THE MESOLITHIC OF THE ATLANTIC FAÇADE: PROCEEDINGS OF THE SANTANDER SYMPOSIUM

MANUEL GONZÁLEZ MORALES & GEOFFREY A. CLARK, EDITORS

CONTRIBUTIONS BY

ESTÉBAN ALVÁREZ-FERNÁNDEZ
PABLO ARIAS CABAL
EMILI AURA TORTOSA
GEOFFREY N. BAILEY
CLIVE BONSALL
GEOFFREY A. CLARK
ALAN CRAIGHEAD
LOUIS DETRAIN
MIGUEL-ÁNGEL FANO MARTÍNEZ
JUAN FERNÁNDEZ-TRESGUERRES
PASCAL FOUCHER
PILAR GARCÍA-ARGÜELLES
MANUEL R. GONZÁLEZ MORALES
PETER ROWLEY-CONWY
CRISTINA SAN JUAN
LAWRENCE G. STRAUS
CHRISTOPHER TOLAN-SMITH
ALAIN TURQ
JOSÉ MANUEL VÁZQUEZ-VARELA
BRADLEY J. VIERRA
PETER WOODMAN
JOÃO ZILHÃO
MAREK ZVELEBIL

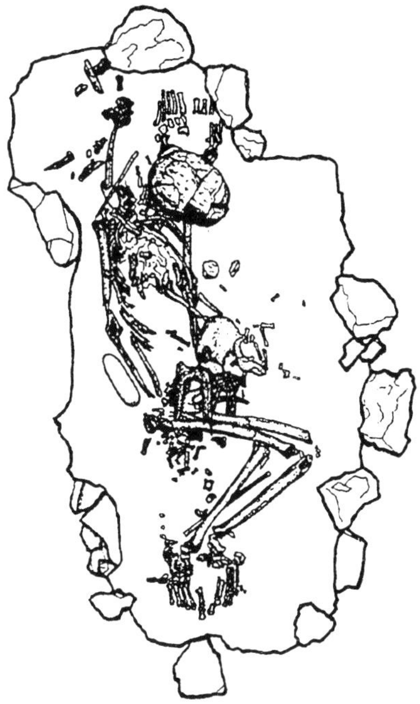

2004
ARIZONA STATE UNIVERSITY
ANTHROPOLOGICAL RESEARCH PAPERS NO. 55

About the Editors . .

Born in Cuba in 1950, Manuel R. González Morales is a Professor in the Department of History, and a former Dean of the Faculty of Philosophy and Letters at the University of Cantabria, Santander, Spain. He received the Ph.D from the University of Oviedo in 1980. A recognized specialist on paleolithic art, the Mesolithic-Neolithic transition, and coastal foragers adaptations, Prof. Morales has long been involved in research on the Paleolithic, Mesolithic and early Neolithic in the north Spanish Principality of Asturias, and the Autonomous Region of Cantabria. immediately to the east. He is the author of *El Asturiense y Otras Culturas Locales,* published by the Centro de Investigaciones y Museo de Altamira (1982), and, with César González Sainz, *La Prehistoria en Cantabria* (Tantín, 1986).

Geoffrey A. Clark is a Regents' Professor in the Department of Anthropology at Arizona State University, Tempe, Arizona. A University of Chicago Ph.D (1971), current interests turn on the logic of inference in the various aspects of modern human origins research *(Conceptual Issues in Modern Human Origins Research,* co-edited with Cathy Willermet, Aldine de Gruyter [1997]) and on applications of neo-Darwinian evolutionary theory in archaeology and human paleontology *(Rediscovering Darwin: Evolutionary Theory in Archaeological Explanation,* co-edited with Mike Barton, American Anthropological Association (1997).

The logo for *Anthropological Research Paper No. 55* is Burial II, the nearly- complete skeleton of a young male from Cueva de los Canes, in eastern Asturias (see pg. 232).

Library of Congress Cataloging in Publication Data

González Morales, Manuel R. (b. 1954)
and Clark, Geoffrey A. (b. 1944)
The Mesolithic of the Atlantic Façade: Proceedings of the Santander Symposium

(Arizona State University Anthropological Research Papers No. 55)

Includes preface, bibliographies
1. Man, Prehistoric - Mesolithic Period. 2. Man, Prehistoric - Atlantic Europe. 3. Man -Prehistoric - Iberia (Portugal, Spain). 4. Hunting and Gathering Societies - Western Europe. 5. Archaeology - Research - Western Europe. 6. Excavations (Archaeology). 7. Europe - Antiquities. I.Title. II. Series.

2004
Library of Congress Control Number 2004117496
ISSN 0271-0641
ISBN 0-936249-16-1

The Mesolithic of the Atlantic Façade

Manuel R. González Morales & Geoffrey A. Clark, Editors

Table of Contents

List of Figures

List of Tables

He visto con mis ojos cosas que Vds. jamás imaginarían.
Todos esos momentos se desvanecerán con
el tiempo, como lágrimas en la lluvia.

The Mesolithic of the Atlantic Façade
is dedicated to the memory of
our friend and colleague

Victoria Cabrera Valdés
1951 – 2004

Con gran tristeza nos hacemos eco del fallecimiento inesperado de la Dra. Victoria Cabrera, una de las arqueólogas más significantes del panorama peninsular. Vicky ha dirigido las excavaciones del yacimiento de la Cueva del Castillo, abordando la problemática de la transición del Paleolítico Medio al Superior en la norte de la Península Ibérica. Sus publicaciones son de los más relevantes que existen para este período en la Península, y su tabajo se enmarca y tiene trascendencia en el ámbito internacional. Desde aquí queremos transmitir nuestro más sentido pésame a su marido el Dr. Federico Bernaldo de Quirós y a toda su familia, así como a su equipo de El Castillo, y a sus estudiantes y colegas en la Universidad Nacional de Educación a Distancia (UNED) de España.

Preface

Most of us would agree that Mesolithic scholars tend not to meet regularly in regional symposia that serve as forums for the exchange of ideas and experiences, or to publicize more widely new data and methodological advances pertinent to their areas of expertise. In the Fourth Mesolithic Symposium (Leuven, 1990), some of the participants were of the opinion that the size and/or scale of the Edinburgh and Leuven meetings, each with more than 75 participants, precluded meaningful discussion of more specialized, regional topics. There was also considerable discussion of what to do during the 5-year intervals between the big, international symposia. These conversations eventually focused on the idea of having smaller, more manageable regional meetings that would permit more informal and direct discussion of current regional issues in a way that would not be possible at the much larger international symposia.

In informal conversations about these problems at the Leuven symposium, we decided that a meeting dedicated to current issues and problems in coastal Mesolithic research along the Atlantic Façade could constitute one such geographical focus, and that a reasonable venue would be the city of Santander on the north coast of Spain – a region that, by virtue of its long tradition of coastal Mesolithic research, would be an ideal setting for such a meeting. Thus, the First International Conference on the Mesolithic of the Atlantic Façade came into being. Generously funded by the Fundación Marcelino Botín, the University of Cantabria, and the municipal government of Santander, it took place in facilities provided by the Fundación Botín from 6-9 July, 1994. The exceptionally long delay in publication was the result of serious problems with the printer in Santander – so serious, in fact, that the conference organizers eventually severed all connection with them, and decided (May, 2003) to publish the proceedings in the *Arizona State University Anthropological Research Papers* (ARP) monograph series. Included here are updated versions of manuscripts originally submitted in 1995 and 1996, and entirely new essays. We are pleased to report that most of the original participants are represented, and thank them for their extraordinary patience.

From the beginning, we planned for a good geographic and intellectual 'mix', with Spanish, Portuguese, French, British, Irish, Scottish and American scholars in attendance. Punctuated by excursions to Mazaculos and La Riera in eastern Asturias, the El Perro rockshelter (Santoña), and the megalithic necropolis of Alto Guriezo-Hayas (Ampuero), the meetings were divided geographically into sessions on interregional perspectives (Straus, Rowley-Conwy, Bailey), the Atlantic Façade (Bonsall, Smith, Kayser, Foucher and San Juan, Turq and Detrain, González Morales, Vázquez Varela, Clark, Arias), Portugal (Zilhão), the Mediterranean world (García-Argüelles, Aura) and the view from northern Europe (Zvelebil). The eminent Cambridge prehistorian and Mesolithic scholar Paul Mellars delivered the keynote address. Slightly modified and considerably expanded, this book reflects in its organization similar geographical subdivisions: *Part I – Northwest Coasts and Islands* (Chapters 1-5; papers by Rowley-Conwy, Bonsall, Zvelebil, Woodman, Tolan-Smith), *Part II – Aquitaine, Charente and the Spanish 'Corners'* (Chapters 6-10; papers by Turq and Detrain, Foucher and San Juan, Aura, Vázquez Varela, García-Argüelles), *Part III – The Atlantic Coasts of Spain and Portugal* (Chapters 11-17; papers by Zilhão, Vierra, Fernández-Tresguerres, Fano and González Morales, Bailey and Craighead, Clark, Arias), and *Part IV – The Mesolithic of the Atlantic Façade* (Chapter 18, overview by Straus).

Set against the backdrop of rather different intellectual traditions (i.e., those of the United Kingdom, Ireland; the anthropologically-oriented approaches of American workers; those of Latin Europe), certain themes recurred over the course of the meeting. One was the sheer diversity of Mesolithic adaptations when compared with those of the final paleolithic and epipaleolithic. Striking changes in environment coincided with the tardiglacial/post-glacial transition and resulted in new vegetation and faunal configurations, to which the foragers of the late Upper Paleolithic had to adapt. Those adaptations were latitudinally distributed to some extent, but also regionally and temporally quite variable, as foragers responded to accelerated change by population movements and technological innovations that comprised a temporal and spatial mosaic with relatively few commonalities beyond the resource intensification and diversification that typically ensues with population-resource imbalances.

A second theme is a 'patchy' distribution of evidence for growing social complexity, seemingly driven ultimately by unequal access to resources. To judge from the archaeological record, most Mesolithic groups were small, fairly mobile foragers practicing logistical subsistence strategies in environments with limited resources. A few, however, evidently maintained defended territories and were relatively sedentary, thus setting in motion the reduced birth-spacing and accelerated rates of population growth usually associated with established agropastoral economies. Along the Atlantic Façade, the best evidence for precocious social complexity comes from rare Mesolithic cemeteries and settlements near the mouths of the Portuguese estuaries and from coastal Brittany and northern Europe. Although a reduction in mobility has well-documented demographic consequences, it does not inevitably lead to complexity, perhaps better explained as an adaptation to the kinds, quantities and spatial distributions of key resources in the environment.

Implicit in many papers, and explicit in others, is the nature of the relationship between the Mesolithic and the Neolithic. For a long time the Mesolithic was considered an inconsequential (and often degraded) survival of a glorious paleolithic age that only assumed importance insofar as it shed light on the origins of domestication economies. While assessing its relationship to the appearance of the Neolithic in any given region is undeniably an important goal (it was much earlier in Mediterranean Spain and France than along the Atlantic Façade), the Mesolithic is now seen as an autonomous research domain, with a set of problems and issues that are unique to it. Some of those issues are explored here under the broadly defined ecological framework that has always been characteristic of Mesolithic research, and that seeks to integrate forager lifeways in the early Holocene with a mosaic of rapidly changing environmental and biotic variables.

• • •

Publication of *The Mesolithic of the Atlantic Façade* was greatly facilitated by Marilynn Bubb, who prepared the camera-ready copy, and Marsha Schweitzer and Michelle Fiedler, who were responsible for scanning and manuscript preparation. Michelle in particular deserves special acknowledgment. Without her diligent, careful yet astonishingly rapid copy editing, we would doubtless still be in the production phases of a book that has already gone far too long unpublished. Clark translated all of the papers originally submitted in French and Spanish.

M. R. González Morales (University of Cantabria)
G. A. Clark (Arizona State University)

April 15, 2004

Part I

Northwest Coasts & Islands

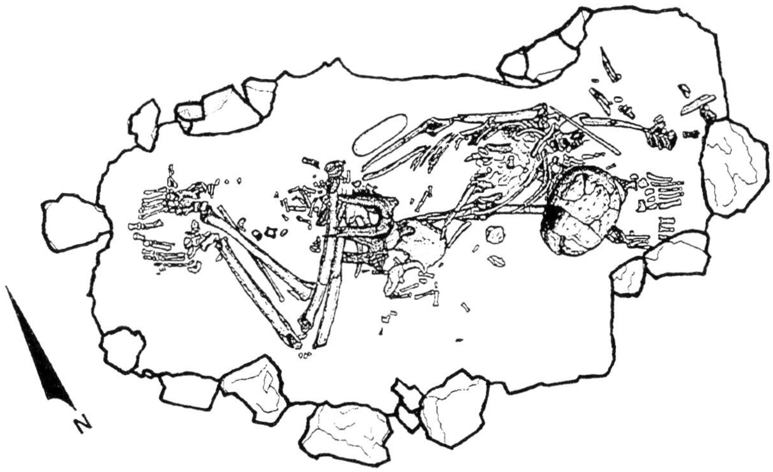

1

Complexity in the Mesolithic of the Atlantic Façade: Development or Adaptation?

Peter Rowley-Conwy
University of Durham

Abstract

This contribution considers socio-economic complexity, and contrasts two views that seek to account for its origins: development and adaptation. The developmental view implies an incremental trend through time towards greater complexity and ultimately agriculture, while the adaptationist view implies a more opportunistic and less directional trajectory based on local ecological conditions. A definition of complexity is put forward, and the importance of territoriality and sedentism are stressed; it is argued that both occur as adaptations to prevailing environmental conditions. Criticisms of developmental views are offered under three headings: population increase, social factors, and non-exploitation of marine resources. It is argued that none of these are likely to inhibit the appearance of complexity, as selection will act on behavioral variability. The implication is that complex hunter-gatherers may have existed at much earlier times than the late Mesolithic. The Upper Paleolithic and early Mesolithic of Europe is examined for supporting evidence.

Introduction

This contribution will examine why socio-economic complexity occurs in Mesolithic societies. Current views may broadly be divided into two opposing camps, based on what are here termed development and adaptation. Not all archaeologists have sufficiently distinguished between these in the past. The position taken here, however, is that we must distinguish very clearly between them, because they embody very different views of change in human societies. By pursuing this question, Mesolithic studies may be in a position to add substantially to our understanding of socio-economic change. As will become clear, the Mesolithic of the Atlantic Façade is one of the best arenas in which to consider the question.

The developmental view seeks to derive socio-economic change from some dynamic inherent within human societies. The environment is at most a passive backdrop, within which various socio-economic formations are equally possible (Ingold 1981). As a result, the intensification of wild resource procurement and ultimately the switch to agriculture are seen as the outcome of non-environmental influences coming from within society. Two main sources of such influence are put forward: (1) a social impetus towards increased production to service alliance networks or feasting (e.g., Bender 1978, Hayden 1992), or (2) a demographic impetus resulting from the long, slow increase of human population levels until they neared environmental carrying capacity and thus caused pressure (e.g., Cohen 1977). In this view socio-economic complexity is a late development in hunter-gatherer societies, signaling that the transition to farming will be the next development.

The adaptationist view on the other hand seeks to link some socio-economic aspects to ecological factors. Not all socio-economic formations are seen as equally viable in a given environment. Some will be more effective than others, and given the individual behavioral variability among humans there is scope for selection of behavior that will tend to move average behavior towards the optimal (e.g., Renouf 1991, Rowley-Conwy 1983, 1998). In this view socio-economic complexity is an adaptation to particular environmental conditions, which may therefore occur at any time and without implying any directional trend towards agriculture.

The adaptationist view will be supported in what follows. Some of its implications for anthropology and archaeology will be examined, and data from the Upper Paleolithic and Mesolithic of the Atlantic Façade will be used in support. It is appropriate that the Cantabrian region has itself been the focus for discussions for and against the views summarized above for some time (Clark & Straus 1983, Straus & Clark 1983, *contra* the adaptationist view of Bailey 1983).

Hunter-Gatherers: What is Complexity?

The uniformitarian model of hunter-gatherer behavior presented by Lee and DeVore (1968) began to break down in the early 1980s under criticism from anthropological approaches stressing greater diversity (Binford 1980, Woodburn 1980). More specifically archaeological views soon followed suit (e.g., Rowley-Conwy 1983, Price & Brown 1985). Various aspects are stressed in discussions of complexity: (1) increased sedentism, (2) increased local group size and population density, (3) food storage, (4) a delayed return economy, (5) logistic organization, (6) cemeteries, (7) territoriality, (8) wealth accumulation and (9) social differentiation (for recent discussions see, e.g., Keeley 1988, Binford 1991, Renouf 1991, Rowley-Conwy 1998.).

For the purposes of this discussion, non-tropical hunter-gatherers are divided into a four-fold typology (Fig. 1.1):

Type 1 - Serial specialists moving from resource to resource in sequence, practicing little or no logistic movement of resources and little or no storage. The Netsilik approximate this strategy (Balikci 1970).

Type 2 - Logistic groups that do not defend territories. The Nunamiut fall into this category (Binford 1980).

Type 3 - Logistic groups that do defend territories. Australian groups along the Murray River fall into this category (Pardoe 1988).

Type 4 - Sedentary groups, who invariably defend territories and store resources. Some Northwest Coast groups fit this category (Richardson 1982).

Many of the aspects of complexity mentioned above are found in Types 2, 3 and 4, and it is these groups that are usually referred to as 'complex'. This definition will be followed here, with the four types forming a continuum from noncomplex (Type 1) to most complex (Type 4).

The non-sedentary logistic category is here divided between Types 2 and 3, depending on their territorial behavior, because this has important implications with regard both to complexity and to archaeological visibility. Territorial groups are found in socially exclusive descent groups (Saxe 1970, Goldstein 1981, Pardoe 1988). Members of other groups are excluded from the defended territories of such groups. Ethnographically, resource patch ownership may be vested in an individual, an extended family, a kin group, a house group, or a village (Richardson 1982). Whatever the case, variation in productivity between resource patches will result in some groups being able to harvest a surplus more often than others. These better off groups will be able to convert their surpluses into social status, particularly if the surpluses are stored. The basis for a ranked society is thus present (Rowley-Conwy & Zvelebil 1989).

Ranking may be taken further among sedentary societies. The additional factors involved are (1) increased storage, (2) more specialized technology and (3) continuous social proximity. Storage as an option can be extended through more of the year if necessary, so the storage linked factors mentioned above may be enhanced. Technology devoted to mass capture of resources may be stored on a permanent settlement and used to maximize returns from even relatively minor resources; this is less feasible when large technological items must be transported from site to site. Surpluses may thus be maximized by sedentary groups. Continuous social proximity leads to a particular type of problem: dispute resolution within the community that does not involve some individuals 'voting with their feet' and moving elsewhere. Such dispute resolution involves high status individuals who can provide an adjudication accepted by both parties. These three factors will therefore enhance complexity as here defined among sedentary groups (Renouf 1991, Rowley-Conwy 1983, Rowley-Conwy & Zvelebil 1989). Territoriality and sedentism are therefore among the most important factors influencing degree of complexity. The circumstances under which they occur are discussed in the next section.

The Environmental Basis for Territoriality and Sedentism

It will be argued here that both territoriality and sedentism are adaptations to particular environmental parameters. Territoriality has been considered anthropologically by a number of writers. The key factors that lead to individual or group territoriality involve crucial resources. Such resources must be (1) productive in a particular localized area, (i.e., not subject to marked spatial variation), and (2) reliable, that is not subject to marked temporal (= interannual) variation. When resources have these characteristics, it becomes worthwhile for humans to undertake territorial defense of the resource point in question (Dyson-Hudson & Smith 1978, Thomas 1981, Richardson 1982). If resources are spatially and/or temporally unreliable, then territorial defense is too rigid a strategy and it is not worth defending a territory. In such cases a fluid and mobile response is preferable.

As mentioned, territorial hunter-gatherers usually live in socially exclusive unilineal descent groups. Where resources are unreliable, groups counteract resource fluctuations with a fluid population structure allowing the relocation of people and the flow of information. Where resources are reliable, a socially and territorially exclusive group emerges. Such exclusive groups justify their ownership of territory by invoking their past: they are (or claim to be) descendants of a similar group of ancestors who occupied the same territory. One direct means of using ancestors to claim territory is to bury them in a cemetery; this serves as a visible and ever present token of territorial ownership and is thus evidence of a territorial group when found by archaeologists (Saxe 1970; also Goldstein 1981, Pardoe 1988, Chattopadhyaya 1996, Rowley-Conwy 1998).

Sedentism is an adaptation to a particular spacing of resources. When seasonally available resources appear too far apart for easy transport, the base camp must move to

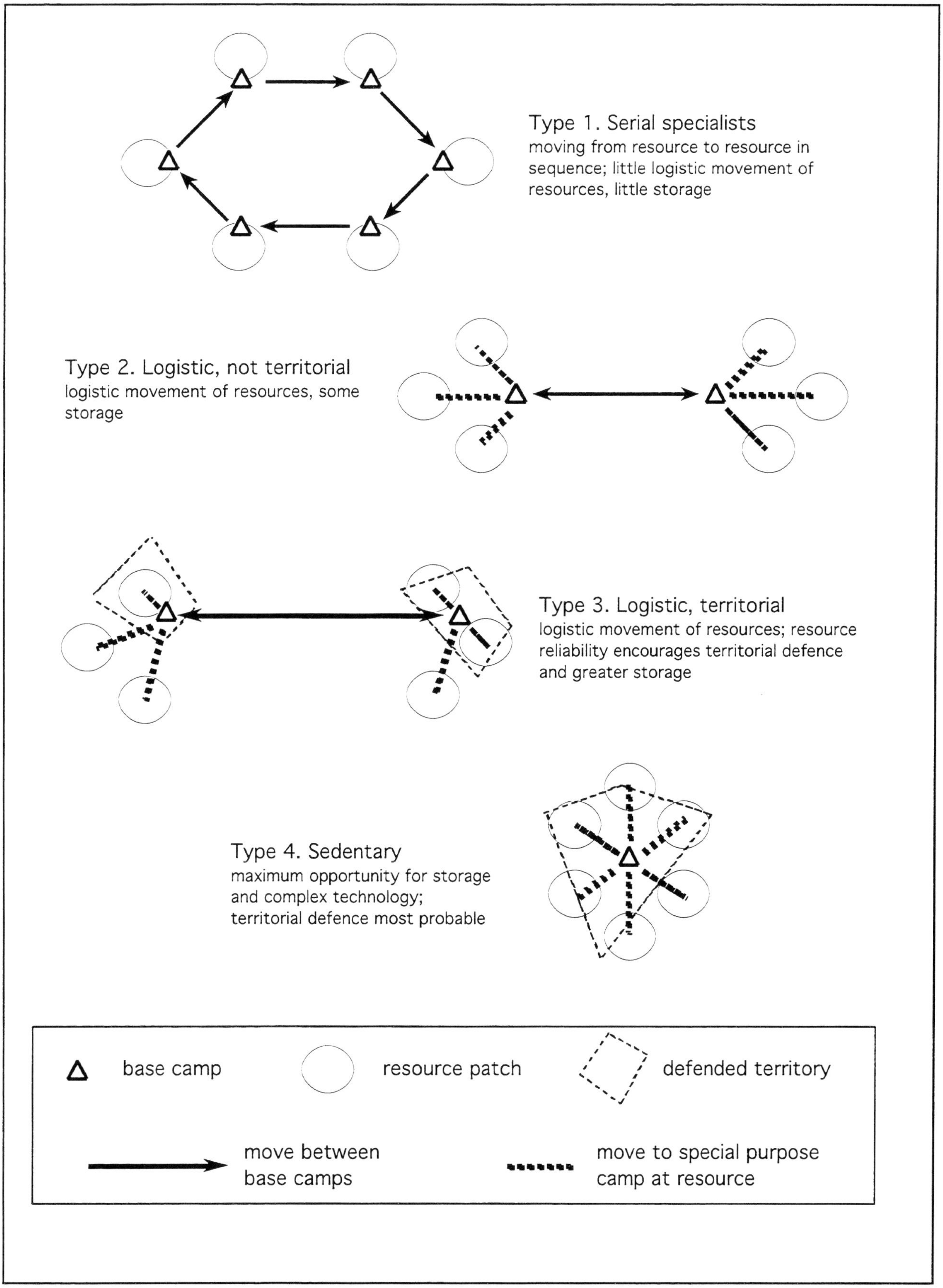

Figure 1.1 Models of hunter-gatherer land use, divided into the four types defined in the text.

the resource (rather than the resource to the base camp). When such resources appear sufficiently close together to allow transport, however, the base camp need not move because all resources are within reach (Fig. 1.2). The increased contributions of storage and mass capture technology to this system were outlined above.

Where is sedentism most likely to occur? Two environmental aspects are important. (1) Multiple seasonal resources are most likely to be found where both aquatic and terrestrial resources are available. Thus some (but not all) coasts, and a few major rivers, are the most likely areas for sedentism. (2) Latitudinal effects are important. Other things being equal, high latitude environments are characterized by few exploitable species, but they occur in large seasonal concentrations. Lower latitudes contrast with this: they are characterized by many more species, which are not usually seasonally concentrated. Thus tropical rain forests have such high resource dispersion that sedentism is most unlikely. Arctic regions have seasonal resource concentrations, but there are usually too few such concentrations available in a given area to make sedentism viable. The temperate zone is intermediate; it is sufficiently far north for many resources to occur in concentrations, but also sufficiently far south for their to be a wide range of resource species.

Territoriality thus requires localized productivity and reliability, and sedentism requires spatial proximity, on the part of resources. The temperate zone, because of its intermediate latitude, will be an area where localized and productive yet reliable resources are often found. Coastlines, because of the juxtaposition of marine and terrestrial biotopes, are areas where spatial proximity among resources will be more common. Temperate coastlines are therefore identified as the environmental zone most often likely to generate socio-economic complexity. One cross-cultural survey has demonstrated that hunter-gatherers of the temperate regions (not just the coasts) on average (1) have larger group size and population density, and (2) undertake fewer residential moves per year, than those of any other environmental zone (Whitelaw 1990, Table 6.8).

Developmental Views of the Mesolithic of the Atlantic Façade

The previous section has argued that territoriality and sedentism are among the most important bases for socio-economic complexity, and that they occur for environmental reasons. Territoriality, sedentism and socio-economic complexity may thus appear at any point in time, and are not restricted to a late 'stage' in hunter-gatherer development. In this view, changes are seen as adaptations and need not only be from simple to complex.

Others, however, prefer to view the changes as developments, partially or entirely separate from ecological factors, conditioned by either increasing population and consequent resource stress, or by internal social factors. For example, in Norway the best recent reviews (Bjerck 1988, Nygaard 1987) stress the difficulty of distinguishing between causative factors due to the paucity of the archaeological record, but accord a large potential role to both social factors and increasing population.

Mesolithic cemeteries, specifically linked with complexity, have recently been considered by Clark and Neeley (1987). They note that cemeteries appear quite late in the Mesolithic; the mean radiocarbon age for single burials is 7557 BP, while that for cemeteries is 6250 BP (op. cit., 122). The cemetery data are more complex; this "points to greater variety in the social organisations of the late Mesolithic period than appears to have been the case with the Upper Paleolithic and the early Mesolithic" (ibid, 124). The appearance of cemeteries in turn suggests that some sort of demographic threshold was crossed for the first time in some parts of western Europe at around 6500 years ago (ibid. 124).

There is no doubt that cemeteries do concentrate mainly in the later Mesolithic, so that evidence for complexity visibly increases through time. Superficially, the archaeological record thus appears to support the developmental view of incremental complexity. However, this may be illusory; the cemetery record is biased and does not necessarily support the developmental view as claimed.

It was argued above that cemeteries are a physical expression of corporate descent groups, which are themselves the result of territorial behavior. If such behavior is much more likely to occur on temperate sea coasts, then an alternative explanation for the late Mesolithic cemetery concentration is possible. Due to the postglacial rise in sea level (and except in areas with steep continental shelves), only the later Mesolithic coastlines survive to be examined by archaeologists. Coastal groups from earlier periods are usually beyond our reach. It is thus entirely possible that cemeteries of early Mesolithic and Upper Paleolithic date existed, but are no longer available for study. Late Mesolithic cemeteries are almost all located close to the coasts (Zvelebil & Rowley-Conwy 1986, Fig. 9); were a substantial rise in sea level to occur, most would be inundated and invisible.

If we consider the environmental zones available for study by archaeologists from the Upper Paleolithic through to the end of the Mesolithic, the sequence goes from an Arctic continental interior in the Upper Paleolithic, to a temperate region including its coastlines in the late Mesolithic. As argued above, temperate coasts are probably the most likely environmental zone in which to find hunter-gatherer territoriality and cemeteries, while Arctic continental interiors are among the least likely. This is borne out by Keeley's cross-cultural survey of hunter-gatherer complexity: as continentality (a measure of seasonal and inter-annual variability) increases, so the likelihood of complexity decreases (Keeley (1988: 393-394). A comparison between the two time periods does not compare like with like, however, and is consequently invalid.

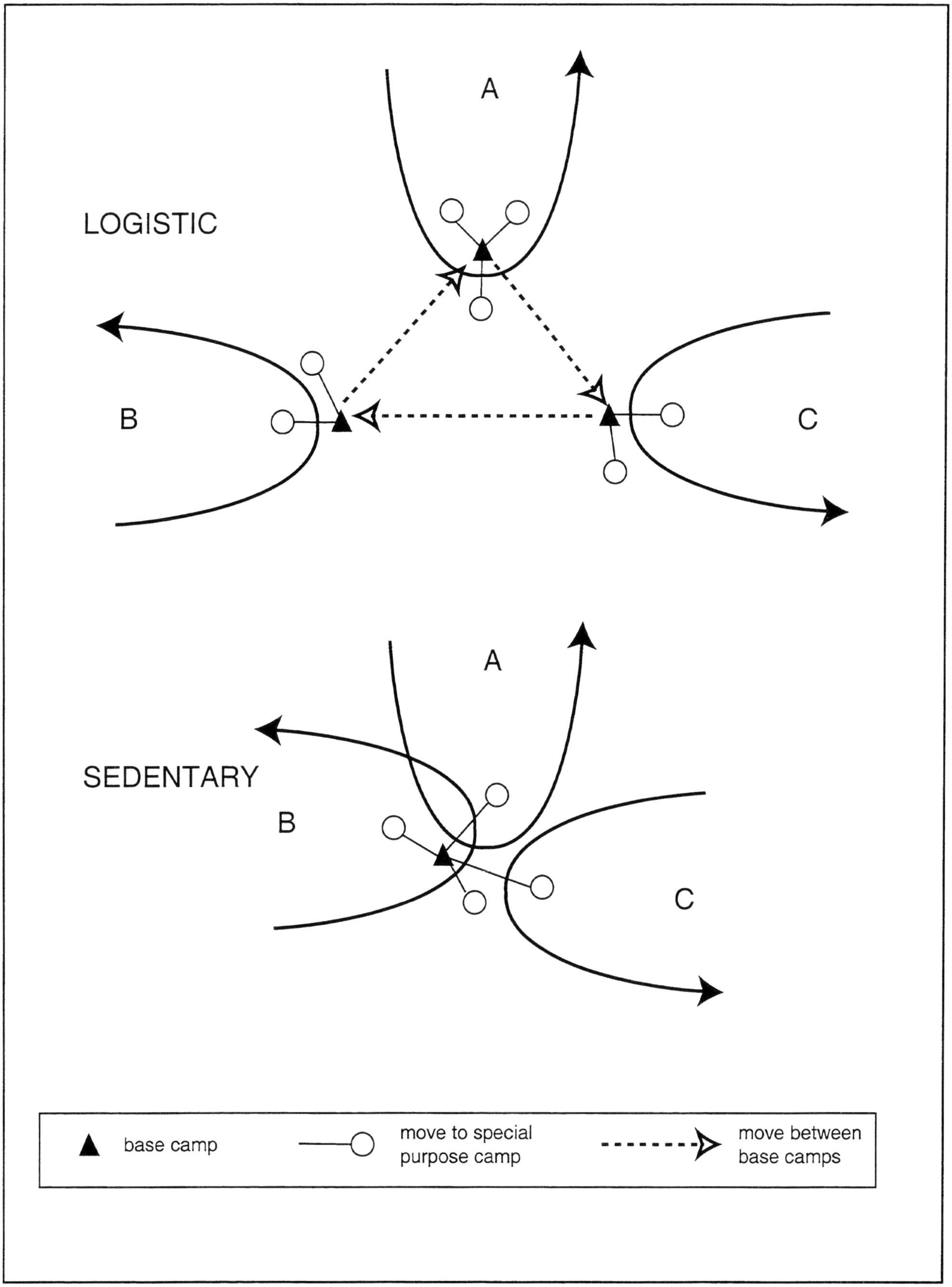

Figure 1.2. Above: seasonal resources A, B and C appear far apart, so the base camp must be moved. Below: seasonal resources appear close together, allowing all to be exploited from a single base camp.

Some Developmental Theory Criticisms

The previous section argues that the archaeological record does not offer support for the developmental view. This section will examine the theoretical basis of three factors invoked to support the developmental view, namely (1) population growth, (2) social forces, and (3) non-exploitation of aquatic resources in earlier periods.

Population growth is sometimes assumed to be an independent variable and is invoked as a causative mechanism. We are conditioned by charts of global population increase to believe that hunter-gatherer populations increase only very slowly (see Fig. 1.3). The clearest proponent of this view is Cohen (1977), who calculates that average population growth on the part of Pleistocene hunter-gatherers would have been on the order of 0.0007% to 0.003% per year. Only after a very long period will population growth thus *cause* intensification and the eventual adoption of domestication economies.

Such models have been criticized as unrealistic (Bronson 1977, Rowley-Conwy 1985). Short term and local oscillations in carrying capacity and human populations would have taken place continuously. Their local impact would have been very much greater than any background trend in the order of magnitude of 0.0007% to 0.003% per year.

It is most unlikely that human populations would, other things being equal, show long slow increases of this sort. Fig. 1.4 shows three different forms of population increase. Theoretically, any population of any species could increase exponentially, creating a J-shaped curve. In practice, environmental resistance increases as carrying capacity is approached, causing a slowdown in population growth. This creates a sigmoidal curve (Boughey 1973). Populations immigrating into a new area are likely to follow the sigmoidal line: when we envisage red deer, for example, moving into northern Europe as the glaciers retreated, we assume a rapid rise in population to carrying capacity followed by population stability. Why, therefore, do we not envisage the same pattern for humans? What sort of "demographic threshold" (cf. Clark & Neeley, quoted above) could not be reached until 6500 BP? If the threshold was reached in southern Scandinavia within 6,000 years of the first postglacial colonization of that region, why was the threshold not reached very much earlier in Iberia, given the continuous presence there of anatomically modern humans from before 40,000 BP? Human population growth could certainly be very rapid. Diamond (1987) points out that an annual growth rate of 1.4% (a rate exceeded in many recorded instances) will turn 100 people into 1,000,000 people in one thousand years. Increase is likely to have been very rapid at times of immigration or favorable environmental change, and zero at times of environmental stasis (assuming a constant technology). There is therefore no theoretical justification for long, slow population increase. A much more likely model involves a relatively rapid rise towards carrying capacity, followed by an oscillation of numbers around a constant level (Layton, Foley & Williams 1991).

It might be objected that culturally determined methods of birth control might inhibit such rapid population expansions; regulating mechanisms would certainly have been available (e.g., Hayden 1972). This would, however, imply that the cultural average rather than the individual was the unit of selection, and would ignore the role of indi-

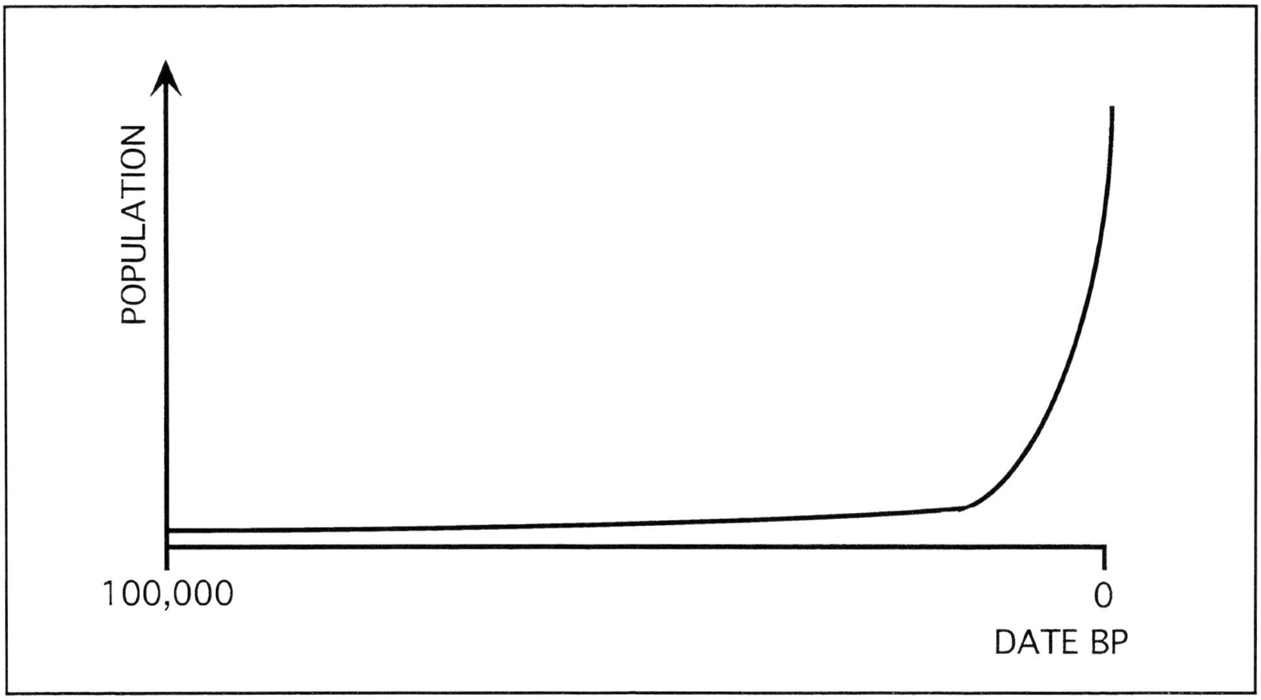

Figure 1.3. Global population curve.

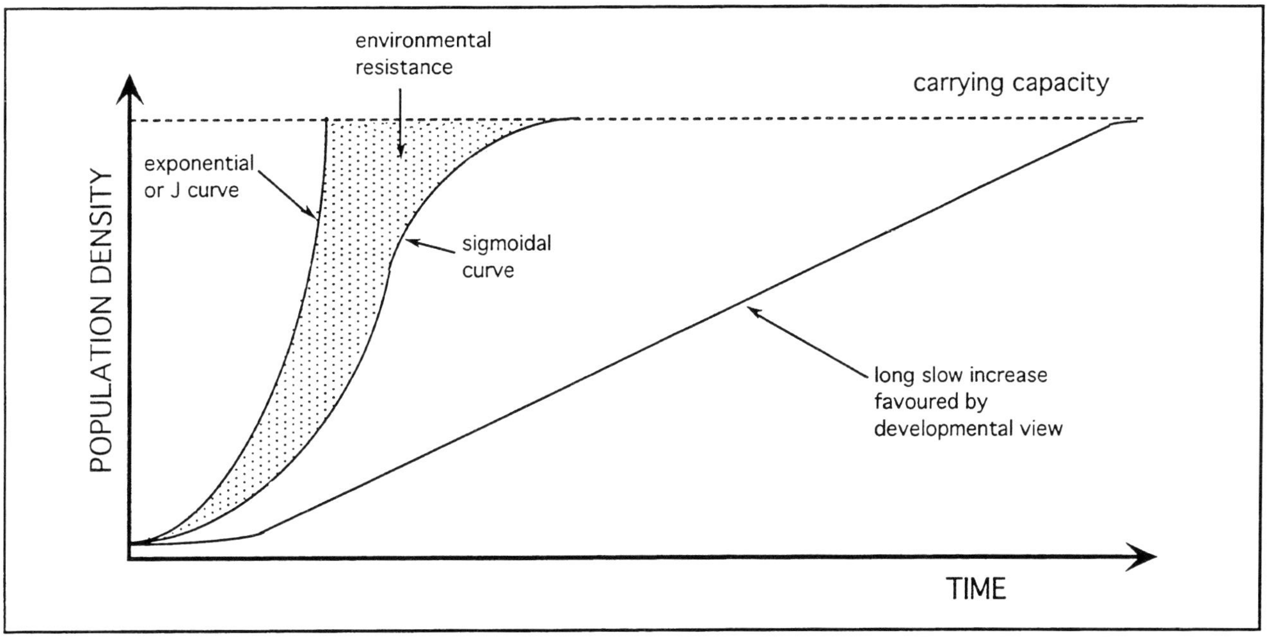

Figure 1.4. Graph illustrating the three kinds of population growth discussed in the text (based partly on Boughey 1973).

vidual behavioral variation. In favorable environmental circumstances, behaviors resulting in more rather than fewer offspring would be selected for, causing the cultural average to change. Culture cannot therefore be argued to 'block' rapid population increases.

Social forces such as the servicing of alliance networks (Bender 1979) or competitive feasting (Hayden 1992) have been invoked as causing economic intensification. Two problems with these views may be mentioned. First, they are somewhat circular. Usually the best argument that social forces were at work to cause a change is the fact that the change subsequently took place. Hindsight is thus crucial, given the difficulty of finding unambiguous archaeological evidence. Second, such scenarios are linked to neither a specific time nor a specific region; there is no explanation as to why the relevant social forces should be active among certain groups at certain time periods, and not in others.

A variety of features such as alliance networks and other methods of maintaining links with distant groups of people are ethnographically recorded among many different hunter-gatherer groups. It is interesting to note that some of the most complex and far-flung systems existed among hunter-gatherers with some of the lowest recorded population densities. These include central desert Aborigines (Yengoyan 1968), north Alaskan Inuit (Minc & Smith 1989) and Kalahari San (Wiessner 1982). An important function of the long-distance contacts is to enable people to move rapidly to food in times of crisis; the systems are thus adaptations to the low carrying capacity environments the groups inhabit (op. cit). There is nothing to suggest that these alliance systems were instrumental in causing resource intensification, or that these groups were in the process of becoming socio-economically more complex. The ethnographic record does not therefore offer any strong support to the linking of alliance networks and socio-economic complexity.

Once again, it is difficult to see how cultural factors could *block* a trend towards complexity. Cultural concepts are not completely uniform; individuals have different perceptions of the best courses of action under particular circumstances. This variability ensures that selection can occur; it is the individual, not the culture, that is the unit of selection. Cultural incapacity or conservatism is therefore no argument against the development of complexity.

Non-exploitation of marine resources until the Pleistocene-Holocene transition has been claimed by Binford (1991): "Pleistocene people..., it is well known, favored terrestrial resources" (op. cit., 134); there was then an *aquatic resource revolution;* the appearance of logistic strategies was dependent on this revolution, and on the development of transport technologies such as water transport and pack animals (ibid., 138). This view is substantially less developmental in outlook than the two previous ones; the appearance of logistic strategies (via aquatic resources and transport mechanisms) is indeed couched in strong selectionist terms (Binford 1991: 139, 146-147). Two questions must, however, be directed at this viewpoint. First, how long may such strong selection be expected to take to cause change? Second, to what extent does the archaeological record support the suggestion of a revolution in aquatic resources and transport at the start of the Holocene?

With regard to the first question, it must be admitted that we do not know how long such changes take to occur. It is largely technological innovation that is under discussion; if specialized technology to procure and transport aquatic resources was not available, this could indeed inhibit the development of a logistic system based on such resources. One implication of Binford's generalization is that selection in North America and Europe would have to cause changes to happen at very different rates because anatomically modern humans occupied Europe much earlier. If things happened so much quicker in North America, then we must admit that they could also have happened equally quickly in Europe but at a much earlier date. Terrestrial hunting in Europe selected for specialized technology at an early date. Tanged Point, Magdalenian and Solutrean technological items certainly merit the term 'specialized' (e.g., Peterkin 1993), as indeed does the multicomponent Aurignacian spear tipped with a split based antler point (Knecht 1993). It seems possible that specialized technology for procuring aquatic resources could have developed just as fast.

With regard to the second question, items like fish traps, canoes and paddles are not known from the Upper Paleolithic, but are found in Mesolithic contexts. However, the old adage that 'absence of evidence is not evidence of absence' must be kept in mind. Two factors may be involved. Postglacial conditions of preservation are far more conducive to the survival of this sort of evidence. Peat bogs in northern Europe preserve quite large numbers of such finds. No Pleistocene peat bogs are available to preserve such items for archaeologists to find; nor, indeed, are they from the postglacial of areas like Iberia or southern Italy, but we do not assume that the Mesolithic of these areas was without these items. In addition, much aquatic resource exploitation would have been coastal, so the rise in sea levels would render much of this invisible to us even if it was preserved archaeologically. River fishing in the Upper Paleolithic is documented from a very early stage of archaeological exploration (Sauvage 1875). The importance of salmon can be disputed (Mellars 1985, Jochim 1983), but that fishing was taking place cannot.

The arguments for a techno-economic 'aquatic resource revolution' at the Pleistocene-Holocene transition are not strong. The alternative view is therefore taken here, namely that no revolution was required to enable anatomically modern humans to exploit aquatic resources. Rather, the selective pressures envisaged by Binford (op. cit.) would probably have brought about such adjustments at much earlier dates, depending on local conditions. The possibility of logistically-organized, complex groups in the Upper Paleolithic cannot be excluded.

Hunter-Gatherer Complexity before the Late Mesolithic?

If the foregoing arguments are correct, hunter-gatherer complexity occurs as an adaptation to a particular set of environmental resources, not as the culmination of a developmental trend. Complexity could thus occur at any time, not just in the late Mesolithic, although the postglacial rise in sea level may obscure much of the evidence; complex groups would have been rare in the Arctic continental interior. This section will briefly examine two Upper Paleolithic cases which may be interpreted as evidence for complexity.

Liguria. The Ligurian topography is extreme. High mountains rise immediately from the coast, and climb to altitudes of >2000 m only a few kilometers inland. The terrain continues to slope steeply down under the sea. As a result, only a few kilometers of land have been submerged since even the height of the last glaciation. Liguria is thus one of the few areas of Europe where the contemporary coast was not far away from the archaeological sites available for study.

The Upper Paleolithic record from Liguria is remarkably rich, containing a variety of important cave sites. Some aspects of the evidence can be argued to indicate Upper Paleolithic complexity. For example, among the famous Balzi Rossi caves at Grimaldi, the site of Barma Grande contained six burials. Although not well dated, they are apparently all Gravettian and predate 18,000 BP (Mussi 1990). Some of these contain substantial quantities of status items, as do the two children buried in the nearby Grotta dei Fanciulli (Grotte des Enfants). According to the definition of Clark and Neeley (1987: 125), the latter would represent inherited high status because the children could not themselves yet have acquired high status. Between 18,000 and 14,000 BP burials in caves are unknown (Mussi op. cit., 133), but more are known in the last millennia of the last glacial. From the cave of Arene Candide a series of burials has been excavated. These date from the last millennium of the last glacial. Represented are 11 adults, one adolescent, four children and two fetuses (Tiné 1984).

With this number of burials, layer M at Arene Candide should be considered a late Upper Paleolithic cemetery. The several burials in the various Balzi Rossi caves might lead one to accord similar status to the caves as a group, but dating from an earlier period. It was argued above that cemeteries are one expression of territorial behavior, and if so then the Arene Candide case at least is evidence of a territorial society. This society was therefore complex, although whether it was of Type 3 (logistic and territorial) or Type 4 (sedentary) is unknown. Interestingly, the presence of a late Upper Paleolithic cemetery on the northern shore of the Mediterranean is matched by the occurrence of several much larger contemporary ones in North Africa (Lubell, Sheppard & Jackes 1984).

Cantabria. Like Liguria, the Cantabrian topography ascends steeply from the coast to high mountains not far inland. Also like Liguria, this steep slope continues under the present sea level, so that sites now above sea level would have been within reach of the Upper Paleolithic shoreline. It has been noted that Upper Paleolithic sites in Cantabria are not evenly distributed, but occur in discontinuous groups which may reflect areas inhabited by different groups. This suggestion has been put forward with regard to the distribution both of large multilayer cave sites (Bailey 1983) and of caves containing rock art (Layton 1987, Straus 1987).

Bailey writes that: " . . the eight major site clusters on the coastal plain are distributed so as to form almost contiguous site exploitation territories. This spacing effect cannot be attributed to the availability of limestone caves. Other cave sites would have been available." (1983: 155).

As an archaeologist, Bailey is not using the word 'territory' in the anthropological sense of a defended territory. Another study does, however, consider the art in this perspective. Layton writes that " . . heavily decorated caves tend to occur at intervals, separated or surrounded by caves containing few or no paintings. Thirty kilometer circles tend to define areas containing one major and several minor caves . . This provides additional evidence to support the hypothesis that Paleolithic art was not simply a record of subsistence activities. It suggests that directly or indirectly the art may have performed some function demarcating territorial units." (Layton 1987: 224). As a social anthropologist, Layton is here using the word 'territorial' to suggest the possibility of defended territories.

Conkey (1980) has argued that Altamira, in Santander, one of the centers identified by both Bailey and Layton, may have been an aggregation site because of the wider range of design elements found there on engraved bone objects. In this argument, outlying sites contain fewer such elements. The site containing most elements not found at Altamira is Cueto de la Mina (Conkey op. cit., 617). This lies some 100 km west of the Altamira center, in the site group identified by Bailey as centred on the multilayered La Riera cave (Straus & Clark 1986), and by Layton as centred on the art site of El Buxú, also in Asturias. Differences in engraved motifs could support the suggestion that separate territorial populations were responsible for the Altamira and Cueto de la Mina engravings. As with Liguria, much remains unsolved. It is unknown whether sedentary sites might have occurred on the coasts, but if the argument for territoriality is correct then a degree of complexity was present. What the visible territories might represent in human terms (e.g., band, clan or mating network) remains to be established.

Conclusions

It has been argued here that complexity among anatomically modern hunter-gatherers is likely to a appear as an adaptation to specific conditions, not as the outcome of independent and inexorable population growth, nor as the result of ill-defined social forces generating development internally. It is suggested that the archaeological record of the Atlantic Façade, while never unambiguous, offers more support for the adaptationist perspective than for the developmental view. The arguments focus attention on those aspects of the archaeological record best able to test and carry further the implications of these ideas. Areas where most postglacial shorelines are above present sea level and accessible to study are of great importance, for coastal adaptations are visible before the late Mesolithic. This bias is thus removed. Areas such as Scotland and Norway fit these criteria due to isostatic uplift of the land. For the Upper Paleolithic no such areas are available, but attention can be directed towards areas like Liguria and Cantabria, which have suffered less land loss than most regions of Europe. The hunter-gatherer archaeological record of these areas assumes a major importance because of their suitability as tests of the ideas put forward here.

Various criticisms can be leveled at the views offered above. One criticism that I have never been able to understand is the assertion that such views are 'determinist'. All constructive approaches to prehistory seek to understand what determined the visible course of events. An argument locating the source of socio-economic change in the ecological sphere is no more deterministic than one which seeks to do so in the demographic or the social spheres. All explanations seek to reduce infinitely complex reality to simplified theories. Such reductionism is to be sought after, not avoided, if we are to further our understanding of our prehistoric past.

Acknowledgements

I wish to thank Bob Layton of the Department of Anthropology, University of Durham, for offering me his comments on an earlier draft of this paper, and for many helpful discussions. I also thank various participants at the conference "Old People and the Sea" in Santander in 1994, in particular Geoff Clark, Paul Mellars, Larry Straus, João Zilhão and Marek Zvelebil, for much useful feedback. None of the above should be held responsible for any views expressed above. And finally, thanks to Manuel González-Morales for organizing such an interesting and productive meeting.

References

Bailey, G.
1983 Economic change in late Pleistocene Cantabria. In *Hunter-Gatherer Economy in Prehistory* (G. Bailey, ed.), pp. 149-165. Cambridge: Cambridge University Press.

Balikci, A.
1970 *The Netsilik Eskimo.* New York: Natural History Press.

Bender, B.
1979 Gatherer-hunter to farmer: a social perspective. *World Archaeology* 10: 204-222.

Binford, L.
1980 Willow smoke and dogs' tails: hunter-gatherer settlement systems and archaeological site formation. *American Antiquity* 45: 4-20.
1991 Mobility, housing, and environment: a comparative study. *Journal of Anthropological Research* 46: 119-152.

Bjerck, L.
1988 Remodeling the Neolithic in southern Norway: another attack on a traditional problem. *Norwegian Archaeological Review* 21: 21-52.

Boughey, A.
1973 *Ecology of Populations.* London: Macmillan.

Bronson, B.
1977 The earliest farming: demography as cause and consequence. In *Origins of Agriculture* (C. Reed, ed.), pp. 23-48. The Hague: Mouton.

Chattopadhyaya, U.
1996 Settlement pattern and the spatial organization of subsistence and mortuary practices in the Mesolithic Ganges Valley, North-Central India. *World Archaeology* 27: 461-476.

Clark, G.A. & M. Neeley
1987 Social differentiation in European Mesolithic burial data. In *Mesolithic Northwest Europe: Recent Trends* (P. Rowley-Conwy, M. Zvelebil & H.P. Blankholm eds.), pp. 121-127. Sheffield: Department of Archaeology & Prehistory.

Clark, G.A. & L. Straus
1983 Late Pleistocene hunter-gatherer adaptations in Cantabrian Spain. In *Hunter-Gatherer Economy in Prehistory* (G. Bailey, ed.), pp. 131-148. Cambridge: Cambridge University Press.

Cohen, M.N.
1977 *The Food Crisis in Prehistory.* New Haven: Yale University Press.

Conkey, M.
1980 The identification of prehistoric hunter-gatherer aggregation sites: the case of Altamira. *Current Anthropology* 21: 609-630.

Dyson-Hudson, R. & E.A. Smith
1978 Human territoriality: an ecological reassessment. *American Anthropologist* 80: 21-41.

Goldstein, L.
1981 One-dimensional archaeology and multi-dimensional people: spatial organisation and mortuary analysis. In *The Archaeology of Death* (R. Chapman, I. Kinnes & K. Randsborg, eds.), pp. 53-69. Cambridge: Cambridge University Press.

Hayden, B.
1972 Population control among hunter-gatherers. *World Archaeology* 4: 205-221.
1992 Models of domestication. In *Transitions to Agriculture in Prehistory* (A.B. Gebauer & T.D. Price, eds.), pp. 11-19. Madison: Prehistory Press.

Ingold, T.
1981 The hunter and his spear: notes on the cultural mediation of social and ecological systems. In *Economic Archaeology* (A. Sheridan & G. Bailey, eds.), pp. 119-130. Oxford: British Archaeological Reports International Series No. 96.

Jochim, M.
1983 Palaeolithic cave art in ecological perspective. In *Hunter-Gatherer Economy in Prehistory* (G. N. Bailey, ed.), pp. 212-219. Cambridge: Cambridge University Press.

Keeley, L. H.
1988 Hunter-gatherer economic complexity and 'population pressure': a cross-cultural analysis. *Journal of Anthropological Archaeology* 7: 373-411.

Knecht, H.
1993 Early Upper Paleolithic approaches to bone and antler projectile technology. In *Hunting and Animal Exploitation in the Later Palaeolithic and Mesolithic of Eurasia* (G. L. Peterkin, H. Bricker & P. Mellars, eds.), pp. 33-47. Washington: Archeological Papers of the American Anthropological Association No. 4.

Layton, R.
1987 The use of ethnographic parallels in interpreting Upper Paleolithic rock art. In *Comparative Anthropology* (L. Holy, ed.), pp. 210-239. Oxford: Blackwell.

Layton, R., R. Foley, et al.
1991 The transition between hunting and gathering and the specialized husbandry of resources. *Current Anthropology* 32: 255-263.

Lee, R. B. & I. DeVore
1968 Problems in the study of hunters and gatherers. In *Man the Hunter* (R. B. Lee & I. DeVore, eds.), pp. 3-12. Chicago: Aldine.

Lubell, D., P. Sheppard & M. Jackes
1984 Continuity in the epipalaeolithic of North Africa with emphasis on the Maghreb. *Advances in World Archaeology* 3: 143-191.

Mellars, P.
1985 The ecological basis of social complexity in the Upper Paleolithic of southwestern France. In *Prehistoric Hunter-Gatherers - the Emergence of Cultural Complexity* (T. D. Price & J. Brown, eds.), pp. 271-297. New York: Academic Press.

Minc, L. & K. Smith
1989 The spirit of survival: cultural responses to resource variability in North Alaska. In *Bad Year Economics* (P. Halstead & J. O'Shea, eds.), pp. 8-39. Cambridge: Cambridge University Press.

Mussi, M.
1990 Continuity and change in Italy at the last glacial maximum. In *The World at 18,000 BP, Volume I: High Latitudes* (O. Soffer & C. Gamble, eds.), pp. 126-147. London: Unwin-Hyman.

Nygaard, S.
1987 Socioeconomic developments along the southwestern coast of Norway between 10,000 and 4,000 BP. In *Mesolithic Northwest Europe: Recent Trends* (P. Rowley-Conwy, M. Zvelebil & H.P. Blankholm, eds.), pp. 147-154. Sheffield: Department of Archaeology & Prehistory.

Pardoe, C.
1988 The cemetery as symbol: the distribution of prehistoric Aboriginal burial grounds in southeastern Australia. *Archaeology in Oceania* 23: 1-16.

1993 Lithic and organic hunting technology in the French Upper Paleolithic. In *Hunting and Animal Exploitation in the Later Palaeolithic and Mesolithic of Eurasia* (G.L. Peterkin, H. Bricker & P. Mellars, eds.), pp. 49-67. Washington: Archeological Papers of the American Anthropological Association No. 4.

Price, T. D. & J. A. Brown
1985 Aspects of hunter-gatherer complexity. In *Prehistoric Hunter-Gatherers: the Emergence of Cultural Complexity* (T. D. Price & J. A. Brown, eds.), pp. 3-20. New York: Academic Press.

Renouf, P.
1991 Sedentary hunter-gatherers: a case for the northern coasts. In *Between Bands and States* (S.A. Gregg, ed.), pp. 89-107. Center for Archaeological Investigations, Occasional Paper No. 9. Carbondale: Southern Illinois University Press.

Richardson, A.
1982 The control of productive resources on the Northwest Coast of North America. In *Resource Managers: North American and Australian Hunter-Gatherers* (N. Williams & E. Hunn, eds.), pp. 93-112. Canberra: Australian Institute of Aboriginal Studies.

Rowley-Conwy, P.
1983 Sedentary hunters: the Ertebølle example. In *Hunter-Gatherer Economy in Prehistory* (G.N. Bailey, ed.), pp. 111-126. Cambridge: Cambridge University Press.

1985 The origin of agriculture in Denmark: a review of some theories. *Journal of Danish Archaeology* 4: 188-195.

1998 Cemeteries, seasonality and complexity in the Ertebølle of Southern Scandinavia. In *Harvesting the Sea, Farming the Forest: the Emergence of Neolithic Societies in the Baltic Region* (M. Zvelebil, L. Domanska & R. Dennell, eds.), pp. 193-202. Sheffield: Sheffield Academic Press.

Rowley-Conwy, P. & M. Zvelebil
1989 Saving it for later: storage among hunter-gatherers in prehistoric Europe. In *Bad Year Economics: Cultural Responses to Risk and Uncertainty* (P. Halstead & J. O'Shea, eds.), pp. 40-56. Cambridge: Cambridge University Press.

Sauvage, H. E.
1875 On fishing in the reindeer period. In *Reliquiæ Aquitanicæ; being Contributions to the Archaeology of Périgord and the Adjoining Provinces of Southern France* (E. Lartet & H. Christy, eds.). London: Williams & Norgate.

Saxe, A.
1970 *Social Dimensions of Mortuary Practices.* Unpublished PhD dissertation, Department of Anthropology, University of Michigan.

Straus, L. G.
1987 The paleolithic cave art of Vasco-Cantabrian Spain. *Oxford Journal of Archaeology* 6: 149-163.

Straus, L. G. & G. A. Clark
1983 Further reflections on adaptive change in Cantabrian prehistory. In *Hunter-Gatherer Economy in Prehistory* (G. Bailey, ed.), pp. 166-167. Cambridge: Cambridge University Press.

Straus, L. G. & G. A. Clark (editors)
1986 *La Riera Cave: Stone Age Hunter-Gatherer Adaptations in Northern Spain.* Tempe: Arizona State University. Anthropological Research Papers No. 36.

Thomas, D. H.
1981 Complexity among Great Basin Shoshoneans: the world's least affluent hunter-gatherers? In *Affluent Foragers - Pacific Coasts East and West* (S. Koyama & D. Thomas, eds.), pp. 19-52. Senri Ethnological Studies No. 9. Osaka: National Museum of Ethnology.

Tiné, S.
1984 *I Cacciatori Paleolitici.* Genova: Sagep Editrice.

Whitelaw, T.
1990 *The Social Organisation of Space in Hunter-Gatherer Communities.* Unpublished PhD dissertation, Department of Archaeology, University of Cambridge.

Wiessner, P.
1982 Risk, reciprocity and social influences on !Kung San economics. In *Politics and History in Band Societies* (E. Leacock & R.B. Lee, eds.), pp. 61-84. Cambridge: Cambridge: University Press.

Woodburn, J.
1980 Hunters and gatherers today and reconstruction of the past. In *Soviet and Western Anthropology* (E. Gellner, ed.), pp. 95-117. London: Duckworth.

Yengoyan, A.
1968 Demographic and ecological influences on Aboriginal Australian marriage sections. In *Man the Hunter* (R.B. Lee & I. DeVore, eds.), pp. 185-199. Chicago: Aldine.

Zvelebil, M. and P. Rowley-Conwy
1986 Foragers and farmers in Atlantic Europe. In *Hunters in Transition* (M. Zvelebil, ed.), pp. 67-93. Cambridge: Cambridge University Press.

2

The 'Obanian Problem' - Coastal Adaptation in the Mesolithic of Western Scotland

Clive Bonsall
University of Edinburgh

Abstract

The history of research on the Obanian shell middens of western coastal Scotland is reviewed, and the Obanian is contrasted with microlithic assemblages from the same region. Recent AMS radiocarbon chronologies indicate contemporaneity between the two Mesolithic entities. The author argues that the sharply defined differences between the Obanian and the Mesolithic of western Scotland are attributed to different structural poses within the same economic system. Most, and possibly all, Obanian sites are the remains of special purpose camps associated with the exploitation of fish and shellfish.

Introduction

Direct evidence for the exploitation of coastal resources in the Mesolithic of western Scotland is restricted to a small number of sites on which shell midden deposits are preserved. These sites are generally referred to as 'Obanian' after the initial discoveries around Oban Bay in Argyll (Fig. 2.1).

The Obanian sites share a number of features:(1) faunal assemblages dominated by the shells of marine molluscs but including crustacean remains and bones of fish, sea birds, sea mammals, and wild land mammals such as red deer and pig; (2) lithic assemblages of flint and quartz in which retouched tools are often rare or absent and bipolar cores and débitage well represented; and (3) a distinctive range of implements made from antler or bone which include bevelled tools, points ('awls') and, occasionally, harpoon heads and perforated mattock-heads.

A contrast is often drawn between the artifact assemblages from the Obanian middens and the microlithic industries recovered from a large number of Mesolithic sites in western Scotland. The latter are invariably surface sites where soil conditions are unsuitable for the preservation of organic materials and Mesolithic activity is indicated mainly by the occurrence of one or more lithic scatters. Compared to Obanian toolkits, the lithic scatters are often characterized by a higher incidence of platform cores and blade débitage and the presence of microliths, scrapers and other retouched tools regarded as typical of Mesolithic sites. This contrast in assemblage types is at the heart of long-standing debate in Scottish archaeology between those workers who believe the evidence justifies the recognition of an Obanian culture or phase within the west Scottish Mesolithic, and those who consider that the assemblage types are most readily explained in functional terms.

This paper examines the background to this debate and offers a possible solution to the 'Obanian problem'.

Obanian Shell Middens: The Development of Ideas

Early work on the Obanian between 1881 and 1926 focused on small caves and rock-shelters around Oban Bay and a series of open-air sites on the islands of Oronsay and Rigsa.

Anderson, in his account of the excavations at Druimvargie Rockshelter in Oban, was in effect the first to suggest that the Obanian sites were sufficiently distinctive to warrant their isolation as a distinct 'culture' (Anderson 1898) although the term 'Obanian culture' did not come into general usage until several decades later (Movius 1940).

Most subsequent workers have followed this interpretation, although there have been disagreements over the origins of the Obanian. Anderson (1898), Breuil (1922) and Clark (1956) interpreted the style of the barbed points as evidence of a link with the Azilian, while Movius (1942) and Lacaille (1954) favoured an origin in the Irish Mesolithic.

Further work on the Oronsay middens was undertaken by Paul Mellars in the 1970s designed primarily to recover information on the economic basis of the Obanian occupation of the island (Mellars 1987). Mellars' excavations also provided the first radiocarbon dates for Obanian sites multiple dating of each of the five excavated sites indicated that the Oronsay middens had been deposited between *c.* 6100 and *c.* 5400 BP. Radiocarbon determinations for microlithic industries available at that time ranged from *c.* 8200-6000 BP.

From this evidence, several authors have concluded that the Obanian sites are a very late manifestation of the Mesolithic in western Scotland, post-dating the sites with microlithic industries and possibly overlapping in time the earliest Neolithic sites. Woodman (1989) argued that the use of microlithic armatures in hunting equipment had been abandoned in favour of the production of barbed projectile points of bone and antler (see also Woodman 1990). Jacobi (1982) suggested that the adoption of barbed points, mattock-heads and other elements of the Obanian toolkit was the result of contact with the earliest farming communities of the region. Mellars (1972: 92) reserved judgment on the relationship between the Obanian and the microlithic sites, but suggested that the Obanian artifact assemblage represented "a highly specialized toolkit adapted to the intensive exploitation of coastal economic resources."

The Obanian in Perspective

Most previous interpretations of the Obanian are underpinned by five basic assumptions:

1. The Obanian has a limited geographical distribution, occurring only in coastal areas of Argyll.
2. The Obanian middens belong to a very late stage in the development of the Mesolithic in this part of western Scotland, after *c.*6500/6000 BP.
3. The manufacture of microliths was not characteristic of the human groups responsible for the Obanian middens.
4. Certain artifact types found in Obanian assemblages are not characteristic of other coastal Mesolithic sites in Scotland.
5. Obanian sites are found only on coasts where sites with microlithic industries do not occur, and vice versa.

Results of recent research on shell midden and microlithic sites in western Scotland, coupled with a reassessment of existing evidence, raise doubts about the validity of most, if not all, of these assumptions.

It has long been recognized that certain artifact types regarded as characteristic of the Obanian are not confined to shell midden sites on the west coast of Scotland. Isolated finds of flat harpoon-heads and perforated mattock-heads are known from a number of coastal locations in northern Britain (Movius 1942, Mellars 1972), while the east coast of Scotland produced bevelled tools of bone similar to those from the west coast middens (Coles 1971). Stone bevelled tools have a wide distribution along the western seaboard of Britain, and have recently been reported from a microlithic site at Kinloch on the island of Rhum in western Scotland (Wickham-Jones 1990).

Bone tools of typical Obanian form have been found recently at An Corran, on the Isle of Skye, in apparent association with a lithic assemblage which contains microliths, scrapers and other retouched tools (Rees et al. 1994, Saville & Miket 1994), and at Ulva Cave on an island off the west coast of Mull in conjunction with platform cores and blade débitage (Bonsall et al. 1994). The An Corran site also extends tile distribution of known Obanian middens along the western seaboard to significantly north of the 'core area' in Argyll.

Knowledge of the chronology of the Obanian has been vastly improved by the use of AMS ^{14}C dating to obtain direct dates on individual antler and bone artifacts from sites that were previously undated (Bonsall & Smith 1990, 1992; Bonsall et al. n.d.). A large series of AMS and conventional radiocarbon determinations are now available for sites with Obanian artifacts (Table 2.1). These dates range from *c.* 8350 to *c.* 4750 BP and can be compared with (reliable) dates for sites with microlithic industries of between *c.* 8600 and *c.* 5450 BP (Morrison & Bonsall 1989, Hedges et al. 1989). The radiocarbon evidence therefore demonstrates a substantial chronological overlap between Obanian and microlithic assemblage types in western Scotland. The dates from Carding Mill Bay I (the latest of the Obanian sites) are similar to those for the earliest dated Neolithic sites in western Scotland and suggest that the economic system represented by the Obanian middens persisted until, and possibly after, the adoption of farming in the region.

Until recently, no microlithic sites were known to occur in close proximity to any of the known Obanian shell middens. At Lón Mór, near Oban, microlithic and shell midden sites have been recorded on the margins of a former marine embayment (Bonsall & Robinson 1992; Bonsall et al. 1993), while research on the island of Colonsay has revealed several sites with microlithic industries within 6-8 km of the Oronsay shell middens (Mithen & Finlayson 1991).

Results of recent research, therefore, do not support the view that Obanian and microlithic assemblage types were the product of populations that were differentiated culturally or chronologically. Other possible explanations need to be considered. It can be argued that much of the variability exhibited by Mesolithic sites in western Scotland is a consequence primarily of the conditions of preservation at the sites and the nature of the activities undertaken there.

Shell Middens and Site Function

Shell middens will accumulate where refuse from shellfish gathering is deposited on a regular basis. They will only survive, however, under favourable circumstances. In western Scotland shell midden deposits are preserved mainly in two geological situations in coastal caves and rockshelters the entrances to which were sealed by talus accumulations, and on open-air sites where the middens were subsequently buried beneath blown sand deposits.

Among ethnographically known shellfish gatherers, shellfish are rarely the principal source of food. But shellfishing is a regular activity, often undertaken by women and children, which normally is integrated with other economic tasks such as fishing, hunting (of land and sea mammals) and plant gathering.

Site	Lab Ref	Stratigraphic Context	Material	^{14}C Age BP	cal BC age	Age range (1σ)	Age range (2σ)
Druimvargie Rockshelter (Oban)	OxA-4608	Midden	Bone artifact	8340±80	7423	7488–7291	7533–7067
	OxA-4609	Midde	Bone artifact	7890±80	6645	6696-6598	7032–6478
	OxA-1948	Midden	Bone artifact	7810±90	6598	6696–6474	7001–6428
Ulva Cave	GU-2704	Midden (residual, cave interior)	Soil matrix (humic acid)	7800±160	6595, 6573, 6569	6993–6453	7043–6252
	GU-2600	Midden (entrance): lower	Marine shells (*Patella* spp.)	7655±65	6457	6477–6415	6595–6372
	OxA-3738	Midden (entrance): upper	Antler artifact	5750±70	4581	4712–4505	4780-4457
	GU-2602	Midden (entrance): upper	Marine shells (*Patella* spp.)	5685±65	4512	4583–4459	4711–4381
An Corran (Skye)	OxA-4994	Midden, C36	Bone artifact	7590 ±90	6416	6483–6267	6552–6189
MacArthur's Cave (Oban)	OxA-1949	Midden	Antler artifact	6700±80	5583	5621–5522	5693–5441
Cumstoun	OxA-3735	Unassociated find	Antler artifact	6665±70	5574, 5543, 5528	5594-5479	5638–5439
Morton (Fife)	Q-981	Midden: lower, T50/T57	Charcoal	6382±120	5290	5435–5226	5564-5059
	Q-988	Midden: lower, T50	Charcoal	6147±90	5062	5219–4940	5266–4839
	Q-928	Midden: upper, T50/T59	Charcoal	6115±110	5046	5215–4869	5268–4780
	OxA-4612	Midden: upper, T59	Bone artifact	5790±80	4681, 4635, 4624	4771–4534	4833–4461
	OxA-4611	Midden: lower, T59	Bone artifact	5475±60	4338	4356–4255	4455–4165
	OxA-4610	Midden: T65	Bone artifact	5180±70	3977	4038–3950	4221–3799
Caisteal nan Gillean I (Oronsay)	Q-3008	Midden: Tr. C, layer 4 (base)	Charcoal	6190±80	5201, 5176, 5136	5229–5003	5275–4927
	Q-3007	Midden: Tr. C, layer 4 (base)	Charcoal	6120±80	5051	5206–4933	5250–4836
	Q-3009	Midden: Tr. C, layer 4 (upper)	Charcoal	6035±70	4932	5036–4839	5196–4780
	Q-3010	Midden: Tr. C, layer 3	Charcoal	5485±50	4341	4358–4264	4453–4238
	SRR-1458b	Midden: (redeposited)	Marine shells	5485±81	4341	4440–4250	4468–4108
	Q-3011	Midden: Tr. C, layer 3	Charcoal	5450±50	4331	4348–4248	4361–4162
Risga	OxA-2023	Midden	Antler artifact	6000±90	4905, 4872, 4865	4963–4787	5196–4711
	OxA-3737	Midden	Antler artifact	5875±65	4773	4809–4695	4907–4571
Meiklewood	OxA-1159	Unassociated find	Antler artifact	5920±80	4793	4905–4719	4952–4591
Priory Midden (Oronsay)	Q-3001	Midden: CT, layer 19	Charcoal	5870±50	4770	4795–4711	4896–4604
	Q-3000	Midden: CT, layer 19	Charcoal	5825±50	4714	4774–4612	4800–4540
	Q-3002	Midden: CT, layer 18	Charcoal	5717±50	4539	4684–4470	4712–4458
	Q-3003	Midden: CT, layers 9/10	Charcoal	5510±50	4348	4435–4333	4458–4252
	Q-3004	Midden: CT, layer 7	Charcoal	5470±50	4337	4352–4257	4448–4230
Shewalton	OxA-1947	Unassociated find	Antler artifact	5840±80	4719	4794–4593	4906-4504
Cnoc Coig (Oronsay)	Q-3006	Pre-midden	Charcoal	5675±60	4502	4549–4458	4685–4361
	Q-3005	Pre-midden	Charcoal	5650±60	4466	4536–4401	4672–4353
	Q-1353	Midden: Tr E, unit 6	Charcoal	5645±80	4465	4544–4385	4690–4340
	Q-1354	Midden: Tr E, unit 6	Charcoal	5535±140	4536	4502–4245	4711–4008
	Q-1351	Midden: Tr E, unit 2	Charcoal	5495±75	4344	4445–4257	4466–4160
	Q-1352	Midden: Tr E, unit 3	Charcoal	5430±130	4323, 4281, 4262	4433–4086	4520–3973
Cnoc Sligeach (Oronsay)	GX-1904	Midden	Mammal bone fragments	5755±180	4587	4805–4368	5048–4245
	Birm-465	Midden (redeposited)	Marine shell (*P. maximus*)	5605±156	4456	4597–4330	4792–4049
	BM-670	Midden: Tr B, layer 7	Charcoal	5426±159	4321, 4284, 4261	4452–4042	4574–3949
	GX-1903	Midden	Marine shell (*O. edulis*)	4610±214	3360	3637–2929	3897–2703
Caisteal nan Gillean II (Oronsay)	Q-1355	Midden: Tr B, layer 4 (base)	Charcoal	5460±65	4334	4353–4245	4464–4151
	Birm-347	Midden: Tr B, layer 4 (base)	Charcoal	5450±140	4331	4453–4091	4546–3973
	Birm-348	Midden: Tr B, layer 4 (base)	Marine shells (*Patella* spp.)	5445±313	4329, 4271, 4269	4592–3958	4946–3636
	Birm-346	Midden: Tr B, layer 3 (upper)	Charcoal	5150±380	3965	4354–3535	4788–2926
Carding Mill Bay I (Oban)	OxA-3740	Midden: lower, XV	Antler artifact	5190±65	3981	4211–3948	4229–3791
	GU-2796	Midden: upper, XIV	Charcoal	5060±50	3930, 3875, 3808	3950–3788	3970–3713
	GU-2899	Midden: lower, XV	Marine shells (*Patella* spp.)	5035±65	3894, 3887, 3798	3946–3729	3972–3693
	GU-2898	Midden: upper, XIV	Marine shells (*Patella* spp.)	5005±73	3785	3938–3702	3966–3646
	GU-2797	Midden: lower, XV	Charcoal	4980±50	3772	3892–3701	3939–3653
	OxA-3739	Midden: upper, XIV	Bone artifact	4765±65	3621, 3580, 3533	3639–3383	3691–3367

Table 2.1 Radiocarbon determinations for Obanian shell middens and sites with antler and bone artifacts of Obanian form. The ^{14}C ages quoted are as reported by the laboratory, with errors expressed at the ± one sigma level of confidence. Dates on marine shell have been 'adjusted' for the marine reservoir effect by subtracting 405±40 years (Harkness 1983). All dates have been calibrated using the CALIB 3.03 program Macintosh version, 12/03/93 (Stuiver & Reimer 1993).

In her study of the Anbarra of the north coast of Australia, Meehan (1982) described three types of site at which shellfish remains accumulate: home bases, processing camps and 'dinnertime' camps. Processing and dinnertime camps are normally located near to the shell beds; the former being special-purpose sites where shellfish are gathered and cooked, and the flesh carried back to the home base; and the latter "small campsites used during the middle of the day while people are engaged in [collecting] trips away from their home base" (Meehan 1982: 26), where food procured up to that time is cooked and eaten some of the food prepared at these sites may also be carried back to the home base for later consumption.

The existence of special-purpose 'processing camps' has been reported for other contemporary shellfish gatherers in different parts of the world (Waselkov 1987), and they may be assumed to have archaeological counterparts in a similarly wide range of coastal situations.

From ethnographic evidence, archaeological 'processing camps' can be hypothesized to exist where the main habitation sites are some distance from the shell beds, and to occur close to the contemporaneous shoreline; they are also likely to lack evidence of substantial structures, and to be characterized by a more restricted range of cultural equipment than at residential base camps (cf. Waselkov 1987: 145).

Ethnographically known 'dinnertime camps' are typically small and were often used on only a single occasion; it follows that their prehistoric open-air equivalents are likely to be less visible archaeologically than either base camps or processing sites. However, if dinnertime camps were located in coastal caves, the chances of survival and discovery would be greatly enhanced; and if such locations were used repeatedly over a long period of time, then significant amounts of food refuse can be expected to have accumulated there.

Shell Middens in the Mesolithic of Western Scotland

These observations drawn from ethnographic data can serve to guide interpretation of Mesolithic shell middens in western Scotland. Mesolithic coast-dwelling communities in western Scotland had access to abundant and varied shellfish resources, and it is probable that most groups were in some degree shellfish gatherers. It is also likely that women and children played an important role in this economic activity. Shellfish on the rocky coasts that characterize much of western Scotland constitute a highly dispersed resource (see below). It is unlikely, therefore, that shellfishing activities would have been confined to base camps. Thus, it may be suggested that in addition to residential camps the use of processing camps and dinnertime camps was a feature of the Mesolithic settlement-subsistence system, and that individual middens will relate to one or other of these site categories.

The quality of the data from the Obanian sites is highly variable, but a number of general features can be discerned which have a bearing on the question of site function:

1. All of the known shell middens (whether within caves or in the open air) appear to have been deposited in proximity to the contemporaneous shoreline; at two sites, MacArthur's Cave (Oban) and Cnoc Sligeach (Oronsay), the midden deposits were interstratified with storm beach gravels (Anderson 1895, Mellars 1987).

2. The sites often occur on open, relatively exposed coasts, and six of the 12 known sites are on small islands (Oronsay, Risga) that could not have sustained resident populations of deer or other large herbivores, and where access to land-based resources generally would have been limited. Conversely, such locations maximize access to shellfish and other shore based resources.

3. In contrast to Mesolithic coastal shell middens in some parts of Europe, the Obanian sites are very small, rarely exceeding 25 m across or 1.5 m in thickness. Yet radiocarbon dating and stratigraphic evidence from several recently excavated sites suggest that they were used repeatedly over long periods; on Oronsay this use appears to have followed a seasonal pattern (Mellars & Wilkinson 1980, see below). At Druimvargie Rockshelter in Oban and Caisteal nan Gillean 1 and Priory Midden on Oronsay midden accumulation occurred over several hundred radiocarbon years, while the Ulva Cave midden (only 8 m across) was deposited over 2000-3000 RC yr (Russell, Bonsall & Sutherland 1995).

4. Although the faunal evidence indicates that seals, otters and sea birds were sometimes procured from the sites, the processing of shellfish and fish appears to have been the main economic activity undertaken. Bones of wild land mammals (principally red deer and pig) occur in all of the middens, but there is no unequivocal evidence that these were procured from the sites, or that major processing of their carcasses or hides occurred there (but see Finlayson n.d.). From a study of the Cnoc Coig assemblage, Grigson concluded that antlers and bones of red deer and bones and tusks of wild boar had been introduced to the site primarily as a source of raw material for tool manufacture (Grigson & Mellars 1987).

5. None of the sites has produced convincing evidence of dwelling structures or storage facilities. The only site with extensive structural remains is Cnoc Coig on Oronsay, where two stake-hole settings and a series of hearths, small pits and concentrations of heat-affected stones were associated with a former land surface underneath the midden (Mellars 1987). These features appear to have resulted from frequent reuse of the site, and their occurrence with burnt shells and fish bones suggests that they were connected essentially with food preparation activities.

6. Many of the artifacts recovered from the Obanian middens appear to reflect 'expediency' in design, manufacture and raw material procurement. Reduction strate-

gies applied to flint and quartz frequently involved use of the bipolar technique. At the majority of sites nearly all of the lithic artifacts, and most of those made from bone, antler or shell, are simple, easily made (and hence 'disposable') items which were probably manufactured at the sites for immediate use, as such they offer an interesting parallel to the 'situational gear' of the Nunamiut which tends to be well represented at their special-purpose sites (Binford 1979)[1]. The frequent absence of retouched pieces, such as microliths and scrapers, commonly found at other Mesolithic sites in western Scotland, implies that activities (production, use, maintenance) involving such artifacts were not normally undertaken at the shell midden sites. Occasionally, elaborate artifacts such as harpoon- and mattock-heads made of bone or antler are represented in the middens. These are items that can be expected to have been 'heavily curated' (cf. Binford 1979). In some cases, they may have been brought to the sites for use there. In many instances, however, they were probably introduced as broken items intended for recycling. Some of the broken harpoon heads from MacArthur's Cave, for example, appear to have been converted into bevelled tools.

From the similarities in their location and in the character of their artifactual and faunal assemblages, it can be argued that the known Obanian sites fulfilled a specific role in the Mesolithic settlement-subsistence system of the region. The absence of major architectural features and the lack of certain categories of industrial and food refuse that can be expected to have been discarded inside residential camps, suggest that the middens do not represent the disposal of refuse from nearby base camps. From the available evidence it can be inferred that they mark the locations of special-purpose processing camps associated with the exploitation of sea food resources, principally fish and shellfish, and that food prepared at the sites was carried back to residential camps located some distance away. As noted above, it is likely that women and children played a dominant role in this economic activity and at some sites, perhaps, an exclusive one.[2]

Processing of shellfish and other food items occurred at all of the known Obanian sites. However, differences in the content of the midden deposits suggest that there was significant variation in the frequency, range, scale and intensity of the processing activities undertaken. The midden in Ulva Cave is substantially smaller than those on Oronsay, is unusually heavily dominated by shells of marine molluscs, and contains comparatively few artifacts or mammalian bones (Russell, Bonsall & Sutherland n.d.) Yet evidently it represents refuse from numerous visits to the site over several millennia, implying that the site was used less frequently than those on Oronsay, or for shorter periods, or that particularly small amounts of material were processed during individual occupations. Such differences suggest that the Obanian sites perhaps represent a spectrum of human activities between dedicated processing camps and repeatedly used dinnertime camps.[3]

Shell Middens and Site Location

It was noted above that Mesolithic shell middens in western Scotland occur in coastal caves/rockshelters and on open-air sites. Among ethnographically known shellfish gatherers the processing of shellfish for food often involves cooking (roasting or boiling) "in, over, under or around open fires" (Waselkov 1987: 100). If climatic conditions in Scotland during the Mesolithic were similar to today with, high rainfall and strong winds, then processing activities would necessitate the frequent use of some form of shelter. It follows that where caves existed near to the shell beds they would have been used occasionally for processing activities (Russell, Bonsall & Sutherland n.d.), and that construction of artificial shelters is likely to have occurred on open-air processing sites. The existence of stakehole settings, hearths and pits in association with shell refuse at Cnoc Coig on Oronsay (Mellars 1987) is consistent with this hypothesis.

The shell middens on Oronsay and around Oban Bay occur at intervals along the shoreline that was formed at the maximum of the Main Postglacial Transgression, *c.* 6000-7000 BP (Fig. 2.1). Radiocarbon dating and stratigraphic evidence indicate that most of the sites post-date or are contemporaneous with the maximum of the transgression. Since the middens are likely to have been deposited in close proximity to the contemporaneous shoreline, it is highly probable that many sites predating the maximum of the transgression (and possibly some that were contemporaneous with it) are likely to have been destroyed during that event (Bonsall & Sutherland 1992). Therefore, the known sites are most probably only a partial record of shellfish-gathering activities along those coasts during the Mesolithic.

From a study of the length distributions of otoliths in the Oronsay middens, Mellars & Wilkinson (1980) have shown that fish were procured from the different sites at specific times of the year. Equivalent seasonality data are not available for the shellfish, but it seems reasonable to assume that they were collected and processed at the sites at the same times of year. Although shellfish probably contributed less to the overall food supply than either fish or sea mammals, it may be suggested that they exercised a significant influence on site location. Oronsay (like many coasts in western Scotland) has a rocky shoreline. As Waselkov (1987: 115) has observed, shellfish associated with such shorelines constitute a dispersed resource which can be exploited most effectively at numerous points. Thus, the choice of location for the processing sites on Oronsay may have been determined primarily by the need for seasonal scheduling of the shellfish resources.

From the ethnographic and archaeological evidence discussed, it can be predicted that remains of Mesolithic processing camps will occur at numerous points around the coasts of western Scotland, around or above the level of the Main Postglacial Shoreline. Unlike Oronsay, many

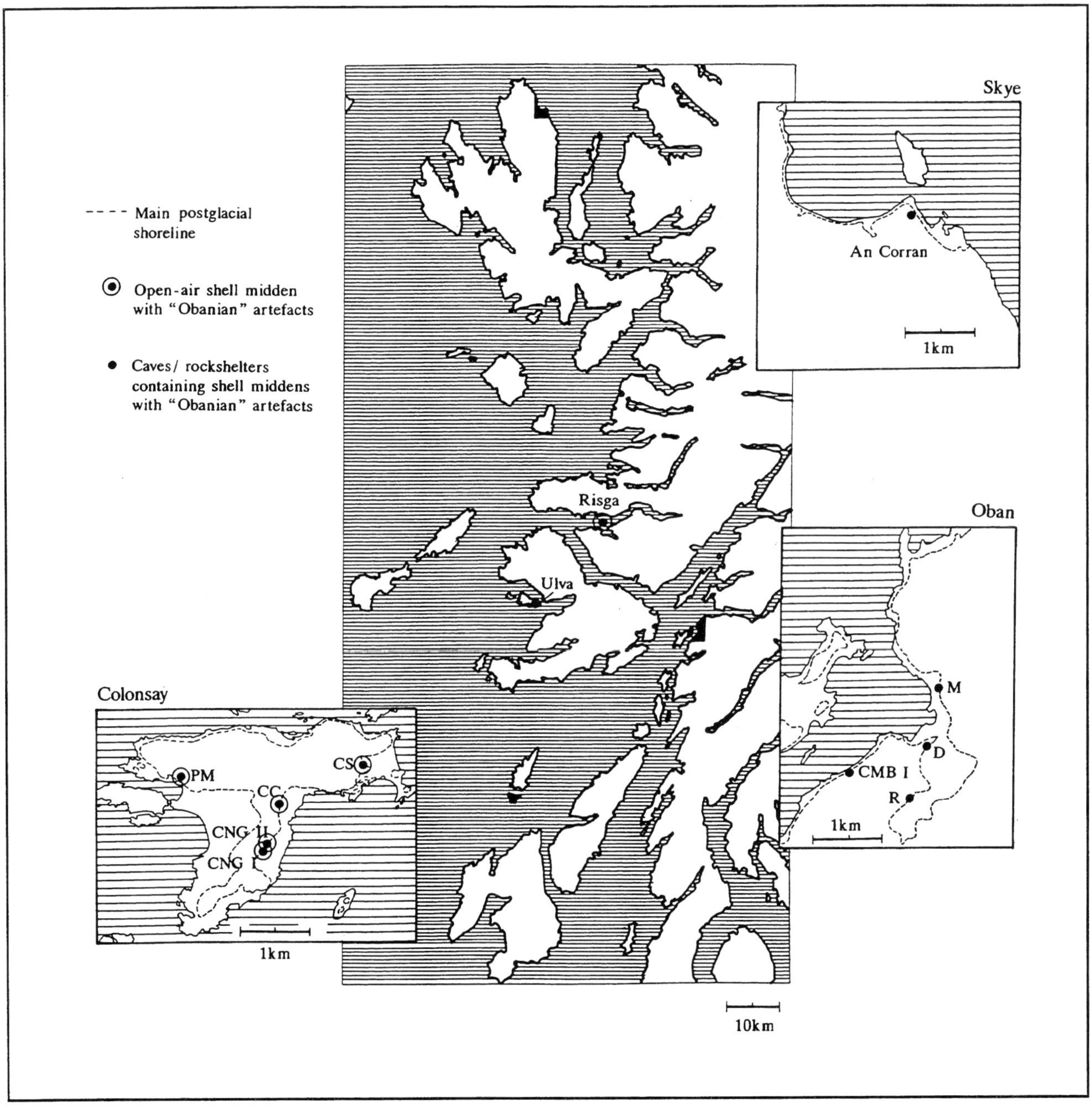

Figure 2.1. Obanian sites in western Scotland (after Bonsall & Sutherland 1992, with revisions): CC. Cnoc Coig; CMB. Carding Mill Bay; CNG. Caisteal nan Gillean; CS. Cnoc Sligeach; D. Druimvargie Rockshelter; M. MacArthur's Cave; PM. Priory Midden; R. Raschoille Cave.

coasts have no blown sand deposits which could preserve open-air shell middens. The former locations of such middens, however, can be expected to be represented by residual scatters of lithic artifacts similar in character to those from known shell midden sites. Such sites are likely to be of limited extent, and hence difficult to locate. Moreover, in the absence of microliths or other diagnostic artifacts, it is doubtful if they would be immediately identifiable as Mesolithic. In a recent survey of the Isle of Ulva, several such scatters were recognized (Russell, Bonsall & Sutherland 1995: Fig. 2.1). None of the sites, however, has been investigated further and their ages are unknown.

Conclusions

In this paper it has been argued that the contrast between Obanian sites and sites with microlithic industries evident in the Mesolithic of western Scotland cannot be explained satisfactorily in cultural or chronological terms, but is primarily a reflection of differences in the conditions of preservation at the two types of site and the nature of the activities undertaken there.

A case can be made for interpreting most, and possibly all, of the known Obanian sites as remains of special-purpose processing camps associated primarily with the exploitation of shellfish and fish. The use of such logistic camps appears to have been an integral part of an economic system that operated in western Scotland (if not more widely in northern Britain) throughout the greater part of Mesolithic and into period after *c.* 5000 BP and the widespread adoption of farming in the region. The model proposed explains a number of 'unusual' features of Obanian sites, such as their frequent occurrence on open coasts and on small islands which lacked resident populations of large land mammals, and the prevalence in the middens of expediently designed and manufactured tools.

The existing evidence, however, has obvious limitations. Although there has been a marked increase in research activity over the past 25 years, with new excavations at sites on the islands of Oronsay, Skye and Ulva, and on the mainland around Oban Bay, the full results of this research are not yet available. Consequently, there is a lack of detailed information on the faunal composition of the majority of Obanian sites and on functional and technological aspects of the Obanian toolkit. There is also a need for additional research, including studies to determine the seasonality of shellfish gathering at individual sites like those conducted by Margaret Deith (1983, 1986) at sites in eastern Scotland and northern Spain, and for more extensive excavations to expose areas adjacent to middens in order to test for associated architectural features and to recover information on site structure in general. The interpretation of the Obanian sites presented in this paper, therefore, should be viewed as a 'working hypothesis', which can be tested, and modified, as new data become available.

Acknowledgements

The author wishes to thank Bill Finlayson for citation of an unpublished paper, Douglas Harkness for advice on radiocarbon dates and calibration, and László Bartosiewicz and Nicola Murray for their comments on an earlier draft of this paper.

Endnotes

1. The principal categories of artifacts with an 'expedient' aspect occurring in Obanian sites are unretouched flakes of flint and/or quartz often produced by the bipolar technique, bevelled tools, and small bone points. There has been little systematic research on the possible uses of these artifacts. Finlayson (in Connock et al. 1993) has observed that a bipolar reduction strategy applied to small pebbles of flint or quartz produces numerous flat, sharp flakes which could be used without further modification as 'knives' for processing fish. Experimental studies of bevelled tools and bone joints are being undertaken at the University of Edinburgh. Pending the results of this research, the present writer favours the view advocated by some previous workers (e.g., Grieve 1885, Bishop 1914) that the bevelled tools were used in the processing or collection of limpets. (In this context it is worth noting that 'Asturian picks' from Mesolithic shell middens along the north coast of Spain, usually interpreted as 'limpet hammers', often exhibit a degree of bevelling at the tip.) The small bone points which occur in large numbers in some Obanian middens may also have been used in the processing of shellfish, e.g. for removing the soft parts of periwinkles and dog-whelks from their shells, and for piercing shellfish/fish meat so that it could be strung up for drying.
2. The age/sex composition of the task group occupying a site is likely to have some influence on the character of the resulting artifact assemblage. The incidence of bipolar débitage is likely to be greater where women and children were more actively involved than men in the use of the site, on the assumption that generally they were less skilled flint knappers. Similarly, items traditionally made and used by men (e.g., remains of hunting equipment) ought to be poorly represented at sites where men and teenage boys were not present or only occasionally participated in the activities of the site.
3. The midden at An Corran (Skye) is unusual in possessing what Saville & Miket (1994) have termed a 'conventional' Mesolithic assemblage with blade débitage, microliths and other retouched artifacts. But the site is also unusual in respect of access to lithic resources. Compared to other Obanian sites a wider range of materials was available locally, including large pieces of high quality volcanic rock; a significant proportion of the microliths and blade débitage are apparently made from this volcanic material. In addition to 'normal' processing activities, therefore, An Corran was possibly important for raw material procurement, and stages in the production of lithic items related to the manufacture or repair of personal gear may have been undertaken there. Such tasks are likely to have involved men, and could have been performed in conjunction with the processing of fish and shellfish or at other times. A similar interpretation can be placed on the structural remains excavated at Morton on the east coast of Scotland.

References

Anderson, J.

1895 Notice of a cave recently discovered at Oban, containing human remains and a refuse heap of shells and bones of animals, and stone and bone implements. *Proceedings of the Society of Antiquaries of Scotland* 29: 211-230.

1898 Notes on the contents of a small cave or rock shelter at Druimvargie, Oban; and of three shell mounds on Oronsay. *Proceedings of the Society of Antiquaries of Scotland* 32: 298-313.

Binford, L.R.

1979 Organization and formation processes: looking at curated technologies. *Journal of Anthropological Research* 35: 255-273.

Bishop, A.H.

1914 An Oronsay shell-mound - a Scottish pre-Neolithic site. *Proceedings of the Society of Antiquaries of Scotland* 48: 52-108.

Bonsall, C. & M.R. Robinson

1992 *Archaeological Survey of the Glenshellach Development Area, Oban: Report to Historic Scotland.* Edinburgh: University of Edinburgh Department of Archaeology.

Bonsall, C., M.R. Robinson, R. Payton, & M.G. MacKlin

1993 Lón Mór, Oban. *Discovery and Excavation in Scotland* 1993: 76.

Bonsall, C. & C. Smith

1990 Bone and antler technology in the British Late Upper Palaeolithic and Mesolithic: the impact of accelerator dating. In *Contributions to the Mesolithic in Europe* (P.M. Vermeersch & P. Van Peer, eds.), pp. 359-368. Leuven: Leuven University Press.

1992 New AMS ^{14}C dates for antler and bone artifacts from Great Britain. *Mesolithic Miscellany* 13: 28-34.

Bonsall, C. & D.G. Sutherland

1992 The Oban Caves. In *The South-West Scottish Highlands: Field Guide* (M.J.C. Walker, J.M. Gray & J.J. Lowe, eds.), pp. 115-121. Cambridge: Quaternary Research Association.

Bonsall, C., D.G. Sutherland, N.J. Russell, G. Coles, C.R.C. Paul, J.P. Huntley & T.J. Lawson

1994 Excavations in Ulva Cave, western Scotland 1990-91: a preliminary report. *Mesolithic Miscellany* 15: 8-21.

Bonsall, C., C. Tolan-Smith & A. Saville

n.d. Direct Dating of Mesolithic antler and bone artifacts from Great Britain: new results for bevelled tools and red deer antler mattocks.

Breuil, H.

1922 Observations on the Pre-Neolithic industries of Scotland. *Proceedings of the Society of Antiquaries of Scotland* 56: 261-281.

Clark, J.G.D.

1956 Notes on the Obanian with special reference to antler and bone work. *Proceedings of the Society of Antiquaries of Scotland* 89: 91-106.

Coles, J.M.

1971 The early settlement of Scotland: excavations at Morton, Fife. *Proceedings of the Prehistoric Society* 37: 284-366.

Connock, K.D., B. Finlayson & C.M. Mills

1993 Excavation of a shell midden site at Carding Mill Bay near Oban, Scotland. *Glasgow Archaeological Journal* 17: 25-38.

Deith, M.R.

1983 Molluscan calendars: the use of growth-line analysis to establish seasonality of shellfish collection at the Mesolithic site of Morton, Fife. *Journal of Archaeological Science* 10: 423-440.

1985 Seasonality from shells: an evaluation of two techniques for seasonal dating of marine molluscs. In *Palaeobiological Investigations: Research Design, Methods and Data Analysis* (N.R.J. Fieller, D.D. Gilbertson & N.G.A. Ralph, eds.), pp. 119-130.

1986 Subsistence strategies at a Mesolithic campsite: evidence from stable isotope analysis of shells. *Journal of Archaeological Science* 13: 61-78.

1995 Complexity in the Mesolithic of the western Scottish seaboard. In *Man and Sea in the Mesolithic. Coastal Settlement Above and below Present Sea Level* (A. Fischer, ed.). Oxford: Oxbow Books/Danish National Forest & Nature Agency.

Grieve, S.

1885 *The Great Auk or Garefowl (Alca impennis L.): Its History, Archaeology and Remains.* London: Jack.

Grigson, C. & P.A. Mellars

1987 The mammalian remains from the middens. In *Excavations on Oronsay* (P.A. Mellars, ed.), pp. 243-289. Edinburgh: Edinburgh University Press.

Harkness, D.D.

1983 The extent of natural ^{14}C deficiency in the coastal environment of the United Kingdom. In *Proceedings of the First International Symposium*

^{14}C and Archaeology (W.G. Mook & H.T. Waterbolk, eds.), pp. 351-354. Strasbourg: Council of Europe.

Hedges, R.E.M., R.A. Housley, L.A. Law & C.R. Bronk
1989 Radiocarbon dates from the Oxford AMS system: Archaeometry datelist 9. *Archaeometry* 31: 207-234.

Jacobi, R.M.
1982 Lost hunters in Kent: Tasmania and the earliest Neolithic. In *Archaeology in Kent to AD 1500* (P.E. Leach, ed.), pp. 12-24. London: Council for British Archaeology Research Report No. 48.

Lacaille, A.D.
1954 *The Stone Age in Scotland.* Oxford: Oxford University Press.

Meehan, B.
1982 *From Shell Bed to Shell Midden.* Canberra: Australian Institute for Aboriginal Studies.

Mellars, P.A.
1972 The Paleolithic and Mesolithic. In *British Prehistory: a New Outline* (C. Renfrew, ed.), pp. 41-99. London: Duckworth.
1987 *Excavations on Oronsay. Prehistoric Human Ecology on a Small Island.* Edinburgh: Edinburgh University Press.

Mellars, P.A. and M.R. Wilkinson
1980 Fish otoliths as indicators of seasonality in prehistoric shell middens: the evidence from Oronsay (Inner Hebrides). *Proceedings of the Prehistoric Society* 46: 19-44.

Mithen, S.J. & B. Finlayson
1991 Red deer hunters on Colonsay? The implications of Staosnaig for the interpretation of the Oronsay middens. *Proceedings of the Prehistoric Society* 57: 1-8.

Morrison, A. & C. Bonsall
1989 The early post-glacial settlement of Scotland. In *The Mesolithic in Europe* (C. Bonsall, ed.), pp. 134-142. Edinburgh: John Donald.

Movius, H.L.
1940 An early post-glacial archaeological site at Cushendun, County Antrim. *Proceedings of the Royal Irish Academy* 46: 1-84.
1942 *The Irish Stone Age. Its Chronology, Development and Relations.* Cambridge: Cambridge University Press.

Rees, T., G. Kozikowski & R. Miket
1994 Investigation of a shell midden at An Corran, Staffin, Isle of Skye. Unpublished report.

Russell, N.J., C. Bonsall & D.G. Sutherland
1995 The exploitation of marine molluscs in the Mesolithic of Western Scotland: evidence from Ulva Cave, Inner Hebrides. In *Man and Sea in the Mesolithic. Coastal Settlement Above and Below Present Sea Level* (A. Fischer, ed.), pp. 273-288. Oxford: Oxbow Books/Danish National Forest & Nature Agency.

Saville, A. & R. Miket
1994 An Corran rockshelter, Skye: a major new Mesolithic site. *PAST* 18: 9-10.

Stuiver, M. & P.J. Reimer
1993 Extended ^{14}C data base and revised CALIB 3.0 ^{14}C age calibration program. *Radiocarbon* 35:215-230.

Waselkov, G.A.
1987 Shellfish gathering and shell midden archaeology. In *Advances in Archaeological Method and Theory, Vol. 10* (M.B. Schiffer, ed.), pp. 93-210. New York: Academic Press.

Wickham-Jones, C.R.
1990 *Rhum. Mesolithic and Later Sites at Kinloch. Excavations 1984-86.* Edinburgh: Society of Antiquaries of Scotland Monograph No. 7.

Woodman, P.C.
1989 Ireland and Scotland: the Mesolithic of Europe's western periphery. Paper presented at the Conference on Scottish-Irish Links, Edinburgh 1989.
1990 A review of the Scottish Mesolithic: a plea for normality! *Proceedings of the Society of Antiquaries of Scotland* 119: 1-32.

3

Social Structure and Ideology of the Late Mesolithic Communities in North Temperate Europe

Marek Zvelebil
University of Sheffield

Abstract

This contribution deals with the problem of understanding social structure and ideology of Stone Age communities in north western and northern Europe. More specifically, it focuses on (1) differences in social organisation among modern hunter-gatherers, (2) indications of social elaboration amongst the Mesolithic hunter-gatherers of North Temperate Europe, and (3) ideology of Mesolithic hunter-gatherers and its relation to their social structure.

Modern Hunter-Gatherers: The Current Debate in Anthropology

In recent years a debate has been raging in anthropology which calls into the question the very existence of hunter-gatherers as a discrete social type, and of hunting and gathering as a self-contained way of life (e.g., Bender & Morris 1988, Bettinger 1991, Bird-David 1988, 1992a,b; Kent 1992, Wilmsen & Denbow 1990, Solway & Lee 1990, Lee 1992). The current debate includes those who, while recognising the cultural variability and flexibility of hunter-gatherers, argue that they share common traits that define them as a group (Lee 1992, Solway and Lee 1990, Bettinger 1991, Kent 1992, 1993; Ingold 1983, 1986, 1988; Woodburn 1982, 1988; Bird-David 1992a,b). As a group, hunter-gatherers are economically dependent on hunting, fishing and gathering. Their social relations typically reflect the personal skill of individual hunters and gatherers in using the resources occurring in the environment. Prestige and status are dependent on personal ability, rather than on land or resource ownership. Ecological relations are dominant and set limits to resource utilisation; social relations are selective in specifying how the environment is utilised. Such hunter-gatherers maintain the practice of sharing, embedded in resources owned communally, 'ownership' conferring the right to distribute food in return for social prestige.

Within such a hunter-gatherer 'mode of production' (Ingold 1988), or 'band society' (Lee 1990, 1992), the ideologically-sanctioned practice of sharing plays a key role. But this is the case only within the context of an ideology which is intolerant of ownership of resources, private capital, status differences, and social hierarchies (Kent 1993, see also Lee 1990).

Ingold goes on to differentiate between hunting-gathering as a mode of subsistence and as a mode of production (Fig. 3.1), where the "essence of hunting and gathering as production lies in the intentional control of procurement behaviour by the self-conscious human object" (1988: 273). That includes not only labour, the technology and the targeted resources, but also the social relations of production. Among hunter-gatherers, these typically include open access to resources, sexual division of labour and sharing. Ingold sees the underlying ideology of hunter-gatherers in terms of ritual appropriation of resources (1988: 277, 1986: 243-256, 1983: 555), but emphasises the practice of sharing as constituting the determinant "social relations of the hunter-gatherer mode of production" (1988: 283).

Woodburn (1982, 1988) takes the argument a step further by making a distinction between *immediate-return* and *delayed-return* hunter-gatherers. An immediate-return system is oriented towards procuring food for immediate consumption using simple technology by groups with an ideology oriented to sharing and to the present, where "distinctions of wealth, power and status are systematically eliminated" (1988: 34), while in delayed-return systems, activities are more diachronically oriented, and people hold rights over four types of assets (Woodburn 1988: 32): (1) facilities requiring advance labour investment (fish traps, weirs), (2) processed and stored food, (3) wild products improved by human labour, and (4) rights held by men over their female kin.

In contrast to these workers, there are those who emphasise the historical perspective and regard contemporary hunter-gatherers as products of their own (modern) history. Many egalitarian 'simple' foragers are seen as a

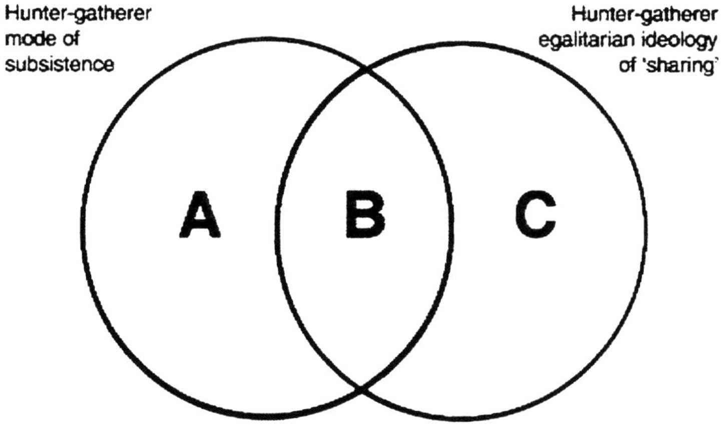

A Hunter-gatherer groups with hunter-gatherer 'mode of subsistence': hunting, fishing, gathering but with non-egalitarian ideology allowing for some inequalities and ranking.

B Hunter-gatherer groups with 'mode of production': based on hunting, fishing and gathering and egalitarian ideology and sharing.

C Hunter-gatherer groups with some farming but with ideology of sharing, who perceive themselves as hunter-gatherers.

Figure 3.1. Mode of production and mode of subsistence in hunter-gatherer societies.

product of resistance to 'encapsulation' by farming societies (Schrire 1984, Bender & Morris 1988, Pedersen & Waehle 1988, Headland & Read 1989), or as a result of alternating foraging and farming lifestyles (Griffin 1984, Denbow 1984, Gordon 1984, Wilmsen & Denbow 1990). The very existence of hunter-gatherers as a distinct and cohesive unit is questioned (Schrire 1984, Bender & Morris 1988, Wilmsen & Denbow 1990). In summary, modern hunter-gatherers are regarded as socially and economically heterogenous societies whose common features are specified, modified or over-ridden by historical contingency: they are primarily a product of their own historical development and of contact with food-producing communities. Within this point of view, the egalitarian social structure of many foragers can be regarded as a form of resistance to their 'encapsulation' by farming societies (Fig. 3.2).

To my mind, these and similar studies have demonstrated three matters of great significance: first, the historical contingency of hunter-gatherer societies; second, the existence of historically evolving relationships between foragers and farmers, and third, the persistence of the egalitarian ideology of sharing among some forager communities, despite the long duration of such foraging-farming contacts.

The general conclusion which can be drawn from this is that although economic and technological elaboration has broadly consonant analogues in social elaboration (e.g., Woodburn 1988), hunter-gatherer modes of subsistence and their mode of production are not always co-eval, allowing for social elaboration to occur which goes beyond the hunter-gatherer mode of production (as in delayed-return societies), while elements of farming can be incorporated within its mode of production, without the abandonment of egalitarian ideology. Only some hunter-gatherers share both the hunter-gatherer mode of production and mode of subsistence (Fig. 3.1).

The North Temperate European Mesolithic

I would now like to turn my attention to the late Mesolithic period in the basin of the Baltic Sea in Northern Europe. The period in question dates between ca. 6000–3000 BC. Returning to Woodburn's model of delayed-return hunter-gatherers, it seems that in the peri-Baltic coastal zones of North Temperate Europe, late Mesolithic communities included all the features of delayed-return hunter-gatherers defined by Woodburn (1988: 34):

(1) Valuable technical facilities used in production are represented by boats, nets, weirs, and fish traps, all of which require considerable labour investment, and which acted as delayed-return facilities in the North European Mesolithic at least since the mid-seventh millennium BC, as shown by the early net find at Antrea on the Karelian Isthmus (Clark 1975, Andersen 1987, Zvelebil 1995b with references, Fisher 1995, Pedersen 1995, Pedersen et al. 1997) (Fig. 3.3).

(2) Processed and stored food materials have been inferred from ceramic jars filled with plant food as at Orimattila in Finland, seal fat as on Aland, in niches in houses as at Sarnate; and from the presence of putative drying/curing/smoking equipment on settlements dating

from the mid-fifth millennium BC (Vankina 1970, Zvelebil 1981, Rowley-Conwy & Zvelebil 1989, Zvelebil 1994, Nuñez 1990 with references).

(3) Human activities which would lead to an increase in the productivity of the environment included coppicing and clearing of woodland, evidenced in Denmark, southern Sweden and northern Poland (as well as Britain) (Zvelebil 1994, Pedersen 1995), selective culling strategies and possibly management of wild pig evidenced by their selective cull, and apparently transport and confinement (Zvelebil 1994, 1995b with references, but see Rowley-Conwy 1995).

(4) Finally, Woodburn notes "assets in the form of rights held by men over their female kin who are then bestowed in marriage on other men" as the fourth type of asset characterising a delayed-return society (1988: 32). Although claims have not been made for the existence of rights held by men over their female kin, claims have been made for the existence of descent-based, or unilineal social structure in the late Mesolithic (see below).

Our understanding of social structure and ritual in the late Mesolithic is based principally on evidence from burials and I should now like to review briefly this evidence.

Major burial areas are shown in Figure 3.4; individual burials are not shown. In northern and eastern Europe, five clusters can be recognised: the Western Baltic, the Eastern Baltic, the Ladoga-Onega region, the Middle Dneiper region, and the Danube Gorges. This distribution reflects not only the intensity of research, but also the favourable ecological conditions of these areas for hunter-gatherer settlement: all clusters occur in coastal or major lacustrine or riverine zones, marked by the concentration of predictable aquatic resources (Rowley-Conwy 1983, Zvelebil 1986, Zvelebil & Dolukhanov 1991).

Between them, the burial grounds cover the entire Mesolithic period, from ca. 8000 BC to the beginning of the second millennium. Some are cemeteries in that the interments are grouped in burial grounds marked exclusively for ritual and burial; others are locations where the prevailing practice was to bury the dead within the houses or within settlements. At a closer inspection, we find three key areas: Ladoga-Onega, the East Baltic and the West Baltic.

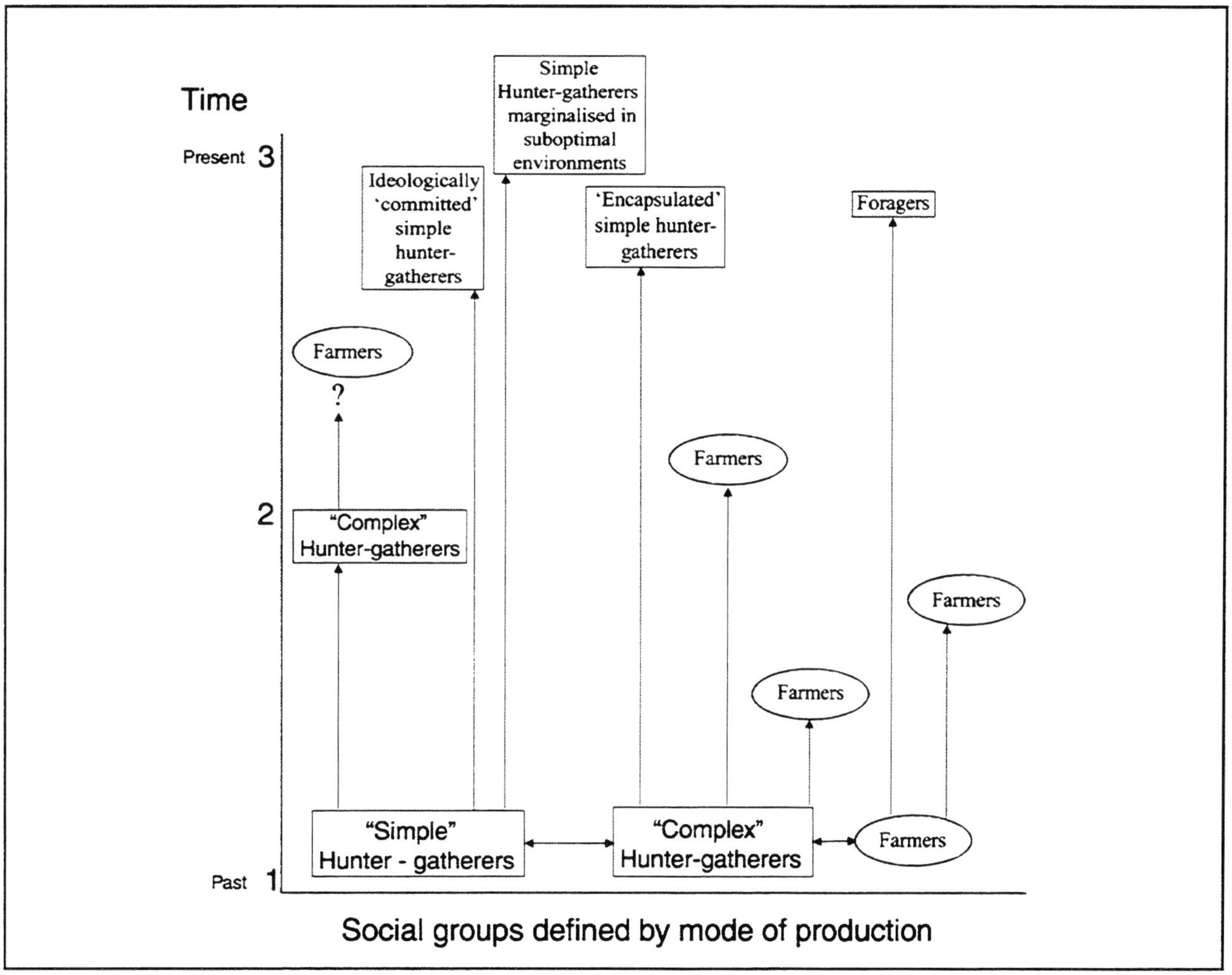

Figure 3.2. Socio-economic developments among hunter-gatherer societies through time.

Ladoga-Onega

In the Ladoga-Onega region, Olenii Ostrov is the largest of the burial grounds. Dated to the mid-sixth millenium BC (Price & Jacobs 1990), the cemetery revealed the existence of at least seven social dimensions, expressing band membership, age, sex, personal wealth, and three specialised ranks (O'Shea & Zvelebil 1984, Zvelebil 1993, but see Jacobs 1995). It seems that two groups, possibly lineages or clans, were using Olenii Ostrov; this is evident from two spatial clusters within the cemetery, and from the association of elk sculptures with the northern cluster, and of snake and human effigies with the southern cluster.

Sex distinctions were expressed by articles placed in the graves of the deceased. Bone points, bone harpoons, axes, flakes and slate knives were associated with males. An absence of implements, ornaments and perforated beaver incisors was associated with females.

Perforated tooth pendants exhibited a clear hierarchical order relative to each other, which corresponded with the number and variety of other goods found in the graves. Graves with bear tusks denoted the most wealthy, followed by graves containing either elk or beaver incisors, and finally those graves with no pendants. These 'wealth ranks' varied with age, so that the adults possessed the greatest quantities, while the young and the old exhibited lower quantities. Such age-dependant change was less pronounced with females, indicating possibly that female wealth markers were obtained through either affinal or consanguinal ties to males.

Special status was accorded to individuals interred in a standing or leaning position in vertical shaft graves. In contrast to others, these graves were oriented westwards. They included two males, two females, one of which was a young adult. Three of the four individuals possessed grave assemblages of the highest wealth level. By reference to ethnographic analogies provided by recent hunter-gatherers in Siberia (Czaplicka 1914, Levin & Potapov 1956, Aleksenko 1967, Resketov 1972, Dioszegi & Hoppal 1978, Ingold 1986), the special shaft graves at Oleneostrovski Mogilnik become comprehensible as graves of ritual specialists, possibly shamen. First, while everyone else was facing east, the western orientation of these graves can be explained as facing the entrance to the Lower World, the domain of spirit ancestors of shamen and of the rulers of the underworld. The souls of the rest of the group inhabited the Upper World, entered from the east. The presence of beaver incisors, a category of pendants normally associated with females, in the shaft graves irrespective of sex, can also be explained in terms of the shaman's role as a spiritual mediator with the underworld on behalf of both sexes, his or her robe consequently retaining symbols of both genders. Finally, beaver mandibles, found in the shaft graves are items ethnographically forming a part of shamens' ritual equipment.

Two other special ranks are present at the cemetery: males interred with bone projectile points only, and interments, both male and female, buried with carved effigy figures. The apparent relationship between the effigy figures and the spatially separate bands would suggest than these individuals held a position central to the ritual identity of the clans or lineages. The range of ages among office holders – adolescent to old – and the independence of personal wealth may also suggest an ascriptive or hereditary component to this social position.

In summary, we seem to be dealing with a descent-based society organised into sequential hierarchies (Aldenderfer 1993), which are linked to ritual (i.e., shaman, effigy holders, possibly bone-point holders), but which operate independently of one another and which are divorced from the general wealth ranks. While the ritual roles can be inherited (i.e., child/juvenile effigy holders), the wealth cannot: it tends to decline in old age. Both men and women could accquire a high-status position, although men tended to acquire higher rank more often than women. Sex-specific variation in the grave goods and the age and sex associations in collective burials suggest that we may be dealing with a patrilineal, or dual-descent social structure.

East Baltic Sites

Olenii Ostrov is only one among several burial grounds in Eastern Europe displaying similar variation. In the East Baltic, the two major cemeteries are Zveinieki in northern and Abora in eastern Latvia. At Zveinieki, the cemetery was used intermittently over three thousand years between ca. 4800 and 1800 BC (Zagorskis 1987). Throughout this period, we find the use of the same wild-animal symbolism as at Olenii Ostrov, as well as differences in status similar to those at Olenii Ostrov. As at Olenii Ostrov, we have individual and collective burials. Some graves were marked by gravestones, or equipped with stone lining: a feature characteristic also of the Jäkärlä Culture graves in southern Finland dated to the early fourth millennium bc (Edgren 1966).

The end phase of the Zveinieki cemetery is contemporary with burials at Abora, where 61 interments were placed in the central part of a fairly permanent hunter-gatherer settlement (Loze 1979). Single, dual and collective burials, perforated tooth pendants, sculptures of waterbirds (8), elk (1), beaver (1), bear (1) and snake (1) attest to the same range of burial practices and symbolism seen already at Olenii Ostrov. The presence of amber beads and the lack of any orientation of burials marks a difference from the other burial grounds. The absence of pottery is striking, since the Abora community belongs among ceramic-using hunter-gatherers. So the same symbolism, the same ideology appears to have lasted in the East Baltic until the mid-second millennium BC.

West Baltic Sites

The burial practice in the West Baltic shows both differences and similarities with the East Baltic. The main concentrations of burials are at Skateholm, southern Swe-

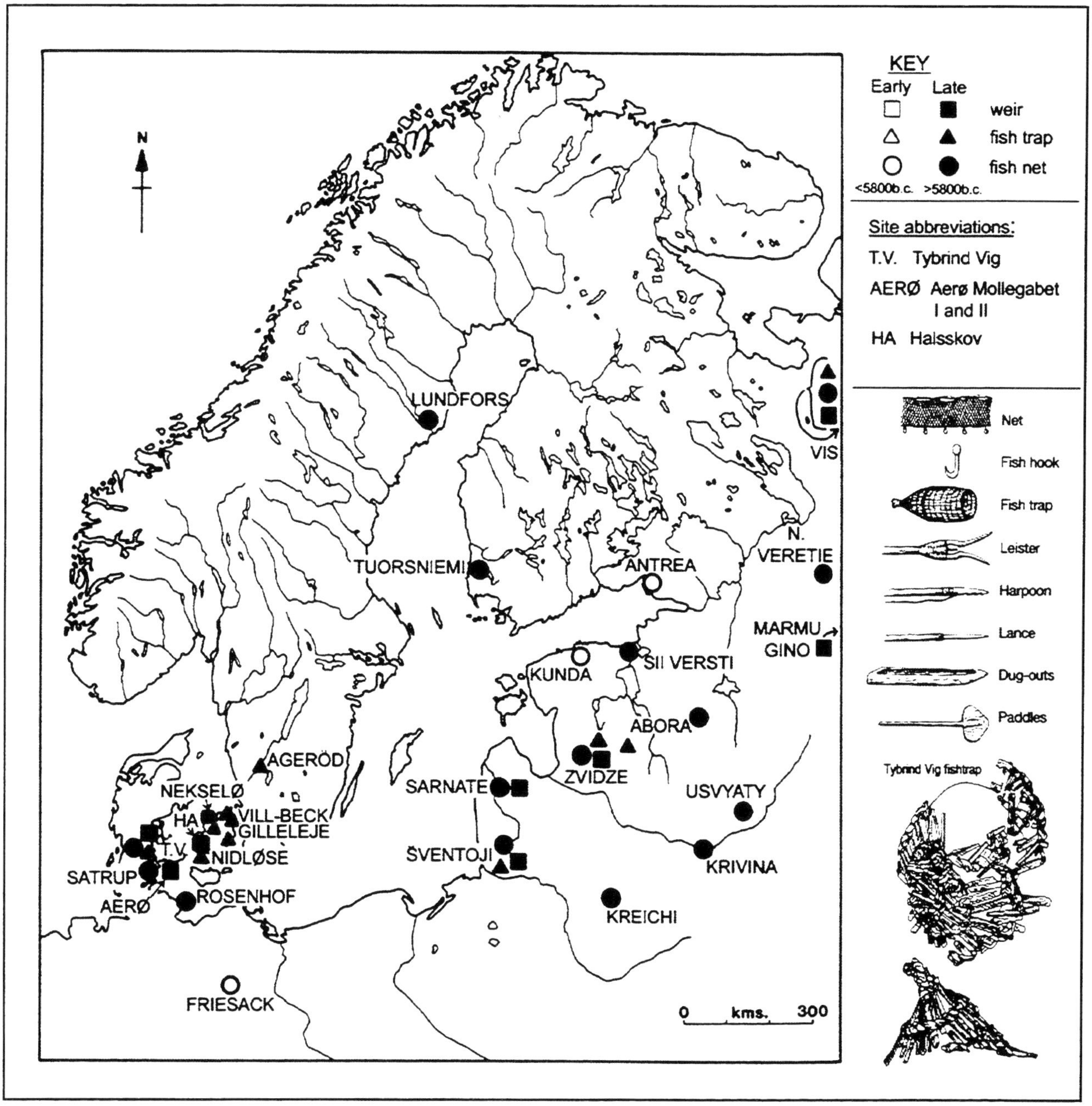

Figure 3.3. The distribution of Mesolithic capture facilities in the Baltic Sea basin. Since the compilation of this figure, a number of additional fixed capture facilities have been discovered, paricularly in Zealand and the surrounding regions of the west Baltic. These include fishing structures, eel baskets and netting equipment (see Pedersen 1995 and Pedersen et al. 1997 for further details).

den and Vedbaek-Bøgebakken in Zealand (Denmark); both date to the Early and the Middle Ertebølle period, ca. 4500 – 3800 BC. While the two burial complexes at Skateholm (I and II) were true cemeteries, the Vedbaek complex consists of burials interred in settlements which dotted the ancient coastline of the Vedbaek fjord (Clark & Neeley 1987, Larsson 1989, 1990a, 1990b, 1993).

At Skateholm, two cemeteries, both located on islands and containing some 85 graves, were excavated by Lars Larsson (1993). In brief, it is again possible to distinguish three wealth or status ranks, with younger females and older males receiving the richest grave goods (Larsson 1989). Burials include cremations, interment in sitting position, double graves containing both women and men with children, rich child graves and dog burials. Timber structures were built over two graves at Skateholm I, while Skateholm II had a mortuary house. With their exceptional grave goods and burial ritual (dog sacrifice in SII - 8, head-dress covering a young man SII-15), the sitting burials can be linked to the shaft graves at Olenii Ostrov, while the rich child graves again suggest some inherited social position.

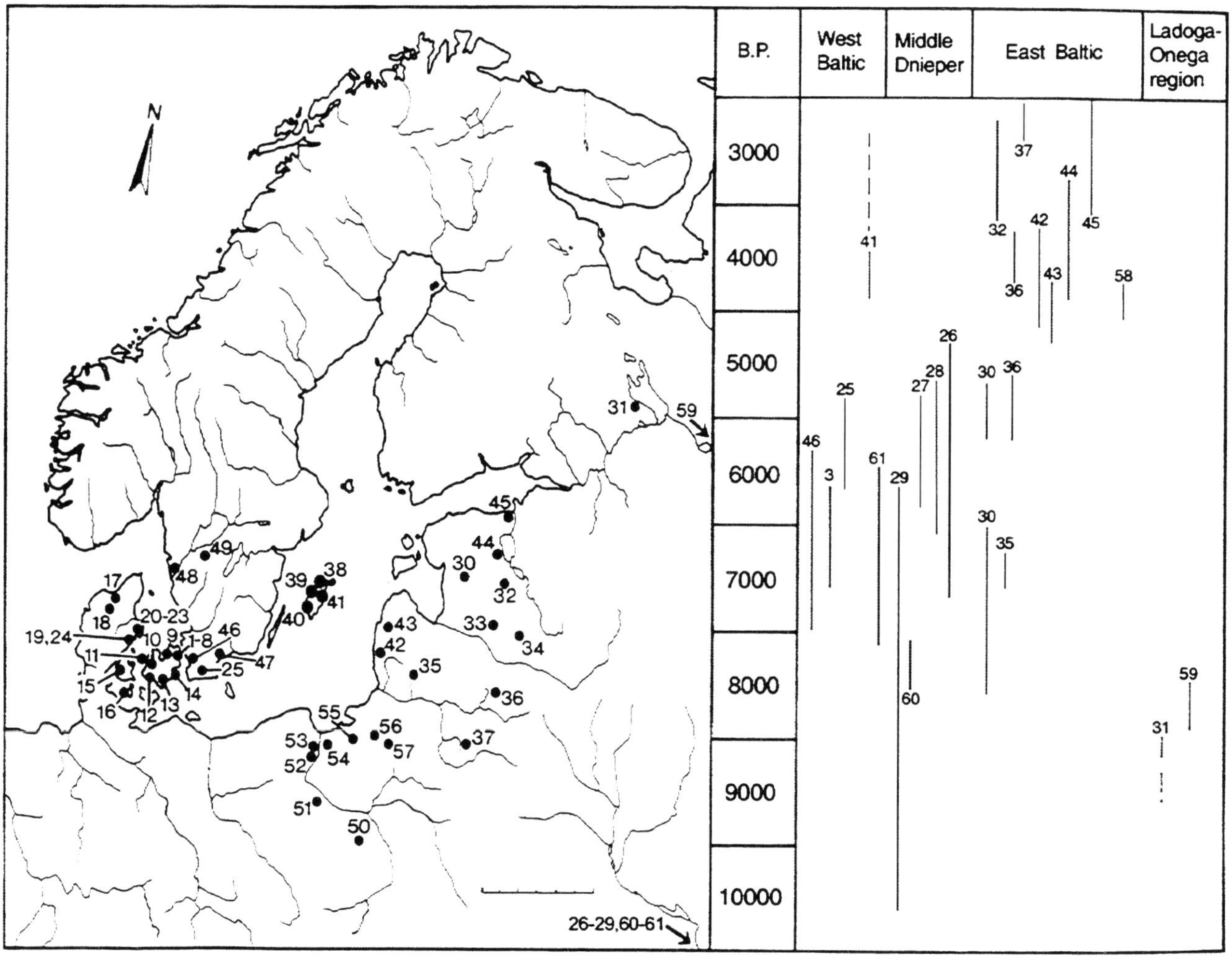

Figure 3.4. The distribution of Mesolithic cemeteries and burial grounds in Europe.

In an effort to understand the division of labour, Condstandse-Westermann and Newell (1989) have investigated markers of stress caused by physical labour: limb laterisation and joint degeneration at Skateholm I and II. They conclude that higher-ranking females show consistently less laterisation than those in poorer graves, while the males show the opposite pattern: stronger laterisation in higher strata. Ethnographically, such a status-related pattern is consistent with the distribution of work-related stress in socially stratified societies. Moreover, the results suggest that women at Skateholm gained their status through means other than physical work: as child-bearers, through descent (suggesting a matrilineal society) or by association with high-ranking males.

At Vedbaek, the grave goods divide the burials again into three groups, and again, as at Olenii Ostrov and Skateholm, there are rich juvenile graves, while some of the oldest individuals are buried without any personal equipment (Albrethsen & Brinch Petersen 1976, Larsson 1990b). Similar situation can be detected at other burial sites in Denmark. In terms of grave-goods, some of the richest graves are those of women (e.g., at Nederst 5, Dragsholm, Bogebakken 8), and a number of artefacts have gender-specific connotations: for example, deer-antler axe was a male accessory, while slotted bone point was a female item. (Kannegaard Nielsen & Brinch Petersen 1993).

As a cemetery – a formal burial area – Skateholm has been interpreted as a territorial marker of a unilineal descent group claiming rights to resources through ancestors (Larsson 1993, Rowley-Conwy 1998). The internal groupings within Skateholm I may suggest discrete lineages, while the notion of territoriality receives further support in the distribution of functionally analogous artifact types such as axes, outlining discrete territories (Vang-Petersen 1984, Larsson 1989, Rowley-Conwy 1998).

In anthropological and ethnographic terms, then, many of the late Mesolithic communities in the coastal areas of the Baltic Sea would fall among 'delayed-return' foragers, whose social structure was, if anything, more hierarchically ranked than was the case among the delayed-return foragers in the ethnographic record. Status distinctions along the major, horizontal social dimensions of age, sex, and achieved status are discernible in general (Clark & Neeley 1987: 125), while evidence for vertical differentiation occurs at Skateholm, Zveinieki, and Olenii Ostrov as discussed in this paper, as well as at other cemeteries along the Atlantic Façade in France and Portugal (Clark & Neeley 1987: 125-126, Schulting 1996). This being the case, such late Mesolithic forager communities would fall within the hunter-gatherer mode of subsistence, but not of production: in other words, we might expect that the ideology of sharing would be at least partly replaced by a cognitive framework which would permit social differentiation and accumulation of wealth.

Ideology

What do we know about the ideology of prehistoric hunter-gatherers in the north of Europe? The essential components of the boreal foragers' belief system, abstracted by Ingold (1986) from Siberian and North American ethnographic data, focus on two things:

(1) A division of the universe into three horizontal layers: sky, earth and the underworld, which correspond to air, land and water respectively. These layers are linked by a 'cosmic pillar' (Ingold 1986), or 'cosmic river' (Anisimov 1963), symbolised in the shaman's *turu* or a tree often placed in the centre of the shaman's tent.

(2) A division of humans and animals into a physical self, a body-soul, and a free-soul. Human beings and those animals who are masters of their animal charges, such as the bear, possess all three substances; wild animals normally possess the physical self and the body soul (their collective 'free soul' residing in the animal master), while among domestic animals 'the spirit of the domestic animal is the soul of man, controlling the animal from without' (Ingold 1986: 255).

Within this cognitive framework, elk (or other deer), bear, and waterbirds play clearly defined roles as guardians of other animals and as channels of communication with the other, non-terrestrial worlds. The 'heavenly elk', for example, is an inhabitant of the heavens, and a central actor in myths of revival and regeneration, and in the mediation between the worlds of the spirits and of humans (Vasilevitch 1963, Anisimov 1963, Tilley 1991). The bear plays an analogous but somewhat different role as the chief guardian of wild animals and a mediator betwen animal beings and human beings. Waterbirds are perceived as the messengers between the other-world and the earth, guarding, in some myths, the entrance to the lower world, and acting as guides to the 'sea of the deceased', to the 'burial beyond the water' in others.

We must, of course, be wary of the possibility of inadvertently mapping the ethnography of modern hunter-gatherers onto the past without careful scrutiny of the archaeological evidence. So how far is the ethnographically derived belief system, outlined above, reflected in the archaeological record?

In the hunter-gatherer prehistory of Northern Europe, the symbolism of rock-carving sites, of carved objects in the material culture, and of the ritual context of burials, clearly relates to this system of beliefs. Material representations include sculpted terminals of wooden household utensils, such as spoon-bowls and ladles, zoomorphic axes and maceheads, rock carvings and zoomorphic ornamentation on pottery. Elk, bear and waterbirds are the most common designs. Within this symbolic context, the meaning of ritual sites can be comprehended by reference to this ideology.

For example, the ritual distinction between land and water, the association of waterbirds with the dead, and the burial of the dead 'beyond the water' in the mythical past of epic poems such as Kalevala (Lönnrot 1963) is reflected in the location of all true cemeteries – Olenii Ostrov, Skateholm I and II, Zveinieki – on islands or promontories, i.e., 'beyond the water'. The waterbird design is a frequent symbol in lake-side petroglyphs and on containers, such as bowls and pottery. Waterbirds are the sole bird remains found in burials; the interment of a child on a swan's wing at Vedbaek is particularly pregnant with symbolism (Nielsen & Petersen 1993).

Elk and bear-headed terminals, which are depicted carried [around] on sticks or poles not only at Namforsen, but also in the rock carvings on the shores of Lake Onega (where Olenii Ostrov is located (Gurina 1956, Savvateyev 1973, Maula 1990) find a direct parallel in the shaman's *turu*, a ritual rod used to mediate between the natural and supernatural worlds.

There are scenes of copulation between hunters and and elk/deer among petroglyphs of Lake Onega and elsewhere (Gurina 1956, Savatyeev 1973). They may refer to the role hunters are held to play in the reproduction of wild animals, which involves symbolic copulation between hunters and their prey. Based on the accounts in the ethnographic record, through such acts of symbolic copulation, hunters return the animal soul of the killed animal to the animal master, thereby ensuring its physical reproduction in the future (Ingold 1986).

Through similar referential reasoning, based on perceived differences between wild and the tamed animals in the ethnographic record of Siberian hunter-fishers and reindeer herders, we can comprehend the canine burials at Skateholm and at Vedbaek. Here the buried dogs are either treated as humans (i.e., with separate burial, grave goods), or are interred as sacrifices (Larsson 1990a).

Animal sacrifice, according to Ingold (1986: 253) can only apply to domestic species as the only animals subordinate to humans. This suggests that hunter-gatherers in the West Baltic made a categorical distinction between the wild and the tame, and that only dogs were considered domesticated. Palaeozoologically, dogs are the only fully domesticated species in the Mesolithic of Northern Europe. The human treatment accorded to some dog burials may symbolise surrogate burial of the human master, since the spirit of the domestic animal and the soul of its human master are the same.

The meaning of special features of the 'vertical' graves at Olenii Ostrov as shamen's graves has already been suggested (see above). This interpretation, too, makes sense only by reference to the structure of the belief system observed in the Siberian ethnographic record. Such understanding is, of course, based on a speculative recognition of patterns spanning vast geographical areas and several thousand years. If true, they suggest direct links between the hunter-fisher ideological system of the Mesolithic and recent foraging groups, and offer a language for decoding the belief systems of prehistoric past. Although there is variation in symbolism across the circum-Baltic zone, for example the use of red deer, elk or reindeer bones and antlers to symbolise the role of elk-deer, I suggest that all such communities were using the same language of symbolism, even though the medium of expression varied. Deer antlers as symbols of revival and regeneration can serve as a good example (Bradley n.d.).

Discussion

How can I justify such literary use of direct historical, or relational, analogy spanning seven thousand years? At the broadest level, it is undeniable that spatial order and social order are both mediated by distinctive principles of a still wider, cognitive order (Levi-Strauss 1968, Bourdieu 1977, Giddens 1984; for archaeological case studies, see Hodder 1990, Chapman 1993). There can be little doubt that the elaboration of material culture in foraging societies is an indication of elaboration in social relations, since ties between people can be objectified through things (Tilley 1991: 152, Chapman 1993: 109). Elaboration in grave goods, burial rites, and in symbolic aspects of material culture, then, should reflect elaboration of social structure in the language of some wider, cognitive order, or conceptual structure.

In the present case, the use of ethnographic analogy might seem valid because we are dealing with a direct historical analogy: the societies in question are historically linked and they operated in similar ecological and economic conditions (see, e.g., Tilley 1991). However, the validity of such analogic reasoning is considerably strengthened if we perceive the historical continuity and change between the hunter-gatherer societies in Northern Europe/Western Siberia in terms of structure and agency (Bourdieu 1977, Giddens 1984, Layton 1985, 1991). In my view, prehistoric and recent hunter-gatherers shared the same organisational structures. Two types of structures are apparent:

(1) Practical structure conditioned by the ecology of the resources, climate and geomorphology of the environment. Seasonal regimes of resource use, organisation of practical time, technological traditions and their relation to the division of labour and social organisation belong to this category of structures.

(2) Ideological-cosmological strucure arising from the hunter-gatherer belief system. This system is based on the perception of nature as a giving environment (Bird-David 1990), which emphasised timelessness and circulation of animal resources in regenerational cycles by means of possessing three substances: physical self, body soul and free soul, managed by human and animal masters in an act of collective appropriation (Ingold 1986, Peterson 1993). Consequently, single individuals could not be owners of wild resources, and the practice of sharing as a normative policy was ideologically justified. Based on these principles, such a belief system would establish a pervading temporal framework, perhaps analogous to the Braudelian concept of *longue durée,* which makes its remarkable temporal continuity more comprehensible.

However, when we consider the evidence for social differentiation in Mesolithic contexts, we can see that ideology and burial practice appear to be in contradiction. How do we reconcile the different messages symbolised by the burial contexts, which show a degree of social hierarchy, and the use of the symbols of the egalitarian hunter-gatherer ideology?

The social (and economic) organisation of the societies in question, embedded within the practical and the cognitive structural framework, will change through the operation of dynamic factors: agency. By agency I understand historically situated negotiation for power, control, or the attainment of goals between different segments of society, which is played out at different scales of organisation, starting with individuals, and then moving on to households, kinship groups, communities and larger units. The use and meaning of symbols will change as a part of this process of negotiation - within the ideological structure which provides the frame of reference for changes in interpretation. In my opinion, although agency modified the use of the symbols, the ideological structure itself did not change until the corpus of symbols associated with hunter-gatherer societies was replaced by those associated with farming, or with contemporary ideologies (i.e., not until about 3000 BC in southern Scandinavia, not until ca. 1000 BC in Finland and East Baltic, not until ca. 500 AD in Karelia, and not until the start of the present century among some groups, such as the Kets in Western Siberia).

Within this broad cognitive framework, however, forces of agency operated to challenge its basic ideological principles. Increasing familiarity with farming practices, and contact with farming communities in the south, contributed significantly to the erosion of these principles, particularly the ideology of sharing. This historical explanation is based on the premise that the ideology of sharing, although adhered to nominally, increasingly did not reflect social practice in the late Mesolithic. The tension betwen ideological prescription and and practice increased with technological innovation, increase in territoriality and, later, contact with farming societies. Territoriality and delayed-return technologies imply restriction of access to resources, and encourage resource ownership: the appropriation of resources appears no longer collective, but a product of human labour.

It is the organisation of labour which is crucial here. In this respect, descent-based social structures, controlled by elders through alliance and exchange, created the structural conditions for social dependency (Bender 1978, 1985, Tilley 1991, Petersen 1993). This may have been achieved through competitive gift-giving and feasting (Hayden 1990), the control of sexual reproduction and of ritual knowledge (Tilley 1991), and may have led to the concentration of social power in the hands of a restricted number of individuals. Ritual specialists were in the best position to assume the control of lineages on the ritual basis of their moral authority (Aldenderfer 1993), and in so doing **to** enhance their power base to create 'simultaneous' rather than 'sequential' hierarchy (Johnson 1982). These would be the individuals represented by the elaborate graves at Olenii Ostrov and other places.

It is important to note that in the prehistoric situation, both men and women enjoyed high status. Moreover status does not seem to accrue with age: we have young individuals, even children, with elaborate burial, suggesting some degree of ascription to their status. Finally, at Olenii Ostrov and later sites, status positions appear to be independent of one another. This would suggest an absence of simultaneously held positions of status; rather we are dealing with sequential hierarchies which were present already in the mid-sixth millennium BC; in other words, additional developments needed to occur, in addition to ritual ranking and control of descent groups, to institute social change.

Such additional developments may have resulted from relations with farming communities and the incorporation of hunter-gatherer societies into broader exchange networks of the time. The impact of the agricultural frontier would have been felt in many aspects of hunter-gatherer social life: particularly if relations assumed a competitive, rather than cooperative character. Farming goods and products, such as polished stone axes, contained a component of added value, arising from their exotic origin and prestigious – subversive perhaps – ideological association. As we know, stone axe imports were found in Denmark and Sweden, and imitations were found at Skateholm. Unless re-interpreted in the context of hunter-gatherer ideology, such exchange goods were bound to promote the establishment of social elites.

The proximity of farming communities and the effects of the agricultural frontier may have had other disruptive effects, leading to the disruption of the hunter-gatherer economic system and social organisation (Zvelebil 1995a). What consequences would this have for the hunter-gatherer belief system? In the archaeological record, we can, perhaps, identify both ideological censure and transformation.

Several people have noted the incompatibility of foraging and farming symbolic codes: Chapman, for example, argues that products which symbolised farming were excluded from Lepenski Vir (1993). Tilley (1991) describes how the fishing and hunting communities at Nämforsen accommodated their economy to the demands of an exchange system controlled by farmers, and how they restructured their symbolic system in the process. In both cases, however, farming appears to have been rejected symbolically and in practice. The subsequent cultural simplification, evident at Lepenski Vir, as well as during the final phases of hunter-gatherer settlement in some parts of Scandinavia (i.e., the Ertebølle context between 3500-3000 BC) suggests a sort of 'prehistoric encapsulation' which followed.

Alternatively, the tension between the ideology of sharing and the practice of accummulation, promoted by contact with farmers as well as earlier technological and economic developments, would be resolved by ideological transformation, allowing for ownership of domesticated resources, formalised status differentiation, and accumulation of wealth. After all, the social meaning of domestication is in the ownership of resources (Ingold 1986). Yet, it seems to me, in the Early Neolithic, we can also identify attempts to perpetuate the old belief system based on the communal use of resources and sharing.

Hunter-gatherer communities in the circum-Baltic zone adopted farming at very different rates (ranging from 100-300 years in the West Baltic to 500-1000 years in the East Baltic). It is commonly assumed that social and ideological structures were transformed in the process. But how much change was actually involved? The emphasis on communal territories and communal ritual, characteristic of the Early Neolithic in north-west Europe, emphasises the continuation of Mesolithic traditions. Other features common to both the Mesolithic and Neolithic include deposits of food and antler in graves, dog burials, mortuary houses, circulation of human bones, votive deposits in aquatic locations and the use of stone in grave architecture (for further evidence and discussion, see van Gijn & Zvelebil 1997). At the same time, there is evidence for social elites and exclusion from ritual practice, the extent of which, to my mind, has not been agreed upon. Could it be that, in some ways, communal ritual, symbolised by causewayed enclosures and megalithic tombs, served as an elaborate ideological practice intended to develop and safeguard the practice of sharing and communal identity by moving it from a personal and universal ideology (embedded in hunting-gathering) to a different level of social organisation: to a level of public ritual? This was perhaps the transformation which facilitated, ideologically, the transition to farming.

Conclusions

In this contribution, I have contrasted the evidence for social structure and ideology of late Mesolithic hunter-gatherers in the circum-Baltic area of the North Atlantic façade. I have argued that the late Mesolithic societies fall within the range of ethnographically known delayed-return hunter-gatherers, that their ideology formally adhered to egalitarian principles, while in practice egalitarian ideological strictures were being eroded and subverted through competition for – and partial achievement of – differentiated social status. This suggests a tension between between ideology and social practice which was eventually resolved by the adoption of farming.

But quite a few questions remain: Was an ideological shift necessary for the adoption of farming? How much continuity and change is there in the transition between Mesolithic and Neolithic ideology and social structure? Is domestication a state of mind or a state of subsistence – and which constitutes the Neolithic?

References

Albrethsen, S.E. & E. Brinch Petersen
1976 Excavation of a mesolithic cemetery at Vedbaek, Denmark. *Acta Archaeologica* 47: 1-28.

Aldenderfer, M.
1993 Ritual, hierarchy and change in foraging societies. *Journal of Anthropological Archaeology* 12: 1-40.

Aleksenko, E.A.
1967 *Kety.* Moskva: Nauka.

Andersen, S.
1987 Tybrind Vig: A submerged Ertebolle settlement in Denmark. In *European Wetlands in Prehistory* (J. Coles & A. Lawson, eds.), pp. 253-281. Oxford: Clarendon Press.

Anisimov, A.
1963 Cosmological concepts of the people of the north. In *Studies in Siberian Shamanism* (H. Michael, ed.) Toronto: University of Toronto Press.

Bender, B.
1978 Gatherer-hunter to farmer: a social perspective. *World Archaeology* 10: 204-223.

1985 Emergent tribal formations in the American mid-continent. *American Antiquity* 50: 52-62.

Bender, B. & B. Morris
1988 Twenty years of history, evolution and social change in gatherer-hunter studies. In *Hunters and Gatherers I: History, Evolution and Social Change* (T. Ingold, D. Riches & J. Woodburn, J., eds.), pp. 4-13. Oxford: Berg.

Bettinger, R.L.
1991 *Hunter-Gatherers: Archaeological and Evolutionary Theory.* New York: Plenum Press.

Bird-David N. H.
1988 Hunters and gatherers and outsiders. In *Hunters and Gatherers I: History, Evolution and Social Change* (T. Ingold, D. Riches & J. Woodburn, J., eds.), pp. 17-72. Oxford: Berg.

1990 The giving environment: another perspective on the economic system of gatherer-hunters. *Current Anthropology* 31: 189-196.

1992a Beyond 'the hunting and gathering mode of subsistence': culture-sensitive observations on the Nayaka and other modern hunter-gatherers. *Man* 27: 19-44.

1992b Beyond 'the original affluent society': a culturalist reformation. *Current Anthropology* 33: 25-47.

Bourdieu, P.
1977 *Outline of a Theory of Practice.* Cambridge: Cambridge University Press.

Bradley, R.
1997 Domestication as a state of mind. In *Ideology and Social Structure of Stone Age Communities in Europe* (A. van Gijn & M. Zvelebil, eds.), 13-19. Leiden: Analecta Praehistorica Leidensia No. 29. University of Leiden Press.

Chapman, J.
1993 Social power in the Iron Gates Mesolithic. In *Cultural Transformations and Interactions in Eastern Europe* (J. Chapman & P. Dolukhanov, eds.), pp. 71-122. Aldershot: Avebury Press.

Clark, G. A. & M. P. Neeley
1987 Social differentiation in European Mesolithic burial data. In *Mesolithic Northwest Europe, Recent Trends* (P. Rowley-Conwy, M. Zvelebil & H. P. Blankholm, eds.), pp. 121-130. Sheffield: University of Sheffield Press.

Constandse-Westermann T. & R. Newell
1989 Limb laterisation and social stratification in Western Mesolithic societies. In *People and Culture in Change* (I. Herskovitz, ed.), pp. 405-434. Oxford: BAR International Series No. 508.

Czaplicka, M. A.
1914 *Aboriginal Siberia. A Study in Social Anthropology.* Oxford: Clarendon Press.

Denbow, J.
1984 Prehistoric hunters and foragers of the Kalahari: the evidence for 1500 years of interaction. In *Past and Present in Hunter-Gatherer Studies* (C. Schrire, ed.), pp. 175-193. New York: Academic Press.

Dioszegi, V. & M. Hoppal (editors)
1978 *Shamanism in Siberia.* Budapest.

Edgren, T.
1966 *Die Jäkärlä Gruppen.* Helsinki: Finska Fornminnesföreningens Tidskrift No. 68.

Fisher, A. (editor)
1995 *Man and Sea in the Mesolithic.* Oxford: Oxbow Monograph No. 53.

Giddens, A.
1984 *The Constitution of Society.* Berkeley: University of California Press.

Gordon, R.
1984 The !Kung in the Kalahari exchange: an ethnohistorical perspective. In *Past and Present in Hunter-Gatherer Studies* (C. Schrire, ed.), pp. 195-224. New York: Academic Press.

Griffin, P. B.
1984 Forager resource and land use in the humid tropics: the Agta of northeastern Luzon, the Philippines. In *Past and Present in Hunter-Gatherer Studies* (C. Schrire, ed.), pp. 95-121. New York: Academic Press.

Gurina, N.N.
1956 *Oleneostrovskii Mogilnik.* Moskva: Materialy i issledovania po arkheologii SSSR No. 47.

Hayden B.
1990 Nimrods, piscators, pluckers, and planters: the emergence of food production. *Journal of Anthropological Archaeology* 9: 31-69.

Headland, T. & Reid, L.
1989 Hunter-gatherers and their neighbours from prehistory to the present. *Current Anthropology* 30: 43-66.

Hodder, I.
1990 *The Domestication of Europe.* Oxford: Blackwell.

Ingold, T.
1983 The significance of storage in hunting societies. *Man* 18: 553-571.
1986 *The Appropriation of Nature.* Manchester: University of Manchester Press.
1988 Notes on the foraging mode of production. In *Hunters and Gatherers I: History, Evolution and Social Change* (T. Ingold, D. Riches & J. Woodburn, J., eds.), pp. 269-285. Oxford: Berg.

Ingold, T., D. Riches & J. Woodburn (editors)
1988 *Hunters and Gatherers I: History, Evolution and Social Change.* Oxford: Berg.

Jacobs, K. 1995 Returning to Olenii ostrov: social, economic, and skeletal dimensions of a boreal forest Mesolithic cemetery. *Journal of Anthropological Archaeology* 14: 359-403.

Johnson, G.
1982 Organisational structure and scalar stress. In *Theory and Explanation in Archaeology* (C. Renfrew, M. Rowlands & B. Seagraves, eds.), pp. 389-421. New York: Academic Press.

Kannegaard Nielsen, E. & E. Brinch Petersen
1993 Burials, people and dogs. In *Digging Into the Past – 25 Years of Archaeology in Denmark.* (S. Hvass & B. Storgaard, eds.), pp. 76-81. Aarhus: The Royal Society of Northern Antiquaries & The Copenhagen & Jutland Archaeological Society.

Kent, S.
1992 The current forager controversy: real versus ideal views of hunter-gatherers. *Man* 27: 45-70.
1993 Sharing in an egalitarian Kalahari community. *Man* 28: 479-514.

Larsson, L.
1989 Ethnicity and traditions in Mesolithic mortuary practices of southern Scandinavia. In *Archaeological Approaches to Cultural Identity* (S. Shennan, ed.), pp. 210-218. London: Unwin Hyman.
1990a Dogs in fraction – symbols in action. In *Contributions to the Mesolithic in Europe* (P. Vermeersch & P. Van Peer, eds.), pp. 153-160. Leuven: Leuven University Press.
1990b The Mesolithic of southern Scandinavia. *Journal of World Prehistory* 4: 257-309.
1993 The Skateholm project: late Mesolithic coastal settlement in southern Sweden. In *Case Studies in European Prehistory* (P. Bogucki, ed.), pp. 31-62. London: CRC Press.

Layton, R.
1985 The cultural context of hunter-gatherer rock art. *Man* 20: 434-453.
1991 Figure, motif and symbol in the hunter-gatherer rock art of Europe and Australia. In *Rock Art and Prehistory* (P. Bahn & A. Rosenfeld, eds.), pp. 23-39. Oxford: Oxbow Monograph No. 10.

Lee, R. B.
1990 Primitive communism and the origin of social inequality. In *The Evolution of Political Systems* (S. Upham, ed.), pp. 225-246. Cambridge: Cambridge University Press.
1992 Art, science, or politics? The crisis in hunter-gatherer studies. *American Anthropologist* 94: 31-54.

Levi-Strauss, C.
1968 (1958) *Structural Anthropology.* London: Routledge & Kegan Paul.

Levin, M. G. & L. Potapov (editors)
1956 *Narody Sibirii.* Moskva: Nauka.

Lönnrot, E.
1963 *The Kalevala or Poems of the Kaleva District.* Cambridge, MA: Harvard University Press.

Loze, I. A.
1979 *Pozdnii Neolit i Rannaya Bronza Lubanskoi Ravniny.* Riga: Zinatne.

Maula, E. (editor)
1990 *Swansongs: Rock Art from Lake Onega 4000–2000 BC.* Tartu: Society of Prehistoric Art.

Nielsen, E. K. & E. B. Petersen
1993 Burials, people and dogs. In *Digging Into the Past – 25 Years of Archaeology in Denmark* (S. Hvass & B. Storgaard, eds.), pp. 82-97. Aarhus:

The Royal Society of Northern Antiquaries & The Copenhagen & Jutland Archaeological Society.

Nuñez, M.
1990 On subneolithic pottery and its adoption in late Mesolithic Finland. *Fennoscandia Archaeologica* 7: 27-52.

O'Shea, J. & M. Zvelebil
1984 Oleneostrovski Mogilnik: reconstructing social and economic organisation of prehistoric hunter-fishers in northern Russia. *Journal of Anthropological Archaeology* 3: 1-40.

Pedersen, L.
1995 7000 years of fishing: stationary fishing structures in the Mesolithic and afterwards. In *Man and Sea in the Mesolithic* (A. Fischer, ed.), pp. 75-87. Oxford: Oxbow Monograph No. 53.

Pedersen, L., A. Fischer & B. Aaby (editors)
1997 *The Danish Storebaelt since the Ice Age.* Copenhagen: A/S Storebaelt Fixed Link.

Pederson, J. & Waehle, E.
1988 The complexities of residential organisation among Efe and the Bamcombi. In *Hunters and Gatherers I: History, Evolution and Social Change* (T. Ingold, D. Riches & J. Woodburn, J., eds.), pp. 75-90. Oxford: Berg.

Peterson, J.
1993 Demand sharing: reciprocity and the pressure for generosity among foragers. *American Anthropologist* 95: 860-874.

Price, T. D. & K. Jacobs
1990 Olenii Ostrov: first radiocarbon dates from a major Mesolithic cemetery in Karelia. *Antiquity* 64: 849-53.

Resketov, A.
1972 *Okhotniki, Sobirateli, Rybolovy.* Moskva: Nauka.

Rowley-Conwy, P.
1983 Sedentary hunters, the Ertebølle example. In *Hunter-Gatherer Economy in Prehistory* (G. N. Bailey, ed.), pp. 111-126. Cambridge: Cambridge University Press.

1992 Wild or domestic? On the evidence for the earliest domesticated cattle and pigs in South Scandinavia and Iberia. *International Journal of Osteoarchaeology* 5: 115-126.

1998 Cemeteries, seasonality and complexity in the Ertebølle of southern Scandinavia. In *Harvesting the Sea, Farming the Forest: The Emergence of Neolithic Societies in the Baltic Region* (M. Zvelebil, R. Dennell & L. Domanska, eds.), pp. 223-241. Sheffield: Sheffield Academic Press.

Rowley-Conwy, P. & M. Zvelebil
1989 Saving it for later: storage by prehistoric hunter-gatherers in Europe. In *Bad Year Economics* (P. Halstead & J. O'Shea, eds.), pp. 40-56. Cambridge: Cambridge University Press.

Savatteyev, V.
1973 *Petroglify Karelii.* Riga: Voprosy Istorii No. 6.

Schrire, C. (editor)
1984 *Past and Present in Hunter-Gatherer Studies.* Orlando: Academic Press.

Schulting, R.
1996 Antlers, bone pins and flint blades: Mesolithic cemeteries of Brittany. *Antiquity* 70: 335-350.

Solway, J. S. & R. B. Lee
1990 Foragers, genuine or spurious? *Current Anthropology* 31: 109-145.

Tilley, C.
1991 *Material Culture and Text: the Art of Ambiguity.* London: Routledge.

Van Gijn, A. & M. Zvelebil (editors)
1997 *Ideology and Social Structure of Stone Age Communities in Europe.* Leiden: Analecta Praehistorica Leidensia No. 29, University of Leiden Press.

Vankina, L.V.
1970 *Torfyanikovaya Stoyanka Sarnate.* Riga: Zinatne.

Vang-Petersen, P.
1984 Chronological and regional variation in the Late Mesolithic of eastern Denmark. *Journal of Danish Archaeology* 3: 7-18.

Vasilevitch, G.
1963 Early concepts about the universe among the Evenks. In *Studies in Siberian Shamanism* (H. Michael, ed.). Toronto: University of Toronto Press.

Wilmsen, E. N. & J. R. Denbow
1990 Paradigmatic history of San-speaking peoples and current attempts at revision. *Current Anthropology* 31: 489-524.

Woodburn, J.
1982 Egalitarian societies. *Man* 17: 431-451.

1988 African hunter-gatherer social organization: is it best understood as a product of enscapulation? In *Hunters and Gatherers I: History, Evolution and Social Change* (T. Ingold, D. Riches & J. Woodburn, J., eds.), pp. 31-64. Oxford: Berg.

Zagorskis, F.
1987 *Zvejnieki Akumens Laikmeta Kapulauks.* Riga: Zinatne.

Zvelebil, M.
1981 *From Forager to Farmer in the Boreal Zone.* Oxford: BAR International Series No. 115.

1986 Mesolithic societies and the transition to farming: problems of time, scale and organisation. In *Hunters in Transition* (M. Zvelebil, ed.), pp. 167-187. Cambridge: Cambridge University Press.

1993 Concepts of time and 'presencing' the Mesolithic. *Archaeological Review from Cambridge* 12: 51-70.

1994 Plant use in the Mesolithic and the implications for the transition to farming. *Proceedings of Prehistoric Society* 60: 95-134.

1995a The agricultural frontier and the transition to farming in the circum-Baltic region. In *The Origin and Spread of Agriculture and Pastoralism in Eurasia* (D. Harris, ed.), pp. 323-345. London: UCL Press.

1995b Hunting, gathering or husbandry? Management of food resources by the late Mesolithic communities of Temperate Europe. In *Before Farming* (D. Campana, ed.), pp. 79-104. Philadelphia: MASCA Research Papers in Science and Archaeology No. 12 (Supplement).

Zvelebil, M. & P. Dolukhanov
1991 Transition to farming in Eastern and Northern Europe. *Journal of World Prehistory* 5: 233-278.

4

The Exploitation of Ireland's Coastal Resources – a Marginal Resource Through Time?

Peter Woodman
University College, Cork

Abstract

Ireland has been an island for more than 10,000 years and is the only large island in the Atlantic to have remained surrounded by water throughout the whole of the Holocene. Yet curiously there appears to be in Ireland relatively little evidence of continual maritime exploitation throughout much of prehistory and the early medieval period. Is the archaeological record providing an accurate reflection of past activities there, or has Ireland's preoccupation with agriculture coloured attitudes and created priorities in Irish archaeology in which maritime concerns are not take adequately into consideration? This chapter will consider the evidence for coastal exploitation during the Mesolithic, whether evidence for the use of coastal resources in later periods represents a continuation of earlier traditions or a different kind of marine exploitation, and whether there are lessons to be learned from the study of Mesolithic coastal adaptations that might shed light on those of later periods.

Introduction

One of the most interesting periods in the history of Irish archaeology was the exploration of the numerous sand dune sites from round the whole of the Irish coastline. In the period 1880 to 1901, William Knowles published a series of papers in the *Proceedings of the Royal Irish Academy* (e.g., Knowles 1901). While these papers were primarily communicating the discovery of a range of artifacts which Knowles and his colleagues had recovered, they also contained certain references to the shell middens and faunal remains which could be found at these sites; the latter were usually identified by recognized zoologists.

This project began a period of coastal investigations that included the first Masters thesis in Irish archaeology at the National University of Ireland (Ms. J. Holland of University College Cork, 1912: *The Shore-Dwellers of Ireland*, published as Brunicardi [1914]) and a further coastal survey by the Reverend Hewson in the 1930s (e.g., Hewson 1938). These surveys identified over 60 localities along the coast of Ireland and the material recovered ranged from the medieval period to the stone age, although virtually none of the localities produced Mesolithic artifacts. A high percentage of the locations investigated were either sand dune sites and/or shell middens (Fig. 4.2) which contained a range of later artifacts and in some cases, faunal remains. The dominant shell type in each midden (see, e.g., Brunicardi 1914) was usually noted while the absence of faunal remains from a number of sites could simply indicate an early or casual and unscientific investigation of a locality. Nevertheless the range of information recorded can be surprisingly good.

Unfortunately in recent decades Irish archaeology placed emphasis on the identification of field monuments and, perhaps somewhat belatedly, there has been a recognition that many forms of settlement are not signaled by upstanding monuments. Teams such as the Wetland Unit (Maloney 1993) have been created about 15 years ago while inter-tidal surveys (O'Sullivan 2001) and the production of definitive lists of shipwrecks have existed for less. Excavations of sand dune sites where coastal erosion is taking place has also recently restarted but again, priority is given to localities where visible stone structures are at risk, for example Omey Island (O'Keefe 1994).

In this intervening period the older antiquarian coastal collections have remained virtually unstudied. Surveys such as the Sites and Monuments Record simply note the existence of shell middens and, on occasions, other coastal settlements. An examination of the literature shows a range of sites which vary from small heaps of shells no more than a few metres across and often containing just one species, to complex sites where it is difficult to separate different phases of activity, to massive middens, particularly of oysters, which tend to occur in the inner, low energy bays such as Cork Harbour (Power 1930), Galway Bay (Knowles 1901) or the Knocknarea area of Co. Sligo (Burenhult, 1984). In the case of the Cork Harbour several are severely damaged while some of those found in Galway Bay have been destroyed. Most of the sites round the Irish Coast still remain undated. In areas such as the Iveragh Peninsula in

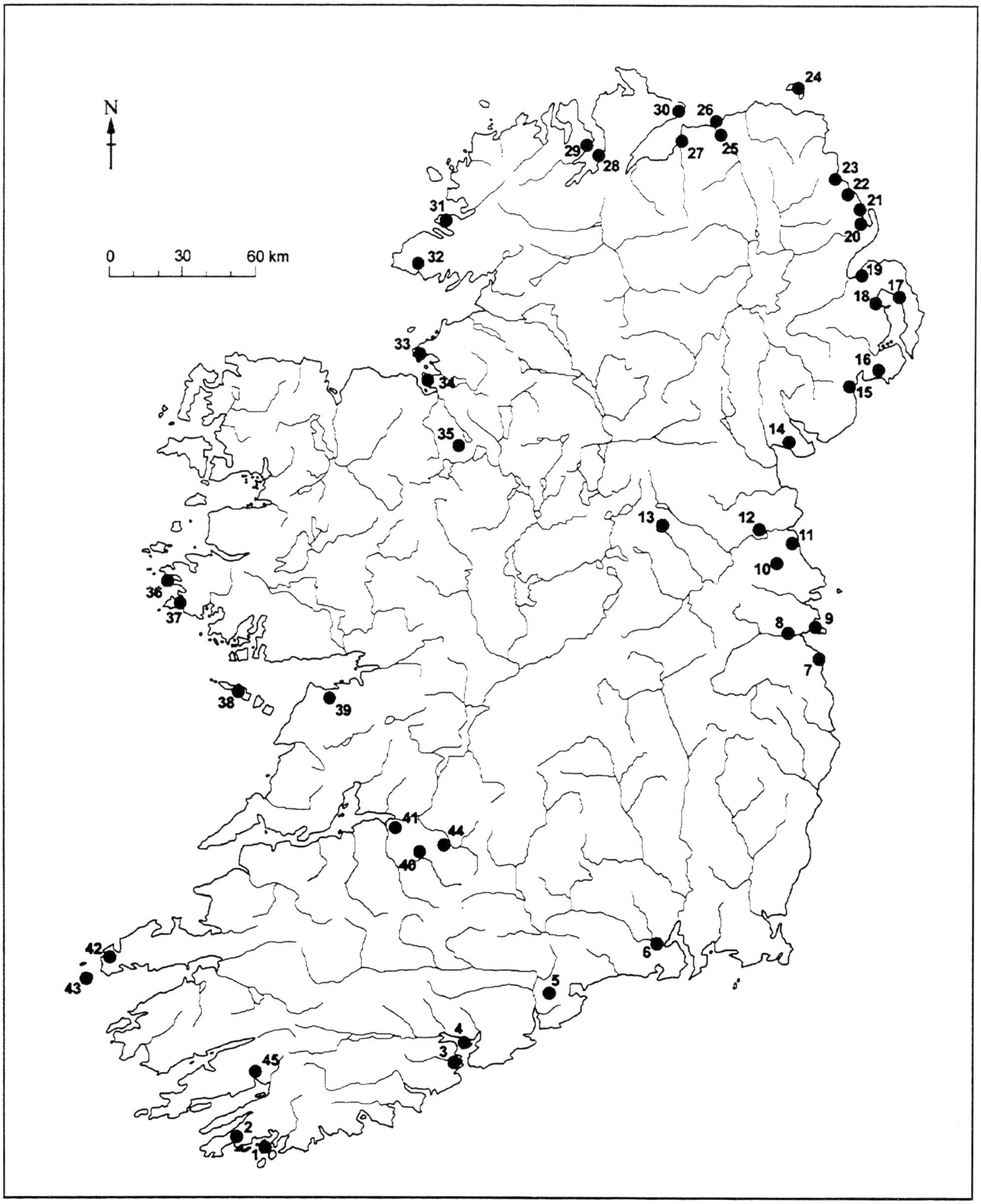

Figure 4.1 Location map of sites referred to in the text: 1. Skeam West, 2. Altar Wedge Tomb, 3. Raheens, 4. Cork Harbour, 5. Kilgreany, 6. Waterford, 7. Dalkey Island, 8. Phoenix Park, 9. Sutton, 10. Fourknocks, 11. Bettaghstown, 12. Newgrange and Knowth, 13. Lough Crew, 14. Rockmarshall, 15. Dundrurn, 16. Rathmullin 17. Temple midden, 18. Rough Island, 19. Glendhu, 20. Glynn, 21. Larne, 22. Ballygalley, 23. Glencloy, 24. Rathlin Island, 25. Mount Sandel, 26. Portstewart, 27. Magilligan Sand Dunes, 8. Baylet and Inch Island 29. Kinnegar Bay, 30. Eleven Ballyboes, 31. Narin, 32. Bavan, 33. Raghly, 34. Culeenamore, Knocknarea, Carrowmore, 35. Carrowkeel, 36. Omey Island, 37. False Bay, 38. Dun Aonghasa, 39. Poulnabrone, 40. Lough Gur, 41. Carrigdirty, 42. Ferriter's Cove, 43. Blasket Islands, 44. Killuragh Cave, 45. Poulgorm.

Figure 4.2. Early R.J. Welch photograph of shell midden at Rosapenna, Co. Donegal (© Ulster Museum).

Co. Kerry, for example, a number of new shell middens have been identified, but it has not been possible to date most of them (O'Sullivan & Sheehan 1996).

Throughout much of the 20th century research, particularly research associated with the Mesolithic period, was concentrated on raised beach sites that could be regarded as little more than traps for artifacts. The one exception was the work lead by Frank Mitchell with his excavations of shell middens at Rockmarshall (Mitchell 1949), Sutton (Mitchell 1956, 1972) and under David Liversage (1968) at Dalkey Island. It has only been at the end of the 20th century that research in coastal settlement has been rekindled. This includes Murray (1999) and others in the Galway Bay area, as well as the Lough Swilly Archaeological Project (Kimball 2000), the Inch Island Prehistoric Landscapes Project (IIPLP) in Donegal (Milner & Woodman 2001) and the work in Strangford Lough (MacErlean et al. 2002) (Fig. 4.1).

The general impression gained from the surveys is of a continuous and consistent exploitation of marine resources but with, on the one hand, little clear evidence of the extent to which fish and sea mammals were exploited and, on the other hand, equally little indication that the sea was exploited in different ways at different periods. In fact the overall assumption has always been that the sea was of marginal significance and a resource that was only used in times of stress after the arrival of farming. There is even today a common 'folk' perception that shell middens in any locality are a product of the Irish Famine of 1845-1847. However as Woodman and Milner have noted there are numerous instances in the recent past such as in the Ballinskelligs area of Co Kerry where Kirby (1990) fondly remembered the collection of shell fish at exceptional low tides as being special occasions. Similarly, Harte (1867) records the visit of the " Mountaineers" and their families from Barnesmore in South Donegal who visited Donegal Bay – turned their horses loose, drank whiskey and consumed mussels! The idea of middens developing as a result of feasting, as suggested by Cross (2001) is therefore quite possible.

The Use of the Sea in Pre-Farming Societies

The tendency to equate marine exploitation with the Mesolithic (hunters/fishers) may have resulted in an unintended juxtaposition of opposites - Mesolithic equals fishing while later periods equal farming. This tendency can be seen in the editorial choice of title for Milner and Woodman (2001) namely *Mesolithic Middens.* This has created a situation where there is little critical analysis on the role of the sea during the Mesolithic while, conversely, there is an assumption that marine resources played a limited role in later periods.

In the study of the Irish Mesolithic[1] the marine element can be considered in terms of three questions.

1. Is there evidence for Early Mesolithic coastal settlement?

2. What marine resources were exploited?

3. How were marine resources exploited?

Evidence for Early Mesolithic Coastal Settlement

Most evidence for Mesolithic coastal settlement comes either from (1) disturbed secondary contexts, such as the mid-Holocene raised beaches where much of the original Irish Mesolithic research was carried out (e.g., Curran Point, Larne, Co. Antrim (Movius 1953), or (2) localities which, by chance, were localities placed well above areas of transgression, or (3) the final phases of the Mesolithic where relative shoreline retreated.

There are numerous sites from the Later Mesolithic which can be placed in these categories. What remains of Ferriter's Cove, Co. Kerry (Woodman et al. 1999) has survived because it is on a ridge overlooking a small bay which may have been created late in the Mesolithic, while shell middens at sites such as Rough Island, Co. Down (Movius 1940) and some of those at Rockmarshall, Co. Louth (Mitchell 1949) survived because the sea had started to regress by the time they were created. Indeed Mallory (pers. comm.) would now believe that the Rough Island Midden may be primarily Neolithic in age. This therefore creates the impression that the sea became more important towards the end of the Mesolithic. However it should be apparent that the greater occurrence of late coastal settlement is based on a natural bias in survivability. It is highly probable that the sea was exploited as extensively in the Early Mesolithic.[1]

In common with many regions on the Atlantic edge, Mesolithic coastal settlement from the earliest stage of the Holocene is difficult to find. In the case of Ireland this may be due to a complex interrelation between sea volume change and isostatic uplift. Even in the northern part of the island most of the early Holocene shorelines have been lost. Here the shorelines advanced to their highest relative point, about 8 m above OD, *c.* 6000 BP. Therefore most early coastal settlements adjacent to the seashore have either been eroded away or buried below marine deposits. Along the south coast of Ireland, which appears to have been less subject to isostatic uplift, the relative shoreline continued to advance until perhaps as late as 4000 BP, when it stabilized at approximately its present level. The absence of Mesolithic material from the coastal sand dunes can therefore be explained by the fact that many dune systems in the northern part of the island only began to form at or after 5000 BP while along the southern coast the sand dunes usually formed in their present location after 4000 BP. Wilson (1995) refers to a ^{14}C date from Grangemore, Co. Derry sand dunes (5315 ± 135 BP) as the earliest date from within a sand dune.

A few coastal localities have produced Early Mesolithic material, usually in a derived context. A particularly large concentration of Early Mesolithic artifacts, including numerous flake axes, was recovered from the edge of a bay at Glynn, Larne Lough, Co. Antrim (Woodman 1977a). This material was recovered during the creation of terraces for rugby pitches and was found in secondary contexts in a series of gravel deposits that had been thrown up at the time of the *Littorina* maximum transgression. Similarly a local amateur, Brian Mc Naught, has found a very significant concentration of Early Mesolithic material washed up on the present shoreline of a Bay in Eleven Ballyboes Townland in Co Donegal (Costa et al.2001) Although the material is in derived contexts this type of site shows that Early Mesolithic communities placed some of their habitations in coastal locations. Mount Sandel, Co. Derry (Woodman 1985) on the other hand was probably located in the upper reaches of the estuary of the River Bann but survived because it was placed on a cliff edge 30 metres above the present estuary. There is every reason to believe that these are little more than indicators of an extensive coastal exploitation that took place throughout the whole of the Mesolithic period.

Marine Resources Exploited

Where material has been preserved *in situ* Irish Mesolithic sites illustrate the exploitation of a full range of marine resources. At Mount Sandel, several kilometres from the coast, the soil was quite acidic but numerous calcined fragments of fish bones were recovered. While these were dominated by the salmonid family and the bones of eel *(Anguilla anguilla)* remains of flounder *(Pleuronectes flesus)* and sea bass *(Dicentrachus labrax)* were identified (van Wijngaarden-Bakker 1985). Flounder *(Pleuronectes flesus)* could of course have penetrated up the estuary but it is likely that sea bass were caught in the open sea. One clear example of a bevel-ended pebble was also recovered. These have usually been associated with shell middens and so this example could indicate the exploitation of the rocky open coastline which lay a few kilometres to the north.

During the later part of the Mesolithic there is little problem in identifying the exploitation of marine resources but at the same time it should be noted that less than ten excavated Later Mesolithic sites are represented by shell middens. However any study of middens must take into consideration the observations of Bailey (1978) that shellfish represent an obvious source of food but even within the middens they may represent a significant but not necessarily the most important source of sustenance. Andersen (1995) has noted that in the Ertebølle of Jutland there is a false assumption that each coastal Mesolithic site is to be associated with a shell midden. He suggests that perhaps only one in six sites would be expected to have a midden. This might be somewhat similar to what has happened in certain parts of Strangford Lough where numerous scatters of Later Mesolithic artifacts have been found but where it can be difficult to identify large numbers of middens of Mesolithic Age. In certain areas such as Strangford Lough middens that would survive could only have been created during the second half of the Mesolithic when the sea had inundated the Lough to a establish viable environment where extensive fishing and shell fish collecting could take place. However the real problem in certain parts of the north of Strangford Lough there are so many middens (see MacErlean 2002, Fig. 7:2) that it is impossible to separate out that the Early Prehistoric examples from later middens. One fact which is becoming apparent from excavations and surveys as part of the Inch Island Prehistoric Landscapes Project (IIPLP) is that certain key localities can be used repeatedly through time for shell fish collection and in the case of Baylet midden in Donegal use may have begun well back in the later Mesolithic and continued throughout much of the Neolithic period although as will be discussed below the use of the lo cality may have been very different. It is surprising that no Mesolithic shell middens have been identified in Co. Antrim particularly given the extensive exploitation of this area for flint (Woodman & Johnson 1996).

Until recently there has been little evidence of substantial middens such as those found in Denmark or Scotland at some of the Obanian sites (Mellars 1987). With the exception of Sutton (Mitchell 1956, 1972), which may have been 100 metres in length, all other published shell middens are quite small. Rough Island, Reagh Island and Temple midden were each perhaps less than 20 metres across; Rockmarshall (Mitchell 1949) had several middens all less than 20 metres across while those found at Dalkey Island, Co. Dublin (Liversage 1968) were even more ephemeral. Other small shell middens may still exist on Dalkey Island. Windle also appears to have uncovered a small Later Mesolithic shell midden at Kinnegar Bay, Co. Donegal

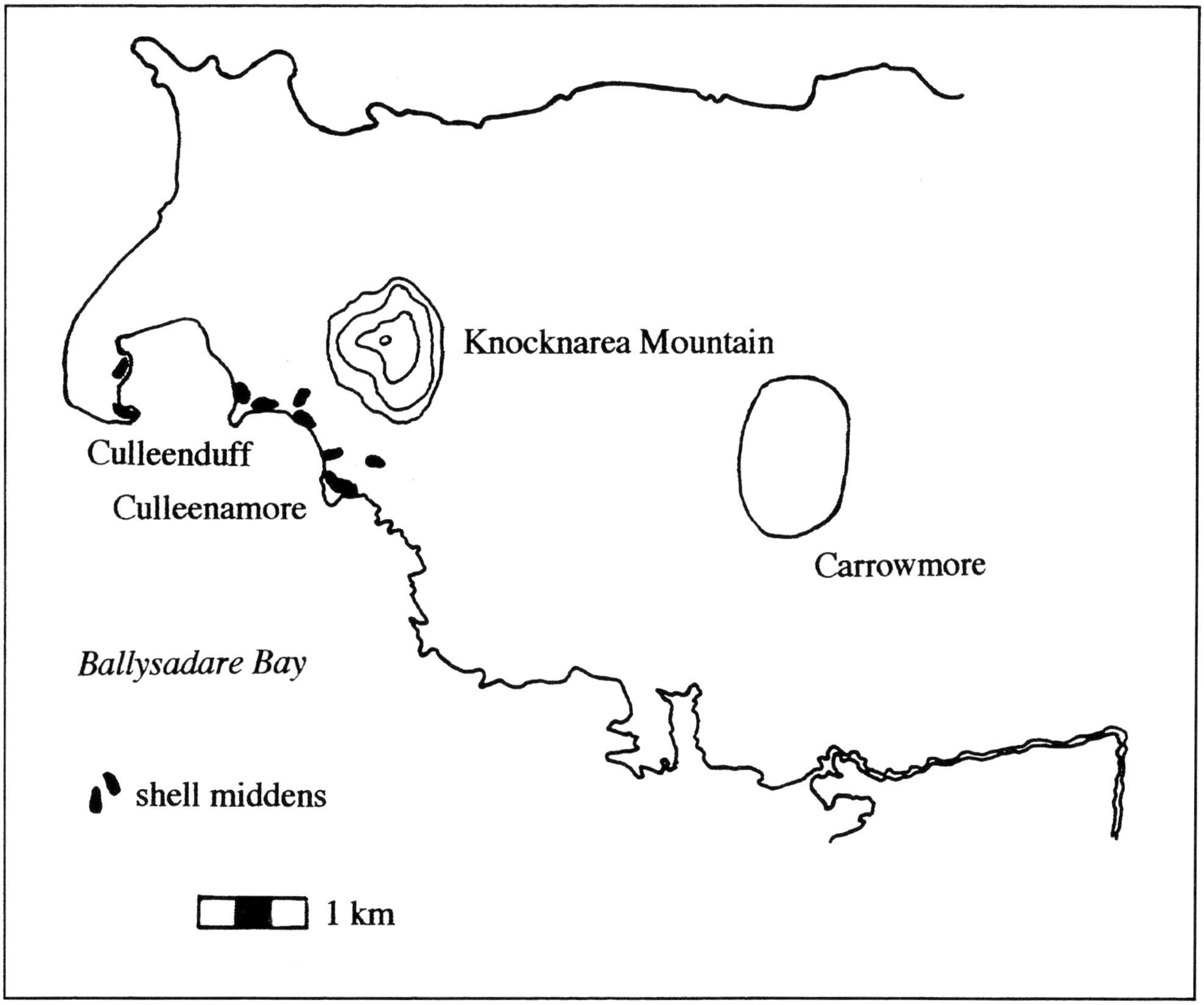

Figure 4.3. Map of shell middens on the Knocknarea peninsula, in the vicinity of Culeenamore, Co. Sligo (after Burenhult 1984).

(Brunicardi 1914). However, the recent research of the IIPLP has re-identified several large prehistoric shell middens including some of Harte's middens (Harte 1866). In some cases these exceeded 75 metres in length and some had their origins in the Mesolithic.

In contrast at Ferriter's Cove, Co. Kerry (Woodman et al. 1999) and Glendhu, Co. Down (Woodman 1985a), it is more a question of small groups of shell on Later Mesolithic settlement sites (Fig. 4.3). It would seem probable that the limited nature of many of the middens may be a function/feature of the duration of exploitation. Preliminary analysis for indicators of seasonality from oyster shells recovered at Glendhu and Rockmarshall carried out by Milner (pers. comm.) at the University of Newcastle upon Tyne (see Milner 2001 for a description of the technique used) has shown that there is no simple collection strategy as the small mound at Glendhu appears to have been a product of a short period of collection in the spring while collection at Rockmarshall may have taken place over a longer period of the year. In Denmark at sites such as at Ertebølle (Andersen & Johansen 1987) concentrations of settlement debris (e.g., ash layers, fire places) have been observed within the middens but in Ireland aside from some evidence of extensive burning of shells at Rockmarshall occupation debris within the midden is quite rare it must be noted that until recently only at Sutton, the largest midden, was a trace of sub midden occupation observed. The recent explorations at Baylet has, however, also shown the same trend.

At one level where Mesolithic shell middens exist in Ireland they appear to simply reflect the availability of particular shellfish in the area. Those in Strangford Lough and on Carlingford peninsula were found in areas that would have encouraged the collection of oysters *(Ostrea edulis)*. However at both Rough Island and at Baylet small clean heaps of periwinkles up to 1 metre across can be found within the oyster midden. Where periwinkles in these small heaps primarily collected for bait?

Only at Dalkey Island were the middens dominated by more open rocky shore species, in particular limpets, and at Ferriter's Cove where the smaller shell middens were usually made up of periwinkles *(Littorina littorea)* and whelk *(Nucella (Purpura) lapillus)*. This would suggest that Mesolithic communities were using an encounter harvesting approach to shellfish collection. However the manner in which shell middens developed and the nature as well as the reasons behind the collection of shellfish appear to be very different. As noted earlier Milner has suggested that there might be evidence for collection at different seasons at different sites and sites also vary dramatically in size. Therefore the simple seasonal model and explanations put forward by Woodman (1978) need to be re evaluated At Ferriter's Cove most of the shell heaps may represent the waste from food. It is noticeable that the limpet shells often occurred as burnt fragments as if they had been roasted to make them more palatable as in the traditions of the Blasket Islands where the 'Bairnach' was roasted. It is also possible that some of the small groups of whelks and periwinkles from Ferriter's Cove were collected as bait.

The other important aspect of the Mesolithic shell middens is that in areas where the soils were acid they provided an environment where other organic remains (i.e., bones) could survive. While sieving was not generally applied, the excavations of Mitchell and Liversage provided faunal remains that were, until recently the only source of information about fishing and sea mammal exploitation. One aspect of these faunal remains is that they are not as common as might have been expected, even allowing for the lack of sieving, as, with the exception of the Dalkey middens, mammalian remains are rare. Obviously fish bones would be expected to be scarce but only 12 bone fragments of mammals and birds were found at Sutton while only 4 were found at Rockmarshall.

The results from Ferriter's Cove were very different. The implementation of a sieving policy resulted in the identification of 18 fish species which, to date, represent the largest number of fish species identified from a Later Mesolithic coastal site – over 2000 identifiable bones (McCarthy 1999). The species are dominated by Labridae - wrasse and whiting. The latter was identified by, the recovery of both vertebrae and otoliths, most of which were very small suggesting that the whiting were immature when caught. Young whiting *(Merlangus merlangus)* tend to stay close inshore and the remains at Ferriter's Cove may conceivably represent a single catch. In contrast, the only fish which may have been caught in any numbers in more open waters was the tope *(Galeorhinus galeus)*. Kimball (2000a) has suggested that extensive use was made of fish traps but McCarthy (1999) feels that most of the fish was caught by inshore line fishing. Overall the Ferriter's Cove fish remains provide a more reliable list of species but also confirm that the same range of fish were being caught in the vicinity of other middens.

Although bones of marine mammals were found elsewhere in Irish Mesolithic contexts, for example at Dalkey, somewhat surprisingly Ferriter's Cove produced no evidence for sea mammal hunting. The nearby Blasket Islands always had a strong tradition of seal hunting (O'Croughan 1937). After the islanders abandoned their permanent settlement on Blascoaod Mór large seal colonies were re-established on 'An Trá Bán' and it is reasonable to assume that colonies would have been present in similar localities during the Mesolithic. In fact the common seal has often been noted today in Ferriter's Cove.

How Were Marine Resources Exploited?

The most remarkable aspect of the Irish Later Mesolithic coastal sites is that they lack any identifiable fishing or sea mammal hunting equipment. Many Danish Mesolithic sites have produced a range of fishing equipment (Skaarup 1993) while Brinkhuizen (1982) has documented a range of equipment found in North Western Europe. Several were found in the exploration of the Danish Storebaelte notably at Halsskov (Pedersen et al. 1997). In Ireland fish traps certainly existed in the River Bann during the Late Mesolithic, but no earlier evidence of fixed facilities have been found in coastal contexts. Unlike the Obanian sites (Mellars 1987), no bone harpoons or antler mattocks have ever been recorded from Ireland, either from excavations or chance finds. The general absence of red deer from Irish Mesolithic sites suggests that any large hunting equipment manufactured from bone would require alternative raw materials and perhaps the large silt stone spear-points (Fig. 4.4) from Ferriter's Cove may have

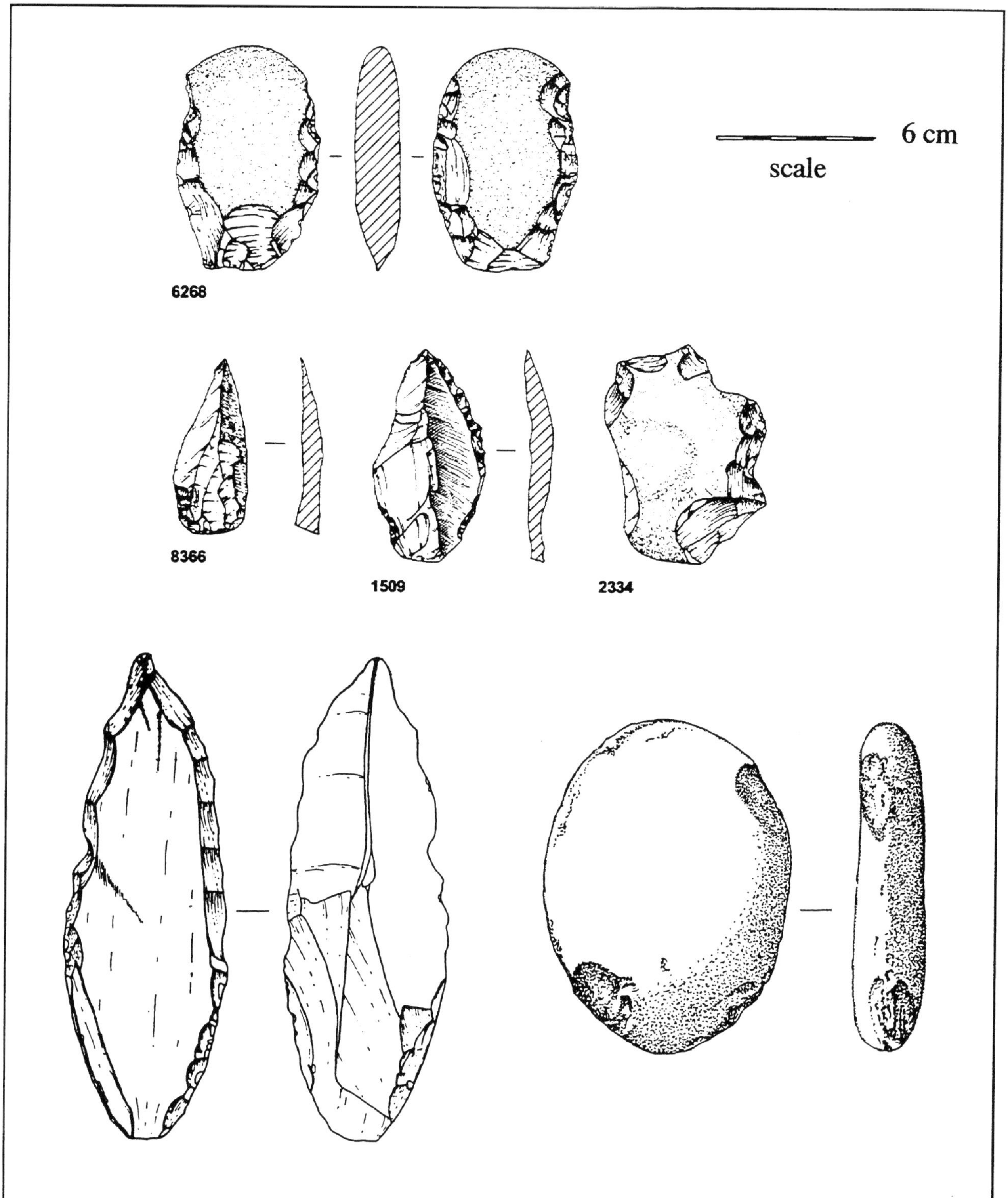

Figure 4.4. Range of stone tools from Ferriter's Cove.

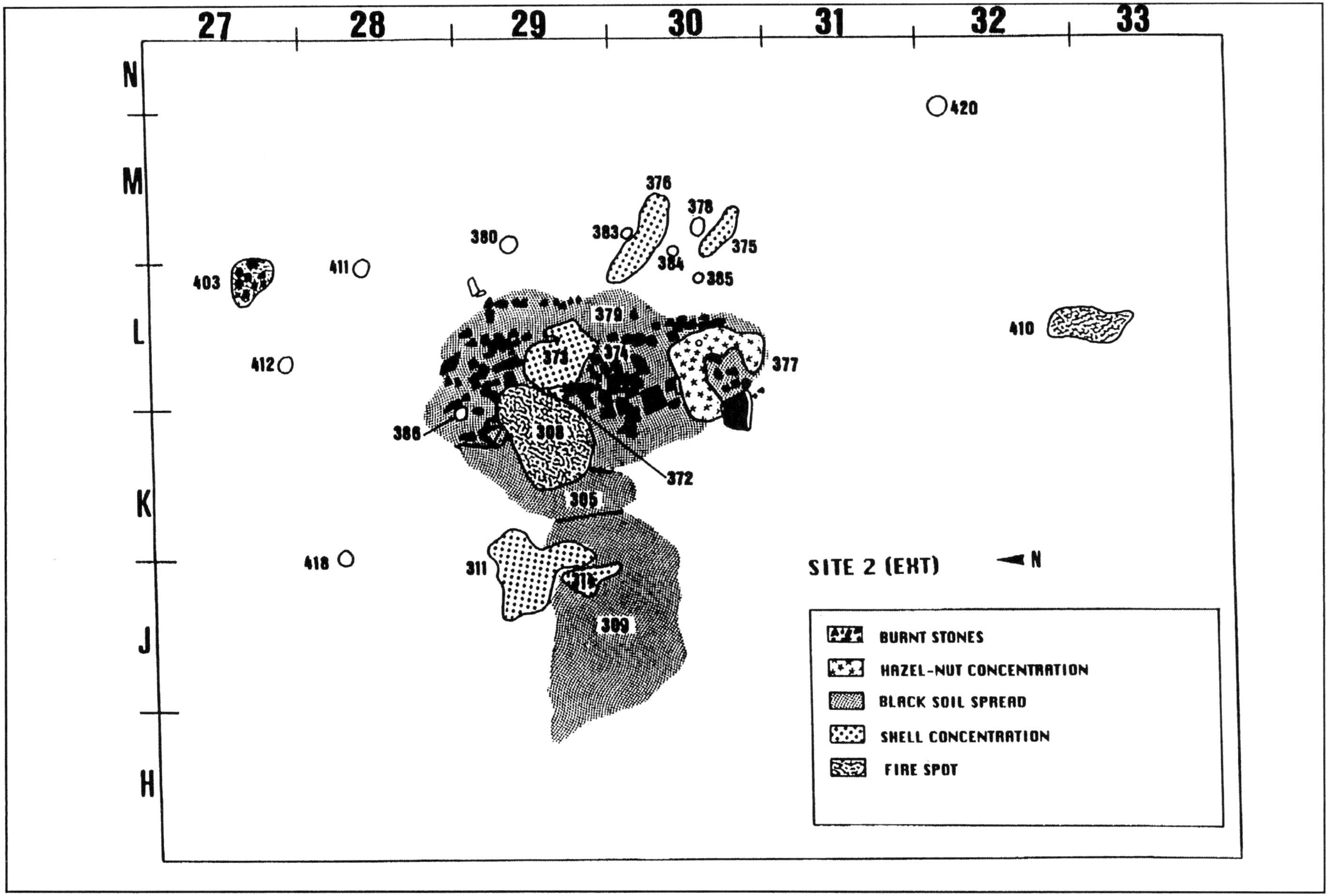

Figure 4.5. Areas of shells around hearth C308 at Ferriter's Cove, Co. Kerry.

served the purpose of hunting spears while some of the large flakes from the same site could have been used for processing sea mammal carcasses.

As on the Obanian sites and the Morton middens, the Irish sites have also produced no trace of fishing equipment. While bone fishhooks have been recovered elsewhere, the Irish sites, like the Scottish ones, lack evidence for both fishhooks and gorges. Unlike some Norwegian Mesolithic sites (Bjerk 1986), net sinkers are also absent. Both rilled and notched pebbles are absent from Irish sites. In spite of a careful examination of the material from Ferriter's Cove no convincing examples could be found. A small number of notched pebbles might be net sinkers. The frustration of using the artifactual remains to establish the extent of marine exploitation is that in Ireland, the same range of equipment occurs in coastal, riverine and lakeshore contexts where some differences could be expected. Instead the same range of large blade implements, notched flakes, axes and picks occur throughout all localities. The one exception is the bevel ended pebbles which appears to be more common where limpet collection has taken place. The other chipped pebbles of various types from Ferriter's Cove are so non-specific it would be dangerous to presume that they were primarily worked to make net sinkers. It is of course equally possible that beach or river cobbles were simply tied to nets with strips of birch bark. This type of net sinker is still used today in traditional net fishing on the Neidin River in North Norway. In a coastal context with less than perfect preservation, on an archaeological site, these sinkers would be unidentifiable. There is of course no reason to believe that these communities did not have access to boats and so it is of interest that one dug out canoe of Mesolithic age from the southern shores of Lough Neagh has been dated.

In conclusion, it is probable that shellfish were collected on numerous shorelines. As noted earlier the reasons and the role of shell fish collection may have varied quite significantly. Bailey (1978) has also noted that, while shellfish were collected in large quantities, they create a very visible form of rubbish that leaves an exaggerated sense of their dietary importance. Fish exploitation can be seen very clearly at Ferriter's Cove. As is being shown in the Baylet excavations, it is probable that if sieving had been employed at the other sites, more substantial evidence of the catching of small fish would have been found (see also McCarthy 2000) and therefore the surviving fish remains from some of the older excavations represents a limited and biased picture.

Two aspects of the Irish Mesolithic prevent the assumption that the sea had a limited role in the Mesolithic period. The concentration of material in low-lying riverine, lacustrine and coastal locations is one indication. The very limited range of large terrestrial mammals and freshwater fish (Woodman et al. 1997) would of course encourage a higher concentration of the use of the sea. It may be important to take into consideration the role of the coast and beaches as a source of raw materials for stone artifacts. While tool manufacturing may have exaggerated the coastal presence, particularly in Co. Antrim where flint was so abundant, there is no doubt that coastal resources of regions such as Strangford Lough and Lough Swilly were used extensively.

A second indicator of the role of marine resources is that the highest ratios of $\partial^{13}C$ recorded from human remains in Irish prehistory come from Mesolithic contexts in particular at Ferriters Cove, Co Kerry (Woodman et al. 1999) – 14.0 and 14.1‰ and to a lesser extent from Rockmarshall, Co. Louth – 18.1‰. These at least suggest that coastal communities living during the later Mesolithic exploited the sea. However there are indications from the earlier Mesolithic that marine exploitation may not necessarily have been intense throughout the Mesolithic.

The Impact of Farming on Coastal Exploitation

The transition to agriculture remains one of the most difficult areas in European prehistory. Therefore the role of marine resources in later prehistory is obviously viewed with certain preconceived theoretical perspectives that colour research. I would argue that agriculture only made a significant impact in Ireland when farming communities introduced a tried and tested mixed domesticate and arable farming economy. While this relied primarily on the keeping of cattle and sheep as well as the growing of cereals, it did not exclude the use of other resources. However later prehistoric archaeological sites have consistently produced domesticates and little trace of the hunting of wild species. In some cases, such as the use of the uplands in Glencloy where transitory Neolithic sites are found at 1100 ft above sea level, it is therefore impossible to know whether they are hunting sites or summer pastures for cattle and sheep (Woodman et al. 1991-92).

There is also no doubt that marine resources continued to be used. Burenhult (1984) documented the existence of several large middens in the Knocknarea area, Co. Sligo. Each of these is 2 metres thick and several hundred metres in length. These middens are made up of oyster shells to the virtual exclusion of anything else. A series of ^{14}C dates showed that the Culeenamore midden began to accumulate by 3000 cal BC. Burenhult felt that earlier phases been eroded away. He also assumed, based on the absence of domesticates in the midden, that the keeping of cattle did not begin until late in prehistory. Instead it was assumed, as there were numerous seal colonies in the area, that seal hunting played a key role in the economy of prehistoric communities of this region. The fact that no seal bones were found during the excavation was explained as due to their tendency to decay at a differentially faster rate than other bones. However seal bones have survived on other Irish middens, and on many mesolithic sites in other countries. Therefore on the Knocknarea peninsula, the seal hunting explanation appears untenable while the oyster middens would seem to be a very specialized component of an economy that is based elsewhere. This could be described as 'target harvesting'.

Unfortunately few Irish sites provide evidence that documents the transition to farming. In the case of Ferriter's Cove at least one area has produced a shell midden with associated charcoal dating to around or just after 4000 cal BC but there is no clear association with a Neolithic economy or material culture. As noted earlier, Mallory feels that the Rough Island shell midden which lies on top of the raised beach may post date the Mesolithic and it contains both cereal grains and pottery. Similarly the IIPLP in Lough Swilly is documenting the continued use of coastal resources from the Later Mesolithic into the Neolithic. Much of this evidence is at the moment based on the sequence in the Baylet Shell middens. These indicate the following possible sequence:

(1) a settlement presence overlooking the beach well before 5000 cal BC

(2) the development of a shell midden in the final stages of the Mesolithic

(3) phases of shell fish collecting taking place throughout much of the Neolithic

In many Irish sand dune locations, such as Dundrum, Co. Down (Collins 1952, 1959) and Portstewart, Co. Derry (May & Batty 1948) there is surprisingly little evidence of the development of shell middens during the Early Neolithic or that fish and sea mammals were exploited. These sites were in environments where organic material might have been expected to survive but in coastal locations where Neolithic material has been found in sand dunes, no substantial Early Neolithic shell middens have been noted. The present broken nature of the coastal sand dunes with their ridges and blowouts gives a false impression of the environment that may have existed in prehistory. It is probable that for much of prehistory a good soil and vegetation cover had developed on a layer of sand. Therefore these regions were probably more than areas for short term squatting and could have maintained more than transitory populations. Therefore the reason for the use of the dune regions was probably more than coastal exploitation. They were regions of light soils which happened to be close to the sea and which could have been used as pasture.

Similarly at Dalkey Island, which was used repeatedly during the Neolithic, it would appear that Neolithic material was cut into the middens at Sites II and IV (Liversage 1968).

Two sand dune sites have produced shell middens with artifacts that belong to the latest phases of the Neolithic. At both these sites, Raghly, Co. Sligo (D'Evelyn 1904) and Narin, Co. Donegal (Knowles 1901), *petit tranchet* derivatives were found in association with shell middens.

In general, few Neolithic sites with faunal remains produce traces of coastal exploitation and where the evidence exists it is possible that the resources of the sea were used in a different manner than during the Mesolithic. Even inland fishing during the Neolithic is difficult to document. The massive faunal assemblage from the final Neolithic phases at Newgrange did not produce a fish or sea mammal bone and Newgrange lies on a major river, the Boyne, reasonably close to the coast (van Wijngaarden-Bakker 1974). McCormick (1986) has noted that domesticates mostly make up the mammalian fauna found in Irish court tombs.

The sea is also a source of raw materials for ornaments and items used in exchange systems. While numerous shellfish were found in the Carrowmore passage tomb cemetery (Burenhult 1984, Herity 1974), their occurrence on a marine peninsula where shellfish exploitation had already been documented is not surprising. In this case they may be the byproduct of consumption. While evidence may be limited through the use of cremation, personal ornament is known to have existed in the Neolithic. A number of sites have produced stone beads (e.g. Bavan, Co. Donegal [Flanagan 1965]) while numerous small hammer pendants have been found in passage tombs such as Newgrange. At the megalithic cist in Phoenix Park, Co. Dublin, two necklaces of shells were recovered (Herity & Eogan 1976) while a *Pecten Maximus* shell as well as *Nerita littoralis* and oyster *(Ostrea edulis)* were found further inland at Fourknocks, Co. Dublin (Herity 1974). At Loughcrew, cairn II, Co. Meath, 200 seashells were recovered along with a portion of an intercostal bone from a whale. Actual pendants made from shells are rare. A *Nautica catena* was found at Carrowkeel, cairn M (Herity 1974) while at Kilgreany Cave, Co. Waterford, Tratman (1928) found three *Littorina obtusata* which had been pierced to be used as necklaces. However, there is little other evidence of seashells being used as ornaments or exchange items. The extensive excavation of a Neolithic settlement site at Ballygalley, Co. Antrim, within 500 m of the coast has also produced numerous beads, but none of them are made from shell (Simpson pers. comm.). Furthermore they are lacking in the Lough Gur area where the limestone bedrock would allow for their preservation and where numerous beads occur (Eogan & Grogan 1987). A similar absence of shells as personal ornament would seem to occur in the graves of the earlier Bronze Age (Waddell 1990).

Therefore there are some limited indicators that the sea continued to provide a component of the resources in Neolithic economies. It is also probable that throughout the Neolithic different communities used the sea in different manners but usually in a very limited fashion and probably as a supplement to a land based economy (see below)

The Evidence and Lessons from Later Period Coastal Sites

Often the study of marine resources is seen too much from teleologi cal perspective of studying the event of the introduction of agriculture rather than from a perspective of how communities have used the sea through time. Indeed, besides being a valid research topic in its own right, the study of later sites can provide lessons for the study of earlier periods. One of the major difficulties in assessing the exploitation of coastal resources is that so many of the middens, and sand dune sites are either undated or undateable. Many of these locations have not been examined in over 50 years and in some cases 100 years (see Murray n.d. for overall survey of these sites). While some sand dune locations in the north east of Ireland in particular were mined for archaeological artifacts, no contexts were recorded. Many other localities round the coast of Ireland produced shell middens that contained little else but shells. Therefore in many instances assessing the dates of midden sites is difficult. In the absence of artifacts three simple observations can be made:

(1) If a midden is within a sand dune rather than below it, it is likely to postdate the Mesolithic.

(2) If it contains cattle bones it probably postdates the Mesolithic.

(3) If it contains horse bones it probably postdates the Neolithic.

Unfortunately many sites contain neither cattle nor horse bones and most are of uncertain age. In Ireland, the virtual absence of red deer before the Neolithic means that the occurrence of red deer bones in a midden does necessarily indicate a Mesolithic date. In recent years, due to problems of coastal erosion, some rescue excavations have been carried out and as a result limited but firm evidence of dating is becoming available. It is apparent that Irish shell middens continued to accumulate throughout the later prehistoric and Early Medieval periods. At Omey Island, Co. Galway, investigation of an extensive occupation site that had been buried in sand dunes revealed the existence of a Bronze Age shell midden (O'Keefe 1994). The Culeenamore shell midden continued to accumulate until the Iron Age when a child burial was placed within it (Burenhult 1984). Investigations of what was hoped to be a series of stone age shell middens at Poul Gorm, Co. Cork, produced Iron Age ^{14}C dates (McCarthy & Woodman). Excavations at Omey Island have also shown that extensive shell middens can build up at monastic settlements (O'Keefe 1994). This pattern can be documented at a number of coastal and island monastic sites.

Investigations of the variation between marine ^{14}C contents and its terrestrial equivalent and how the ratio might change through time has included the ^{14}C dating of limpets and bone or charcoal from several middens in the west of Ireland (Reimer et al. 2002). These again confirm an extensive use of shellfish in the Bronze Age in the Clifden area of west Co. Galway (McCormick et al. 1996) and provide an interesting lesson. The False Bay shell midden is a good example (McCormick 1995). Murray has noted (1996) that in this case no fish bones have been recovered and that analysis of food residues adhering to food vessel sherds suggested a land based diet. Therefore, once again a shell midden does not necessarily indicate a highly marine based economy but conversely it is important to note that at Dun Aonghasa, two hundred metres above sea level on the cliffs of Inis Mor, quite significant concentrations of fish bones were uncovered – more than 5000 (McCarthy 2000). It is probable that fishing played a much more important role than casual investigations would suggest, yet in the context of modern excavations, their absence from Poul Gorm (Iron Age), False Bay (Bronze Age) and Culeenamore (Neolithic to Iron Age) may be significant. If that is the case then many later prehistoric shell midden sites may represent little more than very specialized components which have been created by a single activity. Evidence of fishing would not necessarily be found on midden sites.

A somewhat similar pattern can be seen in Historic periods. One investigation in Lough Foyle at Magilligan sand dunes revealed a series of small shell middens that dated to the Iron Age. Again, in spite of careful sieving, no traces of fish bones were recovered (Mallory & McCormack 1988). This does not of course mean that did take place. At Oughtymore, again in the Magilligan sand dunes of Lough Foyle, some fish bones were recovered from an Early Medieval context (Mallory & Woodman 1984). This is a site where the excavations indicated that long term settlement had taken place. Fish bones have also been found at the settlement associated with the monastic site of Drumcliffe, Co. Sligo (McCarthy pers. comm.). This again is evident at the raised rath and motte of Rathmullen, Co. Down. This site lies about 1.5 km from the sea but has produced quantities of shellfish and fish bones, many of them cod bones (Lynn 1981-82).

Although there is little documentary evidence, it is possible that in later periods sea mammals were also exploited. However, whether this was an opportunistic use of bones from stranded animals or as a result of hunting is unclear. Lynn (1981-82), in discussing fragments of a whale bone from Rathmullen, Co. Down, notes a reference in *The Annals of the Four Masters* for the year 739 in which a whale cast ashore in Boirche is slaughtered.

At Altar wedge tomb, Co. Cork, in the Iron Age, a ritual deposit of 'fruits of the sea' (i.e., shellfish, fish and cetacean bones) had been placed in the entrance to the tomb (O'Brien 1993). The bones of cetaceans have been found on other sites. At Raheens ring fort, overlooking Cork Harbour, cetacean bones were found in an Early Medieval *souterrain* (McCarthy 1994). Knowles (1901) has also noted the occurrence of a paddle made from whalebone in sand dunes in Co. Donegal.

It is also apparent that the medieval period saw an intensification of the use of the sea. With development of Viking and Norman coastal ports, trade and marine contacts with Ireland increased. Urban excavations have produced large numbers of imported pottery while large quantities of fish such as hake (*Merluccius merluccius*) and ling (*Malva malva*) have been recovered from excavations in Waterford city (O'Shaughnessy pers. comm.) while excavations in Cork city produced quantities of shells from within the medieval town walls.

The exploration of some of the later shell middens can also provide important lessons for prehistory. In the Cork Harbour area a series of large and medium sized shell middens have been recorded since the middle of the 19th century (Atkinson 1874). These vary from the Ballintubrid East Midden which was originally over 150 metres in length to sites such as the Rathcoursey midden which is much smaller. Although on occasions Atkinson documented traces of burning bones or artifacts were rarely found within the middens that usually consisted of massive quantities of oysters. A limited programme of radiocarbon dates of oyster shells was carried out. A regional marine effect correction of adjusting the dates by approximately 400 years younger has been used. In the two larger middens samples from near the top of the midden were used. In total 10 middens were sampled with a date from the base of large Ballintubrid midden suggesting that it had already begun to form in the Iron Age. However, as can be seen from Figure 4.6, one group of middens seems to have been in existence up to 1100 AD while a second group were predominantly of post medieval and early modern age. The latter group of dates confirms suggestions made by O'Kelly (1955) that some middens in the harbour area were quite late in date. It is of interest that there are fewer midden which are contemporaneous with the first few centuries of the Anglo Norman phase of the City of Cork yet numerous excavations have revealed layers of oyster shells within the city. It would seem probable that at a certain period of time oyster consumption continued but the oysters were not necessarily processed on the shore adjacent to where they were collected but were instead brought in their shell to the city. Therefore in a shellfish rich area we cannot presume at any period the absence of middens does not necessarily mean that shell fish were not a part of the diet. In contrast it is of interest that in the discussions on Harte's Donegal middens reference was made to the fact that in the 18th century barrel loads of oysters without their shells were transported along the south coast of Ireland from near Kenmare.

Assessing the Role of the Sea

As noted earlier, the major problem is that so many coastal sites and middens in Ireland remain undated. The sample of coast sites also appears to present a biased picture. More Mesolithic and Neolithic settlement is known from middens and sand dunes in the north of the country while middens around the south coast often lack stone age material. This is almost certainly due to the differing rates of isostatic recovery. The shoreline reached its highest relative point round the north of Ireland before 4000 cal BC or earlier. However, in parts of the extreme south where isostatic recovery never outstripped the increase in sea volume, strandlines stabilised at their present levels by about 2500 cal BC (Devoy 1983). So far there is little evidence from all these middens that there are changes in shell types or shell sizes at certain periods in the past. Instead the shell middens usually reflect the range of shellfish available in a particular area.

This problem is compounded by the difficulty in assessing the actual impact of marine resources on the human diet. Shell middens constitute an obvious and highly visible presence in the archaeological record but they may not always provide an accurate reflection of the importance of the sea in the diets of past populations. Even in later periods it would be unwise to assume that shell middens are automatically associated with the loci of consumption. One major debate centres on the role of the dog whelk *(Nucella lapillus)*. It is probable that, in addition to being a food source, the occurrence of concentrations dog whelk could be due to their use in the extraction of a purple dye. Bigger (1895) drew attention to numerous shell middens in the north of Ireland where large numbers of dog whelk shells occur, and argued that there is good historical evidence for their use as a source for dye (Murray 1996). Obviously Irish archaeology can only make use of information collected over the past 100 years, but as has already been observed, the major flaws tend to be absences of information. These are the absences of dates for most known coastal shell middens and the lack of good excavations leading to an uncertainty about the role of fishing. As has been shown by Mellars (1987) in the case of the Oronsay sites, the older excavations missed the presence of numerous saithe bones. It is highly probable that fish bones were either not observed or that only the larger, more distinctive species were collected.

One potential source of information on the role of the sea is based on $\partial^{13}C$ results which have on occasions been obtained from human remains usually associated with radiocarbon dating programmes (e.g., Tauber 1981, Price 1989). There are of course still problems with this data not least the matter of the freshwater reservoir effect and the lack of accuracy of some of the results that were documented by Lanting and van der Plicht (1998). In spite of these problems there is still a robust pattern for Irish data.(many though not all of the results used were taken from Brindley and Lanting [1989-90, 1995]; Lanting [1996] and Hedges et al. [1993]).

The major problem with $\partial^{13}C$ analysis in Ireland is that skeletal remains of Mesolithic and later periods from coastal contexts are rare. Due to a combination of poor soil conditions and cremation practice a very limited sample of skeletal remains are available for analysis. In this analysis remains from within 10 km of the sea coast have been used. In terms of chronology the samples are not evenly spaced through time. In particular the Later Bronze Age and Early Iron Age cremation practices have left a gap in the chronological sequence. However a certain pattern can be seen. Figure 4.7 displays the results of $\partial^{13}C$ analysis on 30 coastal (0-10 km) and 13 insland samples (>10 km from coast). The inland samples are consistently between -20 and -23%, while the coastal samples are more variable.

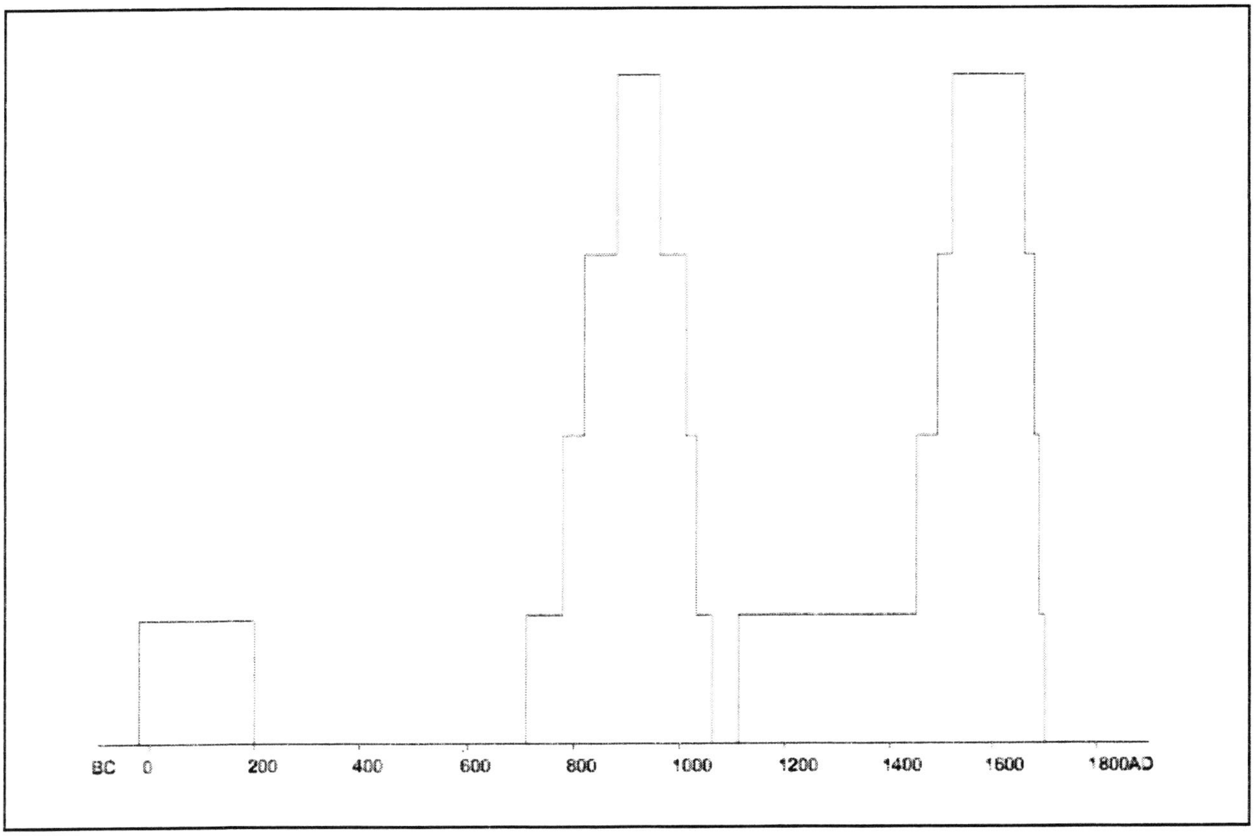

Figure 4.6. Adjusted radiocarbon dates from a series of shell middens in Cork Harbour, Co. Cork.

Unfortunately there are no coastal human remains from coastal locations in the Early Mesolithic. The only uncremated remains are those from Killuragh Cave overlooking the Mulkear River and which lies about 30 km from the Shannon Estuary (Woodman & O'Shaughnessy n.d.). These remains date to 7000-6546 and 7194-6658 cal BC with ∂^{13}C levels of –19.95‰ and –20.86‰. This could, as suggested by Lanting and van der Plicht 1998, be due to the consumption of C3 plant remains and freshwater fish. It should be noted that no fish bones were recovered from the site and that eels would be the only Irish freshwater fish that could have been consumed. However, irrespective of how these results are interpreted, there is, as yet, no evidence that marine resources played a significant role in their diet. This is, of course, likely to be the result of the absence of coastal sites and the range of evidence now being uncovered from many Scandinavian Mesolithic sites (see Fischer 1995) would suggest that the sea was used during the earlier part of the Mesolithic.

Some determinations from Later Mesolithic sites have already been noted (Fig. 4.6). The Ferriter's Cove samples are similar to those obtained at Vedbaek, Denmark (Albrethsen & Brinch Petersen 1976) which would suggest that marine resources dominated their diet while the Rockmarshall result also suggests that marine resources were important. The fact that, in contrast, samples from the more inland Killuragh Cave appear to have a rather more terrestrial signature raises certain questions. The number of samples is limited and in certain areas the transfer of high quality flint artifacts from the coast could be interpreted as significant evidence of mobility between the coast and inland areas. Material can be moved through limited contact or exchange rather than large scale movement of peoples therefore the disparity between the Ferriter's Cove results and Killuragh could suggest that general coastal/inland models may not always be appropriate even in an area such as Ireland where there are few large mammals.

In recent years the very significant dietary shift away from marine resources at the beginning of the Neolithic and the fact that this can be quite sudden has been noted in a number of areas (see most recently Schulting [1998], Schulting & Richards [2002]). In Ireland ∂^{13}C results based on Neolithic human remains from coastal areas are rare. This is to some extent a reflection of research interests in recent years especially since ∂^{13}C analysis became standard. However the results from the Phoenix Park Tomb in the Liffey Estuary, a human bone of Neolithic date from Kilgreany which lie just under 10 km from the sea are strongly suggestive of a land based economy. Some analyses of undated human remains from Poulnabrone that again lies about 10 km from the sea has produced results of a similar range. It is assumed that these remains belong to the earlier half of the Neolithic period (Lynch 1988). In Ireland a Neolithic land based economy need not necessarily be a cereal based economy as Power (1993) has suggested there are some indicators in the skeletal evidence

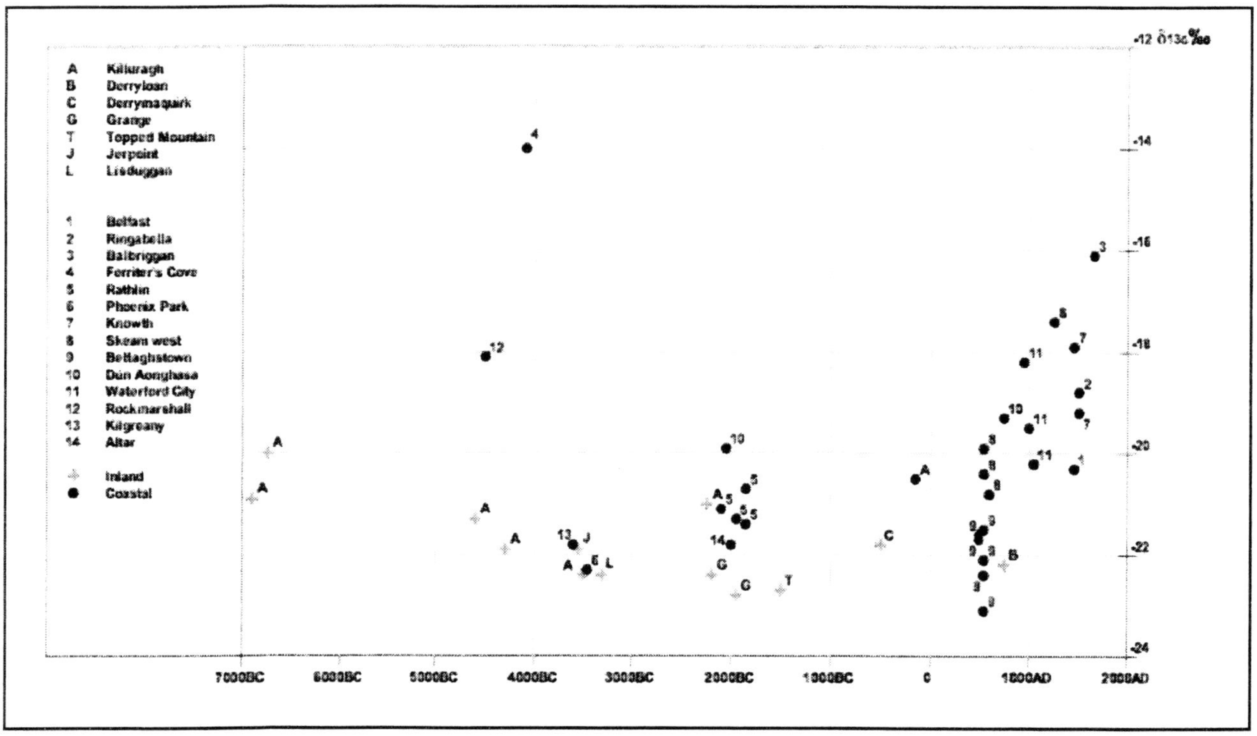

Figure 4.7. $\partial^{13}C$ determinations from coastal (0-10 km) and inland (>10km) locations.

from the Irish Neolithic that could suggest a meat based economy. This does not mean that we should exclude foraging or harvesting wild resources. The location of Neolithic material in the Shannon Estuary at Carrigdirty (O'Sullivan 2001) clearly demonstrates the use of the sea and the river estuaries but they do not appear to be the basis for Neolithic economies. These few results from coastal areas do not differ significantly from those obtained at sites such as Jerpoint West that lies in the centre of Ireland.

This pattern continues into later periods. Although determinations from later periods tend to clump in time, certain obvious patterns emerge from the $\partial^{13}C$ figures. Indeed, in general, after the Mesolithic and until the Medieval period there is no significant difference through time for changes in the $\partial^{13}C$ determinations. Within Ireland, after the Mesolithic period $\partial^{13}C$ determinations tend to be below –20‰. While, as noted earlier, they may not have the same accuracy as tests carried out specifically for $\partial^{13}C$, the scarcity of results below –20‰ from coastal localities provides a strong indication that human diets in Ireland after the Mesolithic tended to be based primarily on terrestrial resources.

In the Bronze age the role of marine resources as reflected in human remains from coastal burials on Rathlin Island and at Altar Wedge Tomb would appear to be limited. Indeed the Rathlin Island results are closer to those obtained from equivalent inland sites (Fig. 4.8) while there is a slightly higher marine signature from Altar.

In the early medieval period the human skeletal remains from the coastal church site of Skeam West or the skeleton from Dun Aonghasa (Cotter1993, 1996) again have signatures that suggest that marine resources formed a small proportion of the diet while in the case of results from the east coast at Bettaghstown there is an even less evidence of a marine content in the diet. In contrast, samples from around 1,000 AD and later (i.e., from the Viking period onwards) begin to show an increased marine presence. This is apparent in some remains from a late burial in Skeam West, in Waterford City as well as in the Balbriggan and Ringabella Skulls (Delaney & Woodman n.d.) and at Knowth where it is reflected both in late dog as well as a human remains. As Lanting and van der Plicht (1998) have shown the development of more industrialized fishing and trade may have had a significant impact on diet in the Medieval and Post Medieval periods. It is also possible of course that islands might have been special places for burials and so the presence of skeletons on an island should not be taken as direct evidence that the island provided their complete livelihood.

Therefore, in Ireland, it would appear that while the sea did continue to be used through time, for a significant time period these resources were not used to the extent that might have been expected. Indeed, in coastal locations, there is often more of a difference between later prehistoric/early medieval $\partial^{13}C$ signatures and medieval/post medieval signatures than there is with contemporaneous signatures from inland sites.

This trend in coastal communities relying less on marine resources is not particular to Ireland. Iron Age human remains from the Obanian caves have produced $\partial^{13}C$ results ranging from 21 to 22‰ (Saville & Hallen 1994). In northern Norway, Johansen and colleagues (1986) have noted on the Lofoten Island, Flakstadøy, that there is a significant shift between determinations on stone age burials with mean values of 13.4±0.8‰ and Late Iron Age burials, where the mean values are 17.0±1.0‰. Again 11 determinations from an Early Medieval cemetery from the very isolated island of Traena has produced a mean of 16.9±1.2‰. While these $\partial^{13}C$ values do not approach the Irish figures, they indicate that in later periods in coastal areas alternative resources probably played a more important role than had been suspected, thus the probable importation of cereals onto Traena (Johansen 1986).

Lessons to be Learnt

Obviously the sea would have played on important role in the Irish Mesolithic but it would be a mistake to assume that this role can be explained by assuming that there was one simple strategy for its use. The highly visible shell middens from Irish later prehistory and later periods may only represent a limited use of marine resources and aside from the Mesolithic it is only with the Viking and Anglo-Norman periods that coastal communities began to exploit the sea in an intensive manner. From the evidence in later sites as well as that from the Mesolithic, certain general observations can be made.

(1) The actual surviving material culture of the Irish Mesolithic itself provides little evidence of marine exploitation. This has to be inferred from the scanty surviving organic data.

(2) As has been noted by many researchers shellfish in middens may not by themselves represent evidence of an extensive reliance on marine resources and, as Milner (2002) has shown, shellfish can be subject to over simple interpretations of use and they may have been collected for other purposes such as bait or even ritual.

(3) Even where there is evidence of midden development, this should not be taken as being the total evidence of marine exploitation. This can be seen in the differences such as the presence and absence of fish remains between Bronze Age sites such as False Bay and Dun Aonghasa in the West of Ireland or between shell middens in Magilligan sand dunes in Co. Derry.

(4) On the basis of the distribution of shell middens through time in areas such as Cork Harbour, one should note that shell fish may not always be processed in the vicinity of the shoreline as they may be removed to settlement sites for consumption and discard. Therefore the absence of middens does not necessarily mean the absence of shellfish consumption.

(5) There is a danger that we might over generalize about models of marine exploitation, especially when the evidence is based on only one of several indicators. Therefore, whether studying the Mesolithic or Later Periods we need to ensure that all potential sources of information are used.

(6) In the case of the Irish Mesolithic with its low-lying distribution, it would be tempting to continue with the assumption that all Mesolithic communities used the sea but this assumption is no longer tenable.

Thus, although Ireland may have been an island for the last 10,000 years, it is by no means certain that the seas which cut it off were always used to their full potential.

Acknowledgements

I would like to thank the late Liz Anderson for many fruitful discussions on the role of the sea and all those who have helped me notably in recent times especially Nicky Milner and Emily Murray. Hugh Kavanagh has provided the drawings and the photograph is from the Welch Collection in the Ulster Museum Belfast.

Endnote

1. The Irish Mesolithic is divided into two phases. An Early Mesolithic phase (8000 to possibly 6500 cal BC) which is associated with microliths and a Later Mesolithic, which had fully developed before 6000 cal BC. This phase is associated with a hard hammer technology and the production of larger blades that produced the blanks for a series of blade tools, including the distinctive butt trimmed forms.

References

Albrethsen, S.E. & E.B. Petersen
1976 Excavations of a Mesolithic cemetery at Vedbaek, Denmark. *Acta Archaeologia* 47: 1-28.

Andersen, S.H.
1995 Coastal adaptations and marine exploitation in the late Mesolithic with special reference on the Limfjord region. In *Man and the Sea in the Mesolithic* (A. Fischer, ed.), pp. 41-68. Oxford: Oxbow Monographs No. 53.

Andersen, S.H. & E. Johansen
1987 Ertebølle revisited. *Journal of Danish Archaeology*. 5: 31-61.

Atkinson, G.M.
1874 Kitchen middens in the estuary of Cork Harbour. *Journal of the Royal Society of Antiquaries of Ireland* 13: 258-61.

Bailey, G.N.
1978 Shell middens as indicators of post glacial economies: a territorial perspective. In *The Early Post Glacial Settlement of North West Europe* (P. Mellars, ed.), pp. 37-64. London: Duckworth.

Bigger, F.J.
1985 Prehistoric settlement at Portnafeadog in the Parish of Morus, Connemara. *Proceedings of the Royal Irish Academy* 3: 727-732.

Bjerck, H.
1986 The Fosna-Nostvet problem – a consideration of archaeological units and chronozones in the Norwegian Mesolithic period. *Norwegian Archaeological Review* 19: 103-121.

Brinkhuizen, D.C.
1983 Some notes on recent, pre- and protohistoric fishing gear from northwestern Europe. *Palaeohistoria* 25: 7-54.

Brindley, A.L. & J.N. Lanting
1989-90 Radiocarbon dates for Neolithic single burials. *Journal of Irish Archaeology* 5: 1-7.
1995 Irish bog bodies: the radiocarbon dates. In *Bog Bodies* (R.C. Turner & R.G. Scaife, eds.), pp. 133-136. London: British Museum.

Brunicardi, J.
1914 The shore dwellers of Ireland. *Journal of the Royal Society of Antiquaries of Ireland* 44: 185-213.

Burenhult, G.
1984 *The Archaeology of Carrowmore, Co. Sligo, Eire.* Stockholm: Theses and Papers in North European Archaeology No. 14.

Collins, A.E.P.
1952 Excavations in the sandhills at Dundrum, Co. Down, 1950-1951. *Ulster Journal of Archaeology* 15: 2-26.
1959 Further investigations in the Dundrum sandhills. *Ulster Journal of Archeology* 22: 5-20.

Costa, L., F. Sternke & P.C. Woodman
2001 The analysis of a lithic assemblage from Eleven Ballyboes, Co. Donegal. *Ulster Journal of Archaeology* 60: 1-6.

Cotter, C.
1993 *Western Stone Fort Project, Interim Report* (J. Wadell, ed.), pp. 1-19. Dublin: Discovery Programme Reports No. 1, Royal Irish Academy.
1996 *Western Stone Forts Project, Interim Report,* pp. 1-15. Dublin: Discovery Programme Reports No. 4, Royal Irish Academy.

Cross, S.
2001 Competitive feasting in the Neolithic. *Archaeology Ireland* 15: 11-13.

Delaney, M. & P.C.Woodman
n.d. Searching the Irish Mesolithic for the humans behind the hatchets. *Festscrift for Professor George Eogan.* Manuscript in possession of authors.

Devoy, R.J.
1983 Late Quaternary shorelines in Ireland: an assessment of their implications for isotatic land movement and relative sea level changes. In *Shorelines and Isostasy* (D. Smith & A.G. Dawson, eds.), pp. 227-254. London: Institute of British Geographers.

D'Evelyn, A.
1904 Prehistoric archaeology report of the 4th triennial conference and excursion held in Sligo, July 12-18th, 1904. *Irish Naturalist* 13: 216-219.

Eogan, G. & E. Grogan
1987 Lough Gur – excavations by Sean P. O Riordain: further Neolithic and Beaker habitation at Knockadoon. *Proceedings of the Royal Irish Academy* 87: 299-506.

Fischer, A. (editor)
1995 *Man and the Sea in the Mesolithic*. Oxford: Oxbow Monograph No. 53.

Flanagan, L. & D. Flanagan
1966 The excavations of a court cairn at Bavan, Co. Donegal. *Ulster Journal of Archaeology* 29: 16-38.

Harte, W.
1866 On the occurrence of *kjökkenmöddings* in the County of Donegal. *Journal of the Royal Geological Society of Ireland* 1: 154-158.

Hedges, R.E.M., R.A. Houseley, C. Bronk Ramsey & G.J. van Klinken
1993 Radiocarbon dates from the Oxford AMS system. *Archaeometry* 35: 305-326.

Herity, M.
1974 *Irish Passage Graves*. Dublin: Irish University Press.

Herity, M. & G. Eogan
1977 *Ireland in Prehistory*. London: Routledge & Kegan Paul.

Hewson, L.M.
1938 Notes on Irish sandhills. *Journal of the Royal Society of Antiquaries of Ireland* 68: 69-90.

Johansen, O.S., S. Gulliksen & R. Nydal
1986 $\partial^{13}C$ and diet analysis of Norwegian human skeletons. *Radiocarbon* 28: 754-761.

Kimball, M.J.
2000a *Human Ecology and Neolithic Transition in Eastern County Donegal, Ireland.* Oxford: British Archaeological Reports No. 300.

2000b Variation and context: ecology and social evolution in Ireland's Later Mesolithic. *New Agendas in Irish Archaeology* (A. Desmond et al., eds.), pp. 31-48. Bray: Wordwell.

Kirby, M.
1990 *Skelligside.* Dublin: Lilliput Press.

Knowles, W.J.
1901 The fourth report on the prehistoric remains from the sandhills of the coast of Ireland. *Proceedings of the Royal Irish Academy* 3: 331-389.

Lanting, J.
1996 The radiocarbon dating. In *Late Viking Age and Medieval Waterford Excavations 1986 – 1992* (M. Hurley & O.M.B. Scully, eds.), pg. 818. Waterford: Waterford Corporation.

Lanting, J. and J.van der Plicht
1998 Reservoir Effects and apparent ^{14}C ages. *Journal of Irish Archaeology* 9: 151-165

Liversage, G.D.
1968 Excavations at Dalkey Island, Co. Dublin 1956-59. *Proceedings of the Royal Irish Academy* 66: 53-233.

Lynch, A.
1988 Poulnabrone, a stone in time. *Archaeology Ireland* 2: 105-107.

Lynn, C.J.
1981-82 The excavation of Rathmullen, a raised rath and motte in Co. Down. *Ulster Journal of Archaeology* 44-45: 65-171.

Maloney, A.
1993 *Survey of the Raised Bogs of Co. Longford (Irish Archaeological Wetland Unit, Transactions Vol. 1).* Dublin: Crannog Publications.

Mallory, J.P. & P.C. Woodman
1984 Oughtymore: an early Christian shell midden. *Ulster Journal of Archaeology* 47: 51-62.

Mallory, J.P. & F. McCormick
1988 Excavations at Ballymulholland, Magilligan Foreland, Co. Londonderry. *Ulster Journal of Archaeology* 51: 103-114.

May, A. McL. & J. Batty
1948 The Sandhills Culture of the River Bann estuary, Co. Londonderry. *Journal of the Royal Society of Antiquaries of Ireland* 78: 130-156.

McCarthy, M.
1994 The animal bone report in A.M. Lennon's excavations of Ringfort Raheens No.2 near Carrigaline. *Journal Cork Historical and Archaeological Society* 99: 47-65.

1999 The faunal remains. In *Excavations at Ferriter's Cove 1983 – 1995* (P. Woodman, E. Anderson & N. Finlay, eds.), pp. 83-95. Bray: Wordwell.

2000 Hunting, fishing and fowling in late prehistoric Ireland. *New Agendas in Irish Archaeology* (A. Desmond, G. Johnson, M. McCarthy, J. Sheehan & E. Shee Twohig, eds.), pp. 107-118. Bray: Wordwell.

McCormick, F.
1985-86 Animal bones from prehistoric Irish burials. *Journal of Irish Archaeology* 3: 37-48.

1995 False Bay in the Bronze Age. *Archaeology Ireland* 9: 12-14.

McCormick, F., M. Gibbons, F. McCormac & J. Moore
1996 Bronze Age to medieval coastal shell middens near Ballyconneely, *Journal of Irish Archaeology* 7: 77-84.

McErlean, T., W. McConkey & R. Forsythe
2002 *Strangford Lough – an Archaeological Survey of the Marine Cultural Environment.* Belfast: Blackstaff Press.

Mellars, P.
1987 *Excavations on Oronsay.* Edinburgh: Edinburgh University Press.

Milner, N.
2001 At the cutting edge: using thin sectioning to determine season of death of the European oyster *(Ostrea edulis). Journal of Archaeological Science* 28: 861-873.

2002 *Incremental Growth of the European Oyster*. Oxford: BAR International Series No. 1057.

Milner, N. & Woodman, P.C.
2001 Mesolithic middens – from famine to feasting. *Archaeology Ireland* 15: 32-35.

Mitchell, G.F.
1949 Further early kitchen middens in Co. Louth. *Journal of the County Louth Archaeological and Historical Society* 12: 14-20.

1956 An early kitchen midden at Sutton, Co. Dublin. *Journal of the Royal Society of Antiquaries of Ireland* 86: 1-16.

1972 Further excavations of the early kitchen midden at Sutton, Co. Dublin. *Journal of the Royal Society of Antiquaries of Ireland* 102: 151-159.

Movius, H.L.
1940 A Report on a Stone Age excavation at Rough Island, Strangford Lough, Co. Down. *Journal of the Royal Society of Antiquaries of Ireland* 70: 111-142.

1953 An historical account of the investigations at Larne. *Ulster Journal of Archaeology* 16: 7-23.

Murray, E.V.
1996 *An Early Bronze Age settlement at False Bay, Connemara.* Unpublished BSc Thesis, Department of Archaeology. Belfast: Queen's University.

1999 *Early Evidence for Coastal Exploitation in Ireland.* Unpublished PhD Thesis, Department of Archaeology. Belfast: Queen's University.

n.d. Molluscs and middens: the archaeology of Irelands 'Savage Races'. In *The Environmental Archaeology of Ireland* (E. Murphy & N. Whitehouse, eds.). Cork: Cork University Press.

O'Brien, W.
1993 Aspects of Wedge Tomb chronology. In *Past Perceptions* (E. Twohig & M. Ronayne, eds.), pp. 63-74. Cork: Cork University Press.

O' Crohan, T.
1937 *The Islandman*. Dublin: Talbot Press.

O'Keefe, T.
1994 Omey and the sands of time. *Archaeology Ireland* 8: 14-17.

O'Kelly, M.J.
1955 A shell midden at Carrigtohill, Co. Cork.. *Journal of the Cork Historical and Archaeological Society* 60: 28-32.

O'Sullivan, A.
2001 *Foragers, Farmers and Fishers in a Coastal Landscape: an Intertidal Archaeological Survey of the Shannon Estuary*. Dublin: Discovery Programme of the Royal Irish Academy.

O'Sullivan, A. & J. Sheehan
1996 *An Archaeological Survey of the Iveragh Peninsula, Co. Kerry*. Cork: Cork University Press.

Pedersen, L.
1997 They put fences in the sea. In *The Danish Storebaelt since the Ice Age* (L. Pedersen, A. Fischer & B. Aaby, eds.), pp. 124-143. Copenhagen: A/S Storebaelt Fixed Link.

Power, C.
1993 Reconstructing patterns of health and dietary change in Irish Prehistoric populations. *Ulster Journal of Archaeology* 56: 9-17

Power, P.
1930 Brick Island shell midden, Cork Harbour. *Irish Naturalist Journal* 3: 123-126.

Price, T.D.
1989 The reconstruction of Mesolithic diets in the Mesolithic of Europe. In *The Mesolithic in Europe* (C. Bonsall, ed.), pp. 48-59. Edinburgh: John Donald.

Reimer, P.J., F.G. McCormac, J. Moore, F. McCormick, & E.V. Murray.
2002 Marine radiocarbon reservoir corrections for the mid to late Holocene in the eastern subpolar North Atlantic. *Holocene* 12: 129-135.

Saville, A. & Y. Hallen
1994 The Obanian Iron Age: human remains from the Obanian cave sites, Argyll, Scotland. *Antiquity* 68: 715-723.

Schulting, R.J.
1998 Slighting the sea: stable isotope evidence for the transition to farming in northwesterm Europe. *Documenta Praehistorica* 25: 203-218.

Schulting, R.J. and Richards, M.P.
2002 Finding the coastal Mesolithic in south west Britain: AMS dates and stable isotope results on human remains from Caldey Island, South Wales *Antiquity* 76: 1011-1029.

Skaarup, J.
1993 Submerged settlements. In *Digging into the Past* (S. Haas & B. Storgaard, eds.), pp. 70-75. Åarhus: Universitetsforlag Åarhus.

Tauber, H.
1981 ^{13}C evidence for dietary habits of prehistoric man in Denmark. *Nature* 293: 332-333.

Tratman, E.K.
1928 Excavations at Kilgreany Cave, Co. Waterford. *Proceedings of the British Historical Society* 3: 109-53.

Van Wijngarden-Bakker, L.H.
1974 The animal remains from the Beaker settlement at Newgrange, Co. Meath: first report. *Proceedings of the Royal Irish Academy* 74: 313-383.
1985 The faunal remains. In *Excavations at Mount Sandel 1973-1977* (P. Woodman, ed.), pp. 71-76. Belfast: Northern Ireland Archaeological Monograph No. 2.

Waddell, J.
1990 *The Bronze Age Burials of Ireland.* Galway: Galway University Press.

Wilson, P.
1995 Sand dunes and archaeology. *Archaeology Ireland* 9: 24-26.

Woodman, P.C.
1977 A narrow blade Mesolithic site at Glynn, Co. Antrim. *Ulster Journal of Archaeology* 40: 12-20
1978 *The Mesolithic in Ireland.* Oxford: British Archaeological Reports No. 58.
1985 *Excavations at Mount Sandel 1973-77.* Belfast: Northern Ireland Archaeological Monograph No. 2.
1985a Excavations at Glendhu, Co. Down. *Ulster Journal of Archaeology* 48: 31-40.

Woodman, P.C., E. Anderson, & N. Finlay
1999 *Excavations at Ferriter's Cove, 1983–1995: Last Foragers, First Farmers in the Dingle Peninsula.* Bray: Wordwell Ltd.

Woodman, P.C., R. Doggart & J. Mallory
1991-2 Excavations at Windy Ridge, Co. Antrim, 1981-82. *Ulster Journal of Archaeology* 54-55: 13-35.

Woodman P.C. & G. Johnson
1996 Excavations at Bay Farm, Carnlough, Co. Antrim. *Proceedings of the Royal Irish Academy* 96:137-235.

Woodman, P.C., N. Monaghan & M. McCarthy
1997 The Irish Quaternary Faunal Project, *Quaternary Science Reviews* 16: 129-159.

Woodman, P.C. & J. O'Shaughnessy
n.d. Killuragh Cave, County Limerick. Manuscript in possession of authors.

5

Radiocarbon Chronology and the Colonisation of the British Sector of the Atlantic Façade

Christopher Tolan-Smith
University of Newcastle

Abstract

A suite of over three hundred radiocarbon dates is used to monitor the spread of human settlement throughout the British Isles during the Late glacial and early Postglacial periods. Examined millennium by millennium, the radiocarbon chronology suggests that the settlement of the British sector of the Atlantic Façade, even allowing for changes in the coastline and the possible loss of evidence, did not begin before the later part of the 10th millennium RCY BP and was mainly a feature of the 9th. This development appears to have coincided with the northward migration of the Polar Front and the establishment of contemporary patterns of oceanic and atmospheric circulation.

This paper is the third in a series addressing the palaeogeography and palaeodemography of northwest Europe during the Lateglacial and early Postglacial. The first (Smith & Openshaw 1990) discussed the principles behind the 'Mapping the Mesolithic' project while the second (Smith 1992a) dealt with the palaeodemographic implications of some of the results. In this paper those results are given a palaeogeographical dimension and I hope to develop further some ideas first expressed at a 'Palaeolithic and Mesolithic Archaeology' seminar at Cambridge in May, 1991, and summarised in Smith (1992b: 159-180).

The results summarised here arise from the analysis of radiocarbon dates from hunter-gatherer contexts in Britain and Ireland, the mean values of which fall within the period between 13000 BP and 6000 BP and which have standard deviations of 250 years or less. A total of 292 dates fulfill these criteria. To these I have added a further 13 dates, the mean values of which lie in the 6th millennium but which at two standard deviations have ranges which extend into the 7th. All dates used are in uncalibrated radiocarbon years, and a full list with laboratory codes is available on request.

I have discussed elsewhere the validity of using only radiocarbon dates in this context (Smith 1992b: 159-61) and the procedures adopted in this research have also been described in the earlier papers (Smith & Openshaw 1990, Smith 1992a). The 305 dates used constitute evidence for a human presence in eighty-nine 10 km by 10 km squares of the British Ordnance Survey National Grid and its Irish equivalent.

The form this analysis takes is a series of maps documenting the spread of settlement in the British Isles between 13000 BP and 6000 BP (Figs. 5.1-5.7). I have used ranges rather than mean values and the maps do not correspond to precisely defined periods of time; instead each shows those 10 km by 10 km squares that have date ranges extending into the millennium in question. It follows from this that some dates contribute to more than a single map while the total, combined range of the dates used to compile each map is larger than the thousand years notionally represented. For example, the thirty-eight dates which have ranges extending into the 12th millennium have a combined total range of 2400 years, from 12700-10300 BP. While this leads to some loss of temporal resolution, it does indicate the maximum extent of settlement in the millennium in question, on the basis of the evidence at present available. My aim is to document a process, rather than precisely date its individual stages and this loss of temporal resolution is considered acceptable.

These individual dated occurrences must imply a more widespread human presence, but there is no way in which this can be demonstrated. However, in order to allow for this, each square has been treated as the focus of a wider area that would have been readily accessible to the groups whose presence is attested in the area. Quite arbitrarily, I have given these wider areas a 50 km radius. I believe this gives a more realistic representation of likely human settlement than would be obtained by simply mapping by 10 km by 10 km squares. The analysis has also proceeded on the assumption that once an area was occupied it was unlikely to have been abandoned during the period covered by the study. Hence the maps document what I considered to be a cumulative process. Lastly, in order to provide an appropriate spatial context I have made some attempt to take account of the changing coastline of the British Isles between the 13th and 7th millennia. This helps establish when, according to the radiocarbon chronology, humans first reached this sector of the Atlantic Façade after the end of the last glaciation.

There are forty radiocarbon dates with ranges which, at two standard deviations, extend into the 13th millennium BP. These dates come from only eight 10 km by 10 km squares, the discrepancy being accounted for by the fact that several well known sites such as Gough's Cave at Cheddar Gorge and the Creswell Crags group have long series of dates. As argued previously (Smith 1992a), once a human presence in an area has been established during the millennium in question, additional dates are redundant in this analysis. This kind of redundancy is a feature of all the periods studied here. The records based on radiocarbon dates for the 13th millennium are illustrated in the first map (Fig. 5.1), to which I have added an additional record for the Hengistbury Head site (Fig. 5.1.1). This site has not been dated by radiocarbon but does have a series of TL dates, the mean value of which has been published as 12500±1150 BP (Huxtable 1992: 60). Although the range given is very wide, this record has been included for the sake of completeness.

The nine records plotted in Figure 5.1 provide only the most slender basis for analysis and interpretation. Leaving aside the isolated date for the Leman and Ower uniserial barbed points (Fig. 5.1.2), the records fall into two groups, a situation that is further emphasised by plotting the 50 km radii focused on each. This spatial separation was first noted by Campbell (1977: 32-33, 159-60) and was commented upon by the present author when I suggested it could be an indication that the recolonisation of Britain had been initiated from two directions, the North Sea Lowlands and the English Channel Plain (Smith 1992b: 165-167). Of interest in the present context is the observation that none of the areas occupied, with the exception of those focused on Kent's Cavern (Fig. 5.1.3) and Three Holes Cave (Fig. 5.1.4), were anywhere near the contemporary coastline. Although the possibility of 13th millennium sites having been lost to eustatic sea level rise cannot be ruled out, the remoteness of known sites from the contemporary coastline implies that the communities who returned to Britain during the Lateglacial Interstadial did not have a particularly maritime way of life.

The second map (Fig. 5.2) plots records for the 12th millennium BP. Thirty-eight dates have ranges extending into this period. These come from thirteen 10 km by 10 km squares, but in only four cases do they document an expansion in the range of human settlement over that accomplished during the 13th millennium. Most of this expansion occurred in the more northerly area and consisted of a single new record from Porth y Waen (Fig. 5.2.1) and a northerly spread represented by records of human occupation at Kirkhead Cave (Fig. 5.2.2) and Victoria Cave, Kinsey Cave and Coniston Dib (Fig. 5.2.3). The fourth new record is from Sproughton in East Anglia (Fig. 5.2.4). As with the records for the 13th millennium, apart from the Kent's Cavern and Three Holes Cave group, none come from near the contemporary coastline.

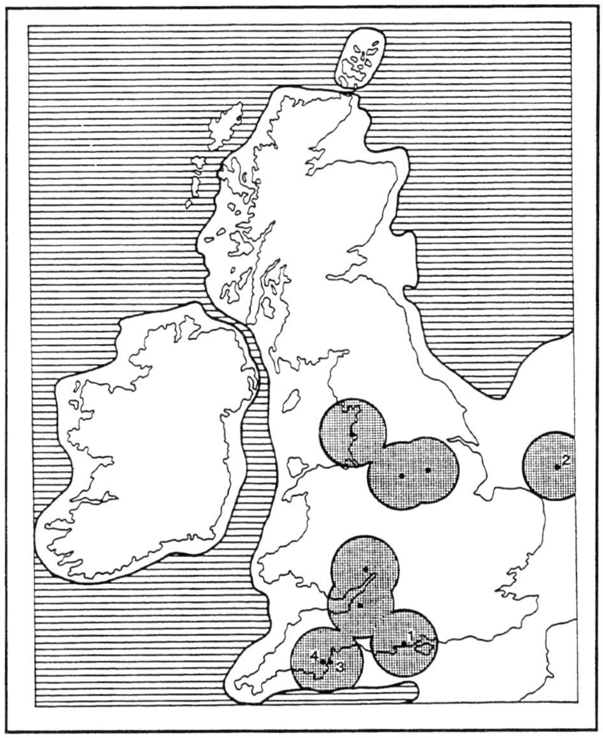

Figure 5.1. 13th millennium.

The thirty-three dates with ranges extending into the 11th millennium distributed among sixteen 10 km by 10 km squares and show a considerable expansion of the area of occupation over that documented for the 12th millennium. This expansion occurred in both the southern and northern areas. In the south we have the earliest evidence for a human presence in the valleys of the Kennet at Marsh Benham and Thatcham (Fig. 5.3.1) and the Thames at Uxbridge (Fig. 5.3.2) and in the Midlands at Earl's Barton (Fig. 5.3.3). In the northwest we have humans present on the fringe of the mountains of North Wales at Kendrick's Cave (Fig. 5.3.4) while to the northeast an expansion into the Vale of Pickering (Fig. 5.3.5) compliments the northwesterly expansion recorded in the previous millennium. This is the period of the Loch Lomond Stadial when substantial parts of northern Britain returned to glacial and periglacial conditions. In light of this, the lack of much northerly expansion is hardly surprising. The coastline recorded in Figure 5.3 is that established by the end the 11th millennium. The areas with a documented human presence are now nearer to the contemporary coast, but settlement on the Atlantic side of Britain still cannot be described as coastal.

It is during the 10th millennium that we see the beginnings of a major phase of expansion into the Atlantic coastal regions of the British Isles (Fig. 5.4). Sixty-three radiocarbon dates, from to twenty-two 10 km by 10 km squares have ranges which extend into the 10th millennium. Some document further expansion in the southeast and northeast but the significant development in this millennium is the occurrence of records in southwest Wales

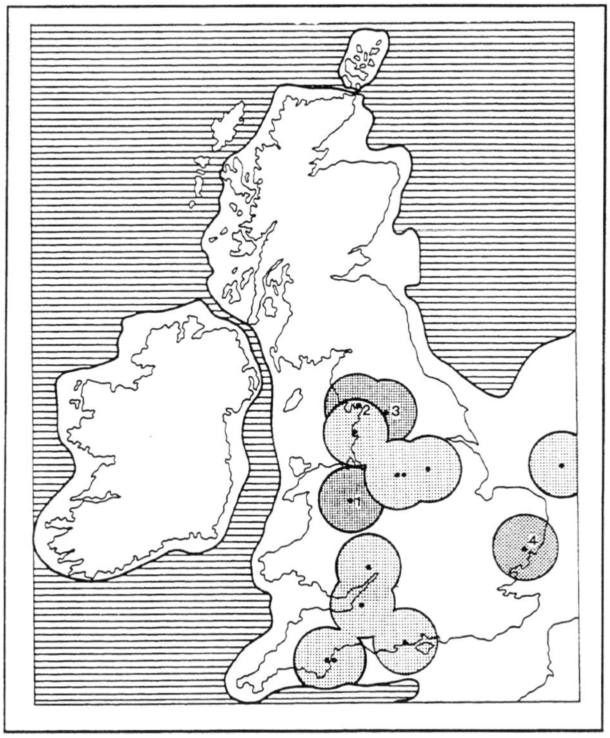

Figure 5.2. 12th millennium.

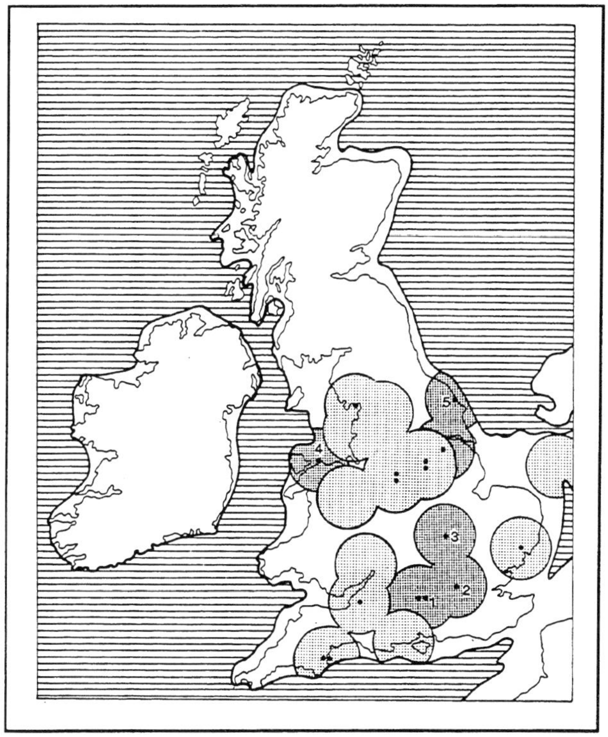

Figure 5.3. 11th millennium.

at Nab Head (Fig. 5.4.1). There are also several dates from northeast Ireland which, at two standard deviations, have ranges just extending, at their extreme lower end, into the 10th millennium. The colonisation of Ireland at this time would have involved a sea crossing, the shortest of which would have been from southwest Scotland. As there are, as yet, no 10th millennium records from this latter area, the colonisation of Ireland, perhaps for the first time in human history, is probably better regarded as mainly a feature of the next phase.

Figure 5.5 maps the evidence from radiocarbon dates for the extent and spread of human settlement in the British Isles during the 9th millennium BP. The forty records used arise from seventy-seven dates with ranges which extend into the 9th millennium. The map shows a considerable expansion of the occupied area over that documented for the 10th millennium, but with the exception of two dates from Lough Boora in the Irish Midlands (Fig. 5.5.1), this was all along the Atlantic Façade. By 8000 BP the documented spread of human settlement had extended up the west coast of Scotland from Red Kirk Point on Solway Firth (Fig. 5.5.2) as far as Kinloch on the Isle of Rhum (Fig. 5.5.3) and included the earliest occupation of Arran and the southern Hebridean islands of Islay and Jura. This could not have been accomplished without an efficient and dependable maritime technology which itself must imply a way of life in which the exploitation of the marine and coastal resources of the Atlantic Façade played a prominent role. It seems evident that by the end of the 9th millennium the hunter-gatherers of western Britain had developed a maritime dimension to their patterns of mobility and subsistence.

It is also interesting to note in Figure 5.5 the areas still without evidence for a human presence. Most of Ireland falls into this category, but the evidence outside the northeast is inadequate for all periods and too much stress should not be placed on its paucity. Other gaps are the Southwest Peninsula, Kent and much of East Anglia. But the most notable blank is provided by Scotland away from

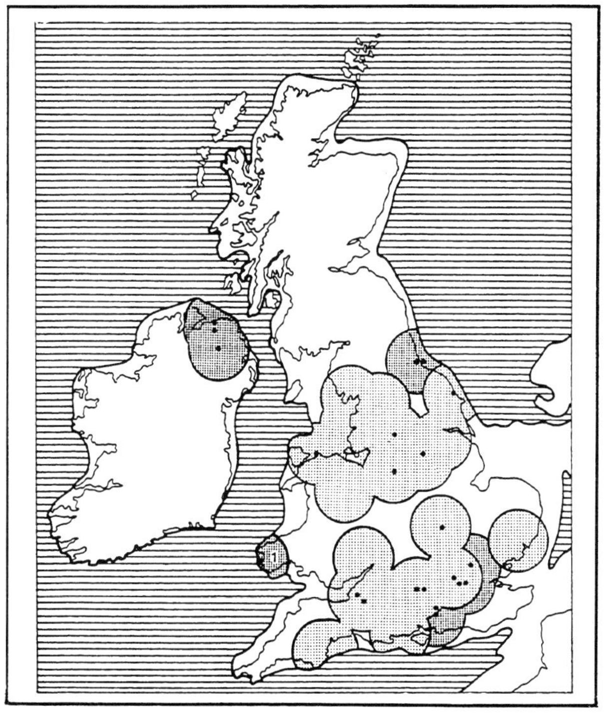

Figure 5.4 10th millennium.

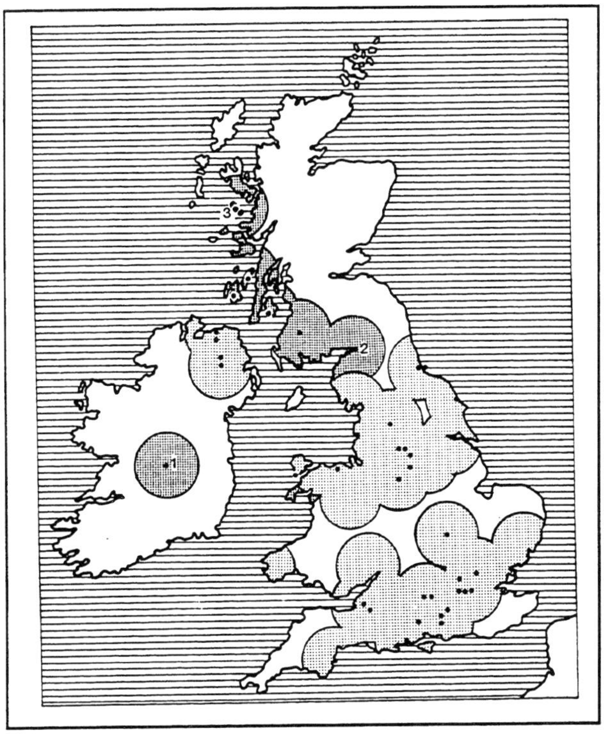

Figure 5.5. 9th millennium.

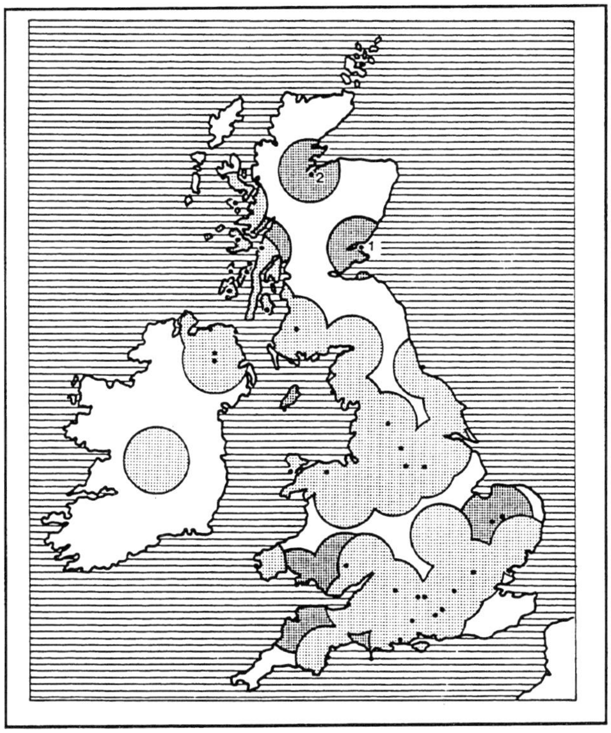

Figure 5.6. 8th millennium.

the west coast. It appears that the rugged and inhospitable north of Britain from the Cheviot Hills to the Flows of Caithness had not yet been brought within the realm of human settlement.

Figure 5.6 documents the next stage in the process and illustrates that the 8th millennium BP was mainly a period of consolidation. The sixty-three radiocarbon dates from forty-two 10 km by 10 km squares document only limited expansion, and newly occupied areas include northern East Anglia, the Isle of Man and the coastal zone of eastern Scotland with records from Morton (Fig. 5.6.1) and Inverness (Fig. 5.6.2). It is during this millennium that we have the earliest dates for the 'Obanian', the quintessential Mesolithic of the British sector of the Atlantic Façade.

The final map in this series (Fig. 5.7) is based on seventy radiocarbon dates from forty 10 km by 10 km squares. Like Figure 5.6 this map mainly documents a phase of consolidation, although the Central Lowlands of Scotland and most of Kent are at last brought into the picture. New, and rather more significant, departures from earlier maps are the first records from the Southwest Peninsula and southwest Ireland, indicating that by 6000 BP virtually the whole of the British sector of the Atlantic Façade had been settled.

In this short paper I have sought to do no more than document a process. While this has been a relatively straightforward matter, explanation of the factors at work behind the observed process is a far more complex and difficult task. It is a task I hope to tackle during the course of continuing research into the palaeogeography of Late Upper Palaeolithic and Mesolithic Britain. By way of conclusion I should like to draw attention to two broad issues arising from consideration of these data.

When considering the first of these maps, I noted that the records fell into northerly and southerly groups. Inspection of the other maps in the series show that this division persists throughout the period studied. The existence of a blank zone between these two groups is obviously an artefact of my decision to demarcate the areas around 10 km by 10 km squares with evidence for human activity by circles of 50 km radius. Had I chosen 100 km circles, no blank zone would have appeared between the two major groupings. However, the fact remains that during the millennia covered by the maps, no occupied 10 km by 10 km square in the southern group was nearer than 110 km from any similarly occupied square in the northern group.

The impression of a blank zone in central southern Britain is real. Its existence is unlikely to have been due to environmental factors, the terrain of southern Britain being relatively homogenous in terms of resources of interest to hunter-gatherers. An alternative explanation must be sought and I should like to refer back to that offered, tentatively as an explanation for the emergence of the pattern in the first place in the 13th millennium BP. If the late glacial recolonisation of Britain was initiated by two discrete populations originating in different parts of Northwest Europe, we should not exclude the possibility that these populations maintained a measure of separateness throughout the early Postglacial. This observation presents a challenge to students of cultural history. Do we have in the archaeological record any evidence for the existence of two such discrete populations and did they survive the Mesolithic?

The other main feature of the series of maps is the dramatic spread of evidence for the settlement of the Atlantic Façade during the 9th millennium BP, although the process may have begun during the 10th and was not actually completed until the 7th. What can be held to account for this? One of the most significant developments in the North Atlantic during the 10th millennium was the northward migration of the Polar Front from the latitude of Galicia to that of Iceland. This produced a fundamental change in patterns of atmospheric and oceanic circulation and brought the warm and biologically productive waters of the North Atlantic Drift to the shores of northern and western Britain. I have said elsewhere that it is unlikely to have been a coincidence that the colonisation of western Scotland, the Inner Hebrides and Ireland began at a time when quantitative and qualitative increases in marine biomass are likely to have encouraged interest in the exploitation of coastal and marine resources (Smith 1992b: 172-3). But a monocausal, environmentally deterministic explanation is unlikely to be the whole answer and the role of other influences should not be ignored. Demographic pressure may have played a part, particularly if the pattern produced by the radiocarbon dates of a southern Britain virtually fully occupied by 10000 BP is genuine (Smith 1992a: 39-40). The exploitation of marine and coastal resources, beyond the collection of shellfish, is an activity involving relatively high levels of investment such as fish traps, nets and, most expensive of all, boats. The high status and prestige of boat owning and handling is well documented ethnohistorically, and coastal hunter-gatherers provide some

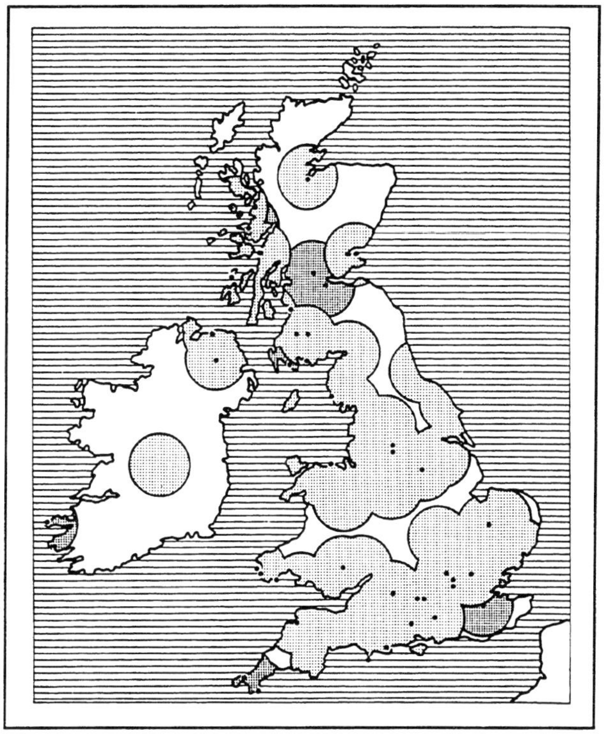

Figure 5.7. 7th millennium.

of the best examples of Woodburn's (1980: 98-99) 'delayed return systems' which tend to be characterised by more complex levels of social organisation than commonly found among hunter-gatherers. If transport and access to resources could be controlled by a relatively small segment of the population we should be prepared to consider that the colonisation of the Atlantic Façade also had a social dimension.

References

Campbell, J. B.
1977 *The Upper Palaeolithic of Britain Vols. I and II.* Oxford: Clarendon Press.

Huxtable, J.
1992 Thermoluminescence (TL) dating of burned artefacts from the Hengistbury Head sites. In *Hengistbury Head, Dorset - Volume 2: the Late Upper Palaeolithic and Early Mesolithic Sites* (R.N.E. Barton, ed.), pp. 60-61. Oxford: OUCA.

Smith, C. & S. Openshaw
1990 Mapping the Mesolithic. In *Contributions to the Mesolithic in Europe* (P.M. Vermeersch & P. van Peer, eds.), pp. 17-22 Leuven: Leuven University Press.

Smith, C.
1992a The population of late Upper Palaeolithic Britain. *Proceedings of the Prehistoric Society* 58: 37-40.

1992b *Late Stone Age Hunters of the British.* London: Routledge.

Woodburn, J.
1980 Hunters and gatherers today and reconstruction of the past. In *Soviet and Western Anthropology* (E. Gellner, ed.), pp. 95-117. London: Duckworth.

Part II

Aquitaine, Charente & the Spanish 'Corners'

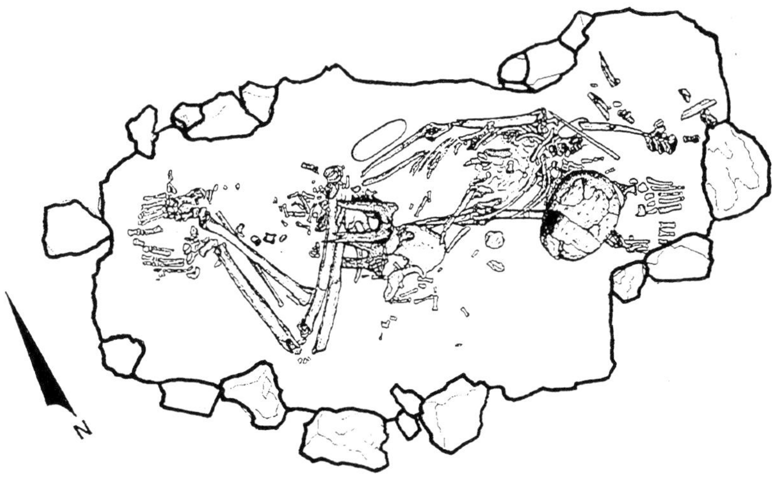

6

Reflections on the Mesolithic of the Aquitaine Basin and its Western Extension

Alain Turq & Louis Detrain
Regional Archaeological Service, Bordeaux

Abstract

Data regarding the Mesolithic of the Aquitane are at present very rare due to a lack of pertinent research. In addition, Mesolithic sites in the region of the Atlantic coastline have been deeply buried or destroyed by geomorphological phenomena. It appears, however, that from the eighth to the beginning of the tenth millennium BC, this zone was occupied by the Sauveterrian.

Introduction

The Aquitaine Basin is a large sedimentary basin with a mosaic of landscapes ranging from low plains to high mountains, from gently rolling relief to abrupt cliffs. It is characterized today by a relatively mild maritime climate, and has been subjected to numerous and profound modification of *c.* 90% of its coastlines, principally in lowland areas. These modifications, which render difficult a clear understanding of the nature of coastline during the Mesolithic, are tied to several factors: (1) the morphological particularities of the region (low plain with a very weak slope extending into a coastal platform), (2) particularly strong coastal currents in this zone of the Gulf of Gascogne, (3) marked variation in sea level, and (4) the influence of westerly winds which are responsible for the creation and displacement of dune lines. The one known fact is that sea levels were lower during the Mesolithic than they are at present. Thus coastal sites that may have existed are now under water or have been destroyed. For these reasons, as well as a lack of pertinent data, we feel that in order to have a general idea of the Aquitaine Mesolithic, it is useful to begin with a brief synthesis of the ensemble of available data regarding the region.

Summary of Available Documentation

Paradoxically, although data from the Aquitaine contributed to the original formulation of a concept of the Mesolithic, current research efforts lag behind those of other regions. Over this vast region (41,407 km^2), relatively credible Mesolithic sites number only about 30. This can be explained by the presence of a dense vegetal covering (e.g., in Landes), the absence for many years of a formal research program devoted to this time period, and the destruction or deep burying of sites. Paleoenvironmental data pertaining to the early and middle Holocene are also uneven and rare.

Moreover, a recent re-examination of the documentation available regarding sites which in the past were the subject of intensive research shows that these data are less than reliable. The material recovered during the 19th and early 20th century excavations is most often incomplete, due either to selective collection or to an absence of systematic water screening for sites which were not totally disturbed, such as Martinet (Kervazo & Mazière 1989), or Cuzoul de Gramat. The nature of the sedimentary matrices (open slope deposits, or sediments rich in organic materials and thus subject to bioturbation) of the archaeological levels cannot assure the homogeneity of the lithic assemblages contained within them, as, for example, at Rouffignac (Barrière 1974). On the other hand, sites without stratigraphic problems, and which were excavated by reliable methods, such as Pöeymau (Laplace-Jauretch 1954, Livache 1983), Ithé (Ebard 1980), and Bignalats (Marsan 1988), have for the most part been only summarily published. Others sites, such as Roc Allan (Turq & Detrain 1993) and Le Gurp (Roussot Larroque 1994), are still in the process of excavation.

The Evolution of Ideas Regarding Mesolithic Cultures

We attribute the earliest work on this theme to L. Coulonges (1935, 1963), who made the still-used bipartite distinction between the Sauveterrian, with few triangles, and the inappropriately named Tardenoisan, with an elevated number of trapezes.

In 1978, J.-G. Rozoy recognized two distinct groups within the first of these ensembles, the Sauveterrian: (1)

the classic Sauveterrian from the Périgord, Gironde, Agenais, and (2) a group found on calcareous bedrock (the Causses Sauveterrian), characterized by a high proportion of microliths (80%). In the second group, the Tardenoisian I and II lithic assemblages are extensions of the recent Sauveterrian, while there is no continuity evident between the Tardenoisian II and III, which last is already Neolithic. According to Rozoy, the Aquitaine was occupied by the classic group.

J. Roussot-Larroque (1985) maintains that Sauveterrian variability cannot be explained by environmental factors alone (geology, topography), but also by the culture histories of the groups in question. Working with the same data as her predecessor, from the sites of Martinet and La Borie del Rey, she confirms the existence of a rupture between the Sauveterrian and the Recent Mesolithic with trapezes which she calls the Roucadourien cycle. This cycle includes the Pre-Roucadourien I (ex-Tardenoisian I), Pre-Roucadourien II (ex-Tardenoisian II) and the Roucadourien (ex-Tardenoisian III). As early as 1981 M. Barbaza recognized nuances in the Sauveterrian scheme, but it was not until the excavation of the Abri de Fontfaurés that he proposed, along with N. Valdeyron (Valdeyron & Barbaza 1989, 1992; Barbaza et al. 1981) the notion of an upper Sauveterrian. In this scheme the classic Sauveterrian and the Causses Sauveterrian are part of the same ensemble but differ from one another due to functional specializations of their tool assemblages correlated with specialized activity sets. Finally, for these authors, the Montclusian corresponds to a middle stage of the Sauveterrian. Recently, through a detailed typological study of armatures from six sites, and a critical examination of the data from 119 other sites, Valdeyron (1984) proposed an evolutionary framework which demonstrates continuity. According to this framework, the changes in question do not necessarily occur synchronously from one to the other end of the Aquitaine Basin.

The Data of the Atlantic Coastal Area

Returning to the Atlantic Façade, the sites are rare, as noted previously, but are all attributable to the Sauveterrian ensemble. They are essentially concentrated in the Gironde along small valleys, around marshes and swamps (e.g., Raysson), and behind present dune lines bordering moors. These localities – humid microenvironments favorable in terms of animal and vegetal resources – were probably selectively chosen for occupation. In fact, they could also correspond to areas with Holocene sedimentary accumulations and to zones of outcrops of these formations, thus favoring preservation and visibility. Most of the sites are known exclusively through unpublished surface collections, with the exception of Saint-Laurent-du-Medoc, located at the foot of the Jacques escarpment in the township of Naujac et Postiac (Rozoy 1978). Saint-Laurent-du-Medoc has yielded a mixed series of deposits from two test pits (Lenoir 1981, 1983; Vadeyron 1995). The first of these corresponds, according to Rozoy, to an early Sauveterrian that he compares with level 5a of Rouffignac. The second is attributed by Lenoir (1981, 1983) to an unspecified Mesolithic and by Valderon to an early Sauveterrian. The only stratified site, Lède du Gurp (located in the township of Grayan et Hôpital), has been known since the end of the 19th century. This site was excavated from 1972 by G. Frugier, and from 1984 to 1994, by J. Roussot-Larroque. The latter has uncovered Mesolithic levels, which for the moment remain largely unpublished. The little information that is available to date regarding these levels is given below.

The excavation has revealed eight Mesolithic levels, alternately composed of sands and peats, which represent genuine occupation floors marked with animal tracks. Wood and vegetal macrobotanical remains are preserved in the peats. The lithic industry is similar to the Sauveterrian but differs in the numbers of armatures and points with concave bases. While ^{14}C dates situate this industry at the end of the Preboreal and in the early Boreal, traits distinctive of the Montclusian (the recent phase of the Sauveterrian of the Midi) are rare. Are these ambiguities due to small sample size, or do they represent a variant of the Sauveterrian in the Atlantic margins of its distribution?

Pollen analyses were done by L. Marambat (Marambat & Roussot-Larroque 1989; Marambat 1995). These show that at the beginning of the sequence there was an herbaceous vegetation rich in grasses, with a low proportion of trees (25%). Arboreal species were dominated by *Pinus silvestris*. Then, there is a development of birch, followed by mixed oak species, notably green oak, and hazelnut. In these very acid environments faunal remains are not preserved. The salvage excavations done by the Service Regional de l'Archéologie (Rouzeau 1993, Roussot-Larroque 1994) have permitted only limited observations regarding the sedimentary context of the site, which is located at the bottom of a large, natural depression of at least 1500 square meters. The original depositional processes have not yet been determined. In the absence of a serious geological study, of the site as well as of its context, and considering the strong slope of the levels (nearly 20%), it is impossible to guarantee the homogeneity of the archaeological levels which could be, at least in part, redeposited. In our present state of knowledge, neither the nature, nor the function, still less the position of the site in relation to the coastline, are well understood.

Conclusions

On the Atlantic coastline, as in the Aquitaine Basin, renewed research efforts focusing on the Mesolithic involve several objectives. The first is rapid publication of recent excavations (Pöeymau, Ithé, Bignalats), and continued research on sites still in the process of excavation. The second, and more important goal, is a systematic revision of sites already known, in particular those which serve as type sites (La Borie del Rey, Rouffignac), as well as a search for new sites with confirmed stratigraphic integrity (homogeneity of lithic assemblages). These research efforts will be used to address the following problems:

(1) The relationship between the earliest Sauveterrian and the late Azilian which appear to co-exist or overlap somewhat, essentially in the Pyrénées (Bignalats, Pöeymau). Is it a question of acculturation of the last Azilians by the earliest Sauveterrians, the origins of the Sauveterrian, or contamination of gravel and rock deposits?

(2) The relationship between the Mesolithic with triangles and the Mesolithic with trapezes, to confirm whether these belong to the same technological assemblage (Sauveterrian), or to two different entities.

(3) The evolution of industries with trapezes, in order to understand in detail changes related to the first indications of horticulture or agropastoral economies.

All of these questions concerning the lithic assemblages must not only be addressed through typological analyses, but also through research on technology and microwear use and damage patterns. We can no longer accept the oversimplistic analytical unit definitions of J.-G. Rozoy.

References

Barriere , C.
1974 Rouffignac– l'archéologie. *Travaux de l'Institut d'Art Préhistorique de l'Université de Toulouse le Mirail* 16: 27-38.

Barbaza , M.
1981 *Recherches sur l'Épipaléolithique en Languedoc et Catalogne*. Thèse de troisième cycle. Toulouse: L'université de Toulouse, École des Hautes Études en Sciences Sociales.

Barbaza, M. & N. Valdeyron
1989 L'Abri de Fontfaurés. Contribution à l'étude du Sauveterrien. *Travaux de l'Institut d'Art Préhistorique de Toulouse le Mirail* 31: 23-38.
1992 Les industries lithiques de l'abri de Fontfaurès à Lentillac-Lauzés (Lot) et la séquence chrono-culturelle sauveterienne. *L'Anthropologie* 96: 163-178.

Barbaza M., N. Valdeyron, T. Andre, Fr. Briois, H. Martin, S. Philibert, D. Allios & E. Lignon
1991 *Fontfaurès en Quercy*. Paris: Archives d'Écologie Préhistorique, École des Hautes Études en Sciences Sociales No. 11.

Briois, Fr.
1991 Économie des matières premières. Économie du débitage. In Barbaza, M. et al., *Fontfaurès en Quercy*. Quercy: Archives d'Écologie Préhistorique, École des Hautes Études en Sciences Sociales No. 11: 97-103.

Coulonges, L.
1935 *Les Gisements Préhistoriques de Sauveterre-la-Lémance (Lot-et-Garonne)*. Paris: Masson et Cie.
1963 Magdalénien et Périgordien post-glaciaires. La grotte de la Borie del Rey (Lot-et-Garonne). *Gallia-Préhistoire* 6: 1-29.

Ebrard, D.
1980 Assurucq, Abri Ithelatseta. *Gallia Préhistoire* 23: 1-417.

Kervazo, B. & G. Maziere
1989 Le Gisement du Martinet à Sauveterre-la-Lémance (Lot-et-Garonne). Nouvelles données et implications. *Bulletin de la Société Préhistorique Française* 86: 263-265.

Laplace-Jauretche, G.
1954 Les couches à escargots des cavernes pyrénéennes et le probléme de l'Azilien de Piette. *Bulletin de la Société Préhistorique Française* 54: 199-211.

Lenoir, M.
1981 Le gisement de Jacques à Naujan et postiac (Gironde). *Revue d'Histoire et d'Archéologie du Libournais* 49: 45-52.

Lenoir, M.
1983 *Le Paléolithique des Basses Vallées de la Dordogne et de la Garonne, I, II.* Bordeaux: L'Université de Bordeaux I, Mémoire de Thèse d'Etat.

Livache, M.
1983 Interprétation des couches á escargots de la grotte du Poeymau á Arudy (Pyrénées-Atlantiques). *Revue Régionaliste des Pyrénées* 1: 97-101.

Marsan, G.
1988 Le gisement préhistorique de la grotte du Bignalats á Arudy (Pyrénées-Atlantiques). deuxiéme partie : Les industries humaines et leur place dans la Préhistoire récente des Pyrénées occidentales. *Archéologie des Pyrénées Occidentales* 8: 31-67.

Marimbat, L. & L. Roussot-Larroque
1989 Paysage végétal et occupations humaines sour la côte atlantique: l'exemple de la Lède du Gurp. *Bulletin de L'Association Française d'Ecologie Quaternaire* 2: 73-89.

Marimbat, L.
1995 *Paysages de la Façade Atlantique Girondine et de la Saintonge au post-glaciaire. L'empreinte de l'Homme*. Paris: Cahiers du Quaternaire du Centre National de la Recherche Scientifique No. 21.

Rozoy, J.-G.
1978 *Les Derniers Chasseurs. L'Épipaléolithique en France et Belgique I-III.* Charleville-Mezières: Numéro special du Bulletin d'Archéologie Champenoise, Tomes 1-3.

Roussot-Larroque, J.
1985 Sauveterre, et après . . . In *La Signification Culturelle des Industries Lithiques* (M. Otte, ed.), pp. 170-202. Oxford: British Archaeological Reports International Series No. 239.

Rouzeau, N.
1993 Grayan-et-l'hôpital, La Lède du Gurp. In *France – Ministère de la Culture et de la Communication. Direction Régionale des Affaires Culturelles Aquitaine. Service Régional de l'Archéologie. Bilan scientifique 1993,* pp. 49-50. Bordeaux: Ministère de la Culture et de la Communication.

Rouzeau, N.
1994 Grayan-et-l'hopital, La Lède du Gurp. In *France – Ministère de la Culture et de la Communication. Direction Régionale des Affaires Culturelles Aquitaine. Service Régional de l'Archéologie. Bilan scientifique 1994*, pg. 55. Bordeaux: Ministère de la Culture et de la Communication.

Turq, A. & L. Detrain
1993 Sauveterre-la-Lémance, Le Roc Allan. In *France – Ministère de la Culture et de la Communication. Direction Régionale des Affaires Culturelles Aquitaine. Service Régional de l'Archéologie. Bilan scientifique 1992*, pp. 100-101. Bordeaux: Ministère de la Culture et de la Communication.

Valdeyron, N.
1994 *Le Sauveterrien, culture mésolithique du Sud-Ouest de la France du Sud durant le Xéme et le IXéme millénaire B.P.* Toulouse: Mémoire de doctorat nouveau régime, Université de Toulouse le Mirail.

7

The Tardiglacial and Initial Holocene on the Charente-Maritime and Vendée Coasts: Status of Current Research and New Perspectives

Pascal Foucher & Cristina San Juan
Regional Archaeological Service, Toulouse

Abstract

This chapter gives a general overview of the Tardiglacial and early Holocene on the Charente and Vendée coasts, as well as the preliminary results of a collective research project focused on the lower Charente Basin and carried out in collaboration between the authors, regional French prehistorians and Spanish researchers working on the Cantabrian coast.

Introduction

The Mesolithic of Charente-Maritime has been relatively poorly studied, although two recent French publications summarize much of what is available through the late 1990s (Marchand 1999, Foucher et al. 2000). Our initial efforts consisted of assessing the state of research in Charente-Maritime and Vendée, recognizing the lack of information concerning the period between the end of the Würm and the beginning of the Holocene, and assembling what few data existed as compiled by various researchers in the region. This allowed us to map the distributions of known sites in the Charente and Seudre basins; it also resulted in the identification of about 20 previously unpublished sites (Fig. 7.1). The discovery of the Pierre Saint-Louis site during the salvage operation preceding the construction of highway A837 (we directed the second phase of this project) afforded us the unique opportunity to conduct research on a mesolithic open-air occupation surface with constructed hearths.

The second phase, which has already begun, includes continuation of surveys (ca. 40 sites recorded since the beginning of 1995) and the excavation of test pits at several sites with the goal of obtaining a significant stratigraphy that would permit establishment of a more complete, regional, chronological sequence (Figs. 7.2, 7.3).

The Geographic Context

Located in the northern part of the Aquitaine Basin, the Charente coast presents exposed sediments dating from the Jurassic Period in Aunis and Ile de Ré, and the Upper Cretaceous in the vicinity of Saintonge and Ile d'Oléron. At Aunis, the topography takes the form of low hills with crestlines at ca. 8 km inland from the present coast, following a SE/N arc. In the region of Saintonge, Cretaceous sediments include, on the one hand, two parallel crests which come to an end at Port-des-Barques and at Bourcefranc and, on the other, a scatter of small, dispersed islands in swampy terrain, some of which connect in a chain to make up the Arvert peninsula. The islands are of very low relief (about tree-level), in consequence of which they are scarcely visible, even up close, from the sea.

Oversimplifying somewhat, there are three major geographical characteristics of the Charente:

(1) *The sea:* The present day coastline exhibits occasional low cliffs, of friable limestones, subject to frequent landslides. Since the continental shelf extends far beyond the present shore, the coastline fluctuated several tens of kilometers from its present location during the Flandrian Transgression.

(2) *A hydrographic system* composed of the watersheds of the Seugne and the Charente basins.

(3) *The coastal marshes:* They consist of 'maritime plains', which are remarkably extensive and flat, varying between 1-4 m above sea level. The soil is composed entirely of *bri,* a fine clay of marine or fluvial origin. The plains were formed during the Holocene, and can be up to 15 or 20 m thick.

The Vendée coast offers a slightly different morphology because its geological substrate is composed of metamorphic rocks; the Breton Swamp occupies a significant portion of the coast.

Knowledge of these geographic units and their evolution over time is of great importance to an understanding

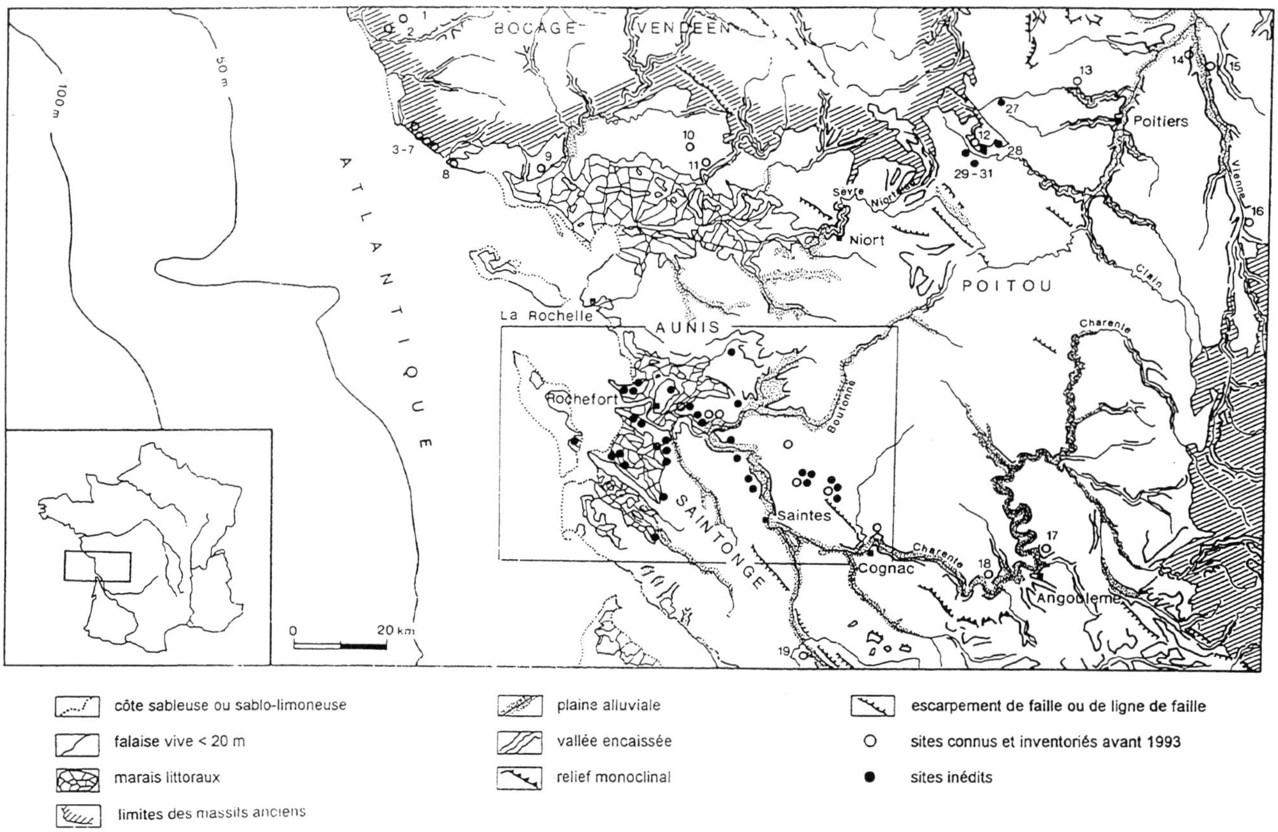

Figure 7.1. Map of the distribution of Epipaleolithic and Mesolithic sites along the Charente and Vendée coasts (based on the geomorphological map of France – scale 1:1000000). Sites known and inventoried before1993: 1 Coex, 2 L'Aiguillon, 3–7 Bel Air, St. Jean d'Orbestier (Le Château d'Olonne), Baie de Caillola, Anse du Corps de Garde, La Mine (Talmont St. Hilaire); 8 Pointe du Payré (Jard-sur-Mer), 9 St Benoit-sur-Mer, 10 Le Fief du Mitan (Petosse), 11 Les Sables de Loi (Auzay), 12 La Bouronnière, 13 La Grande Roche, 14 Le Gué St Mars, 15 Abri de Bellefonds, 16 Bois Ragot, 17 St. Genis d'Hiersac, 18 Garde Epée, 19 Heurtebise, 20 Les Parpaillons, 21 Bel Air, 22 La Sablière, 23 Abri Piphrez, 24 Maison Neuve, 25 L'Autruère, 26 Le Cloux. Unpublished sites 1993-94: 27 Vaulifier (Vasles), 28 Marconnay (Sanxay), 29–38 Le Moulin de Bois, Pouvreau, La Brouinière, L'Aurière, La Boutinière (Ménigoute); 32–34 Fontbelle, Chez Bonpain, Peu Deis (Villars-les-Bois); 35 Chez Chenard (Saint-Sulpice de C.), 36–40 La Brouisse, Chez Gautret, Les Prises, Chez Jouneau, La Garenne (Brizambourg); 41 Le Prieuré (St. Vaize), 42 Bois Leclerc (Port d'Envaux), 43 La Pierre St. Louis, 44 La Vigne à Jacques (Bords), 45–50 Le Buffant, La Grande Chauvinière, La Rabotellerie, L'Alaigne (Tonnay-Charente); 51 Bois des Sablières (Le Breuil Magné), 52 Le Tertre, 53–55 Champon, Les Brochins (St. Laurent de la P.), La Baronnière (Fouras); 56 Pougne (St. Nazaire), 57 Bois du Grand Parc (Soubise), 58 Moulin d'Angle (St. Agnant), 59–61 Le Creux de l'Orage, La Brumauderie, Le Godinet (Saint-Agnant); 62 Pièce de la Vache d'Oron (Le Château d'Oléron), 63 Le Fief Jardin (Marennes), 64 Tête du Grand Fief (Marennes), 65 Ile d'Hérablais (Hiers-Brouage), 66 La Châtaigneraie (Gripperie-St-Symphorien), 67 Champs du Marais du Port (Saujon).

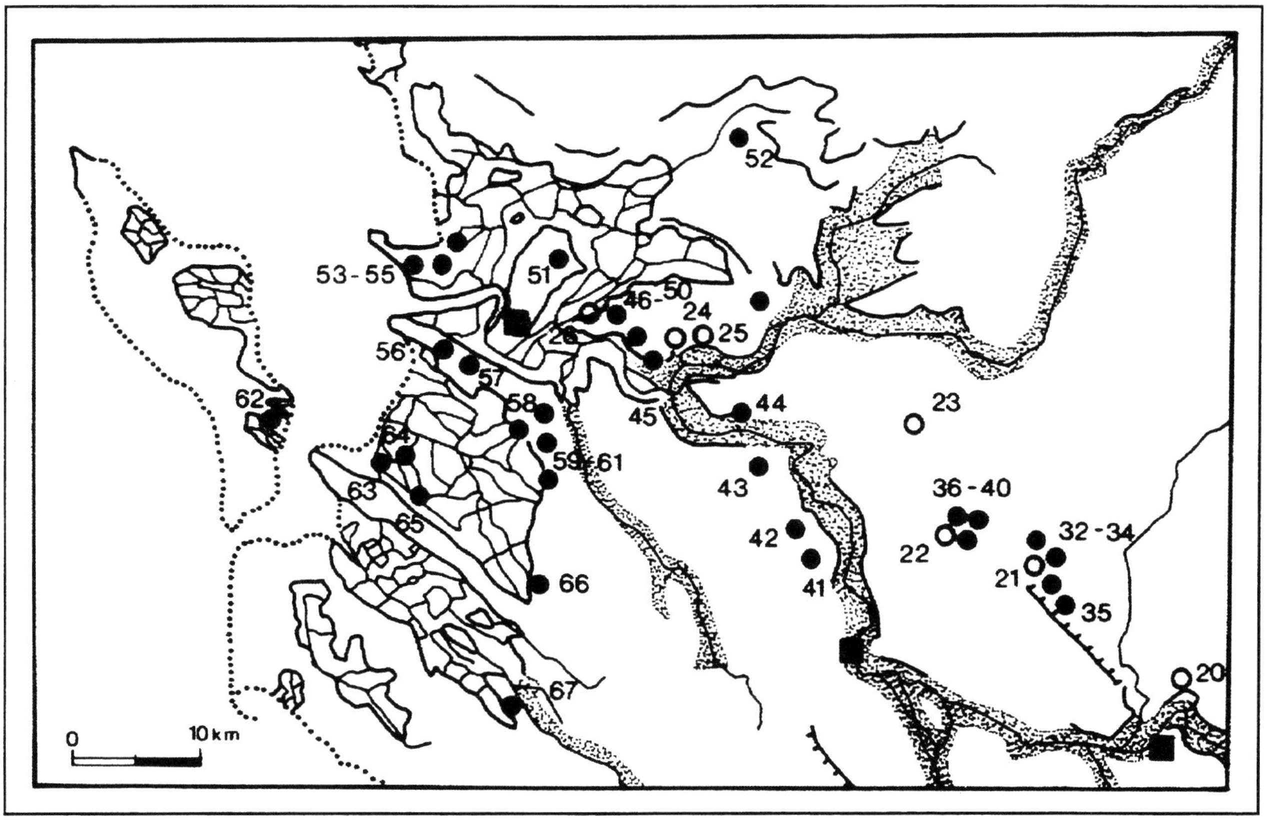

Figure 7.2. Distribution map of the Epipaleolithic and Mesolithic sites along the Charente coast.

of the prehistoric landscape. The landscape varied considerably over time to judge from the Holocene sedimentation in the marshes and valley floodplains, and from the evidence of marine transgression (Fig. 7.4). These changes must have entailed numerous modifications in the territories occupied and used by foragers, where they put their campsites, and the range of natural resources available to them.

State of Regional Research

The Final Magdalenian and Azilian

The Final Magdalenian/Azilian transition is well-defined in Vienne thanks to the stratigraphy of Bois Ragot (Gouex). According to ^{14}C dating, the transition spans the interval 11-8 kyr BP, and there was a progressive, continual evolution from the Final Magdalenian to the Azilian, without any signs of rupture (the ^{14}C date 8800 ± 220 BP for the Azilian level at Bois Ragot, however, seems very late (Chollet 1979). Nevertheless, for the region, and especially for the Vienne drainage, it constitutes the only stratified site to have been even partially published. As for the Final Magdalenian, the sites of Le Roc-aux-Sorciers (Angles-sur-l'Anglin), La Marche (Lussac), La Piscine (Montmorillon), and Le Chaffaud (Savigné) have not yet been fully published. Of these, we have only the information concerning the open air sites at our disposal, the subject of a critical review by the authors (Foucher & San Juan 1991, 1994).

In Charente, the state of our knowledge is little better (Debenath 1979). Final Magdalenian occupations are reported at the site of Montgaudier (Montbron) (Duport 1987) and at La Petite Courrière (Massaud et al. n.d.). The cave of Quéroy provides a stratigraphy which covers the period from the end of Dryas I to Dryas III. In the level corresponding to Dryas III, some Azilian points were recovered, but the site is primarily a paleontological locality without definable occupation surfaces (Tournepiche 1987). The nearest epipaleolithic occupations are in the Dordogne at Rochereil cave and at the Pont-d'Ambon rockshelterat Bourdeilles and, for the Final Magdalenian, at the sites of La Mairie and Le Mège at Teyjat.

In Charente-Maritime, if we are to take at face value the scant indications of LUP occupation, the Magdalenians would hardly ever have resided in the area. The site of Piphrez (Grandjean), which was unfortunately destroyed by the A10 highway in 1979, was the object of a few collections made after the bulldozers had passed (Blanchet & Rouveau 1986). The rockshelters of Heurtebise (Jonzac) produced a lithic assemblage which is probably Final Magdalenian. Because of the number of destructive disturbances that have affected these sites over the centuries (e.g., during the Middle Ages, World War II, clandestine diggers etc.), Gaillard was not able to recover any in situ material (1986, 1987). To find evidence of epipaleolithic occupations, one must look south to the Médoc, and to the lower Dordogne valley (Roussot-Larroque 1979, Lenoir 1979).

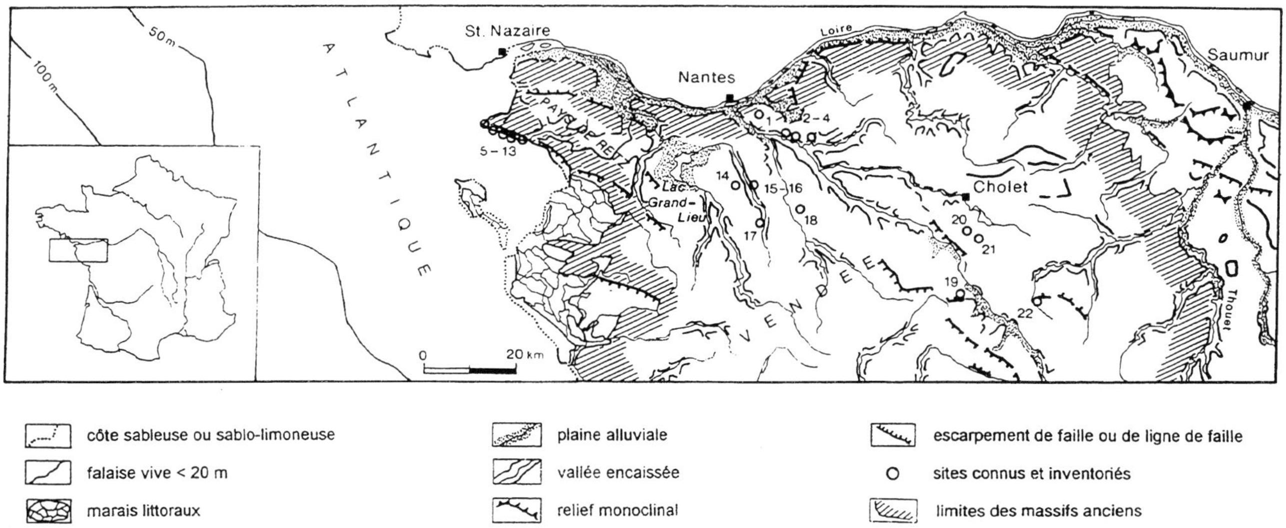

Figure 7.3. Distribution map of the Epipaleolithic and Mesolithic sites along the Vandée coast. 1 Les Cléons (Haute Goulaine), 2 Bégrolles, Guérivière (Haie Fouassière), 3 Le Breuil (Haie Fouassière), 4 La Haie Pallet (Mouzillon), 5–13 Pointe St. Gildas, Anse Sud, Préfailles II, Le Chatelet, L'Etang-Ouest, Biochon, Port aux Goths (Préfailles), La Girardière, Porteau-Ouest (Ste Marie); 14 Les Garennes (Geneston), 15–16 Les Etangs de la Brennière, Les Majoires (4 locus) (Montbert); 17 La Garne (Vieillevigne), 18 Senard (St. Hilaire de Loulay), 19 Chatelliers-Chateaumur, 20 La Plaine (Loublande), 21 Larsicot, Puy Albert, La Robière, Roche Allon (Moulins), 22 Puy Gayard (Pin).

The picture of the end of the Magdalenian and the Azilian in Poitou-Charente remains very sketchy because of the lack of information at our disposal and despite the potential importance of the archaeological heritage in the region.

The Mesolithic

The study of the origin and evolution of the cultural groups assembled under the generic term 'mesolithic' remains very problematic due to lack of publication and the absence of long stratigraphic sequences. The mesolithic of the region has always been considered either in relation to the late Upper Paleolithic (which latter represents the apogee of cultural evolution) or in relation to the Neolithic (in relation to which the mesolithic is seen as a development antecedent to the rearticulation of productive forces characteristic of the Neolithic). The mesolithic is rarely studied as a chronological and cultural entity in its own right. Radiocarbon dates are given in Table 7.1.

Massaud (1962, 1965) is credited with bringing to the attention of the profession the Sauveterrian occupation in the Charente at the sites of Cloux (Tonneay-Charente) and Maison-Neuve (Lussant), and for recognizing Garde-Epée (Saint-Brice) as the first Sauveterrian site in France. This site, published by Delamain (1914), consisted of material collected from a peat. It was also the focus of an excavation undertaken by the author. Grey and yellow sands, stratified under the peat, produced a lithic industry characterized by large blades, endscrapers, scalene and isosceles triangles, and triangular points with retouched bases (neither 'rhomboid', nor 'trapezes') that Delamain attributed to Breuil's 'Azilo-Tardenoisian', and that we consider today to be Middle Sauveterrian.

The two synthetic works which exist for this period are those of Joussaume (1981), which was part of a collected work under his direction - *Les Sites à Microlithes entre Vilaine et Marais Poitevin* (1984), and Perpère (1976). Despite their 'dated' quality, they remain important and complementary studies. The first surveys the coastal region from the south of Brittany to the Seugne estuary and the Charente (Charente-Maritime is partially covered); the second includes west-central France (Poitou-Charentes and the departments to the south of the Loire). The contribution of Fouéré (1991) adds little in the way of new information; it focuses more on the perspectives of prior investigators and the results that we might expect in this region.

Regional studies were conducted with greater frequency beginning in the 1960s by Tessier (Pays de Retz), and in the 1980s by Gourad (Bassin de Grand Lieu). Rozoy used some of these data in his massive synthetic work on mesolithic hunter-gatherers (1978). Based on the information provided in these studies, a preliminary discussion of the current state of research is possible.

The Early Mesolithic

The oldest mesolithic sites are in the Loire valley (Etangs de la Brenière, La Majoire D, La Garne) They are open sites attributed by Gourad to the Preboreal. The industry, gathered from surface collections, seems homogeneous from a chronological point of view and includes a range of types of which the most characteristic are scalene and isosceles triangles, and points (with and without retouched bases) (Figs. 7.5, 7.6). Nevertheless, each site has minor variations in the percentage representation of the different types. In addition, new types appear like the

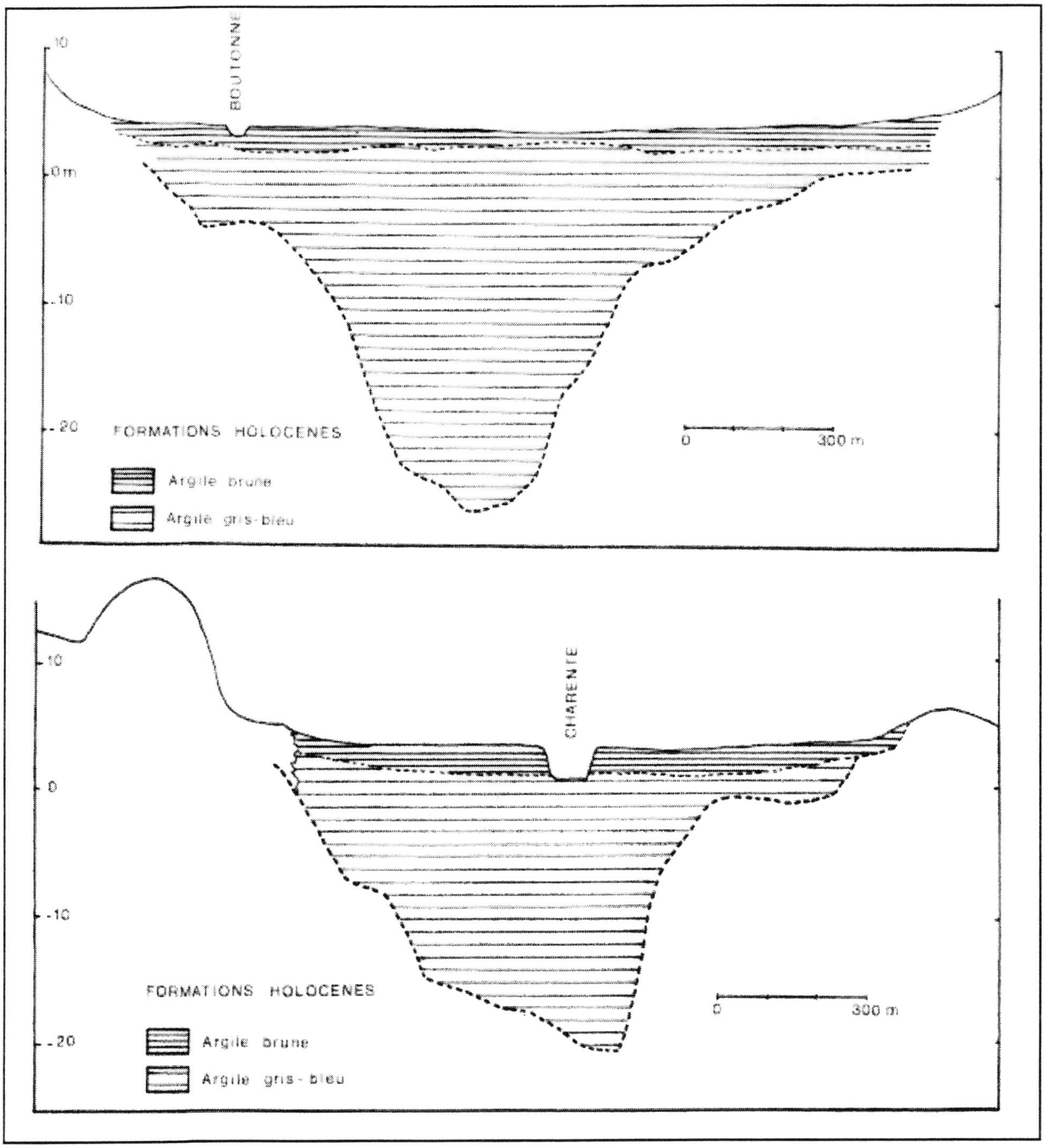

Figure 7.4 . Transverse sections of the valleys of the Boutonne near its confluence with the Charente, and of the Charente in the vicinity of Rochefort (after P. Fouéré 1991).

Majoire point, and the Majoire D 'crescents' (segmentiforms). With the existence of 'diminutive' versions of Malaurie points, Etangs de la Brenière could be the oldest mesolithic site in the region (Gourad 1987, 1992, 1993).

Other open-air sites have yielded epipaleolithic elements in Charente-Maritime and in Deux Sèvres (La Sablière, Bel Air, La Bouronnière). These are inland sites and Azilian points occur in the larger collections. However, more recent mesolithic elements are often found mixed in these assemblages (Blanchet 1987).

The Middle Mesolithic

For the Middle Mesolithic, the data are also imprecise because of the already-noted mixture in assemblages gathered in surface collections. However, it would seem that this stage is (also) characterized by a lithic industry emphasizing scalene and isosceles triangles, and points with and without retouched bases. Percentage differences and a tendency towards more standardization of geometric tool forms, with the introduction of trapezes, could indicate a later phase. It is notable that in all the sites known,

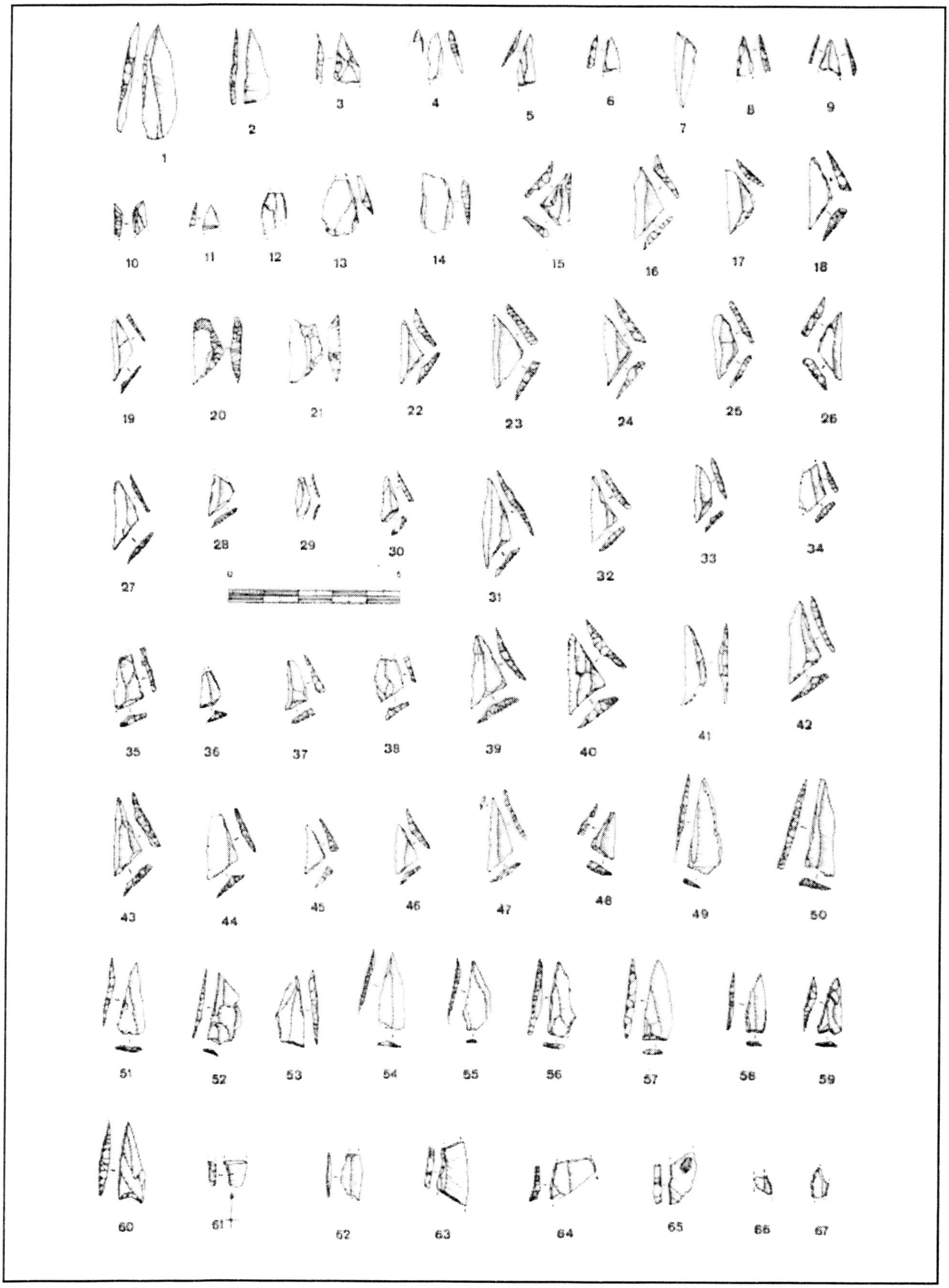

Figure 7.5. Microliths from Etangs de la Brenière (Montbert, Loire-Atlantique) (Gouraud 1992). 1–14 points, 15–29 isosceles triangles, 30–40 scalene triangles, 49–60 transverse base points, 61–67 various microliths.

Sauveterrian points are very rare or non-existent. La Sablière, Bel Air, and Chez Jouneau are representative of this phase in Charente-Maritime (Figs. 7.7-9).

The only reliable chronological indicators for this period are found at the site of La Pierre-Saint-Louis, excavated in 1993 in a salvage operation during the construction of the Saintes-Rochefort highway (A837). This excavation revealed several superimposed occupation surfaces (Middle Mesolithic, Final Mesolithic, Early and Late Neolithic), of which the main one, centered on eight hearths of 'polynesian' type, was determined to be Middle Sauveterrian (ca. 8550 BP). The Late Neolithic dates to ca. 5750 BP (Figs. 7.10-12). These sites are found in the lower Charente basin, but are still rather far from the present coastline (between 20 and 50 km).

Inland, only a single rockshelter site has been excavated and published. It is the Bellefonds site in the Vienne basin, where Patte (1971) defined two large assemblages: (1) a Middle Mesolithic with an industry dominated by scalene triangles (all the types are represented), isosceles triangles, circle segments and triangular (Tardenoisian) points, and (2) a more recent assemblage dominated by trapezes and scalene triangles, amongst which are a single Châtelet point, Sonchamp points and Montbani bladelets. The author notes an intermediate level where the first trapezes appear and above which the percentage representation of the various types is reversed. Not a single Sauveterrian point was found.

The Final Mesolithic

Amongst the sites attributed to the Final Mesolithic, the best known are found along the coast of Vendée. Joussaume proposes a chronocultural outline in two phases: (1) a 'South-Breton' group, independent of the Tardenoisian, and characterized by atypical trapezes (symmetric with concave truncations), numerous scalene triangles with one concave edge, but no isosceles forms (e.g., the site of Point Saint-Gildas). (2) Beginning in the mid-8th millennium BP, another chronocultural group follows the 'South-Breton' group in the same area - the Retzian. This group is characterized by tanged forms *(? - armatures à éperon),* Châtelet points, numerous trapezes and some scalene triangles with their short edges concave (often retouched inversely). Use of the microburin technique is common in Retzian assemblages, and the so-called 'coastal debitage' production mode (flaking on an anvil of small flint pebbles found along the coast) is associated with it. This production modes continues into later periods. The most important Retzian sites are coastal, cliff-edge occupation sites located in Pays de Retz (e.g., Le Châtelet, La Girardière) and on the Vendée coast (e.g., Pointe du Payré) (Figs. 7.13, 7.14).

'Retzian' influence can also be detected at some sites in Deux-Sèvres (e.g., Puy Albert, Lariscot) and Charente-Maritime (Bel Air), signalled by the presence of Châtelet types. Elements attributed to the Late Mesolithic *sensu lato* (principally trapezes) are found at numerous other sites in Charente and Charente-Maritime, but again, these are open sites with mixed industries (some cited above in connection with the Middle Mesolithic) (e.g., Garde-Epée, La Sablière, Bel Air, Chez Jouneau).

The Mesolithic-Neolithic Transition

As for the evolution of the Final Mesolithic toward the earliest phases of neolithization in the region, the Bellefonds rockshelter still remains the reference sequence. According to Patte's excavations (1971), there is a late Mesolithic level there with decorated pottery, seemingly related to the large family of early Neolithic ceramics from southern France and the Iberian Peninsula. The recent resumption of excavations at the shelter by a team headed by Joussaume will, without doubt, provide us with more precise information.

If the presence of potsherds in late mesolithic levels is a rare and equivocal occurrence, some microliths show up in numerous neolithic sites in Vendée, as well as in certain dolmens. Joussaume considers this association in the dolmen of La Pierre Levée (Nieuil-sur-l'Autize) to be the result of mixture after the reoccupation of the mesolithic site by neolithic people. But, more generally, he admits there is a persistence of microlithic technology in the neolithic (Tessier 1984, Joussaume 1984, Joussaume & Tessier 1991). Nevertheless, we do not yet have sufficient, reliable data to define what kinds of microlithic industries continue into the neolithic, nor in what neolithic phases they are represented.

Current Perspectives

During the last 15 years, several regional researchers have conducted surveys in the region of Saintonge and in the Charente basin. We have established contact with them and with M.R. González Morales at the University of Cantabria (Santander, Spain) to try to develop a collaborative research project focused on these periods, and to establish a comparative framework to guide research in France and along the Cantabrian coast (González Morales et al. 1992).

Between 1993 and 1995, about 40 unpublished epipaleolithic or mesolithic sites, of various importance, were recorded. The sites in the maritime zones and those inland present a series of common features. For the most part, they are situated close to water sources (streams, springs), located on low slopes (knolls, hills, former islands) and on well-drained, sandy soils. Only four are located in marsh-edge environments in peats, but given the continuing development of marshy shorelines, and the significant amounts of Holocene deposition, we cannot determine whether these site locations are due to relatively recent advance of the *bri* on earlier, pre-existing sites, or uplifting of underlying sediments caused by draining of the marshes. A program of test excavations should help us to resolve this question.

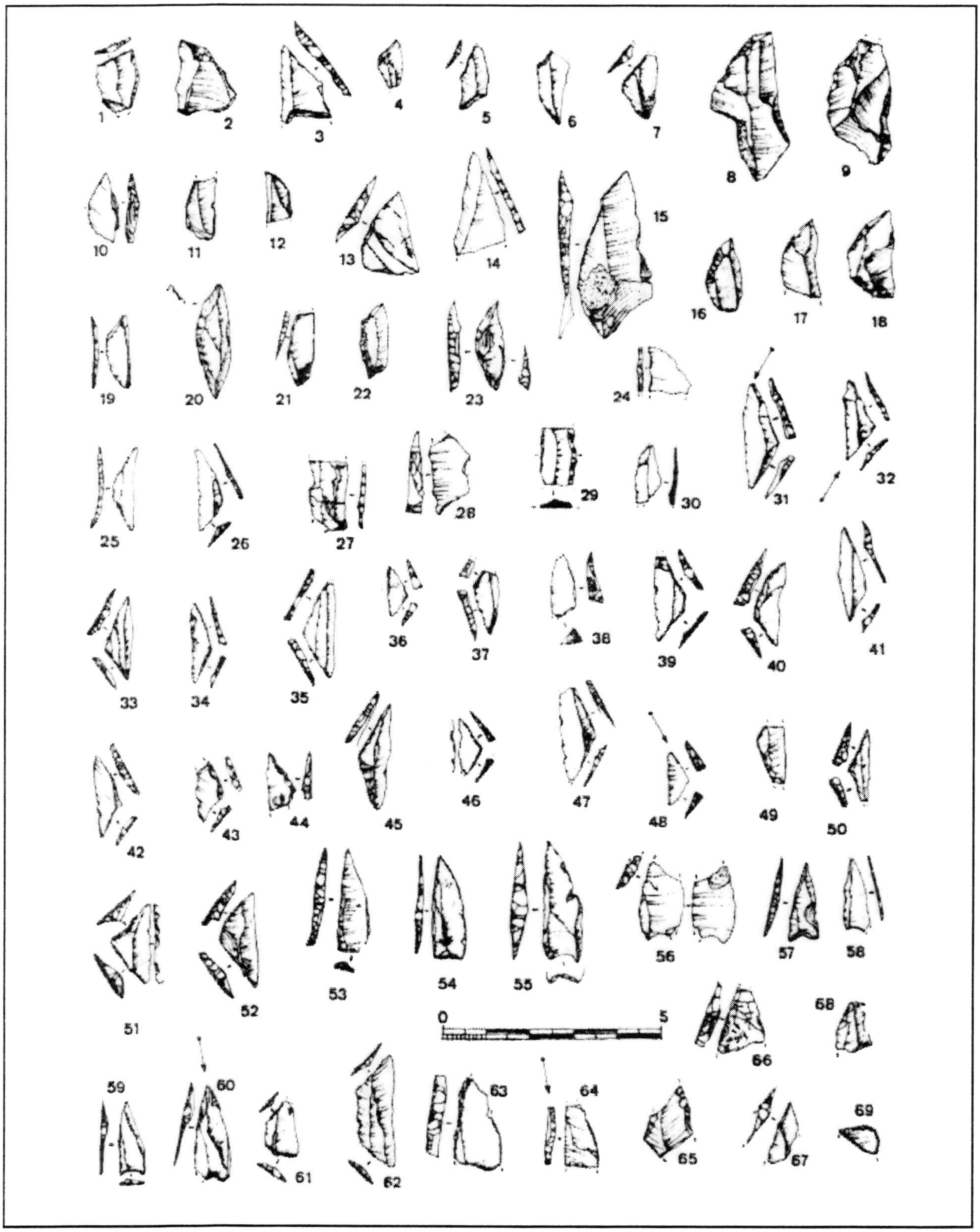

Figure 7.6. Microliths from la Majoire D (Montbert, Loire-Atlantique) (Gouraud 1987). 1–4 broken bladelets on oblique truncations, 5–7, 10 oblique truncated points, 11–15 unilaterally retouched points; 8, 9, 16–18 Chaville points; 19–23 Majoire points; 25, 27 circle segments; 26 asymmetric segment; 24, 28–30 backed bladelet fragments; 31–37 typical scalene triangles; 38, 39 atypical scalene triangles; 40–43 scalene triangles on small concave truncations, 44–52 isosceles triangles, 53 long triangular point, 56–60 concave base Tardenoisian points, 61 short asymmetric trapeze, 62–69 various microliths.

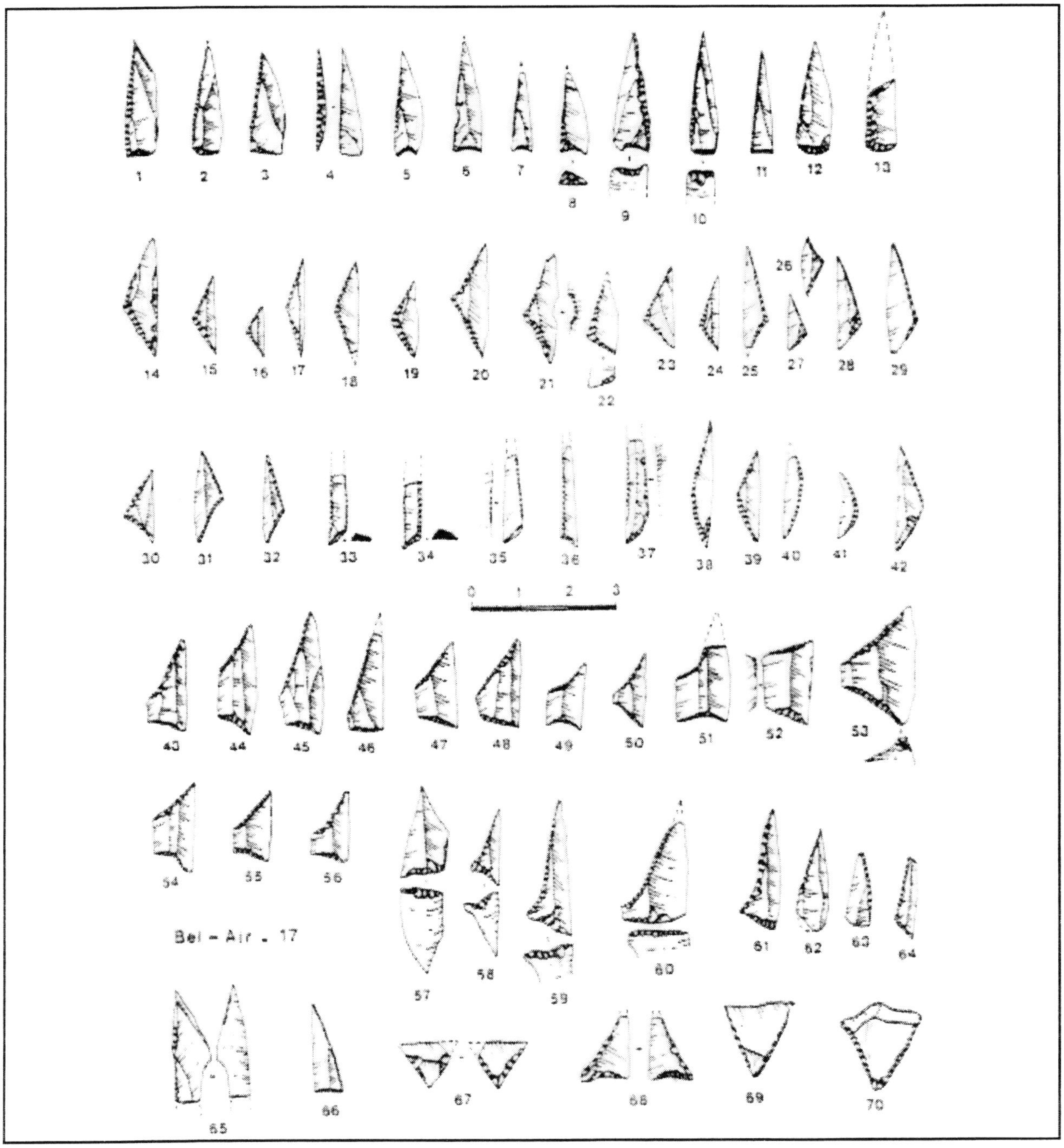

Figure 7.7. Microliths from Bel Air (Villars-les-Bois, Charente-Maritime) (Joussaume 1981). 1–42 pieces typical of the Middle Mesolithic, 43–70 pieces typical of the Final Mesolithic.

Sites closest to the sea or the marshes are located at elevations of 3-27 m, without significant modality in elevation. The inland sites are found at elevations ranging from 20-70 m, with a mode at about 50 m. Most sites face south (S, SW, SE) or east, and are found on tilled land with the exception of those in meadows in the marshes, and the four just mentioned, which were found in commercial sand quarries in wooded areas. Among the surveyed sites, none contained any recognizable structures. They generally consisted of remains dispersed by ploughing (e.g., flint debitage, artifacts, cores etc.).

In the case of the sand quarry sites, the pieces of flint were recovered after sifting the excavated soil, but no occupation surfaces were noted in the stratigraphic sections. It should be noted that the sand quarries concentrate their efforts only on fairly homogeneous sand layers from Cenomanian geological formations, after removing the overlying vegetation, sandy humic and gravel deposits, which are of no commercial value. It is probable that these overlying deposits contain the archaeological remains of interest here.

Most of the sites average 10-200 pieces (various microliths, debitage), but at the Chez Jouneau sand quarry

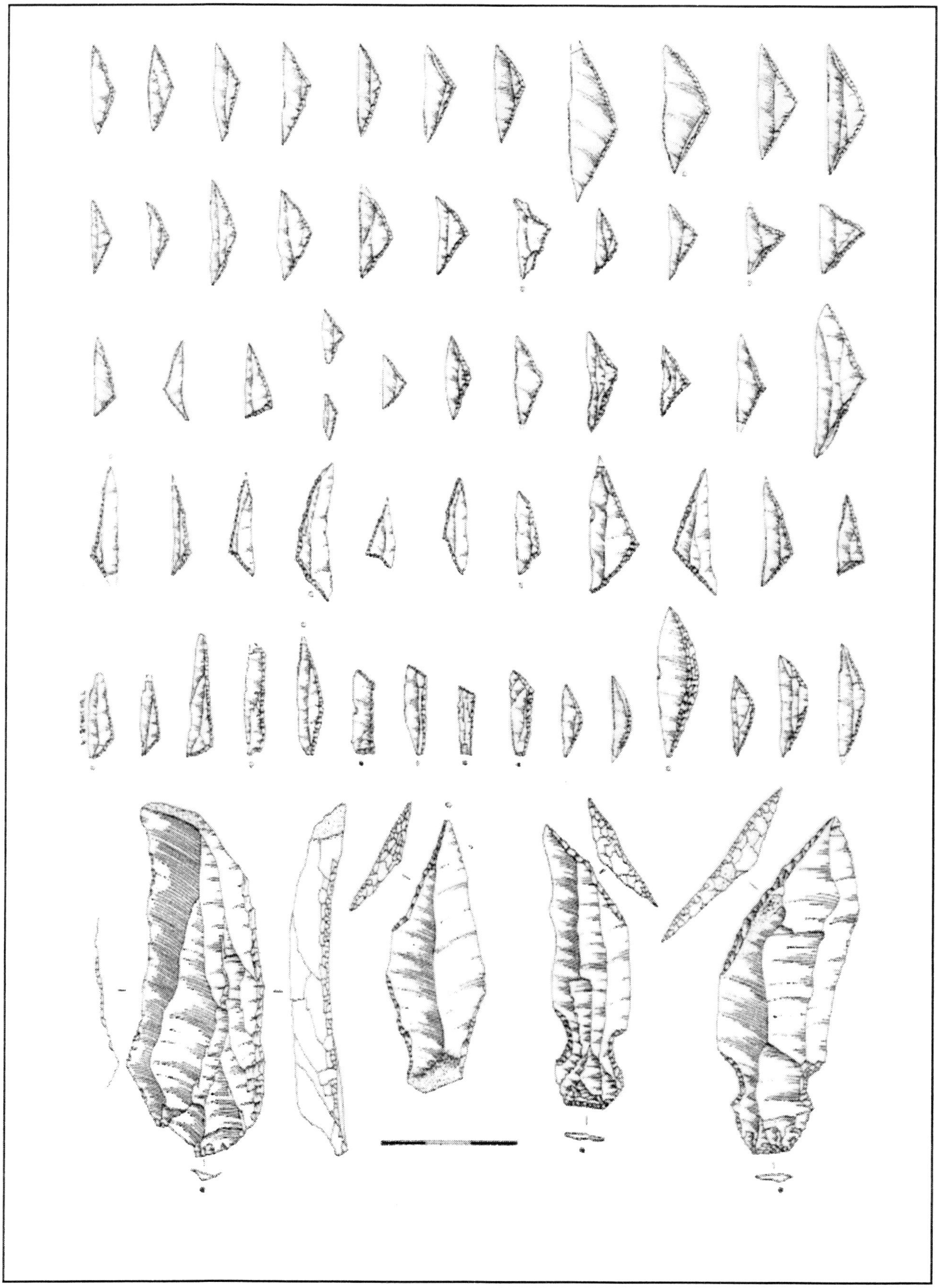

Figure 7.8. Microliths from Chez Jouneau (Brizambourg, Charente-Maritime) (after F. Blanchet). Pieces typical of the Early and Middle Mesolithic.

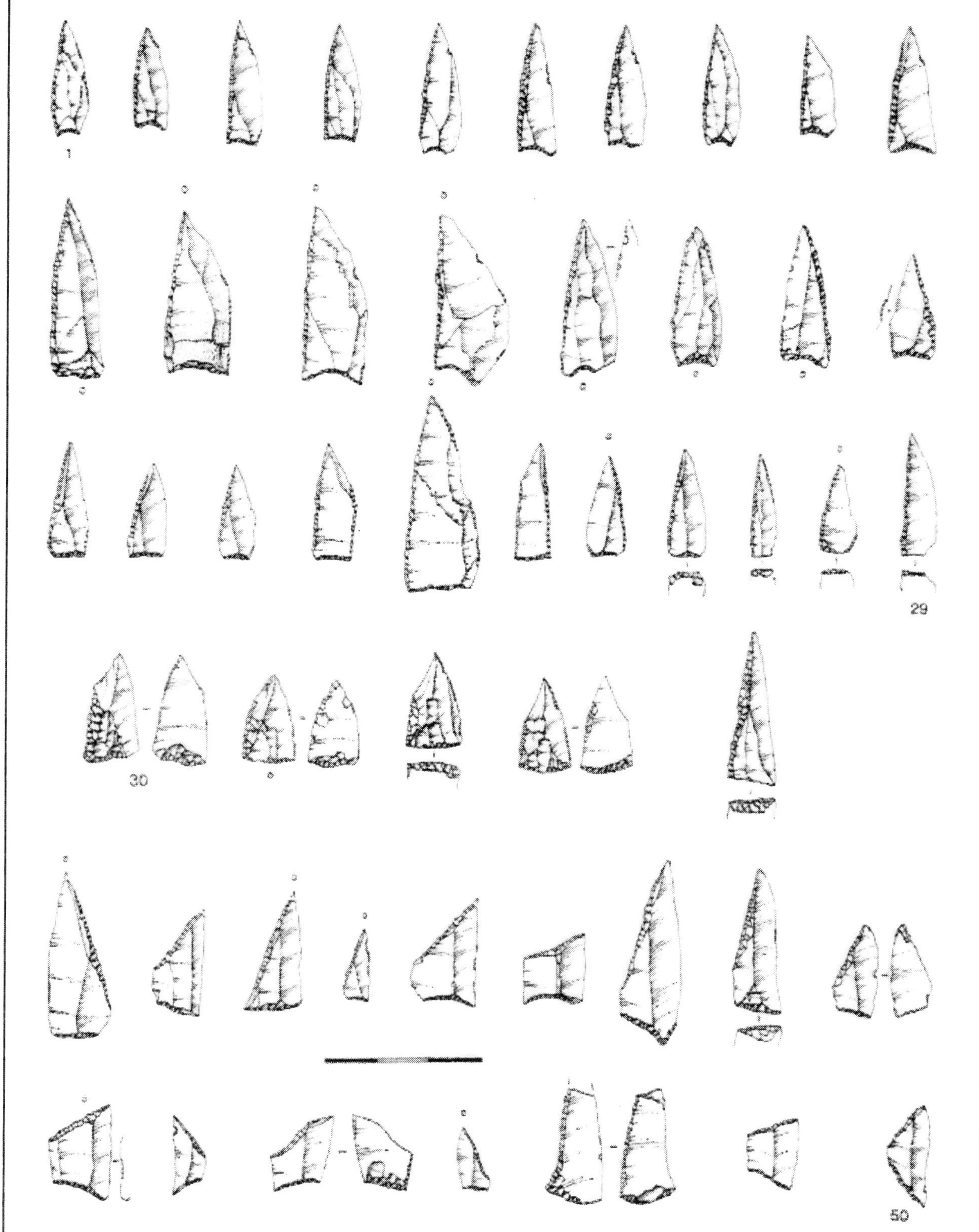

Figure 7.9. Microliths from Chez Jouneau (Brizambourg, Charente-Maritime) (after F. Blanchet). 1–29 pieces typical of the Middle Mesolithic, 30–50 pieces typical of the Final Mesolithic.

Figure 7.10 'Polynesian' type hearth from the site of La Pierre Saint-Louis, st. 2 sect. VI (8420 ± 110 BP) (P. Foucher excavations).

Figure 7.11 'Polynesian' type hearth from the site of La Pierre Saint-Louis, st. 4 sect. VI (7245 ± 70 BP) (P. Foucher excavations).

Figure 7.12 'Polynesian' type hearth from the site of La Pierre Saint-Louis, st. 6 sect. III (8695 ± 95 BP) (P. Foucher excavations).

site (Brizambourg, Charente-Maritime), the Blanchets (1987) found, after screening, some 20,000 flint artifacts (of which 2,000 were microburins and 1,000 various kinds of microliths). The chronological placement of these sites remains a problem. About half the surveyed sites contain artifact types attributed to more than a single period between the Azilian and the Neolithic. In the absence of stratigraphy or datable structures, dating can only be done using typological criteria, with all the ambiguity and imprecision that that involves. Nine sites have yielded Azilian points. Unfortunately, the taphonomic conditions at these open-air sites do not allow for the preservation of bone or antler artifacts (e.g., harpoons, fish hooks) and the rest of the assemblage is not diagnostic (i.e., it cannot be differentiated from the range of Sauveterrian microliths also found at the these sites).

The difficulties just mentioned concerning the identification of Middle Mesolithic sites also apply to current research efforts. At least 10 sites have microlithic assem-

Table 7.1. Radiocarbon dates for the Mesolithic of the Charente-Vendée region.

Saint-Gildas 1b		7 520 ± 140 BP	(Gif 3531)	on shell
Saint-Gildas 1c		6 790 ± 90 BP	(Gif 4847)	on shell
La Pierre St. Louis	St 4 Sect VI	7 245 ± 70 BP	(Ly 147 : AMS)	on charcoal, base of hearth
La Pierre St. Louis	St 2 Sect VI	8 420 ± 110 BP	(Gd 6937)	on charcoal, base of hearth
La Pierre St. Louis	St 6 Sect III	8 695 ± 95 BP	(Ly 116 : AMS)	on charcoal, base of hearth
La Pierre St. Louis	St 15 Sect III	8 765 ± 95 BP	(Ly 117 : AMS)	on charcoal, base of hearth
Bois-Ragot	niv. 3	8 800 ± 220 BP	Azilien (Gif 1588)	charcoal
Bois-Ragot	niv. 4b	10990 ± 160 BP	Azilien (Gif. 3580)	charcoal
Bois-Ragot	niv. 5b	11030 ± 140 BP	Magdalénien final (Gif. 2537)	charcoal
Bois-Ragot	niv. 6	10180 ± 160 BP	Magdalénien final (Gif. 3579)	charcoal

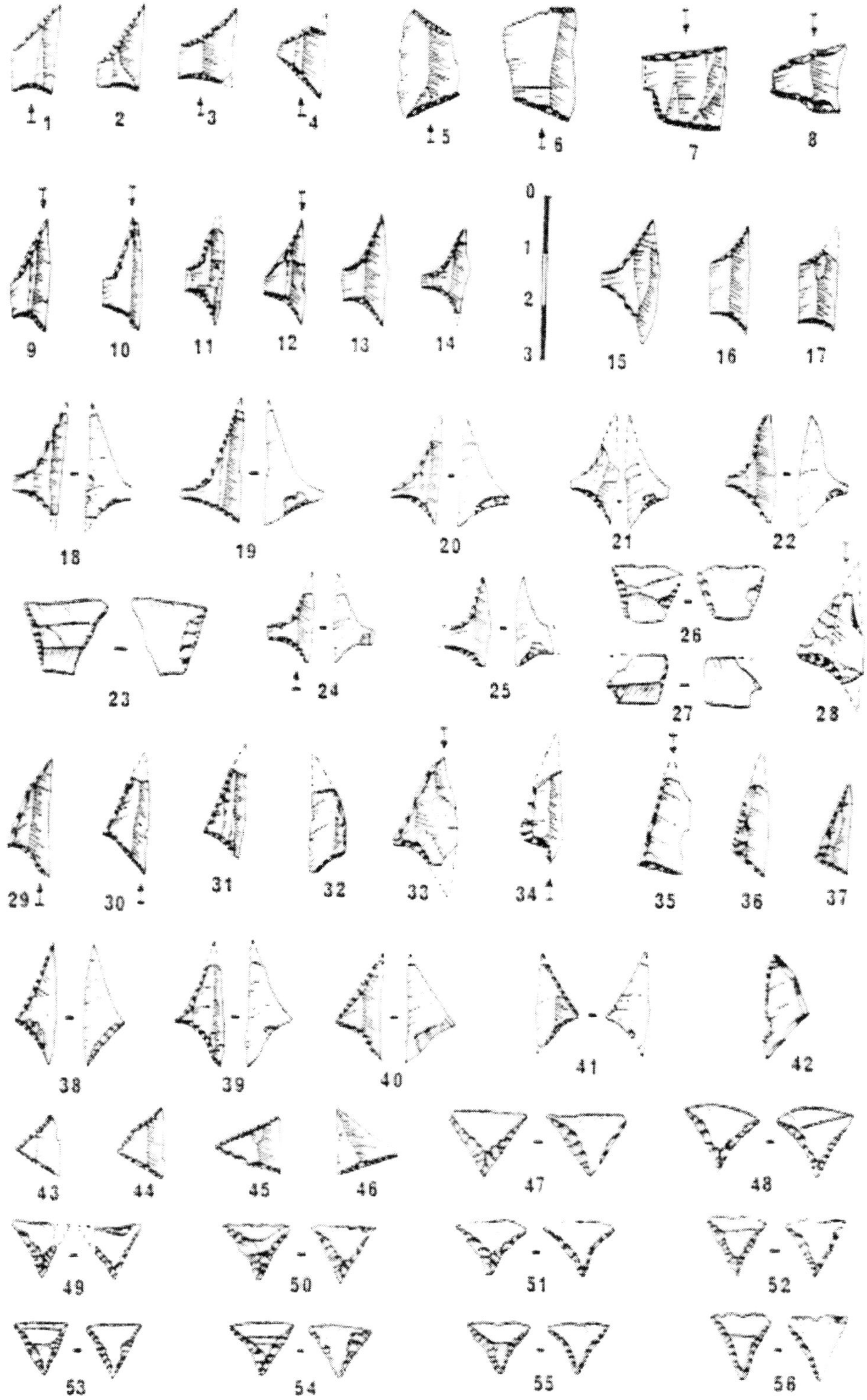

Figure 7.13. Microliths from la Pointe du Payré (Jard-sur-Mer, Vendée) (Joussaume 1981). 1–25 trapèzes, 28–46 triangles; 26, 47–56 Châtelet armatures.

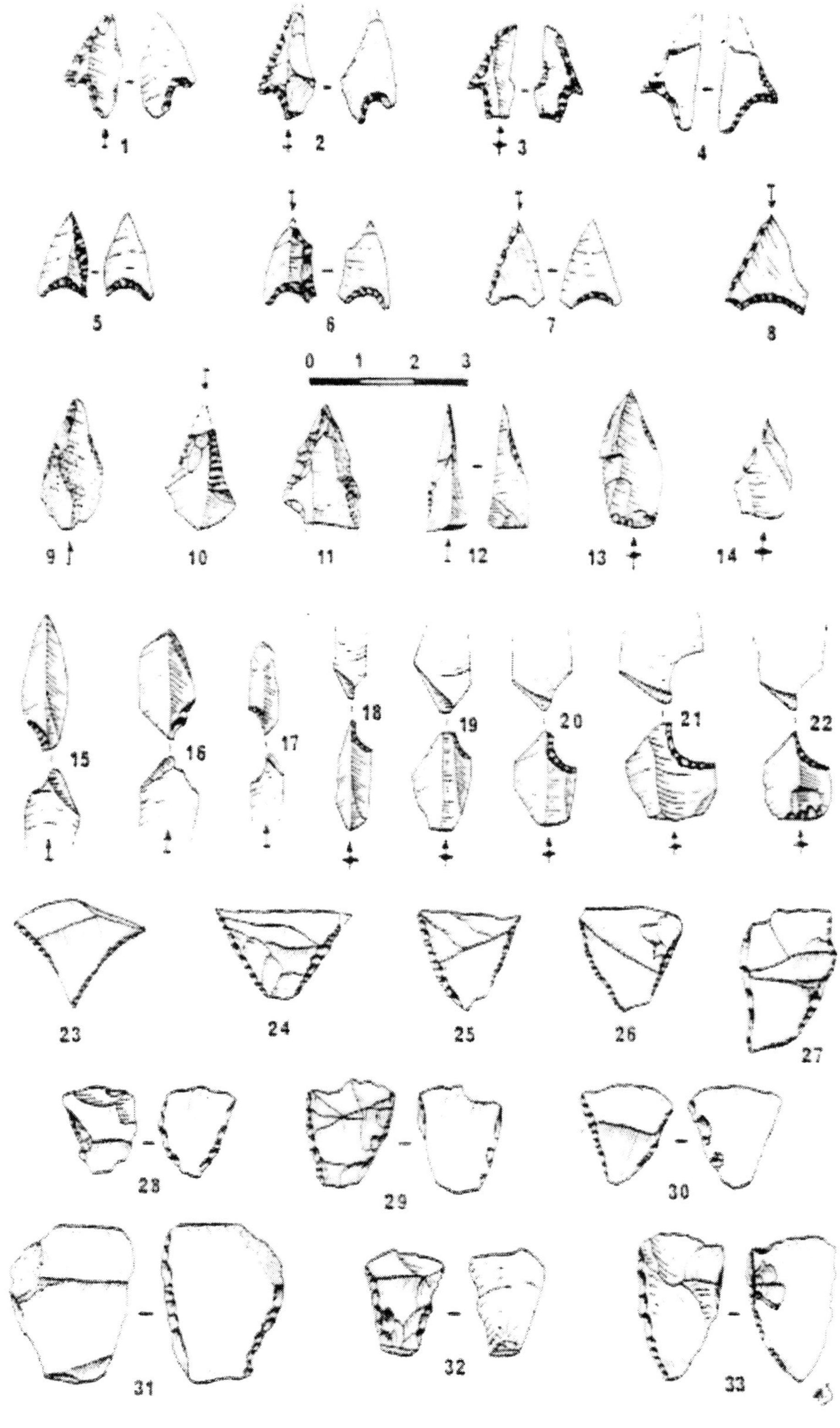

Figure 7.14. Microliths from la Pointe du Payré (Jard-sur-Mer, Vendée) (Joussaume 1981). 1–8 shouldered armatures, 9–14 points 15–22 microburins, 23–33 *armatures à tranchant transversal.*

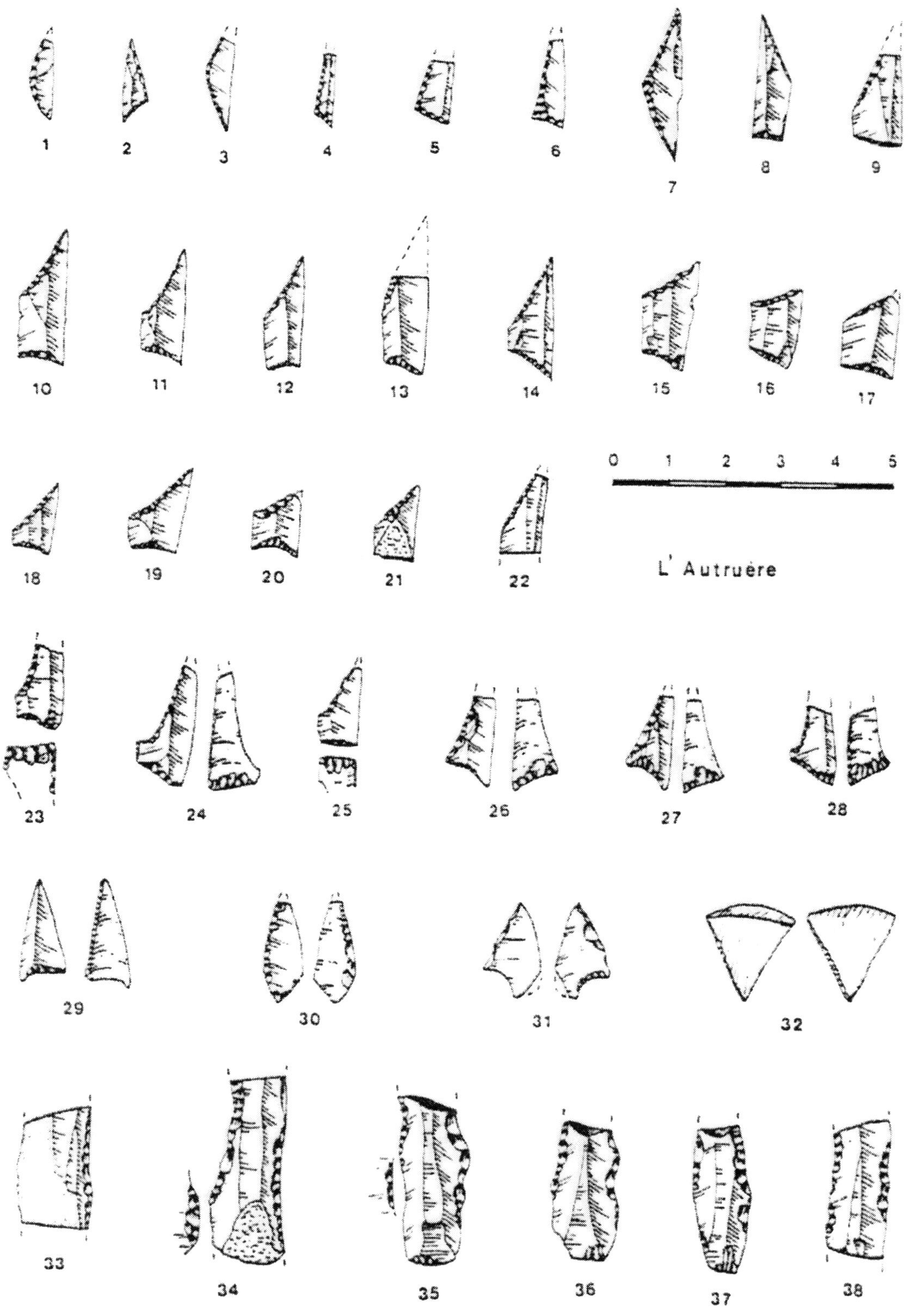

Figure 7.15. Microliths from l'Autruère (Lussant, Charente-Maritime) (Joussaume 1981). Typical pieces from the Final Mesolithic.

blages that can be assigned to the Sauveterrrian, and five others to the Final Mesolithic, but these distinctions rest solely on the diversity of the microliths collected, in particular the fortuitous presence or absence of 'index fossil' tool types: trapezes, Châtelet points, tanged pieces *(= armatures à l'éperon?),* as well as the flat retouch typical of the final phases of the mesolithic). On sites where the assemblages are large enough, we can document all the stages of the *chaîne opératoire* from the selection of nuclei to the production and discard of microliths, and the incidence of microburin technique. At some sites, large amounts of debitage could indicate workshops. All the unpublished sites are presently being written up as monographs, and we will have to await their completion before undertaking a detailed, synthetic analysis of these assemblages.

Conclusions

The results of recent work summarized here document an important human occupation of the maritime zones between Sèvre Niortaise and the Gironde during the tardiglacial and the beginning of the Holocene. The mapping of the distributions of sites known before and after this preliminary study clearly reveal a new 'territory' to explore, and a much higher density of archaeological remains than expected. Some of these sites were undoubtedly damaged by ploughing or destroyed in the extraction of sand from quarries. But others are clearly more valuable from a scientific point of view, and our efforts will be directed toward investigation of those. The major problem confronting us is to locate an in situ stratigraphic sequence in order to study the diachronic evolution of the industries, and to establish paleoenvironmental reference points.

Following this initial effort at reconstructing the archaeological 'facts' (in which we find ourselves at present), our intention is to construct a diachronic, chronocultural and paleoenvironmental interpretive model for the Tardiglacial on the coast of Charente. Once this is accomplished, we should be in a position to compare this model with the situation in the better-studied but topographically-distinctive region of Cantabria, in the western part of the Gulf of Gascony.

References

Blanchet, F. & J. Blanchet
1987 Note sur quelques sites préhistoriques poitevins. *Groupe Vendéen d'Etudes Préhistoriques* 18: 42-50.

Blanchet, J. & M. Rouvreau
1986 *L'Abri Sous-roche détruit de Piphrez chez Guérin (Grand-jean).* Rapport dactylographié, 11 pp., 12 fig.

Bonnet, J.
1963 Gisement mésolithique en Charante. *Bulletin de la Société Préhistorique Française* 60: 23-25.

Callot, G.
1976 *Analyse d'un Système Géo-pédologique Régional: Étude Sédimentologique des Sols et Formations Superficielles sur Plateforme Calcaire peu Déformée.* Montpellier: Thèse de l'Université des Sciences et Techniques du Languedoc.

Chollet, A., P. Boutin, A. Debenath, F. Delpech & J.-C. Marquet
1979 La grotte du Bois Ragot à Gouex (Vienne), industries, géologie, paléozoologie. In *La Fin des Temps Glaciaires en Europe* (D. de Sonneville-Bordes, ed.), pp. 365-380. Paris: Centre National de la Recherche Scientifique, Colloques Internationaux No. 271.

Chollet, A., H. Reigner & P. Boutin
1974 La grotte du Bois Ragot à Gouex (Vienne), note préliminaire. *Gallia-Préhistoire* 17: 286-291.

Collectif
1984 *Les Sites à Microlithes entre Vilaine et Marais Poitevin.* Nantes: Association d'Etudes Préhistoriques et Protohistoriques des Pays de la Loire, Etudes Préhistoriques et Protohistoriques des Pays de la Loire, Vol. 7.

Debenath, A.
1979 La fin des temps paléolithiques dans le Bassin de la Charente. In *La fin des temps glaciaires en Europe.* (D. de Sonneville-Bordes, ed.), pp. 381-385. Paris: Centre National de la Recherche Scientifique, Colloques Internationaux No. 271.

Delamain, R.
1914 Restes d'industrie lithique sur les bords de la tourbière de Garde Epée (commune de Saint-Brice). *Bulletin de la Société Archéologique et Historique de la Charente 3*: 190-207.

Duport, L.
1987 Grotte de Montgaudier, commune de Montbron (Charente): le foyer et les gravures magdaléniennes. In *Préhistoire de Poitou-Charentes. Problèmes actuels.* pp. 37-48. Paris: Comité des Travaux Historiques et Scientifiques.

Favre, M.
1994 Un gisement de tradition tardenoisienne à Puissoteau (Tonnay-Charente, Charente-Maritime). *Bulletin de la Société de Géographie de Rochefort* 14: 232-234.

Foucher, P. & C. San Juan
1991 Les Genêts (Leigné-les-Bois, Vienne), atelier de taille magdalénien dans son contexte régional. *Revue Archéologique de Centre de la France* 30: 7-29.
1992 Les ateliers de taille du silex de l'Est-Châtelleraudais: prospection thématique. *Bilan scientifique 1991 du Service Régional de l'Archéologie de Poitou-Charentes,* p. 74. Poitiers: Ministére de la Culture.
1993 Les ateliers de taille du silex sur la rive gauche de la Creuse: prospection thématique. *Bilan scientifique 1992 du Service Régional de l'Archéologie de Poitou-Charentes,* p. 66. Poitiers: Ministére de la Culture.
1994 L'atelier de taille paléolithique des Marineaux (Leigné-les-Bois, Vienne) et le problème d'attribution chronologique des sites de plein air du Bassin de la Creuse. *Paléo* 6: 175-196.

Foucher, P., J. Musch, R. Guilbert, A. Boguszewski & L. Laporte
1994 La Pierre saint-Louis à Geay (Charente-Maritime). *Bilan Scientifique 1993 du Service Régional de l'Archéologie de Poitou-Charentes,* p.33. Poitiers: Ministère de la Culture.

Foucher, P., J. Wattez, A. Gebhardt & J. Musch
2000 Les structures de combustion mésolithiques de La Pierre-Saint-Louis. *Paléo* 12: 165-200.

Fouere, P.
1991 Mésolithique et processus de néolithisation. in *Les Archéologiques Poitou-Charentes. Bilan et perspectives sur la recherche régionale.* Préactes du Colloque du 23-24 nov. 1991 organisé par le Service Régional de l'Archéologie de Poitou-Charentes et l'AAPC, p. 37-40.

Gabilly, J.
1978 *Poitou-Vendée-Charentes.* (Guides Géologiques Régionaux). Paris: Masson.

Gaillard, J.
1987 Heartebise (Jonzac): rapport de sauvetage. *Bulletin de l'Association des Archéologues de Poitou-Charentes* 16: 33.

González-Morales, M.-R., J.-C. García Codrón & A. Morales Muñiz
1992 El bajo Asón del X al V milenio BP: cambios ambientales, económicos y sociales en el paso a

la prehistoria reciente. *The Late Quaternary in the Western Pyrenean Region*, pp. 333-342. Bilbao: Universidad del País Vasco.

Gourad, G.
1987 Le Mésolithique des Majoires à Montbert (44), la station D. *Groupe Vendéen d'Études Préhistoriques* 18: 14-41.
1992 Le campement mésolithique des Etangs de la Brenière à Montbert (Loire-Atlantique). *Revue Archéologique de l'Ouest* 9: 39-55.

Gourad, G., C. Dugast, J.-M. Jauneau & B. Cousseau
1993 La station mésolithique de la Garne à Vieillevigne (Loire-Atlantique). *Groupe Vendéen d'Études Préhistoriques* 29: 1-19.

Joussaume, R.
1981 *Le Néolithique de l'Aunis et du Poiton occidental dans son cadre atlantique.* Rennes: Travaux du Laboratoire d'Anthropologie-Préhistoire-Protohistoire et Quaternaire armoricain, Université de Rennes I.
1984 Microlithes dans les mégalithes et les camps néolithiques de la Vendée. In *Les Sites à Microlithes entre Vilaine et Marais Poitevin,* pp. 279-283. Nantes: Association d'Études Préhistoriques et Protohistoriques des Pays de la Loire, Études Préhistoriques et Protohistoriques des Pays de la Loire, Volume 7.
1993 La Néolithisation dans le Seuil du Poitou et ses marges: milieux karstiques et alluviaux. Projet collectif. In *Bilan scientifique 1992 du Service régional de l'Archéologie de Poitou-Charentes,* pp. 69-70. Poitiers: Ministère de la Culture.

Joussaume, R. & J.-P. Pautreau
1990 *La Préhistoire du Poitou: Poitou-Vendée Aunis des Origines à la Conquête romaine.* Tours: Editions Ouest-France Université.

Joussaume, R. & M. Tessier
1991 Mésolithique et Néolithisation entre Loire et Marais Poitevin. In *Mésolithique et Néolithisation en France et dans les Régions Limitrophes,* pp. 173-176. Paris: Editions du Comité des Travaux Historiques et Scientifiques.

Leduc, M., C. Boujot & S. Cassen 1987 La grotte du Bois Bertaud. *Archéologie Pontoise* 79: 1-8.

Lenoir, M.
1979 Les industries du Paléolithique supérieur terminal des basses vallées de la Dordogne et de la Garonne. In *La Fin des Temps Glaciaires en Europe* (D. de Sonneville-Bordes, ed.), pp. 401-423. Paris: Centre National de la Recherche Scientifique, Colloques Internationaux No. 271.

Marchand G.
1999 *La Néolithisation de l'Ouest de la France. Caractérisation des Industries Lithiques.* Oxford: BAR International Series No. 748.

Massaud, J.
1962 Présence du Mésolithique dans le nord de la Saintonge, station du Cloux (commune de Tonnay-Charente) et son environnement, *Bulletin de la Société Préhistorique Française* 59: 642-648.
1964 La Sauveterrien dans la Charente. In *Actes du 20e Congrès des Sociétés Savantes du Centre-Ouest*, pp. 108-109. Rochefort: Congrès des Sociétés Savantes du Centre-Ouest.

Massaud, J. & M. Rouvreau
1975 Les microlithes post-néolithiques des Charentes. *Bulletin de la Société Préhistorique Française* 72: 113-116.

Musch, J. P. Foucher, A. Boguszewski & R. Guilbert
1993 *La Pierre Saint-Louis à Geau (Charente-Maritime)-I,II.* Document Final de Synthèse de Sauvetage Urgent 15 janv. 1993-15 juil. 1993. Poitiers: SRA Poitou-Charentes.

Patte, E.
1971 Quelques sépultures du Poitou du Mésolithique au Bronze moyen. *Gallia-Préhistoire* 14: 139-244.

Perpere, M.
1976 Les civilisations de l'Epipaléolithique et du Mésolithique en Poitou-Charentes. In *La Préhistoire Française* (H. de Lumley & J. Guilaine, eds.), pp. 1436-1440. Paris: Centre National de la Recherche Scientifique.

Regrain, R.
1980 *Géographie Physique et Télédéction des Marais Charentais.* Amiens.

Roussot-Larroque, J.
1977 Néolithisation et Néolithique ancien d'Aquitaine. *Bulletin de la Société Préhistorique Française* 74: 559-582.
1979 Stations aziliennes du Médoc et des Landes de la Gironde: documents et problèmes. In *La Fin des Temps Glaciaires en Europe* (D. de Sonneville-Bordes, ed.), pp. 387-400. Paris: Centre National de la Recherche Scientifique, Colloques Internationaux No. 271.
1989 Magdalélien final et Mésolithique du Sud-Ouest de la France. In *Le Temps de la Préhistoire* (J.P. Mohen, ed.), tome 1, pp. 362-364. Baume-les-Dames: Société Préhistorique Française.

Surmely, F.
1991 Le Palèolithique supérieur en Saintonge. *Recherches Archéologiques en Saintonge* 1: 1-12.

Rozoy, J.-G.
1978 *Les Derniers Chasseurs I-III.* Charleville-Mezières: Société Archéologique Champenoise.

Tessier, M.
1984 Les armatures microlithiques des sites néolithiques de la zone cotière du Pays de Retz. In *Les Sites à Microlithes entre Vilaine et Marais Poitevin,* pp. 123-126. Nantes: Association d'Etudes Préhistoriques et Protohistoriques des Pays de la Loire, Vol. 7.

Thibaudeau, J.
1972 *A travers le Canton de Gémozac, Promenades Historiques, Artistiques et Archéologiques (Livre-Guide V).* Saintes: Delavaud.

Tournepiche, J.-F.
1987 L'occupation épipaléolithique de la grotte du Quéroy, commune de Chazelles (Charente). In *Préhistoire de Poitou-Charentes. Problèmes actuels,* pp. 207-221. Paris: Comité des Travaux Historiques et Scientifiques.

Visset, L., D. Voeltzel, E. Maison-Neuve, J. Nikodic, J.-P. Margerel & V. Borne
1990 Paléoécologie holocène des 'Marais du Roucher' en Maillezais (Vendée), dans le Marais poitevin (littoral atlantique-France). *Quaternaire* 2: 11-121.

8

Archaeology of the Early Holocene in the Mediterranean Region of Spain (12,000-7,000 BP)

J. Emili Aura Tortosa
University of Valencia

Abstract

This chapter reviews the archaeological record of the central and southern Mediterranean region of Spain, from the Pleistocene-Holocene transition to the beginning of the Neolithic. The archaeological data are integrated with current information on changes in climatic and paleoeconomy across the region. Ecological, economic and industrial transformations from the end of the Tardiglacial (12-10 kyr BP) to the beginning of the Neolithic underscore elements of continuity and define the character of the Epipaleolithic-Neolithic transition in Spanish Levante.

Introduction

Since the publication of the sequence of epipaleolithic industries developed by Fortea (1973), projects dedicated to the regional Epipaleolithic have been few. Only in the last decade has the tempo of investigation increased, especially regarding paleoenvironmental reconstruction, although research on the Spanish Levante remains uneven and limited in scope. Most projects have approached Holocene transformations in a tangential way, either from the perspective of changes accompanying the final Upper Paleolithic, or from that of the process of neolithization.

This essay seeks to remedy this deficiency by focusing on recent work on hunter-gatherer adaptations in Mediterranean Spain between 12-7 kyr BP. Current data on paleoenvironmental change, chronology, industrial and economic transformations, and settlement characteristics are summarized. A brief commentary on funerary ritual and the exploitation of marine resources is also included. I will consider ecological, economic and industrial transformations from the end of the Tardiglacial (12-10 kyr BP) in an effort to throw into sharp relief elements of continuity with the late paleolithic and to define the character of the regional Epipaleolithic-Mesolithic.

Paleoenvironment

Mediterranean Spain exhibits considerable biogeographic diversity due to its latitude, topography and coastal location (Fig. 8.1). Nevertheless, the temperate effects of low latitude and the proximity of the Mediterranean Sea combine to create a certain environmental unity. Differences in elevation and relief, and in the orientation of coastal hills and mountains are perhaps the most important determinants of subregional microenvironments. The peripheral mountain ranges define a littoral and pre-littoral coastal plain of variable extent upon which most of the archaeological sites are found. The Pyrenees define the northern boundary of Mediterranean Spain; the Straits of Gibraltar its southern one.

Because of topography, interregional communication along the coastal strip, and between the coast and the interior, is not easy. The coast along a N/S axis alternates between stretches with steep cliffs (e.g., Costa Brava, La Nao, Almería-Málaga) and those with low-lying, sandy beaches (e.g., the Ebro delta, the Gulf of Valencia, Murcia) which lie at the mouths of the major river valleys and basins leading to the interior of the Peninsula: the Ebro Valley to the north, the Montesa-Vinalopó in the center, and the Segura-Guadalentín in the south, which connects with the Bética-Guadalquivir depression to the southeast.

The Tardiglacial-Holocene Transition

Continuity seems to be the dominant element in bioclimatic change over the Tardiglacial-Holocene (T-H) transition. The extreme climates of the last glacial maximum had an uneven effect on this ecologically very diverse region. Sedimentological analyses indicate a series of short, cold and dry pulses between 12 and 11 kyr BP, identified as local manifestations of Final Dryas . The transition to the Holocene is marked by erosion, which became accelerated under a climate with marked seasonality and with

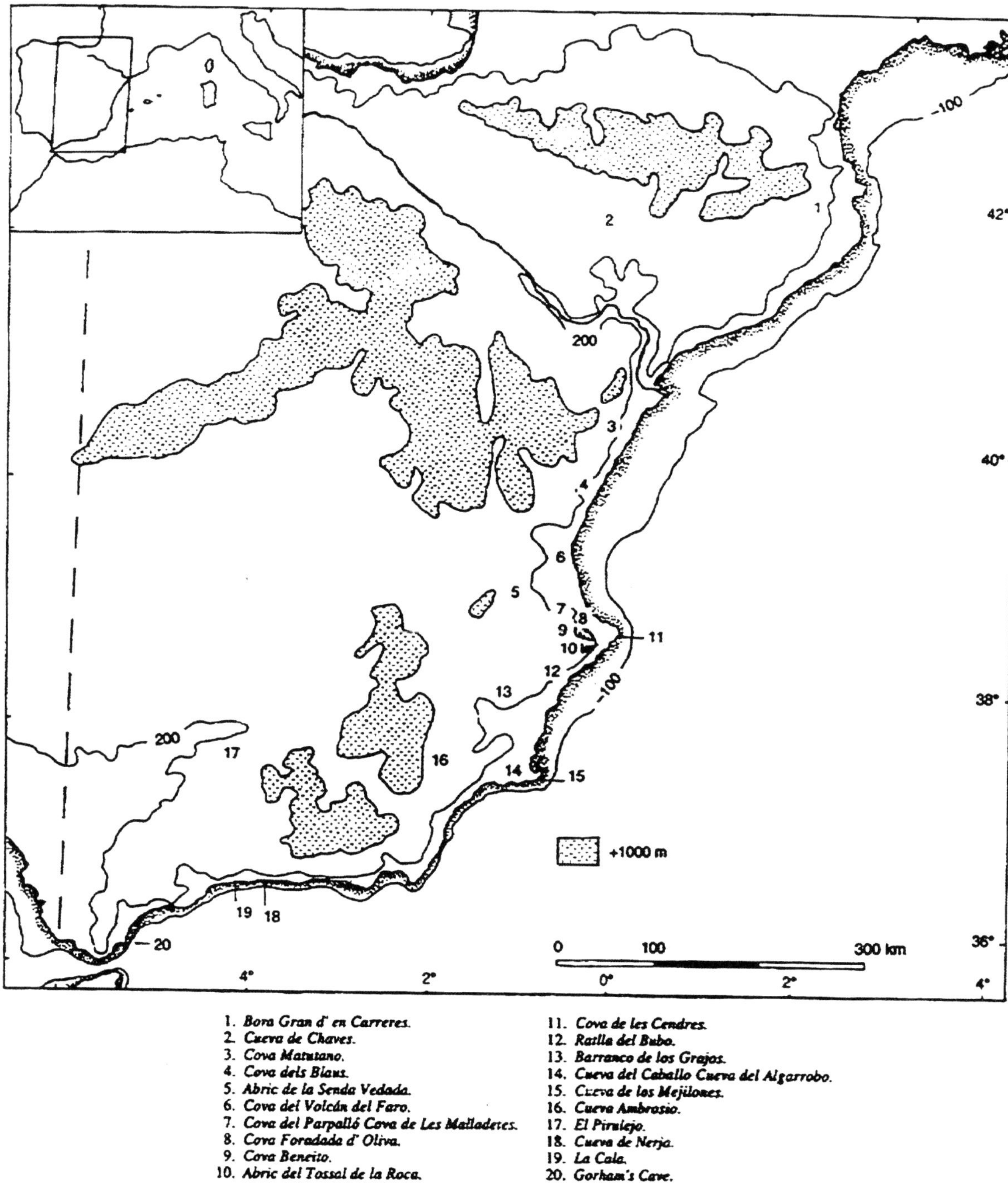

Figure 8.1. Map of Mediterranean Spain showing topographic relief, sites cited in the text.

unevenly distributed rainfall over the course of an annual cycle (Fumanal 1986, Jordá Pardo 1986).

Paleobotanical data indicate a N/S time-lag in the phytogeographic associations identified from the south of France to Gibraltar (Table 8.1). Although it is possible to recognize a continuous, progressive increase in thermophile taxa (AP) after ca. 16 kyr BP (Dupré 1988), both the coast and the interior are characterized by an open landscape, with conifers and junipers, and a dry, more temperate climate south of 40° N. lat. (Vernet & Thiebault 1987, Badal 1991). The Mediterranean deciduous forest does not seem to reach its fullest development until the Boreal/Atlantic transition.

Although, within a region, the anthropogenic faunas by themselves do not shed much light on biostratigraphic changes (i.e., they vary minimally through time), interregional ecological differences are recorded in the paleontological faunas. In northern Catalonia and in southeastern

France, some Euro-Siberian species occur throughout the LGM (e.g., *Rangifer tarandus, Bison priscus, Elephas* sp., *Crocuta* sp.) that are not found south of the Ebro. In the latter region, *Cervus elaphus* (red deer) and *Capra pyrenaica* (ibex), along with some *Equus* sp. (horse) and *Bos primigenius* (auroch) were the basic species after c. 20 kyr BP (Davidson 1989, Aura & Pérez 1992, Villaverde & Martínez 1992). Over the T-H transition, there is a certain increase in forest-adapted species (*Rupicapra rupicapra, Capreolus capreolus* and *Sus scropha,* mainly), which seems to indicate increases in temperature and humidity (Tables 8.2 - 8.4).

There are, however, some cold-climate indicators in the marine and avifaunas from sites in eastern Andalucía: (1) in level B at Gorham's Cave (= Tardiglacial?), *Nucella lapillus* among the mollusca, and *Pyrrhocorax pyrrhocorax* and *P. alpina* among the birds, have been recorded (Baden-Powell 1964, Eastham 1967); and (2) in the Cueva de Nerja (Málaga), *Pinguinus impennis* and *P. pyrrhocorax* are reported in levels post-dating 12 kyr BP (Eastham 1986). Marine fish have also been identified from a level dated to 10.8 kyr BP that today have a boreal distribution: *Pollachius pollachius, Melanogrammus aeglefinus* (Rodrigo 1991).

Some of these species were probably brought into the Mediterranean with an influx of cold Atlantic waters via the Straits of Gibraltar, an observation that seems to be corroborated by analyses of foraminifera from the Alborán Sea, which indicate a cold marine microfauna between 12-10 kyr BP (Pujol & Vergnaud 1989). Conflicting terrestrial (paleobotanical, paleofaunal) and marine (mollusca, fish, forams) bioclimatic surrogates suggest, therefore, that the Tardiglacial deglaciation was not synchronous on the land and sea.

Chronology

One of the important consequences of Fortea's (1973) proposed industrial sequence was the collection of a large number of radiometric dates for the 12-7 kyr BP interval

CRONOLOGIA		N del EBRO (> 40 º)	S del EBRO (< 40 º)
HOLOCENO	6.000		
	7.000	Fase 2: CLIMA SUPRAMEDITERRANEO Quercus cf. pubescens Acer sp. Buxus sp.	Fase 2: CLIMA MESOMEDITERRANEO Pinus sp. Quercus faginea Quercus ilex-coccifera Pistacia lentiscus Olea europaea var. sylvestris
	8.000	Fase 1b: CLIMA OROMEDITERRANEO Juniperus sp. (+) Pinus sylvestris	
	9.000	Fase 1a: CLIMA CONTINENTAL Pinus sylvestris (+) Juniperus sp.	
TARDIGLACIAR	10.200		Fase 1: CLIMA SUPRAMEDITERRANEO Pinus nigra Juniperus sp. Quercus sp.
	12000		
	14.000		

Table 8.1. Bioclimatic phases in the vegetation of Mediterranean Spain: 14,000 – 6,500 BP (Vernet & Thiebault 1987, Badal 1991).

Table 8.2. Record of Tardiglacial and Holocene faunas in Levante by period (fish and molluscs excluded).

	MSM		EMM		EGM	
	NR	%	NR	%	NR	%
Equus + Bos	86	0.3	24	0.1	7	0.3
Cervus elaphus	1.543	5.8	2.220	15.3	407	19.8
Capra pyrenaica	1.227	4.6	1.623	11.2	1.238	60.3
Sus scropha	11	0.0	61	0.4	51	2.5
Rupicapra + Capreolus	1	0.0	46	0.3	11	0.5
Lagomorfa	23.162	86.9	10.179	70.1	299	14.5
Aves	537	2.0	285	2.	14	0.7
Carnivora	64	0.2	75	0.5	24	1.1

Table 8.3. Frequencies of large and middle-sized mammals by period.

	MSM	EMM	EGM
Equus + Bos	2.9	0.6	0.4
Cervus elaphus	52,6	54.8	23.4
Capra pyrenaica	41.8	40.1	71.2
Sus scropha	0.3	1.5	2.9
Rupicapra + Capreolus	0.03	1.1	0.6
Carnivora	2.2	1.8	1.4

Table 8.4. Large and middle-sized mammals at Nerja, Tossal de la Roca and Matutano.

	Nerja		Tossal de la Roca			Matutano	
	MSM	EMM	MSM	EMM	EGM	MSM	EMM
Equus + Bos	0.1	-	-	0.7	0.3	8.7	0.4
Cervus elaphus	11.7	3.3	15.3	27.4	35.2	68.5	88.8
Capra pyrenaica	86.8	92.4	83.5	59.6	59.2	22.7	10.3
Sus scropha	1.3	4.2	1.1	5.7	1.1	-	0.3
Rupicapra + Capreolus	-	-	-	6.4	4.1	-	0.1

(Fig. 8.3 and Table 8.2). The chronological framework derived from these dates tends to confirm the succession of different epipaleolithic complexes established by Fortea (1973). However, some temporal overlap between the various industries is also indicated, along with important axes of variability among contemporaneous industries:

From the 13th millennium (mil.) BP, there are industries known from their typological characteristics as Mediterranean Microlaminar Epipaleolithic (EMM). On the other hand, there are occupations attributed to the Mediterranean Upper Magdalenian (MSM) that date up to the 11th mil. BP. This apparent continuity between the MSM and the EMM - especially as regards lithic industries but also in respect of subsistence and settlement patterns - raises the question of whether they should be considered as two different industrial complexes.

A greater antiquity (10th mil. BP) is recorded for the 'Sauveterroid facies' of the Mediterranean Epipaleolithic (EMM/S) north of the Ebro, while to the south, available data could indicate a long persistence of the EMM. The 'Tardenoisoid facies' of the Mediterranean Geometric Epipaleolithic (EGM) also might be older in Catalonia (e.g., at Balma Margineda), although it seems to develop throughout the 8th mil. BP (García Argüelles et al 1992). The possibility of a persistence of the EMM up to the neolithization should not be discounted. This process should be analyzed from the perspective of possible interaction between groups with EGM industries and those of the Cardial Neolithic (Fortea 1973, Cabanilles 1992, Bernabeu et al. 1993). Despite these observations, the general industrial sequence in Mediterranean Spain is confirmed, and is coherent with the chronology of sequences recognized in neighboring regions:

11-10th millennium BP:	Aziloid facies (= EMM)
10-9th millennium BP:	Sauvetteroid facies (= EMM/S)
8th millennium BP:	Tardenoisoid facies (= EGM)

Lithic Industries: a Regional Model

Investigations developed over the last 20 years on the basis of Fortea's *Los Complejos Epipaleolíticos del Mediterráneo Español* (1973) underscore its signal importance in the intellectual history of archaeology in Mediterranean Spain. The organization of Holocene lithic industries into two industrial complexes (EMM, EGM) departs from the existence of two very different successive (but temporally overlapping) typological entities.

The Mediterranean Microlaminar Epipaleolithic (EMM)

By virtue of their stratigraphic position and their microlaminar components (of Magdalenian derivation), these complexes were considered an 'Early Epipaleolithic' pertaining to a kind of regional 'Azilioid' complex. The elements that differentiate these assemblages from those of the Magdalenian substrate are the scarcity of an industry in bone and antler, and the near-total disappearance of Upper Paleolithic-like portable art. The sites of St. Gregori and Les Mallaetes lend their names to the two recognized EMM facies:

Mallaetes facies (c. 11.5 - 7 kyr BP?): characterized by a significant endscraper index (IG: 33-26%), moderate numbers of burins (IB: 9-4%) and a progressive increase in the microlaminar component. In Mallaetes, this facies is overlain by a Cardial Neolithic, a datum that has not been confirmed in other sequences.

St. Gregori facies (c. 11.5 - 10 kyr BP): with markedly unequal numbers of endscrapers (IG: 60-43%) and burins (IB: 4%), there is an increase over time in the incidence of notches and denticulates, as well as the presence of some geometrics (triangles, segments) which, according to the sequence at El Filador, form a link with a Sauveterroid industry (Fortea 1973, 1985).

The Mediterranean Geometric Epipaleolithic (EGM)

The EGM is also divided into two facies, named after the sites of El Filador and La Cocina:

Filador facies (c. 10 - 8 kyr BP): considered as 'Sauveterroid' (EGM/S in Catalonia and EMM/S in Valencia) because of the presence of geometric elements - triangles, circle segments (Igm: 18%). The geometrics occur associated with a macrolithic component that, in the last phases of the Filador sequence, comes to acquire a 'regressive'character.

Cocina facies (c. 8 - 6 kyr BP?): the 'Tardenoisoid' facies, named after the sequence in the eponymous site (La Cocina) is a typical 'Mesolithic' industries. Of the four phases recognized there (Cocina I-IV), the two latest are have neolithic technoeconomic elements (pottery, ground stone) but without a complete rupture in chipped stone technology (Fortea 1973; Cabanilles 1985, 1990). Its principle typological characteristics are summarized in Table 8.5.

Recent Data

Since 1973, assemblages have been excavated that can only be attributed with difficulty to one of the four facies enumerated above. This situation does not negate the validity of the existing typological structure, but it does open new avenues of debate and discussion. Two central issues characterize the debate in recent years:

(1) the continuity between the MSM and the EMM, and the variability within each; and

(2) the recognition of non-geometric assemblages dated to the 9 - 8th millennia BP.

In respect of the first issue, the basic industrial transformation between the MSM and the EMM is based on an inversion of the principle typological groups, and the loss of bone/antler technology and portable art (Aura & Pérez 1995, Casabó 1995, Aura 2001). These changes in material culture are temporally vectored, become amplified over

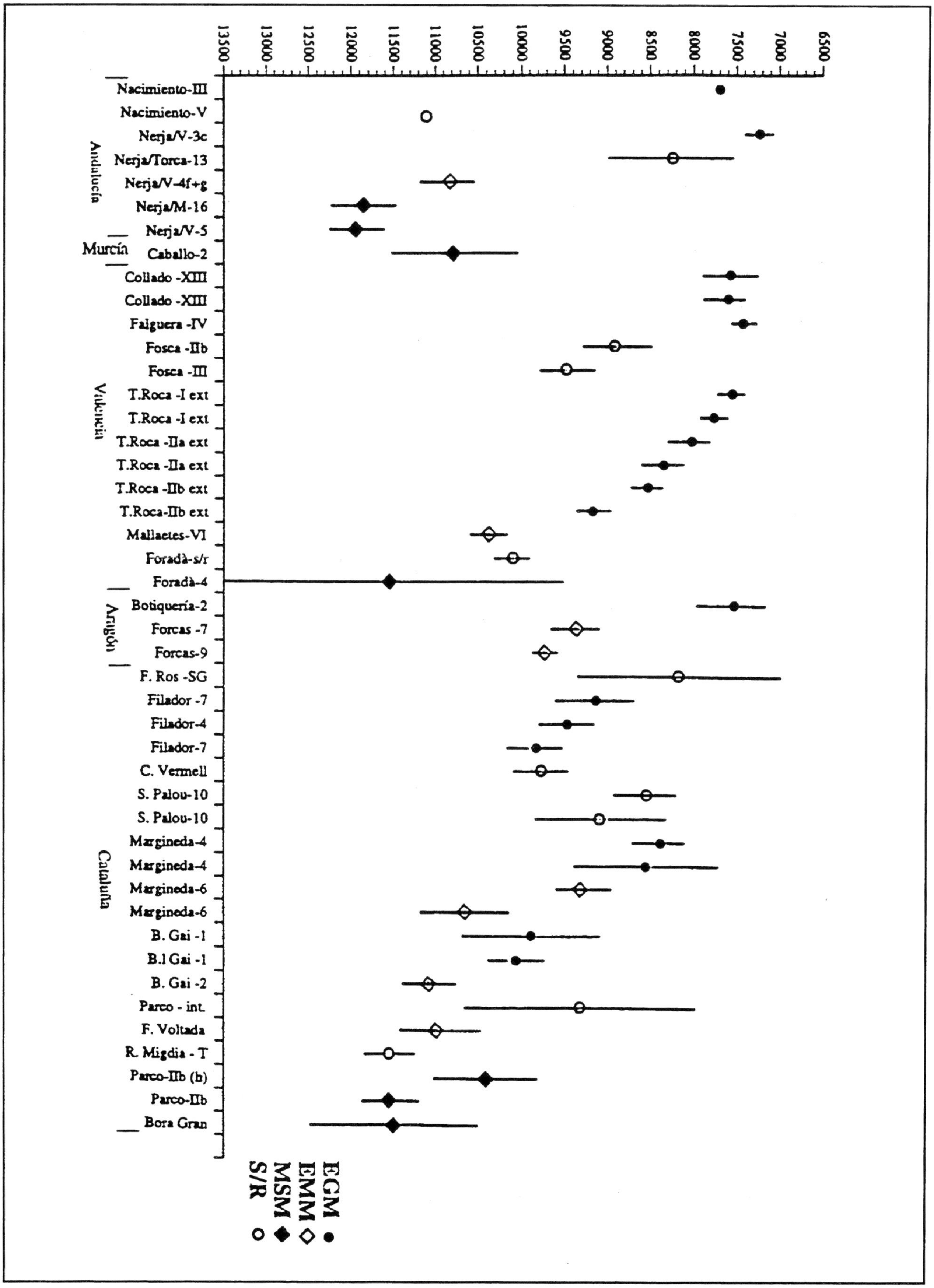

Figure 8.2. Radiometric dates between 12,000 and 7,000 BP (± 2 sd) for Spanish Levante and Catalonia.

Table 8.5. Diachronic evolution of the principal typological groups—assemblages attributed to the MSM and EMM grouped by radiocarbon dates. IG = enscrapers, IB = burins, IL = backed bladelets, IT = truncations, IM-D = geometric microliths.

	IG	IB	IL	IT	IM-D
IXº	29.9	4.7	4.3	2.8	28.4
Xº	30.2	4.3	12.3	5.8	18.9
XIº	25.4	18.0	15.8	6.3	16.5
XIIº					
XIIIº	15.2	18.8	29.6	5.8	13.4
XIVº	14.1	35.8	7.6	1.2	14.1

time, and are especially obvious if we compare MSM assemblages dated to the 13th mil. BP with EMM assemblages dated to the 10th mil. BP (Fig. 8.2). But, across those three millennia, there is significant variability that cannot be explained in sequential terms.

There are two elements that lend a new perspective to the analysis.The first is the regional fragmentation that can be observed in Mediterranean Spain, since the MSM. Although it is a complex endeavor to define territorial facies, one can postulate the existence of three subregions within Mediterranean Spain: (1) the region north of the Ebro Valley (= Catalonia), (2) the central region (= Valencia), and (3) that of the southeast (= eastern Andalucía). Differences in the use of raw materials, in the bone and antler technologies, in the 'style' of stone projectile points, and in the important concentration of paleolithic art sites in Andalucía should be taken into consideration (Aura 1995).

The second is the chronological-cultural versus the functional nature of the EMM facies. Variability in material culture generally has a strong chronological-cultural component. But, in order to explain that component, perhaps we would find our analytical approaches more compelling if we could also take technofunctional variability and site function into account.

These two issues or considerations have generated various equivocal terms to refer to the industries spanning the T-H transition: final MSM (Mediterranean Upper Magdalenian), Epimagdalenian, post-Magdalenian; early, middle and final EMM (Mediterranean Microlaminar Epipaleolithic), even Mesolithic (!). These terms reflect the intrinsic complexity of the interval, in that it is difficult to isolate evolutionarily 'pure' states in what is effectively a continuum. The terms generally carry a chronological, rather than a cultural, loading.

In respect of the second issue, the EMM - EMM/S transition is not evenly recorded throughout Mediterranean Spain. 'Sauveterroid' facies assemblages are dated in Catalonia to c. 10 kyr BP, while south of the Ebro some geometric elements (triangles, segments) are added to the EMM substrate at Tossal II (Cacho et al. 2001) and Cova de Santa Maira (Aura et al. 2000).

EGM industries appear throughout the 8th mil. BP.The later EMM /S and early EGM industries with geometric pieces have large notch and denticulate components (macrolithic components generally) in some cases, sometimes described as 'Campanoid style' (= M-D). They appear at the end of the EMM/S sequence (at Tossal de la Roca and Santa Maira) and at the base of some EGM sequences (at Cocina and Costalena), although they also show up in mesolithic contexts of neolithic age (e.g., Botiquería) (Fortea 1973, Barandiarán & Cava 1989, Barandiarán 1978) (Tables 8.6, 8.7).

A new pattern recognized in recent years complicates the picture still more. Over the past few years (since about 1990), assemblages have been recognized that are made up of notches, denticulates, retouched flakes, and some becs, with scarcely any classic Upper Paleolithic pieces (i.e., endscrapers, burins), nor backed bladelets, nor geometric microliths (= M-D). La Font del Ros, Sota Palou and Cingle Vermell exemplify these assemblages in Catalonia (García Argüelles et al. 1992); Cova de Sta. Maira in Valencia and other sites in the Ebro Basin (Utrilla et al. 1998). All are dated between the 9th and the 8th mil. BP. These apparently more or less contemporaneous 're-gressive' assemblages could indicate different levels of expediency, determined by raw material available, the planned duration of site use and the purpose for which these implements were intended. These trends offer sig-

Table 8.6. Typological Groups from the four phases at Cueva de la Cocina (Fortea 1973). IG = endscrapers, IB = burins, ILF = backed bladelets, IMD = geometric microliths, Pqt = microburins.

	IG	IB	IL	IMD	Pqt
Cocina IV	4,8	0	4,8	7,3	7,3
Cocina III	10,2	0	10,2	10,2	15,3
Cocina II	0	0	2,2	10,8	46,2
Cocina I	4,4	1,4	2,9	29,6	2,2

nificant parallels to those of the Atlantic Façade (e.g., the Asturian) (González Morales 1995).

Settlement Patterns and Characteristics

Two important considerations affect settlement patterns in the region: (1) changes in coastline which, in some cases, inundated lowlying coastal beaches (e.g., the Gulf of Valencia) but which had much reduced effects on more elevated parts of the coastal plain (e.g., Nerja); and (2) the absence of data on open-air sites, due perhaps to erosional processes that resulted in slopewash and denudation, and in valley alluviation (Fumanal & Calvo 1981).

Sites in the interior are found in caves and rockshelters along a strip approximately 50 km from the present coastline. They tend to be concentrated along the contact zone between the coastal plain and the interior valleys, although they also appear in valley headwaters, and in the intervening heavily dissected terrain.

Continuity in settlement location is also evident between the MSM and the EMM. Sites occur from the present coastline up to elevations of 1000 m, in the steep-sided prelittoral interior valleys. Combined use of both environments (coastal plain, interior interfluves and upland plateaux) seems probable and is in fact confirmed ethnographically by numerous studies of transhumance from Greece to Gibraltar (Davidson 1989).

On the other hand, sites that document an EMM/S - M-D - EGM transition are rare. Although they are found in practically the same areas, many caves and rockshelters were occupied or used for the first time during the EGM, with evidence of occupation extending, in some cases, up to the Neolithic. Another element of discontinuity with respect to earlier times is the occurrence of EGM sites along coastal salt-water lagoons and in estuarine brackish water environments, often with genuine shellmiddens (e.g., El Collado). This situation can perhaps be explained by differential preservation caused by changes in coastline.

With respect to settlement pattern changes, the economic reorientation described below is worth keeping in mind. It very likely had consequences for the size of Holocene foraging territories, and might have led to greater territoriality vis à vis conditions during the final Pleistocene.

Subsistence

The major patterns of resource utilization established during the Tardiglacial continued during the early Holocene (Tables 8.2 - 8.4). Along with the specialized hunting of certain ungulates, there is a diversification in the resource base. This diversification - new ungulate species, small mammals, birds, marine resources, plants - and above all the use of new exploitation territories - the *montaña*, estuaries, the sea - dates in Mediterranean Spain at least as far back as the 13th mil. BP.

One of the salient characteristics of forager adaptations in Mediterranean Spain is the exploitation of small game (lagomorphs, avifauna, marine fish) from at least the last glacial maximum (Villaverde, Aura & Barton 1998, Aura et al. 2002b, with references). These resources were quantitatively very important during the late Tardiglacial and the early Holocene (Table 8.2). Riverine fishing is documented from the early Upper Paleolithic (Juan-Muns 1987), although in no case does it seem to be an important subsistence activity. Marine and terrestrial shellfish were utilized as ornaments until the end of the Tardiglacial. Most of the sites have quite specialized faunal spectra, emphasizing red deer (*Cervus elaphus*) and ibex (*Capra pyrenaica*) (Tables 8.2 - 8.4). Specialization seems to be strongly correlated with the immediate site environs. Thus, sites in which ibex are dominant tend to be found in the rocky environments these alpine caprids favour, while those associated with deer tend to occur on more regular terrain (Bailey & Davidson 1984).

The forms that specialized hunting took tend to corroborated by animal mortality profiles (Aura et al. 2002b). Selection for subadult prey is a common element, although certain diachronic tendencies emerge that await future confirmation. Thus, juveniles and subadults are better represented during the MSM and the EMM than in the EGM.

Table 8.7. Geometric component from the four phases at Cueva de la Cocina (Fortea 1973). DB = double bevel technique.

	Triángulos	Triáng./Co	Trapecios	Segmentos	Total Gm	DB
Cocina IV	4,8	0	7,3	9,7	21,9	2,4
Cocina III	15,3	2,5	5,2	20,5	43,5	0
Cocina II	2,8	16	10,2	1,1	30,2	0
Cocina I	0.7	0	34,8	0	35,5	0

This selectivity might be attributed to hunting strategies linked with the growth and maturation cycles of the prey - herd hunting or hunting of isolated individuals - and also to the degree of planning by the hunters, as Clark pointed out for Cantabria a long time ago (1971). However, all the variability among faunal assemblages cannot be accounted for by the orography of site environs. Ecological readjustments and/or diversification in resource use could explain the increase in forest-adapted species: *Sus scropha* (boar), *Capreolus capreolus* (roe deer), and *Rupicapra rupicapra* (chamois). There is a corresponding decrease in the frequency of *Bos primigenius* and *Equus* sp., that confirms tendencies observed after ca. 20 kyr BP (Table 8.2). Vegetal resources (roots, tubers, nuts) were possibly important in low latitudes like this one (Clarke 1976), although few data on plant use are currently available (Table 8.8).

Marine Resources: Cueva de Nerja

Two elements have tended to undervalue the perceived importance of marine resources for hunter-gatherers in the region of Mediterranean Spain:

(1) the misconception of the Mediterranean as a sea 'poor' in resources, due to low primary productivity, the absence of strong tides, high salinity, and a low turnover rate in its continental and oceanic waters. This global view is rather naïve, however, since there are some regions of very high primary productivities vis à vis the mean. This is the case with the Alborán Sea and the southern coast of Andalucía, for example, where there are upwellings of cold Atlantic waters which enrich and refresh both the flora and the fauna (Alcaraz et al. 1985), and (2) the uneven effects of the T-H transgression, and the differential preservation of sites related to the procurement and use of marine resources (Aura et al. 2001).

Exploitation of marine resources is associated in Mediterranean Spain with MSM and EMM assemblages, although there are references to human use of marine resources as far back as the Middle Paleolithic (e.g., Waechter 1964). On the coasts of eastern Andalucía and Murcia, exploitation of marine resources dates back to c. 12.5 kyr BP (Cueva de Nerja, Hoyo de la Mina, Caballo), but for the rest of the region, the dates are much more recent (e.g., the *conchero* at El Collado [Oliva], dated to 7.5 kyr BP). The Cueva de Nerja exemplifies highly specialized marine resource exploitation: ducks and geese, mollusca, fish, equinoderms, crustaceans, tortoises and even seal (*Monachus monachus*) (Pérez Ripoll & Raga 1998). Some ungulates (mainly *Capra pyrenaica*), lagomorphs and plant resources, especially *Pinus pinea* fruits (Badal 1998), can be added to the inventory (Tables 8.2 - 8.4, 8.8).

Counts of fish bones and teeth made as part of a sampling program by Mª J. Rodrigo indicated a ratio of five fishbones for each lagomorph bone in the MSM, and double that (10:1) in the EMM. Along with these quantitative data, there are representations of marine species in the paleolithic art of Andalucía and the fabrication of bone fishing gear - bipointed gorges and harpoons - in the MSM (Aura 1989, Aura et al. 2001).

During the T-H transition, a *Mytilus edulis* (mussel) shell-midden formed in one of the chambers of the Cueva de Nerja; it also contained many fish bones. Dated 10,800 and 10,000 BP, it also produced an assemblage of straight bone fish hooks (gorges), but no harpoons. The lithic assemblage of this unit (N/V4) yielded both retouched pieces and crude, heavy-duty blanks, with the latter numerically dominant (Straus 1979, Clark 1989). At the base of the midden the lithic artifacts showed some typological resemblances to those of the EMM, although with a new element: the presence of worked, and sometimes painted, cobbles. Above the middle levels in the midden, some triangular microliths were recovered and, near the top, in contact with levels containing ceramics, some trapezes and microburins. Use of marine resources continued throughout the Neolithic, and there is a reduction in the average size of fish vertebrae, perhaps indicating overexploitation. In addition to fishing, there is good evidence for ovicaprine and cereal domestication economies, and the changes in

Table 8.8. Use of vegetal resources in Mediterranean Spain between 12,000-7,000 BP (Buxó 1990, Badal 1991).

	Cueva de Nerja	Tossal	Cingle Vermell	Roc del Migdia	Sota Palou	Balma del Gai
Pinus pinea	+		+			
Quercus sp.	+	+	+	+		
Corylus avellana			+	+	+	
Juglans regia			+			
Prunus spinosa						+

butchering practices and in the ages at death that are typical of neolithic societies (Pérez Ripoll n.d.).

The Cueva de Nerja is not an isolated site; in its immediate environs there are a number of penecontemporaneous sites that contain similar sequences and paleofaunal records. Neither is Nerja exceptional in the broader context of the Epipaleolithic-Neolithic transition in the Mediterranean Basin. From Greece to Gibraltar there are well documented sites that are specialized to a greater or lesser degree in the exploitation of littoral resources: Franchthi, Grotta dell'Uzzo, and Châteauneuf-les-Martigues (Payne 1975, Durante 1980, Desse 1987). Despite the overall 'impoverishment' of the Mediterranean Sea, these sites are all found in areas of high ecological productivity. In all these cases, exploitation of littoral resources has been viewed as a preadaptation that facilitated a rapid assimilation of neolithic technoeconomic elements (Jacobsen 1981).

Funerary Ritual

The few interments known in Mediterranean Spain between 12-7 kyr BP indicate an evolution of funerary practices similar to that described for neighboring areas. There is only single case of individual burial: Nerja/T-13 (García Sánchez 1982). This burial had little in the way of grave goods and consisted of a simple pit delimited by stones, with the skeleton lying on its back. The description of the context of discovery and associated materials is not sufficiently detailed to warrant assignment to the EMM (9th mil. BP).

In contrast to this individual burial, a multiple burial comprising 14 inhumations was reported (for the first time in Mediterranean Spain) from a shell-midden at El Collado, dated to the 8th millennium BP (Aparicio 1990). Most of the interments were in simple pits, with the bodies buried in flexed positions. The only mention of funerary offerings is the presence of shell associated with the heads and feet of the deceased (buried, after all, in a shell midden . .). Trace element analyses done on the skeletons indicate a diet rich in marine resources, and to a lesser extent, plant foods (Chimenos et al. 1992).

The change from individual to collective interment can be considered as the final consequence of the economic reorientation that came about during the T-H transition, directed as it was toward the more intensive procurement of marine resources. Within this dynamic, there was likely a need for greater social and territorial cohesion. Perhaps the shift from individual to collective burial can be understood as a prelude to the pattern of collective interment in necropoli characteristic of the Neolithic.

The Neolithization

The neolithization in Mediterranean Spain not take place in a uniform fashion. In the interior this process lasted until ca. 6,000 BP, while on the coastal plain and in coastal river valleys agropastoral systems were in place earlier (i.e., by c. 6,800 BP). This situation can be explained by the unequal levels of Holocene forager development in the area, and their differential receptivity, mediated by cultural options, to contact with neolithic groups with Cardial Ware. According to the dual model proposed recently by Bernabeu et al. (1993), this interaction can be analyzed by means of the study of the different components of cultural systems (technology, economy and settlement), thus differentiating the conceptual framework of general explanation from what might be isolated cases which contributed relatively little to our understanding of the process of Neolithic expansion.

The lithic industries mark a rupture between the EGM tradition and the Cardial Neolithic, with only geometric pieces a common element between them. Comparison of formal and technical characteristics reveals important differences and scant evidence of assimilation of the EGM tradition by Cardial Neolithic agropastoralists. Such assimilation is indicated only by the adoption by Cardial peoples of segments and the double-bevel technique (Cabanilles 1985, 1990).

Finally, the presence of ceramics and ground stone in levels that are chronologically neolithic (but with no evidence of agropastoral economies) at Cocina, does not distance these levels completely from their techno-typological traditions, nor from pre-existing patterns in resource use (Fortea 1973). Variable subregional trajectories were adopted, recognizable so far only in the composition of the geometric microlithic armatures and in their techniques of manufacture.

Postscript

This manuscript maintains the structure and content presented at Santander Conference in 1994. In spite of the time that has passed since this conference, we have preferred to leave the text essentially unchanged while adding additional recent bibliographic citations that reflect current research directions.

Acknowledgements

We want to express our thanks to the volume editors, Geoffrey A. Clark and Manuel R. González Morales, who invited us to participate in the International Conference on The Mesolithic of the Atlantic Façade. Although the region discussed in this manuscript could be considered peripheral to the focus of the conference, the processes described here offer significant parallels to those of the Atlantic Façade. We also want to thank the editors for their persistence and efforts to bring the conference proceedings to publication.

References

Alcáraz, M., M. Estrada, J. Flos, J. Font, J. Romero & J. Salat
1985 *L'Oceanografia I. Introducció a l'Ecologia Marina Mediterrània.* Barcelona : Quaderns d'Ecologia Aplicada No. 8.

Aparicio, J.
1990 *II Seminari sobre el Mediterrani – El Mare Nostrum.* Valencia: Academia de Cultura Valenciana Serie Histórica No. 5: 7-91.

Aura, J.E.
1989 *A Preliminary Report on Marine Resources Exploitation on the Andalucian Coast: the Gorges from the Cave of Nerja (Málaga, Spain).* Fifth Meeting of the ICAZ Fish Remains Working Group, Stockholm.
1995 *El Magdaleniense Mediterráneo: la Cova del Parpalló (Gandía, Valencia).* Valencia: Servicio de Investigaciones Prehistóricas No. 91.
2000 Les coves de Santa Maira (Castell de Castells, la Marina alta- Alacant): primeros datos arqueológicos y cronológicos. *Recerques del Museu d'Alcoi* 9: 75-84.
2001 Cazadores emboscados. El Epipaleolítico en el País Valenciano. In *De Neandertales a Cromañones. El Inicio del Poblamiento Humano en Tierras Valencianas* (V. Villaverde, ed.), pp. 219-238. València: Universitat de València.

Aura, J.E., E. Badal & J. Bernabeu
1992 *Al Oeste del Eden.* Madrid: Síntesis.

Aura, J.E. & M. Pérez Ripoll
1995 El Holoceno inicial en el Mediterráneo español (11,000-7,000 BP). Características culturales y económicas. In *Los Últimos Cazadores,* (V. Villaverde Bonilla, ed.), pp 119-146. Alicante: Instituto de Cultura Juan Gil-Albert.

Aura, J.E, V. Villaverde, M. González Morales, C. Gónzalez Sainz, J. Zilhão & L.G. Straus
1998 The Pleistocene-Holocene transition in the Iberian Peninsula: Continuity and Change in Human Adaptations. In *As the World Warmed: Human Adaptations Across the Pleistocene/Holocene Boundary,* L.G. Straus & B. Eriksen (eds.). *Quaternary International* 49-50: 87-103.

Aura, J.E., J.F. Jordá Pardo, M. Pérez & M.J. Rodrigo
2001 Sobre dunas playas y calas. Los pescadores prehistóricos de la Cueva de Nerja (Málaga). *Archivo de Prehistoria Levantina,* 24: 9-39.

Aura, J.E., J. Jordá Pardo, M. Pérez Ripoll, M.J. Rodrigo, E. Badal García & P. Guillem Calatayud
2002 The Far South: the Pleistocene-Holocene transition in the Nerja Cave (Andalucía, Spain). *Quaternary International* 93/94: 19-30.

Aura, J.E., V. Villaverde, M. Pérez Ripoll, R. Martínez & P. Guillem
2002 Big Game and Small Prey: Paleolithic and Epipaleolithic Economy from Valencia (Spain). *Journal of Archaeological Method and Theory,* 9: 209-262.

Badal, E.
1991 *Aportaciones de la Antrocología al Estudio el Paisaje Vegetal y su Evolución en el Cuaternario Reciente en la Costa Mediterránea del País Valenciano y Andalucia (18.000-3.000 BP).* Valencia: Tésis Doctoral, Universitat València.
1998 El interés económico del pino piñonero para los habitantes de la Cueva de Nerja. In *Las Culturas del Pleistoceno Superior en Andalucía* (J.L. Sanchidrián & M.D. Simón, eds.*),* pp. 287-300. Málaga: Patronato de la Cueva de Nerja.

Barandiarán, I.
1978 El Abrigo de Botiqueria del Moros. Mazaleón (Teruel). Excavaciones Arqueológicas 1974 *Cuadernos de Prehistoria y Arqueología Castrellonense* 5: 49-138.

Barandiarán, I. & A. Cava
1989 *La Ocupación Prehistórica del Abrigo de Costaleña (Maella, Zaragoza).* Zaragoza.
1992 Caracteres industriales del Epipaleolítico y Neolítico en Aragón: su referencia a los yacimientos levantinos. In *Aragón/Litoral Mediterráneo: intercambios culturales durante la Prehistoria (*P. Utrilla, ed.), pp.181-211. Zaragoza: Instituto Fernando el Católico.

Barton, C.M., F. Rubio, C.A.Miksicek & D.J. Donahue
1990 Domestic olive. *Nature* 346: 518-519.

Buxó, R.
1990 *Metodología y técnicas para la recuperación de restos vegetales (en especial referencia a semillas y frutos) en yacimientos arqueológicos.* Girona: Cahier Noir No. 5.

Cacho, C.
1986 Nuevos datos sobre la transición del Magdaleniense al Epipaleolítico en el Pais Valenciano: el Tossal de la Roca. *Boletín del Museo Arqueológico Nacional* 4: 117-129.

Cacho, C. (editor)
1995 El Tossal de la Roca (Vall d'Alcalà, Alicante). Reconstrucción paleoambiental y cultural de la transición del Tardiglacial al Holoceno inicial. *Recerques del Museu d'Alcoi* 4: 11-101.

Casabó, J. & M.L. Rovira
1987-88 El Paleolítico Superior y Epipaleolítico Microlaminar en Castellón. Estado actual de la cuestión. *Saguntum* 21: 47-107.

Casabó, J.
1995 *Las Sociedades Depredadoras entre el Final del Pleistoceno Superior y el Holoceno. Un Ejemplo: la Comunidad Valenciana.* València: Tésis doctoral, Universitat de València.

Clark, G.A.
1971 The Asturian of Cantabria: subsistence base and the evidence for post-Pleistocene climatic shifts. *American Anthropologist* 73: 1241-1257.
1989 Site functional complementarity in the Mesolithic of northern Spain. In *The Mesolithic in Europe* (C. Bonsall, ed.), pp. 589-603. Edinburgh: John Donald.

Clarke, D.
1976 Mesolithic Europe: the economic basis. In *Problems in Economic and Social Archaeology*, G. Sieveking (J.K. Longsworth & K.E. Wilson, eds.), pp. 449-481. London: Duckworth.

García-Argüelles, P., P. Adserias, R. Bartrolí, M. Bergadá, A. Cebrià, R. Doce, J.M. Fullola, J. Nadal. G. Ribé, T. Todón & R. Viñas.
1992 Síntesis de los primeros resultados del programa sobre el Epipaleolítico en Cataluña central y meridional. In *Aragón/Litoral Mediterráneo: intercambios culturales durante la Prehistoria* (P. Utrilla, ed.), pp.269-284. Zaragoza: Instituto Fernando el Católico.

García Puchol. O.
2002 *Tecnología y Tipología de la Piedra Tallada durante el Proceso de Neolitización.* Valencia: Tésis Doctoral, Universitat València.

García Sánchez, M.
1982 El esqueleto epipaleolítico de la Cueva de Nerja (Málaga). *Cuadernos de Prehistoria de la Universidad de Granada* 7: 37-71.

González Morales, M. R.
1995 La transición al Holoceno en la region cantábrica: el contraste con el modelo del Mediterráneo español. In *Los Últimos Cazadores* (V. Villaverde Bonilla, ed.), pp 63-78. Alicante: Instituto de Cultura Juan Gil-Albert.

Chimenos, E., A. Malgosa & M.E. Subirá
1992 Paleopatología oral y análisis de elementos traza en el estudio de la dieta de la población epipaleolítica de El Collado (Oliva, Valencia). *Actas del I Congreso de Paleopatología*, pp. 177-182. *Munibe,* Suplemento No. 8. San Sebastian: Sociedad de Ciencias Naturales 'Aranzadi'.

Davidson, I.
1989 *La Economía del Final del Paleolítico en la España Oriental.* Valencia: Servicio de Investigaciones Prehistóricas No. 85.

Desse, J.
1987 La pêche: son rol dans l'economie des premières sociètès néolithiques en Méditerraneé occidentale. In *Premières Communautes Paysannes en Méditerraneé Occidentale* (J. Guillaine, J. Courtin, L. Roudil & J.L. Vernet, eds.), pp. 281-285. Paris: CNRS.

Dupré, M.
1988 *Palinología y Paleoambiente. Nuevos datos Españoles.* València: Servicio de Investigaciones Prehistóricas No. 84.

Durante, S.
1980 Nota preliminare sulla ittiofauna e sullo sfruttamento delle risorse marino. *Sicilia Archeologica* 42: 65-66.

Eastham, A.
1986 The birds of the Cueva de Nerja. In *La Prehistoria de la Cueva de Nerja (Málaga).* J. Jordá Pardo (ed.), pp. 109-131. Málaga: Patronato de la Cueva de Nerja.

Fortea, F. J.
1973 *Los Complejos Microlaminares y Geométricos del Epipaleolítico Mediterráneo Español.* Salamanca: Universidad de Salamanca.

Fortea, F. J. & Jordá, F.
1976 La Cueva de les Malletes y los problemas del Paleolítico Superior del Mediterráneo español. *Zephyrus* 26-27: 129-166.

Fumanal, Mª. P.
1986 *Sedimentología y Clima en el País Valenciano. Las Cuevas Habitadas en el Cuaternario Reciente.* Valencia: Servicio de Investigaciones Prehistóricas No. 83.

Fumanal, Mª. P. & A. Calvo
1981 Estudio de la tasa de retroceso de una vertiente mediterránea en los últimos 5,000 años. *Cuadernos de Geografía* 29:133-150.

Jacobsen, T. W.
1981 Franchthi cave and the beginning of settled village life in Grece. *Hesperia* 50: 303-319.

Jordá Pardo, J. F.
1992 *Neógeno y Cuaternario del Extremo Oriental de la Costa de Málaga.* Tésis Doctoral, Universidad de Salamanca, Salamanca.

Jordá Pardo, J.F., J.E. Aura & F. Jordá Cerdá
1990 El límite Pleistoceno -Holoceno en el yacimiento de la Cueva de Nerja (Málaga). *Geogaceta* 8: 102-104.

Juan Cabanilles, J.
1985 El Complejo Epipaleolítico Geométrico (Facies Cocina) y sus relaciones con el Neolítico Antiguo. *Saguntum* 19: 9-30.
1990 Substrat epipaléolithique et neolithisation en Espagne: apports des industries lithiques a l'identification des traditions culturelles. In *Rubané et Cardial* (D. Cahen & M. Otte, eds.), pp. 417-435. Liège: ERAUL No. 39.

Martínez Andreu, M.
1989 *El Magdaleniense Superior en la Costa de Murcia*. Murcia.

Olaria, C.
1988 *Cova Fosca*. Castellón: Monografies de Prehistòria i Arqueologia Castellonenques No. 1.
1999 *Cova Matutano (Vilafamés, Castellón). Un modelo ocupacional del magdaleniense superior final en la vetiente mediterránea peninsular*. Castellón: Monografies de Prehistòria i Arqueologia Castellonenques No. 5.

Payne, S.
1975 Faunal change at Franchthi Cave from 20,000 B.C. to 3,000 B.C. In *Archaeozoological Studies* (A.T. Clason, ed.), pp. 120-131. Amsterdam: North Holland Publishing.

Pérez Ripoll, M.
1992 *Marcas de Carnicería, Fracturas Intencionadas y Mordeduras de Carnívoros en Huesos Prehistóricos del Mediterráneo Español*. Alicante: Instituto de Cultura Juan Gil Albert.

Pérez Ripoll, M.
n.d. *Estudio Zooraqueológico de la Cueva de Nerja: La Sala del Vestíbulo*. Manuscrito inédito. 109 pp. and 58 figs.

Pérez Ripoll, M. & J. A. Raga
1998 Los mamíferos marinos en la vida y en el arte de la Prehistoria de la Cueva de Nerja. In *Las Culturas del Pleistoceno Superior en Andalucía* (J.L. Sanchidrián & M.D. Simón, eds.), pp. 251-275. Málaga: Patronato de la Cueva de Nerja.

Pujol, C. & C. Vergnaud
1989 Palaeoceanography of the last Mediterranean deglaciation in the Alborán Sea (western Mediterranean). Stable isotopes and planktonic foraminiferal records. *Marine Micropaleontology* 15: 153-179.

Rodrigo, M.J.
1991 *Remains of* Melanogrammus aeglefinus *(Linnaeus, 1758) in the Pleistocene-Holocene passage of the Cave of Nerja (Málaga, Spain)*. Kiel: Schriften aus der Archäologische-Zoölogische Arbeitsgrüppe Schleswig No. 43.

Rodriguez, J.M.
1979 La Cueva del Nacimiento (Pontones, Jaén). *Saguntum* 14: 33-38.

Straus, L.G.
1987 Upper Palaeolithic ibex hunting in southwest Europe. *Journal of Archaeological Science* 14: 163-178.

Utrilla, P., A. Cava, A. Alday, V. Baldellou, I. Barandiarán, C. Mazo & L. Montes
1998 Le passage du Mesolithique au Neolithique ancien dans le Bassin de l'Ebre (Espagne) d'apres les datations C14. *Préhistoire Européenne* 12: 171-194.

Villaverde, V., J.E. Aura & C.M. Barton
1998 The Upper Paleolithic in Mediterranean Spain: A Review of Current Evidence. *Journal of World Prehistory* 12: 121-197.

Waechter, J. d'A.
1964 The excavation of Gorham's Cave, Gibraltar (1951-1954). *Bulletin of Institute of Archaeology* 4: 189-222.

9

The Mesolithic on the Northwest Coast of the Iberian Peninsula: a Current Perspective

J. M. Vázquez Varela
University of Santiago de Compostela

Abstract

After a brief characterization of its biology and hydrology which underscores the extraordinary subsistence potential of the Galician coasts and estuaries, the current status of the Mesolithic of Galicia and of northern Portugal is reviewed. Modern-era investigations show that many of the lithic assemblages reported by earlier workers are in fact palimpsests of artifacts that accumulated along fossil and modern beaches over a very long period of time. The historically problematic relationship between the Mesolithic Asturian of Cantabria, identified by its characteristic unifacial picks, and the Galician and Portuguese 'Asturian', which has produced similar objects, is evaluated. While there are traces of Asturian-like coastal adaptations that likely date to the Preboreal and Boreal periods, the scarcity of well-excavated sites, the absence of clear stratigraphies, relatively few radiometric dates, and only preliminary understanding of the geomorphological and paleoecological frameworks all combine to make it difficult to arrive at an adequate understanding of early Holocene forager adaptations in the region.

Introduction

The autonomous community of Galicia, in the northwest of Spain, has a surface area of 29,240 square kilometres and 1,300 kilometres of coastline, from the estuary of the River Eo on the boundary with Asturias on the north coast, to the mouth of the Miño, forming the frontier with Portugal in the southwest. This shore faces both the Cantabrian Sea in the north, and the Atlantic Ocean in the west, and has open sections with little relief, alternating with more sheltered stretches with long, shallow estuaries extending far inland. Its morphology includes areas of abrupt cliffs and also sheltered beaches, which are sometimes very long.

For the above reasons, and other biological and hydrological characteristics (e.g., nutrient-bearing currents), most of the Galician coast is extremely rich in marine resources, and these have given rise to fishing, shellfish gathering and seaweed harvesting. The oldest evidence for the first two activities consists of the shell midden on the small island of Guidoiro Areoso in the Arousa estuary in the province of Pontevedra, dated to 2070 ± 40 BC (Rey 1995). Although this date is the oldest known, the presence along the coast of concentrations of Acheulean handaxes, as at Louselas near Ribadeo, or isolated finds of bifaces of this period, often associated with fossil beaches, all along both the Cantabrian coast and the Rías Altas and Baixas of the Atlantic, suggest the possibility that marine resources could have been exploited from remote antiquity (Vázquez 1998, Vázquez & Rodríguez 2001).

Some flint tools collected on the beaches, which might be of Upper or Middle Paleolithic Age, should be considered with great care, as place names and geological, ethnographical and historical evidence all indicate that at least some of them come from the ballast of medieval ships returning from England, or from ports along the Cantabrian coast. This ballast was loaded on the return journey when carrying a lighter cargo than the one loaded for the outward voyage (Vázquez 1996).

Modern archaeological data assigned to the Mesolithic show great discontinuity in their geographical distributions and are much rarer than the evidence of earlier or later prehistoric periods on the Galician coast. The most abundant and accurate information, apart from a few isolated finds, comes from sites in the southernmost part of the province of Pontevedra. The site of O Reiro, on the coast of La Coruña, has also been assigned to the Mesolithic, but this date might not be reliable (Vázquez 1988b, see below) (Fig. 9.1).

The scarcity and geographical discontinuity of Mesolithic sites could be explained by the important rise in sea level during the Holocene (Blanco et al. 2002) and also by the geological and geomorphological character of the coast, which has very little limestone and no caves, in some contrast with the situation along parts of the Cantabrian Coast, where numerous caves and rockshelters have undoubtedly helped to preserve Mesolithic remains.

The Holocene sea level rise, while not fully understood in detail, submerged Chalcolithic sites, dated to ap-

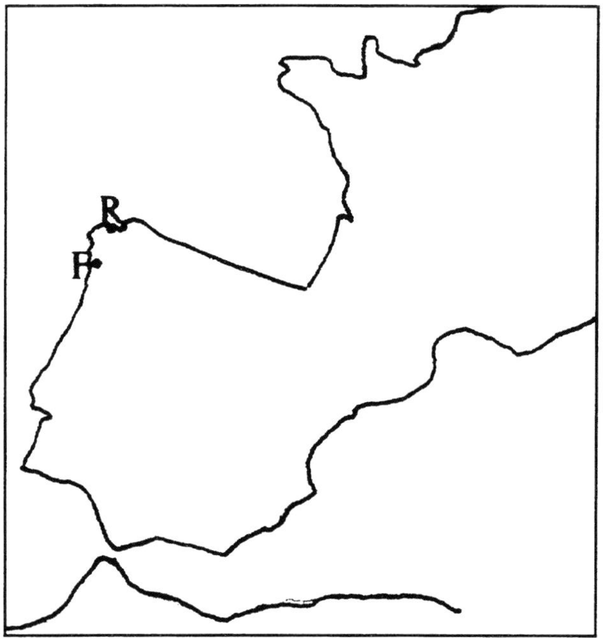

Figure 9.1. Location of the sites of Riero (R) and Fiales (F) on the Atlantic coast, northwest Iberia.

proximately 2000 BC, and some Roman remains from the first centuries AD (Martínez & Vázquez 2002). This could explain the loss of open-air Mesolithic sites located a short distance from the shore.

ARCHAEOLOGICAL DATA

The Site of O Reiro (La Coruña)

The open-air site of O Reiro, which has been attributed to several different cultural periods, was located in a sheltered area between rocks, at the western end of Barrañan beach, to the west of the city of La Coruña. It is now, however, mostly destroyed by recent construction (Vázquez 1988). The black sediment, rich in organic material, overlies the bedrock, and was covered by a dune, in the process of being eroded by the sea at the time of its discovery. During exploratory fieldwork, bones of land mammals and fish were found, as well as a few examples of a non-descript lithic assemblage classified by Ramil (1973) as Mesolithic. Later work provided remains of fish, deer and boar, together with a small collection of artifacts comprising a pebble worked simply on a single face, a backed blade, a notch, and a few doubtful burins, blades, flakes and cores, manufactured in quartz, quartzite and rock crystal (Fernández 2000). Some of these tool types persisted for millennia in the northwest of the Iberian Peninsula. For example, pebbles worked in this way are found in Romano-Galician sites in the first centuries AD. Hence, it is not possible to establish chronological or cultural attributions with such generalized data. The site also had poor quality, undecorated pottery, which cannot be easily assigned to any particular prehistoric period. Pollen analysis of the sediment, dated by ^{14}C to 6590 ± 70 BP (CSIC 580), showed the presence of cereal pollen (Vázquez 1988).

The date supports the original hypothesis that this was a Mesolithic site, as it is contemporaneous with the late phase of the Mesolithic of inland Galicia, and much older than even the oldest megalithic monuments in the same region, at present the only archaeological evidence clearly pertaining to the Neolithic. However, the presence of pottery and cereal pollen show that this could be a site where specialized activities were carried out, at least partially related to the exploitation of marine resources, by a population familiar with pottery and agriculture. As noted above, the lithic elements are not definitive, as similar types are found in Chalcolithic and Bronze Age sites, and cannot be used to confirm the date. On the other hand, inland Mesolithic sites of the same time have a different lithic assemblage (Vázquez 2003).

The disagreement between the date, pottery sherds and cereal pollen becomes even clearer when one considers that the oldest dates for agriculture in the rest of the region, following the pollen record, belong to the fourth millennium BC, in uncalibrated ^{14}C dates (Vázquez 2003). The final assessment of this site should therefore await publication of the full range of studies on the site and its surroundings.

The Southwest Coast of Galicia

The southwest coast of Galicia, with a length of about 30 kilometres, runs from the Silleiro headland, on the south bank at the mouth of the Vigo estuary, to the mouth of the River Miño, marking the frontier with Portugal. The coastline is straight, with very few bays, and has alternating cliffs and beaches of mostly quartzite pebbles and cobbles.

Lithic artifacts have been collected from these beaches since the early 1920s. They have been classified in various chronological and cultural periods, ranging from the Lower Paleolithic to the Romano-Galician culture in the first two centuries AD (Cano 1991).

The interpretation which is most widely supported, and which provoked the most discussion, is that all these lithic finds pertain to the Mesolithic Asturian culture. This idea originated with the Conde de la Vega del Sella's study on the Asturian of the Cantabrian coast (1923), and is based on the presence of numerous picks which are identical in shape, size, raw material, technique and location to the classic picks found in coastal Asturian sites. The great abundance of these objects led some authors to consider that the Asturian began in Galicia and spread from there along the Cantabrian coast.

Since the 1920s, a large number of papers have reflected the research based on speculations about the lithic artefacts, which have been given different ages. A large group of researchers have supported the existence of an Asturian culture in the region, due to the presence of numerous picks identical in every way to those from Asturias: raw material, knapping technique, type, size and coastal distribution. Other tool types have sometimes been grouped in the Asturian for the sole reason that they were found in the same open air sites as the above-mentioned picks, or in shell-middens assigned to the Mesolithic simply because of the presence of tools worked from quartzite pebbles, with the result that the reasoning became circular.

As a reaction to the oversimplification and questionable logic, other trends have been developed to explain the attribution of these artefact types, assigning them to a specific chronological or cultural era. Interpretation has occasionally gone to the opposite extreme and claims have been made that all of the picks belong to a much later time (i.e., the Christian era) and/or to the Castreña culture represented by the sites of Santa Trega (Tecla) and A Forca. A critical review of research between 1923 and 1979, the design of a new analytical strategy, renewed fieldwork, and the excavation of coastal sites carried out since 1984 by the Department of History of the University of Santiago have contributed to make radical changes in the interpretation of the finds.

To summarize, it is clear that the assemblages found in coastal sites are the result of the accumulation of objects from different intervals in the past, from the Acheulian to recent times (e.g., the past 50 years, when sailors worked stone weights from quartzite beach pebbles, sometimes using prehistoric cores or hammerstones). The disturbance of old archaeological levels by the sea, agriculture, and the different uses of the beaches during prehistory and history are the causes of the formation of this polygenetic assemblage (Cano & Vázquez 1986, Cano 1991; see also Clark 1976).

By comparing the open air finds with those from excavated sites such as Fiales, Portecelo from the late Bronze Age, the pre-Roman fort of A Forca, or the Romano-Galician fort of Santa Trega, it can be concluded that these sites represent at least the following culture/stratigraphic or historical units:

Acheulean
Fiales
Portecelo
Pre-Roman Castreño
Romano-Galician
Historical

The Question of the Asturian Picks

Objects identical to Asturian picks, once very common on this coast (in fact, on some beaches, more were collected than in all of Cantabria) are only found in situ in the open air coastal site of Fiales. During the excavation (Cano 1990, 1991), 46 worked items were collected, with the following distribution:

Core artifacts:

Pebble worked unifacially with extraction	1
Pebble worked unifacially simply	4
Pebble worked unifacially extensively	1
Type 1 (Asturian) Pick	1
Pebble worked bifacially simply	1

Flake tools:

Sidescraper	4
Denticulate	3
Bec	1
Endscraper	1
Cleaver	1

Flakes:

Retouched	2
First Order	4
Second Order	21
Third Order	1

Despite having a broken point, the Type 1 pick is comparable with classic Asturian examples, and is the only one found in situ in a definite geochronological context (Fig. 9.2).

Although no absolute date for the organic material at the site was available at the time of the excavations, by comparing its pollen data, sedimentology and geomorphology with sediments on the same coast which had been dated by ^{14}C, the archaeological level at Fiales was placed at between 13,600 BP and 9,800 BP, with a certain degree of probability (Cano 1991). This proposed date coincides with that of assemblages containing this kind of pick in the north of Portugal, according to the geomorphological and assemblage series proposed by J. Meireles (1994, 1996).

A recent and detailed analysis of deposits with organic material in the region, as part of a program studying the climatic and biogeographic dynamics of littoral and sublittoral Galicia during the Upper Pleistocene-Holocene, has revised many of the dates and climatic assignments made so far, including those used for the correlation of Fiales, and this means that the reasoning used to establish the chronology of the site is no longer valid. Recent ^{14}C dates allow us to place the lithic industry at a time immediately preceding 4,760 ± 80 BP. Its presence in low-energy sediments indicates little disturbance or transport of the assemblage (Cano et al. 1997, 1999).

The fact that the Type 1 (Asturian) pick has been found only at this site, despite the low number of tools recovered and the small area dug, whereas they have not been found in later sites (e.g., Portocelo (initial Late Bronze), A Forca (late pre-Roman Castreño), Santa Trega (Romano-Galician) suggests that at least some of the surface finds of this type belong to the same period as Fiales.

There have been many references to the existence of Mesolithic shell middens on this stretch of the coast. This information, which has been reiterated in secondary sources up to modern times (e.g., Clarke 1976) is mistaken, as the middens known at present pertain exclusively to the Castreño culture (late Iron Age), or the Romano-Galician culture of the first and second centuries AD, as shown by diagnostic pottery and metal objects. The confusion springs from early research in the shell-middens, when the fact that they contained tools made from pebbles was enough for them to be included in the culture and chronology of the Asturian. By reviewing published and conserved material, and by examining the middens, it was possible to establish that they contained no Asturian picks, and indeed the lithic types found there are similar to the ones known from hillforts in the region, whether pre-Roman or Romano-Galician. The archaeological material of the shell-middens confirms their use as rubbish dumps with various kinds of remains, including food debris, derived from use of marine resources (Vázquez 1998, Vázquez & Rodríguez 2001).

The North Portuguese Coast

The coast of the Portuguese Minho has geographical, historical and cultural similarities with the southwest coast of Pontevedra mentioned above. In the same way, several of the pebble beaches on the coast between the mouth of the Miño, and the coastal town of Viana de Castelo are known to have lithic assemblages made from worked pebbles with identical characteristics to those from the Galician shore.

Researchers on each side of the Miño considered that the sites formed part of a whole and the history of research on each bank is very similar, which is hardly surprising if, besides the ecological and archaeological similarities, we take into account that the researchers knew the work being carried out in the neighbouring country and often maintained personal and scientific contacts. They have therefore applied the same conceptual frameworks, theories and methods in similar ways in order to interpret the assemblages with more or less the same explanations.

In the late 1970s, J. Maury published an extensive study in which he established a modern typology following contemporary European research, and described the assemblages of the northern Portuguese coast in their geological context (Maury 1977). He proposed that most of the tools found on the beaches formed a homogeneous group, which he classified as Asturian, and which would be of the same approximate age as Asturian assemblages on the Cantabrian Coast. Once more the lack of a geological study in sufficient depth, and the absence of absolute dates for objects found in situ with the guarantee that their analysis would be significant, leave doubts about the true age and affiliation of these assemblages, which can only be hypothetical without the grounding of solid facts.

The doctoral thesis of J. Meireles (1994) is a more ambitious, profound and conscientious attempt to give the Minho coast assemblages a modern typology in their geological context. In this study the author shows how these assemblages, despite temporal variations, always maintain a high percentage of unifacial pebbles, and suggests that they belong to (1) an Acheulean corresponding to the penultimate interglacial, (2) an Acheulean corresponding to the last glacial period, (3) assemblages of particular facies, (possibly Acheulian), and corresponding to the last interglacial, (4) assemblages of unifacial pebbles corresponding to the Würm I and II, and (5) extending to the climatic deterioration of the last glacial cycle, where they are assigned to the 'Ancorian'. The Ancorian represents the continuity of the monofacial pebble assemblage tradition which persists throughout the tardiglacial, and into the postglacial, extending from the Dryas III until the near present.Recently, Brun (1996) has pointed out that it is difficult to establish a precise geological sequence for the evolution of the Miño coast during the Pleistocene and Holocene, as can be seen by the different interpretations made by Granja (1990), Araujo (1991), Meireles (1994) and others. These have clear discrepancies in important aspects, which means that important chronological issues have yet to be resolved. Analytical difficulties are exacerbated by the absence of 'index fossil' tool types, the lack of absolute dates, and the absence a solid geomorphological sequence into which the assemblage series can be fitted. Meireles (1994, 1996) places the Asturian picks in the penultimate phase of his series, together with the different types mentioned above as belonging to this culture in the Miño region.

From the archaeological point of view, it is pretty clear that excavations are needed to locate artifacts in situ, and to establish relations between the different types, and between them and their paleoenvironmental contexts. The construction of evolutionary series with few in situ objects and without absolute dates is not solid enough to be accepted as definitive without due caution. It is also true that unifacially worked artifacts made from pebbles are found in several hillforts near the Minho shore, such as O Coto da Pena, near Caminha.

Assemblages of this type have clear parallels with hillforts on the Galician side of the Miño, ranging from early in the Castreño culture, in the first half of the first millennium BC, to the second century AD within a fully Roman context. Until recent historic times, pebbles were knapped with simple techniques in northern Portugal and in Galicia to make core tools for fishing, especially weights (Cano & Vázquez 1996). The interpretation of the assemblage sequences of material collected in the open air on beaches must therefore take into account the information from archaeological sites and ethnographical data, as has been done to the north of the Miño.

From all this we can conclude that, although we have no certain evidence of the existence of an archaeologically distinctive coastal Mesolithic, and despite the traditional view based on the presence of 'Asturian picks' (sometimes taken as an indicator of the Mesolithic), there is reasonable evidence for an Asturian-like culture with dates which are still indeterminate. The information from Fiales, its position in the sequence of the coast north of the Miño, and the partial agreement of this sequence with that from the south of the river, allow us to infer the existence of a similar way of life to the Asturian of the Cantabrian coast, and at more or less the same time. The question will be tackled on a more solid base only through intensive programs of archaeological excavation, an increase in the number of geological studies, and above all, of absolute dates. Some of these objectives are already being achieved.

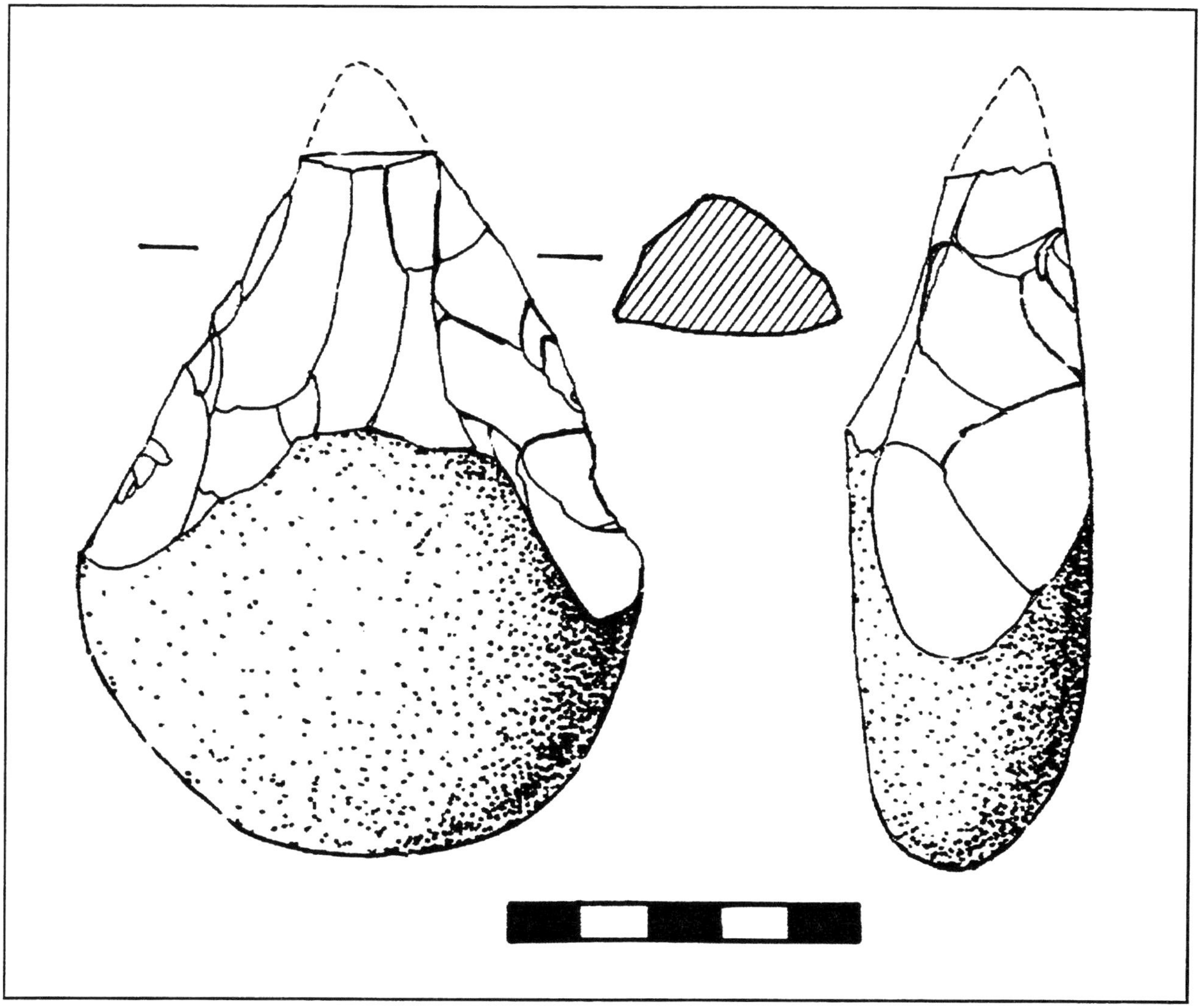

Figure 9.2. Type 1 unifacial pick from the site of Fiales (modified from Cano 1990).

Subsistence Potential of the Coast

The environmental characteristics of the coastal area have placed two constraints on human populations. First, the hilly terrain, granitic soils and humid climate have reduced the potential for agriculture. During historic times farmers systematically used seaweed as fertilizer for the land, and as feed for pigs. Second, the rocky nature of the shoreline, with exposed cliffs and straight pebble beaches, has reduced the kinds of shellfish which can be collected to the molluscs and crustaceans surviving in rocky, wave beaten environments. Fishing can be quite dangerous, and in fact there is only a single modern port, A Guarda, with a difficult entrance and scarce shelter, and a small harbour at Oia. The local inhabitants still use special tools to gather limpets, mussels and periwinkles to complement the animal protein in their diet. They sometimes catch octopus or the fish trapped in pools formed in the intertidal area.

This type of activity has been documented in the shell-middens of the Romano-Galician hillfort of Santa Trega, dated in the first two centuries AD, where mostly mussels and limpets are found, as well as some periwinkles, barnacles and sea urchins (Fernández & Rodriguez 1996). Similarities in marine resource exploitation, and in similar environments, allows us to extend this model into earlier phases of the Holocene. The prehistoric inhabitants of the coast could have complemented their hunting and gathering with the exploitation of marine resources and the raw material necessary for their stone tools, the pebbles found in countless millions on the beaches. Perhaps the use of edible seaweed was another attraction for the human occupation of the narrow coastal corridor. Other parts of the Galician coastline are likely to have prehistoric sites of the same age, but little systematic exploration has been carried out. It should be remembered that the rise in sea level must have destroyed many coastal sites.

Mesolithic and Megalithic

In the present state of research, it seems that the builders of megalithic monuments were responsible for the introduction of food producing economies in Galicia. However, taking into account typological parallels with Neolithic material from Portugal, some pottery found in rock-shelters on the coast of Pontevedra could belong to a Neolithic phase prior to the spread of megalithic tombs throughout the northwest of the Iberian Peninsula. The oldest megalithic dates extend back to the fourth millennium BC, and coincide with the earlier evidence of cereal in the pollen record, with the exception of O Reiro mentioned above (Vázquez 1996).

Tumuli are found from islands in the Rías Baixas (e.g., Guidoiro Areoso in Arousa estuary) to the high mountains in the east, and their distributions are by no means confined to the coastal plains. Their lithic assemblages bear no resemblances to what is normally considered as coastal Mesolithic. This suggests the presence of two distinct cultures (or adaptations), and no relation can be expected between these communities of Holocene hunter-gatherers exploiting the sea, and the first communities of food producers.

Conclusions

Research on the coastal Mesolithic of the northwestern 'corner' of the Iberian Peninsula began in the area around the mouth of the River Miño about 80 years ago, as a result of the early work of Conde de la Vega del Sella on the Asturian of the Cantabrian coast (1923). Since then a number of authors have considered the lithic artefacts collected on the coast as being Asturian, simply because of the presence of picks similar to those found in this type of site in the Cantabria.

Although recent research has not been conclusive, due to the absence of clear stratigraphies and absolute dating, it suggests that some of the shoreline assemblages of worked pebbles, including the picks, belong to a period between the end of the Würm and the early Holocene. The numerous problems still existing will only be solved by carrying out new archaeological excavations, obtaining absolute dates, and developing geomorphological and palaeoecological frameworks more adequate than those currently available.

References

Blanco, R., M. Costa, A. Martínez, A. Pérez & M. Vázquez
2002 Holocene evolution on Galician coast (NW Spain): an example of paraglacial dynamics. *Quaternary International* 93-94: 149-159.

Cano Pan, J.
1990 El yacimiento de Fiales (Oia, Pontevedra). *Gallaecia* 12: 19-28.
1991 *Las Industrias Líticas Talladas en la Costa de La Guardia a Baiona.* Coruña: Excelentísima Diputación Provincial de La Coruña.

Cano Pan, J., M. Fumanal, C. Ferrer, J. Usera, A. Blázquez & J. Olmo
1997 Evolución de la costa meridional de Galicia durante el Cuaternario Superior. *Cuaternario Ibérico* 7: 33-46.
1999 Evolution of the southern coast of Galicia (Spain) over the Upper Quaternary. *Zeitschrift für Geomorfologisch,* N. F. 43: 531-540.

Cano Pan, J. & J. M. Vázquez Varela
1986 Nuevas aportaciones al estudio de las industrias líticas del suroeste de Galicia: la prospección de 1984. *Studia Zamorensia Historica* 7: 267-304.
1996 An ethnographic model for the flaking of quartzite artifacts. In *Non-Flint Stone Tools and the Palaeolithic Occupation of the Iberian Peninsula* (N. Moloney, I. Raposo & M. Santonja, eds.), pp. 131-135. Oxford: Tempus Reparatum.

Clark, G. A.
1976 *El Asturiense Cantábrico.* Madrid: Consejo Superior de Investigaciones Científicas, Bibliotheca Praehistorica Hispana No. 13.

Clarke, D.
1976 Mesolithic Europe: the economic basis. In *Problems in Economic and Social Archaeology* (G. de G. Sieveking et al., eds.), pp. 449-482. London: Duckworth.

Fernández Rodríguez, C.
2000 *Los Macromamíferos en los Yacimientos Arqueológicos del Noroeste Peninsular: un Estudio Económico.* Tésis doctoral, Departamento de Prehistoria. Santiago de Compostela: Universidade de Santiago de Compostela.

Fernández Rodríguez, C. & C. Rodríguez López
1996 Nuevos datos acerca de los recursos económicos del poblado castreño de Santa Trega (A Guarda, Pontevedra). *Gallaecia* 14-15: 359-392.

González Morales, M. R.
1982 *El Asturiense y Otras Culturas Locales.* Santander: Monografías del Centro de Investigación y Museo de Altamira No. 7.

Martínez Cortizas, A. & J. M. Vázquez Varela
2002 El clima en la Galicia romana: una aproximación interdisciplinar. *Revista Real Academia Gallega de Ciencias* 21: 87-104.

Maury, J.
1977 *L'Asturien du Portugal.* Oxford: British Archaeological Reports International Series No. 21.

Meireles, J.
1994 As indústrias líticas pré-históricas do Litoral Minhoto (Portugal) é o seu quadro litoestratigráfico. In *Actas do 1° Congresso de Arqueología Peninsular, Trábalhos de Antropología e Etnología No. 4,* pp. 17-42. O Porto.
1996 *As Industrias Líticas Pré-Históricas do Litoral Minhoto. Contexto Cronoestratigrafico e Paleoambiental.* Braga: Cuadernos de Arqueología No. 6.

Ramil Soneira, J.
1973 Paradero de Reiro. *Cuadernos de Estudios Gallegos* 28: 23-31.

Rey García, M.
1995 Excavación arqueolóxica no illote de Guidoiro Areoso (Vilanova de Arousa, Pontevedra). In *Arqueología – Informes 3: Campaña 1989,* pp. 15-17. Santiago de Compostela: Xunta de Galicia.

Vázquez Varela, J.M.
1988 El Neolítico en Galicia. In *El Neolítico en España* (P. López, ed.), pp. 329-336. Madrid: Catedrático.
1991 El Paleolítico Inferior y Medio: formas de vida. In *Galicia Historia I – Prehistoria e Historia Antigua* (F. Acuña Castroviejo & J. M. Vázquez Varela, eds.), pp. 79-86. La Coruña: Hércules de Ediciones.
1996 Prehistoria. In *Nova Historia de Galicia,* pp. 7-71. La Coruña: Editorial Tambre.
1998 O aproveitamento dos recursos mariños na Prehistoria e a Antigüidade de Galicia. In *Historia da Pesca en Galicia* (C. Fernández Casanova, ed.), pp. 13-50. Santiago de Compostela: Universidade de Santiago de Compostela.
2003 *Prehistoria de Galicia.* Santiago de Compostela: Lóstrego.

Vázquez Varela, J.M. & C. Rodríguez López
2001 El aprovechamiento de los recursos marinos en la Prehistoria y la Antigüedad de Galicia. *Boletín do Museo Provincial de Lugo* 9 (1999/2000): 335-366.

Vega del Sella, Conde de la
1923 *El Asturiense, Nueva Industria Preneolítica.* Madrid: Comisión de Investigaciones Prehistóricas y Paleontológicas, Memoria No. 32.

10

Epipaleolithic Phases in the Northeast of the Iberian Peninsula

Pilar García-Argüelles i Andreu
University of Barcelona

Abstract

This chapter gives an overview of the Epipaleolithic, in the widest sense of the term, in Catalunya, based on the latest data. It reviews the oldest phases: the microblade Epipaleolithic, known from La Font Voltada, Cova del Vidre, Sant Gregori, levels 8-9 of Filador in the Province of Tarragona, layers 8 and 9 of Balma de la Margineda (Andorra), level 1b of Parco Cave (Lleida) and Cova de la Guineu (Barcelona). These are followed by the Barcelona site of Balma del Gai, which links this phase and the next, geometric Epipaleolithic, with the sites of Balma de la Margineda, Filador, and level 1a2 of Cova del Parco. The third phase corresponds to the end of the period, with a few examples in Catalunya. At the same time, there are other sites which are characterized by the use of varied raw materials (Font del Ros, Sota Palou or Roc del Migdia, Balma del Guilanyà) and which belong to these final moments of the Mesolithic.

The study of the Epipaleolithic in the northeast of the Iberian Peninsula raises a series of problems which force us to make a real effort to synthesize the data available for this period, because not all the excavations were carried out with the same precision, and even more importantly, the results have not always been published. There is, in addition, a lack of good radiocarbon series which would allow us to obtain a tighter, more reliable chronocultural sequence.

Because of these problems, the S.F.R.P. (Seminari d'Estudis i Recerques Prehistòriques) at the University of Barcelona has been developing a long-term research project in which a series of important sites have been studied with similar criteria, obtaining the results to be described here, together with the data mentioned above, in order to have an overview of the Epipaleolithic in Catalonia (Fig. 10.1).

In Catalonia the model proposed by Fortea (1973) has the widest currency, and is the one used here. It subdivides the Epipaleolithic into (1) a microblade facies (roughly equivalent to the Azilian of the Pyrenees and Franco-Cantabrian region) and (2) a geometric facies (mostly parallel to the classic French Sauveterrian).

Initial Phases: the Microblade Epipaleolithic

In Catalonia this phase has received the most attention, owing to the number of sites located and included in this phase because of their chronology and characteristics. The first group includes an ample series of sites belonging to the oldest phase, such as the site of La Font Voltada (Sarral, Tarragona), the oldest known with a date of 10,920 ± 240 BP. Archaeological excavations have been able to distinguish three occupation floors with flint industries made up of endscrapers on flakes or short blades (5.3%,), abundant denticulates (17.9%), a high proportion of backed points and blades (3.4%, 22.1% respectively) and burins on unretouched fractures (5.3%). The fauna collected is basically rabbit *(Oryctolagus,* 94.3%), with some ibex *(Capra pyrenaica,* 5.1%) and red deer *(Cervus elaphus,* 0.6%). There are abundant land snail fragments *(Helix spp.).* The vegetation surrounding the site is dominated by pine *(Pinus sylvestris)* and juniper *(Juniperus),* indicating an open, cold-climate cover, dated to the period between 12,000 and 8,000 BP in the Mediterranean region when forest-steppe predominated.

Layers 8 and 9 at Balma Margineda (Andorra) also form part of this microblade Epipaleolithic group, with curved backed points, backed blades, scrapers, bone working tools, and with faunas with a high percentage of ibex *(Capra pyrenaica* - 73% in layer 8, 94% in 9), together with wild boar *(Sus scropha),* red deer *(Cervus elaphus),* carnivores and lagomorphs, and a date of 10,640 ± 260 BP. The tree cover at that time was dominated by conifers (*Pinus sylvestris*, *Juniperus* and *Avies sp.*) with cold dry climatic conditions (Guilaine et al. 1988, Geddes et al. 1990).

More recently, we have publications of another two sites included in this phase. The first is Cova del Vidre (Tortosa, Tarragona), with a lower level belonging to the microblade Epipaleolithic, a date of 10,740 ± 130 BP, and an industry composed of endscrapers, microblades and backed points (Bosch 1993). The second is the rock shelter site of Picamoixons (Tarragona), which was at first thought to be Upper Paleolithic (L.A.U.B.T. 1992) but which, as a later publication shows, could belong to this

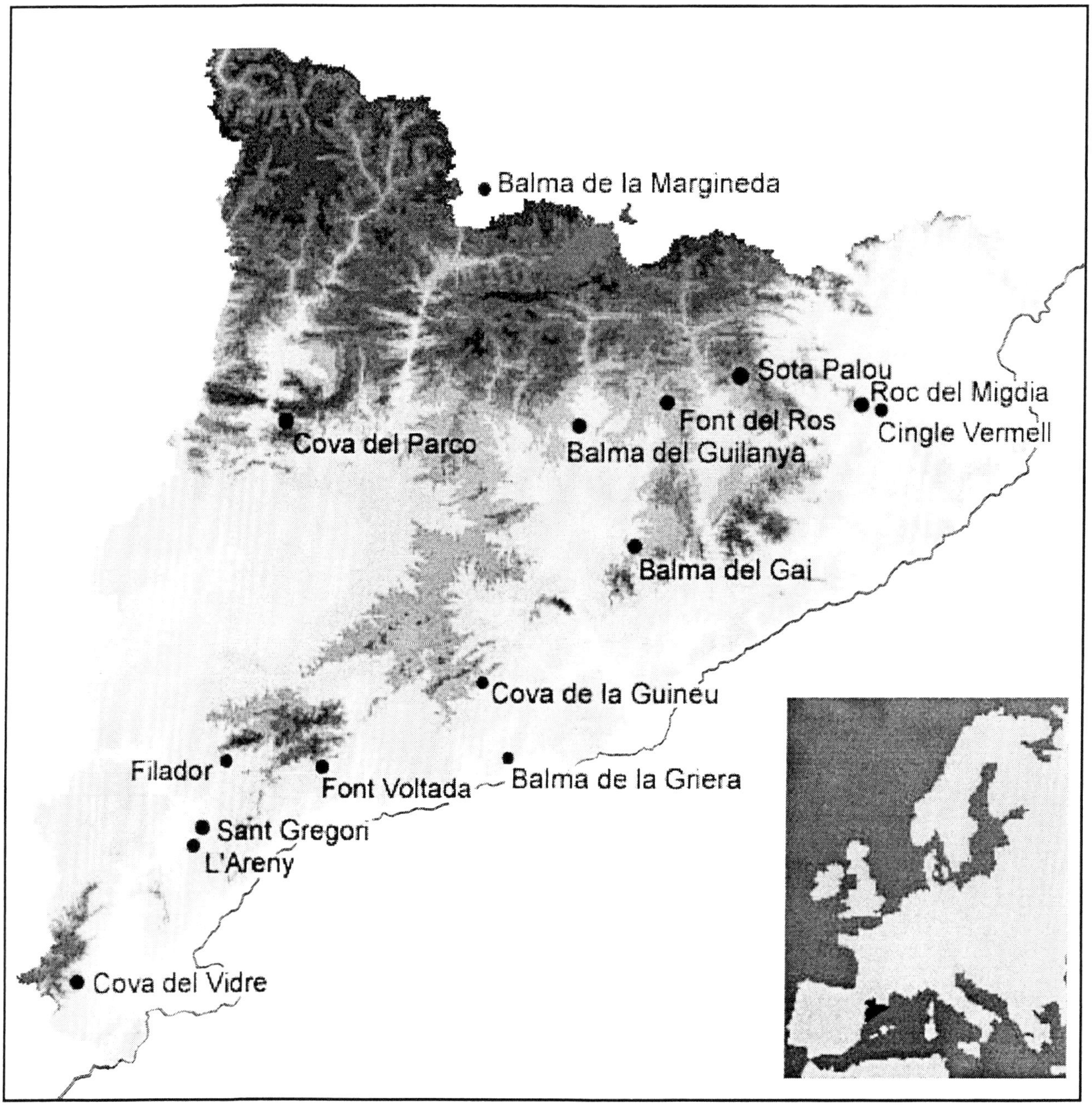

Figure 10.1. Epipaleolithic and Mesolithic sites in Catalunya discussed in the text.

phase. Its industry is comprised of endscrapers (20.7%), burins (18.5%), denticulates (18%), and backed blades and points (15.9%, 9%). The fauna is dominated by *Orytolagus* (91.2%), with small numbers of *Capra sp.* (6.8%) and *Equus* (1.3%). Radiocarbon dates were obtained from bone from group I and the upper part of II, level A (9,170 ± 80, 9,370 ± 95, 10,900 ± 90 BP), while a charcoal sample from group II level B was dated to 11,055 ± 90 BP. The authors do not favour any particular date and it seems that the lithic elements form a unity so that no kind of interior evolution or development can be discerned (Allué et al. 1992) and level 1b of Parco cave.

A second period within this microblade phase is represented by strata 1c of our re-excavation of the rock shelter of Sant Gregori (Falset, Tarragona) (García-Argüelles et al. 1992) with material comparable to the level 2 proposed by S. Vilaseca. We chose to abandon his terminology (we equate stratum 1c with Vilaseca's level 2) as it is possible, from his description of the stratigraphy, the nature of the sediment produced by disintegration of sandstone, and the micromorphological results, that part of the deposits were already contaminated. Strata 1c of Sant Gregori is characterized by an abundance of scrapers, backed elements, the appearance of a few microburins and by a low percentage of burins (2.5%).

Level IIIb of Cova de la Guineu (Font Rubí, Barcelona) also belongs to this phase. Its industry is dominated by endscrapers (26%), backed elements (22.5%) and

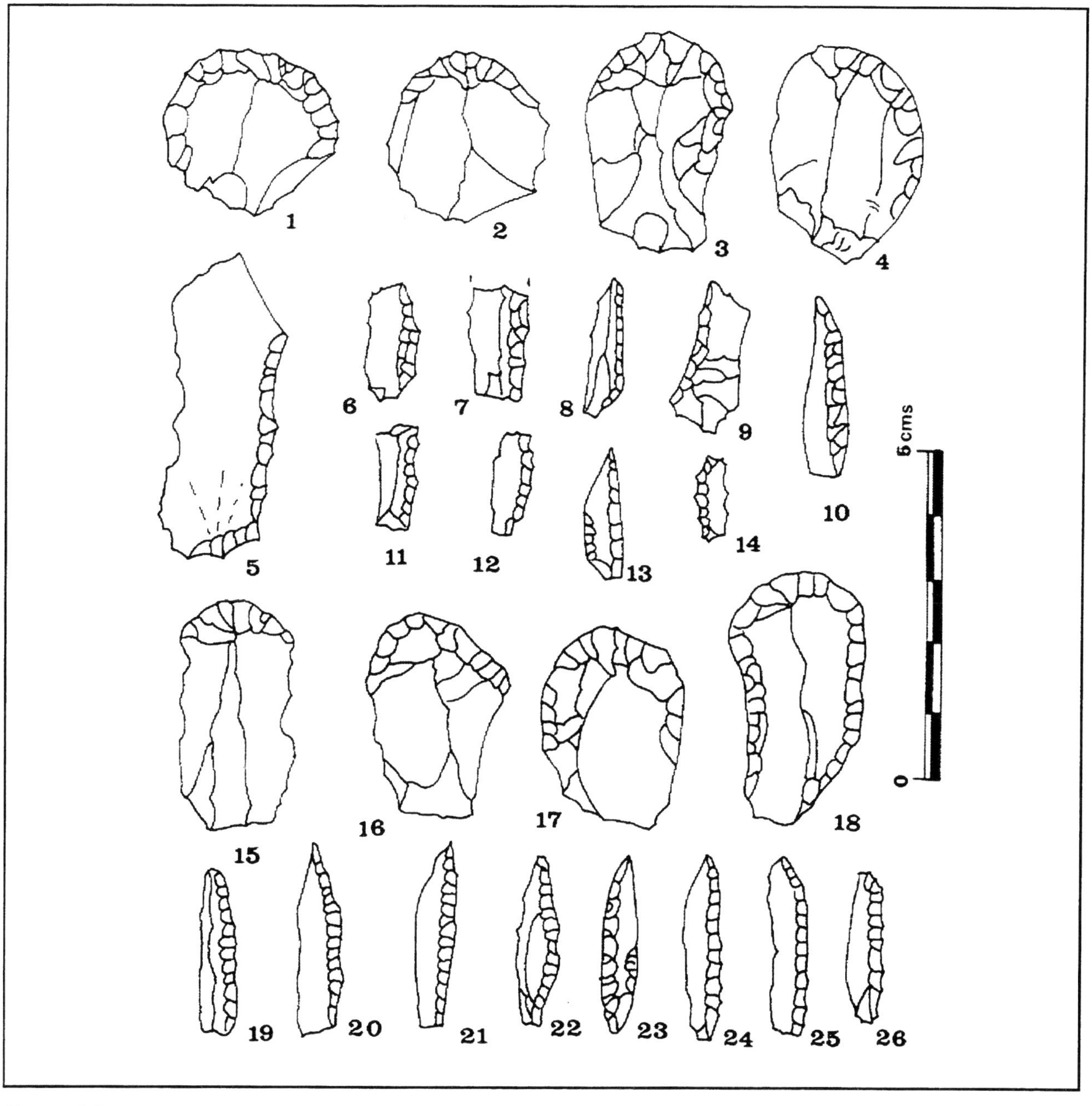

Figure 10.2. 1-14 Guineu Cave, level IIIa; 15-26 El Filador, level 8-9.

burins (65%), and has been dated radiometrically to 9,850 ± 80 BP (Fig. 10.2). The most common faunal element here is *Oryctolagus*, while microfauna studies indicate a typical intermediate mountain fauna, with semideciduous forest and mountain holly oak (Cebrià et al. 1990). We can also include in this period the rock shelter of L'Areny (Vilanova d'Escornalbou, Tarragona) and strata 8-9 of the rock shelter of El Filador, excavated in the mid-1990s, with an assemblage dominated by backed elements, but in which geometric pieces and microburin technique are both absent (García-Argüelles 1990, García-Argüelles et al. 2002).

The link between the microblade and geometric phases is documented at Balma del Gai (Moià, Barcelona), a site with a single archaeological level, subdivided into three layers (1, 2 and 3) by its original excavators (Guilaine et al. 1982, Barbaza 1984). In its upper part it has some geometric forms (scalene triangles and segments), while the lower layers have large quantities of small endscrapers, backed blades and points. The fauna is dominated by *Orytolagus*, with *Capra pyrenaica, Rupicapra rupicapra, Sus scropha, Vulpes vulpes, Lynx,* a few birds and numerous land snails (*Cepraea nemoralis*). This type of fauna is characteristic of other Catalonian sites already mentioned (Font Voltada, Cova de la Guineu, Sant Gregori). Its dates range between 10,030 ± 160 BP and 9,860 ± 400 BP for the upper layer, while layer 2 (the intermediate zone of level 1) has a date of 11,054 ± 160 BP. These dates do not agree with the micromorphological analyses of sediments

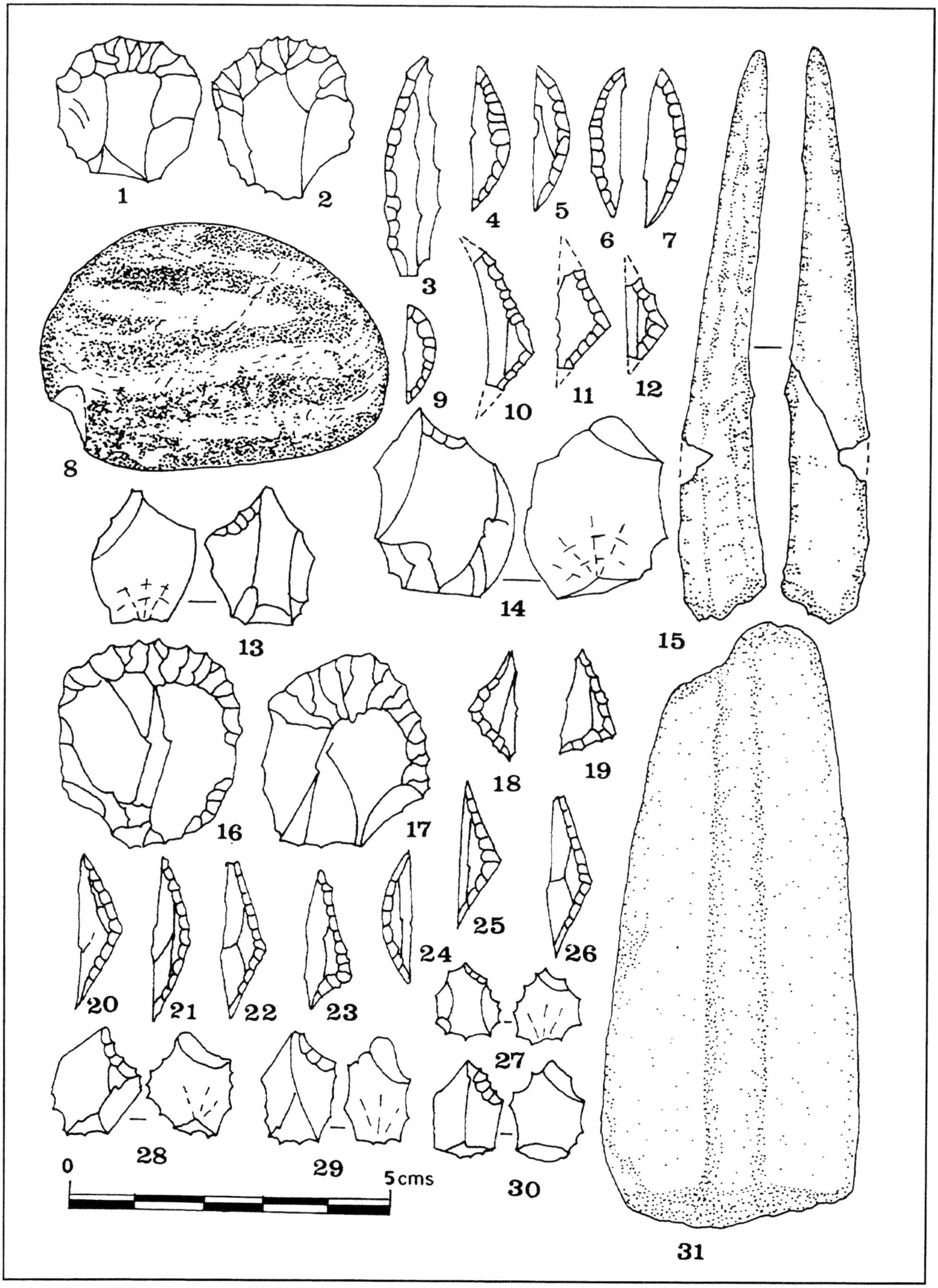

Figure 10.3. 1-15 Filador, level 4; 16-21 Filador, level 7.

obtained from our recent re-excavation of the site, which show a very humid climate for level 1 that does not correspond with the late glacial age indicated by the dates.

Advanced Phases: the Geometric Epipaleolithic

The geometric facies is well represented in the northeast of the Peninsula; in the north, layers 7 and 6 of Balma Margineda are especially important. Dated to 9,250 ± 160 BP, these levels exhibit an industry of scalene triangles, segments, small scrapers and the microburin technique. They also contain some worked bone and a few painted cobbles. The fauna hunted is still *Capra pyrenaica*, *Sus scropha*, *Cervus elaphus*, carnivores and lagomorphs; the tree cover is the same as in the lower layers (*Pinus sylvestris*, *Juniperus* and *Abies*). The climate is cold and dry.

Towards the south, the most representative site is the rock shelter of El Filador (Margalef de Montsant, Tarragona), where strata 4 and 7 have given dates of 9,460 ± 190 BP and 9,830 ± 160 BP respectively (Fig. 10.3). The main characteristic of these levels is the important presence of geometric forms (11%); triangles are dominant over segments in strata 7, whereas the reverse is true in strata 4; microburin technique (32% in strata 7, 43% in strata 4), and a fair proportion of backed elements (24% in strata 7). Like Balma Margineda, this site has also produced worked bone and painted cobbles. The fauna is quite similar in both levels, with a predominance of *Capra pyrenaica, Cervus elaphus* and *Oryctolagus*, and a significant exploitation of land snails (*Cepraea nemoralis*). Archaeozoological studies of strata 7 show a specialization in ibex hunting, and two clearly delimited activity areas. Whereas the first was specialized in butchery, we can suppose that the animals were transported to the other area almost whole. From the results obtained from the study of juvenile dentitions, it would appear that the site was mainly occupied in spring and summer.

Paleoenvironmental studies show the existence of open vegetation (*Pinus sylvestris*, which diminished in strata 4 (*Juniperus sp., Quercus ilex*); there are also species typically found near watercourses (*Salix, Corylus*). In strata 4 there is an increase in thermophile (*Olea, Phyllyrea*) and temperate species (*Ulmus, Abies* and *Betula*). The climate was still quite cold and dry, however, with severe winters but much warmer, drier summers.

The third site representative of this phase is Cova del Parco (Alòs de Balaguer, Lérida), whose level 1a2, situated on top of a boulder collapse which separates it from paleolithic levels, has an apparent living floor with geometric elements (triangles 8.5%, segments 22.8%), backed bladelets (14%) and microburins (25.7%) (Fig. 10.4). *Cervus elaphus* has been identified, while pollen studies describe a vegetation comprised of *Juniperus*, *Quercus ilex*, *Rhamnus spp.*, *Acer sp.* and *Quercus robur*. The absence of pine, together with the appearance of rosemary (*Rosmarinus officinalis*) and *Ephedra* are evidence for dry environmental conditions and poor soils. The oaks and holly oaks would grow in protected locations with higher humidity. The landscape would be quite open – steppe with bushes – with a climate which was not too cold, but dry. The dates obtained situate this level in an early moment of the geometric phase: 10,190 ± 290 BP (Fullola et al. 1998.). The Epipaleolithic level of Cova de la Griera (Calafell, Baix Penedès) should also be included in this phase. Although it had been disturbed by human action, it contained a low incidence of geometric forms. Epipaleolithic radiocarbon dates available through 1993 are given in Figure 10.5 (see also end note).

Mesolithic Phases

The Mesolithic, in fact an Epipaleolithic in the process of economic transformation, is found in level 4 of Balma Margineda (8,530±420, 8,390 ± 150 and 8,210 ± 180 BP), with the presence of trapezes, the disappearance of microburin technique and the first examples of *Ovis aries* in the upper part of the level. This phase at Margineda has its nearest parallels in the French sites of Gazel and Dourgne.

We can also include in this period stratum 3 of El Filador, where there is a change in the size and shape of the geometric forms, with the disappearance of segments and the substitution of scalene triangles by isosceles ones (8.8%), a reduction in the percentages of microburins (26.4%), the appearance of a macrolithic component (denticulates), and the progressive substitution of steep retouching by simple, marginal retouching. The fauna is still wild, dominated by *Cervus elaphus* and *Sus scropha*. Phytolith analyses show a strong presence of *Juniperus sp.*, accompanied by cyperaceous and poaceous species.

Except in Margineda, this period cannot be compared with the final phase of the La Cocina-type geometric complex, as it lacks all of the latter's defining elements. The following period corresponds to the Neolithic, and is present in Cova de la Guineu, in Cova del Parco (6,450 BP), in Balma Margineda (6,670 BP), in Cova del Vidre, with Cardial pottery, and in the upper level of El Filador.

Other Northwestern Assemblages

Independent of this first group, there is a large sector of the northwest of Catalonia with a series of sites which are very distinct from this sequence. Some common denominators include the use of local raw materials, such as quartz, quartzite and limestone, with low percentages of flint; and a predominance of simple retouching mainly on denticulates, continuously retouched pieces and endscrapers.

In this group we have Cingle Vermell (Vilanova de Sau, Barcelona) with a date of 9,760±160 BP, and an assemblage composed of quartz, limestone, *lidita*, and jasper, and with most retouched pieces made on flint blanks.

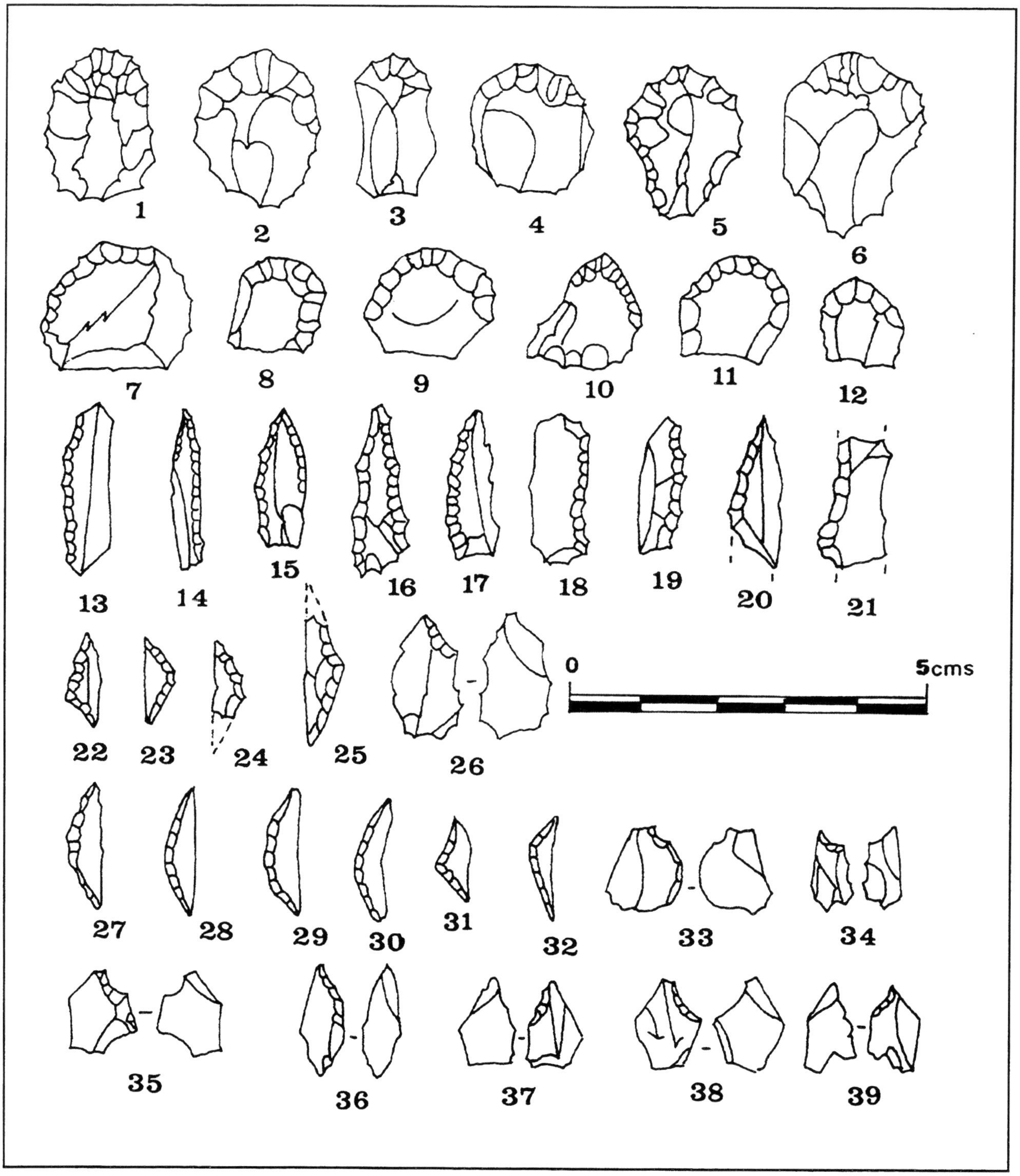

Figure 10.4. 1-25 Balma del Gai; 27-39 Parco Cave, Level 1aZ.

The knapping technique seems to have been largely determined by the raw material. The retouched pieces are mainly endscrapers (21.2%), denticulates (14.8%), and sidescrapers (12.7%); backed blades and points are also present (15.6%) with an important percentage of burins (10%). The fauna consists of *Sus scropha, Cervus elaphus, Capreolus capreolus, Capra pyrenaica, Rupicapra rupicapra, Oryctolagus* and *Cepraea nemoralis*. The vegetation was dominated by deciduous species, and the climate was humid (Vila 1985).

Another site is Sota Palou (Campdevànol, Girona), dated to 9,060±380 BP and 8,540±180 BP. This was a seasonally occupied, open-air camp, located at the boundary between mountain and plain (700 m above sea level), and with an assemblage of quartz, *lidita* and some flint, and mostly simple retouching (denticulates). The fauna is

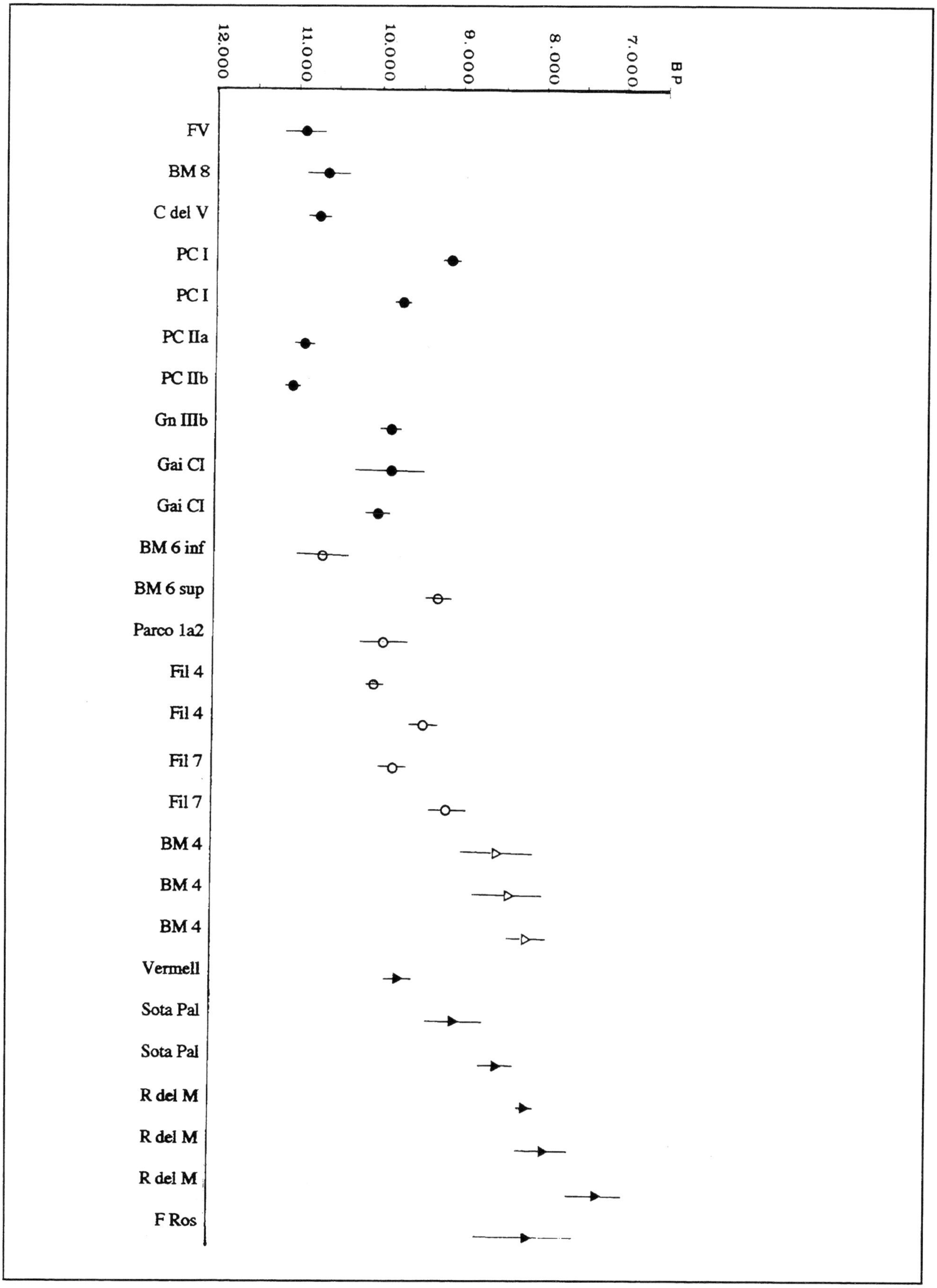

Figure 10.5. Epipaleolithic and Mesolithic radiocarbon dates to 1993 (see also endnote).

restricted to *Capra pyrenaica* and *Sus scropha*, and the vegetation is composed of such species as *Pinus sylvestris, Corylus avellana* and *Quercus* spp. (Carbonell et al. 1985).

Font de Ros (Berga, Barcelona) is second seasonally occupied open-air camp, in a similar situation to the previous site, in an open valley between the Catalonian depression and the Pre-Pyrenees (650 m above sea level). It was dated to 8,150 ± 590 BP, and its assemblage is of flint, quartz, limestone, quartzite and *lidita*. There is a clear predominance of simple retouching (54%), mainly of denticulates (29%) and side scrapers (36.3%) (Terradas 1995). The faunal assemblage is dominated by *Cervus elaphus, Capra pyrenaica* and *Oryctolagus*. Forest vegetation was of a sub-mediterranean type with a full extension of deciduous woods and riparian communities. An upper level yielded some Cardial ware, and other kinds of 'early Neolithic' pottery with impressed decoration which are difficult to date precisely. The excavators do not favour any particular moment within the early Neolithic (Mora et al. 1991, Terradas et al. 1992).

Finally, we must include in this section the new results from the revised study of Roc del Migdia (Vilanova de Sau, Barcelona), whose lower levels have new dates of 7,280±370 BP, 7,950±370 BP and 8,190±32 BP (Rodríguez and Yll 1992: 76). Stone-working was carried out on quartz, flint and limestone, with an absolute predominance of denticulates, suggesting an expedient assemblage. *Cervus elaphus, Capra pyrenaica, Sus scropha* and *Oryctolagus* were hunted, while the vegetation was dominated by *Buxus, Acer* and *Prunus* in the earlier phases, with *Quercus* appearing later. The climate was dry between 8,800 and 8,200 BP, after which time it became humid and cold.

The dates of these sites are perfectly comparable with those of the other site group, and they often fill the time gap that we find between the geometric Epipaleolithic and the Neolithic.

Final Considerations

As we have seen, in the northeast of the Iberian Peninsula we can propose the co-existence of two types of industry between 11,000 and 8,000 BP. On the one hand we have the sites which follow the classic model found over large areas of western Europe, and which has its origin in the Upper Paleolithic tradition, evolves during those millennia exploiting flint as its basic raw material, starts to incorporate new economic elements in the middle of the 9th millenium BP, and later leads to the Neolithic in its full sense. This scheme is broken in Catalonia in the penultimate phase as, so far, we do not know of any site with Fortea's La Cocina-type assemblage (1973). We also do not know how the geometric Epipaleolithic evolved into the Neolithic – the chronology of this phase would roughly correspond to a number of sites which do not follow this line (Font del Ros, Roc del Migdia and Sota Palou).

On the other hand, we have groups who mainly used (rather poor quality) local raw materials to manufacture their tools. The quality of the raw material obscures any historical connection with earlier assemblages. These sites rarely show chronostratigraphic continuity (e.g., Cingle Vermell, Sota Palou, and Roc del Migdia) and only Font del Ros seems to have been (re)occupied fairly intensively.

The development of these groups would doubtless be similar to that of the sites in the first group. The evident differences emphasized here should not hide other equally evident parallels which form a model of neolithization that is less heterogeneous than it might first appear. We think that seasonality of occupations and/or different site functions could help us to understand the differences amongst these assemblages. We conclude by proposing as a working hypothesis that in Catalonia, the last phase of the Epipaleolithic evolves towards the model suggested by Font del Ros, Roc del Migdia and Sota Palou, as happens in other zones in the French eastern Pyrenees.

Endnote

The following radiometric dates have appeared since this manuscript was originally submitted to the editors.

Microblade Epipaleolithic:		
Balma del Gai	layer 3:	10,269±90 BP
	layer 5:	12,240±110 BP
	layer 7:	11,170±160 BP
Cova del Parco		10,810±160 BP
		10,190±290 BP
		10,260±280 BP
		10,720±280 BP
		11,510±170 BP
Filador	level 8-9:	11,880±60 BP
	level 8-9:	11,000±55 BP
Geometric Epipaleolithic:		
Balma del Gai	layer 1:	8,930±140 BP
Filador	level 7:	9,130±230 BP
	level 5-6:	9,988±97 BP
	level 4:	10,020±80 BP
Cova del Parco	level 1a2:	10,190±100 BP
	level 1a2:	10,930±100 BP
	level 1a2:	10,420±110 BP
Mesolithic:		
Balma del Guilanyà		8,970±430 BP

References

Allué, M. et al.
1992 *Picamoixons, un Assentament Estratègic dels Darrers Caçadors – recollectors.* Tarragona: Arqueologia d'Intervenció 1, Museu d'Historia de la Ciutat de Tarragona y L.A.U.T.

Barbaza, M.
1984 L'Epipaléolithique Méditerranéen espagnol. *Travaux de l'Institut d'Art Préhistorique* 25: 3-14.

Bartrolí, R., M.M. Bergadá, A. Cebriá & M. Fontugne
1992 El model microlaminar - geométric a la Catalunya meridional: aportacions del projecte d'investigació de la serra de Font-rubí (Alt Penedés). In *9è Colloqui Internacional de Puigcerdà,* pp. 34-37. Andorra: Institut d'Estudis Ceretans y Servicio d'Arqueologia d'Andorra, Andorra.

Bosch, J.
1993 Cronologia prehistórica al curs inferior de l'Ebre. Primeres datacions absolutes. *Pyrenae* 24: 53-56.

Carbonell, E. et al.
1985 *Sota Palou, Campdevànol: un Centre d'Intervenció Prehistòrica Postglaciar a l'aire llivre.* Girona: Centro de Investigaciones Prehistóricas de la Diputación de Girona, Série Monografías No. 5.

Estevéz, J., A. Vila & E. Vila
1983 Reflexions sur la dynamique économique et industrielle au Nord du Levant ibérique (16.000-7.000 BP). *Rivista di Scienze Preistoriche* 38: 9-19.

Estrada, A., P. García-Argüelles, J. Nadal & C. Arnau
1990 Noves aportacions a l'estudi de la Balma del Gai (Moià). *Modilianum* 2: 3-14.

Fortea, J. 1973 *Los Complejos Microlaminares y Geométricos del Epipaleolítico Mediterráneo Español.* Salamanca: Seminario de Prehistoria y Arqueología, Memoria No. 4.

Fortea, J., J.M. Fullola, V. Villaverde, I. Davidson, M. Dupré & P. Fumanal.
1983 Schéma paléoclimatique, faunique et chronostratigraphique des industries à bord abattu de la région méditerranéenne espagnole. *Rivista di Scienze Preistoriche* 38: 21-67.

Fullola, J.M. & P. García-Argüelles
1996 La formation des complexes mésolithiques de la Peninsule Ibèrique. In *Actas du Colloque XIII sur la Formation des Complexes Mésolithiques en Europe,* pp. 9-15. Forlí: XIIIème Congrés U.I.S.P.P.

Fullola, J.M., R. Viñas & P. García-Argüelles
1990 La nouvelle plaquette gravée de Sant Gregori (Catalogne, Espagne). In *L'art des Objêts au Paléothique, Tome 1 – L'art Mobilier et Son Contexte,* pp. 279-285. Paris: C.N.R.S.

Fullola, J.M., P. García-Argüelles & J. Nadal
1992 Els caçadors-recollectors de l'Holocè a Catalunya: epipaleolítics o mesolítics? In *9è Colloqui Internacional de Puigcerdà,* pp. 20-22. Andorra: Institut d'Estudis Ceretans y Servicio d'Arqueologia d'Andorra.

Fullola, J.M. et al.
1998 Occupation épipaléolithique de la grotte du Parco (Alòs de Balaguer, Catalogne, Espagne). In *Proceedings of the 13th International Congress of the UISPP, Vol. 2, Section 6: The Upper Palaeolithic,* pp. 535-542. Forlí: XIIIème Congrés U.I.S.P.P.

García-Argüelles, P.
1990 *Las Industrias Epipaleolíticas del Sur de Cataluña: Antecedentes, Desarrollo y Evolución hacia Nuevas Formas Neolíticas.* Barcelona: Colección de Tésis Microfichadas No. 625, Publicaciones de la Universidad de Barcelona.

García-Argüelles, P.
1993 Los pulidores de arenisca del Tardiglaciar y Poslglaciar de Europa occidental: síntesis y nuevas aportaciones. *Pyrenae* 24: 35-52.

García-Argüelles, P., M. Adserias, R. Bartroli, M.M. Bergadá, A. Cebria, R. Doce, J.M. Fullola, J. Nadal, G. Ribit, T. Rodon & R. Viras
1992 Síntesis de los primeros resultados del programa sobre Epipaleolítico de la Cataluña central y meridional. In *Aragón – Litoral Mediterráneo. Intercambios Culturales durante la Prehistoria,* pp. 269-284. Zaragoza: Instituto Fernándo el Católico.

García-Argüelles, P., D. Serrat & M.M Bergadá
1993 Las terrazas fluviales del curso medio del río Montsant (Tarragona) y su relación con los asentamientos prehistóricos. In *Actas de la IIª Reunión del Cuaternario Ibérico,* Vol. 1, pp. 493-499. Madrid: I.N.Q.U.A.

García-Argüelles, P. & J. Nadal
1998 The geometric sequence of the Filador rock shelter (Catalonia, Spain). In *Proceedings of the 13th International Congress of the UISPP, Vol. 3, Section 3: The Mesolithic,* pp. 49-54. Forlì: XIIIème Congrés U.I.S.P.P.

Garcia-Argüelles, P., J.Nadal & J.M. Fullola
2002 Vint anys d'excavacions a l'abric del Filador (Margalef de Montsant, Priorat, Tarragona). In

Tribuna d'Arqueologia 1998-99, pp. 71-95. Barcelona: Servei d'Arqueologia, Generalitat de Catalunya.

Geddes, D., J. Guilaine, J. Coularou, O. Le Gall & M. Martzluff
1990 Postglacial environment, settlement and subsistence in the Pyrenees: the Balma Margineda, Andorra. In *The Mesolithic in Europe* (C. Bonsall, ed.), pp. 561-571. Edinburgh: John Donald.

Guilaine, J., M. Barbaza, D. Geddes & J.L. Vernet
1982 Prehistoric human adaptation in Catalonia (Spain). *Journal of Field Archaeology* 9: 407-416.

Guilaine, J. et al.
1988 *Les Recerques a la Balma Margineda (1979-1985).* Andorra: Institut d'Estudis Andorrans.
1995 *Les Excavacions a la Balma de la Margineda (1979-1991).* Andorra: Institut d'Estudis Andorrans.

L.A.U.B.T.
1992 Relaciones depresión central – litoral Catalán durante la prehistoria: el centro de intervención de Picamoixons. In *Aragón– Litoral Mediterráneo. Intercambios Culturales durante la Prehistoria,* pp. 139-151. Zaragoza: Institución Fernando el Católico.

Mir, A. & A. Freixas
1993 La Font Voltada, un yacimiento de finales del Paleolítico Superior en Montbrió de la Marca (La Conca de Barberá, Tarragona). *Cypsela* 10: 13-21.

Mora, R., X. Terradas, A. Parpal, C. Plana, J. Martin, A. Figols & G. Roca
1991 Les ocupacions mesolítica i neolítica de la Font del Ros (Berga, Berguedà). In *Tribuna d'Arqueologia 1989-90,* pp.19-29. Barcelona: Servei d'Arqueologia, Generalitat de Catalunya.

Paz, M.A. & A. Vila
1988 Roc del Migdia, un jaciment mesolític atípic? In *Tribuna d'Arqueologia 1987-88,* pp. 133-143. Barcelona: Servei d'Arqueologia, Generalitat de Catalunya.

Rodríguez, A. & R. Yll
1992 Materias primas y cadenas operativas en el yacimiento epipaleolítico del Roc del Migdia (Vilanova de Sau, Barcelona). In *Tecnología y Cadenas Operativas Líticas,* pp. 73-82. Bellaterra: U.A.B., Treballs d'Arqueologia No. 1.

Terradas, X.
1995 *Las Estrategías de Gestión de los Recursos Líticos del Prepirinéo Catalán en el IX Milenio BP: El Asentamiento Prehistórico de La Font del Ros (Berga, Barcelona).* Bellaterra: U.A.B., Treballs d'Arqueologia No. 3.

Terradas, X., R. Mora, C. Plana, A. Parpal & J. Martinez
1992 Estudio préliminar de las ocupaciones del yacimiento al aire libre de la Font del Ros (Berga, Barcelona). In *Aragón – Litoral Mediterráneo. Intercambios Culturales durante la Prehistoria,* pp. 285-296. Zaragoza: Institución Fernando el Católico.

Tresserras, J.
1993 Nuevos datos sobre la evolución del paisaje y los recursos vegetales en el Abric del Filador (Margalef de Montsant, Tarragona). *Pyrenae* 24: 23-33.

Vila, A.
1985 *El Cingle Vermell: Assentament de Caçadors – recollectors del Xe Milleni BP.* Barcelona: Generalitat de Catalunya, Excavacions Arqueològiques a Catalunya No. 5.

Vilaseca, S.
1934 L'estació-taller de sílex de Sant Gregori. *Memòries de la Real Acadèmia de Ciències i Arts de Barcelona* 23: 415-440.

Yll, E., J. Watson & M.A. Paz
1994 Les darreres excavacions al Roc del Migdia (Vilanova de Sau, Osona): estat de la questió i noves perspectives. In *Tribuna d'Arqueologia 1992-1993,* pp. 15-24. Barcelona: Servei d'Arqueologia, Generalitat de Catalunya.

Part III

The Atlantic Coasts of Spain & Portugal

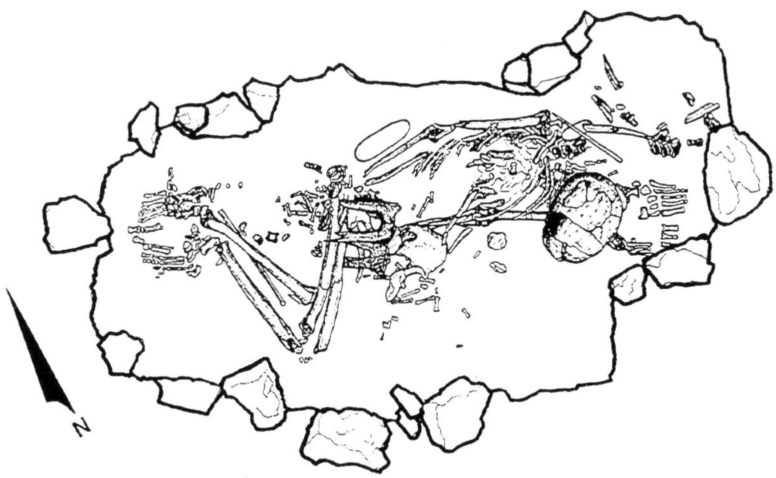

11

Radiocarbon Evidence for Maritime Pioneer Colonization at the Origins of Farming in West Mediterranean Europe[1]

João Zilhão
Max Planck Institute, Leipzig

Abstract

Most radiocarbon dates for the earliest Neolithic cultures of west Mediterranean Europe are on samples of unidentified charcoal. If only results obtained on short lived samples (seeds, shells, and bone) of diagnostic material (domesticates, artifacts, and human remains) are considered, then the dates for the first appearance of the Neolithic package are indistinguishable statistically from central Italy to Portugal and cluster around 5400 calendar B.C. This rapidity of spread, no more than six generations, can be best explained in the framework of a maritime pioneer colonization model.

Introduction

A sound dating of the first appearance of agro-pastoral economies across Europe is a basic prerequisite to the evaluation of how, why, and when hunter-gatherer adaptive systems eventually disappeared from most of the continent in prehistoric times. The task, however, has not proved easy. Before the advent of accelerator mass spectrometer (AMS) radiocarbon dating, bulk samples of charcoal, bone, or shell had to be used, and in many instances the resultant dates were at odds with archaeological expectations based on stratigraphy and typology.

One source of problems was the use of shell samples, providing results biased by reservoir effects. This problem remains largely unsolved except for mid-Holocene marine samples from the western and southern Atlantic seaboards of Iberia. Through the comparison of results obtained by dating different kinds of samples from the same levels (44 shell and 20 charcoal or bone samples from 18 different archaeological contexts), it was possible to establish that the reservoir effect was of 380±30 years for the most common mollusk species accumulated in the shell-middens of the time: *Cerastoderma* sp., *Patella* sp., and *Mytilus* sp. (Monge Soares 1993, and Marine Reservoir Correction database, *http://radiocarbon.pa.qub.ac.uk/marine/*). Thus, Portuguese and western Andalucian late Mesolithic and early Neolithic samples made up of these species can be calibrated now by using the curve for terrestrial samples (Stuiver et al. 1998) after subtraction of that apparent age from the raw before present (BP) result.

Another potential source of inaccuracy was the old wood effect, the impact of which was made apparent over the last decade, because the AMS direct dating of diagnostic material (bones from domestic animals and charred cereal seeds) became possible. In the case of the Linear Bandkeramik culture from central Europe, these AMS results suggested that its earliest phase dated to after ≈6400 BP, whereas dates on bulk charcoal samples suggested ages as early as ≈6900 BP (Hedges et al. 1989, Whittle 1990). AMS dating of domesticates has also shown that the beginning of agriculture in different parts of northwestern Europe was sometimes as much as two millennia later than thought previously on the basis of pollen diagrams and the assumption that appearance of 'Neolithic' artifacts in the archaeological record could be used as a proxy for the introduction of agro-pastoral economies (Rowley-Conwy 1995). Similar problems in the New World and Oceania were reported by Fritz (1994) and Spriggs (1989), respectively. The former noted that AMS dates on remains of the actual domesticated plants from sites in the Tehuacán valley and elsewhere in the Americas suggested the inception of agricultural practices much later than accepted previously on the basis of bulk charcoal samples from the levels containing those remains. The latter showed that 'chronometric hygiene' was required for a correct understanding of the Lapita process and, hence, of the human colonization of Polynesia.

In west Mediterranean Europe (Fig. 11.1), problems inherent to the nature of the samples have been compounded

1. Originally published in the *Proceedings of the National Academy of Sciences,* Vol. 98, No. 24, pp. 14180-14185 (2001); published here with the permission of the National Academy of Sciences.

by the fact that most archaeological evidence comes from caves and rock shelters (Ammerman 2001). Postdepositional disturbance is commonplace in these kinds of sites; hence, when bulk charcoal samples are used, the association between the dated material and the historical events one wants to date cannot be taken for granted before appropriate critical filters are applied to evaluate the association (Zilhão 1992, 1993). When scrutinized from a taphonomic perspective, the very early chronology claimed for the inception of production economies in Mediterranean France and Spain must be rejected. The dates (some as early as ≈8000 BP) for the appearance of domesticates in such Spanish sites as Cueva de La Dehesilla (Acosta & Pellicer 1990) and Cova Fosca (Olaria 1988) were obtained from samples collected in disturbed contexts. The dated material contained in variable proportions late Paleolithic, Mesolithic, and Neolithic bone and charcoal, and the Neolithic artifact assemblages found therein corresponded to Epicardial contexts intruded into the underlying late Pleistocene or early Holocene deposits (Zilhão 1993).

A side issue of the Mesolithic-Neolithic transition as seen from these kinds of sites was the local domestication of goats, suggested for Cova Fosca (Estévez 1988), and the late Mesolithic acquisition of exotic domestic sheep through long distance exchange mechanisms, suggested for Abri Dourgne and Grotte Gazel, among others (Geddes 1983). This was a definitional issue, involving the identification as domesticates of juvenile ovicaprid bones recovered in high mountain sites that might in fact belong to young chamois or ibex. Such identifications created the impression that in these regions the Neolithic did not spread as a complete integrated package. Instead, local hunter-gatherers would have gone through a process of piecemeal independent invention, or adoption, of different elements. However, once the taxonomic status of bones attributed to sheep reported for those Languedoc sites was reevaluated, the evidence for domesticates in the Mesolithic vanished (Guilaine et al. 1993). This made it clear that the Cardial and related impressed ware cultures of west Mediterranean Europe represent the simultaneous appearance in the archaeological record of the whole package of features involved in the earliest agro-pastoral practices of the region: animal and plant domesticates, ceramic vessels, polished stone axes, and village dwelling. None of these have ever been found in secure, undisturbed Mesolithic contexts.

Since originally formulated (Zilhão 1993), these conclusions have remained unchallenged. Their verification removed any empirical foundation for models describing the emergence of the Neolithic in west Mediterranean Europe as resulting from gradual east-west dispersal of economical and technical innovations. Such models dominated the literature for the last quarter of a century (Zvelebil 1986, Whittle 1996, Barnett 2000). They also are difficult to reconcile with the fact that there is no evidence for long distance exchange or trade networks through which indigenous hunter-gatherer groups could have acquired the Neolithic package (or any of its individual components) and transformed themselves into agro-pastoral societies. Moreover, no sound and testable ecological or social explanations have been offered for why local Mesolithic people would have been willing to undertake such a transformation.

These facts suggest that the spread of farming in west Mediterranean Europe was associated with some sort of demic diffusion process such as the "wave of advance" model put forward by Ammerman and Cavalli-Sforza (1973, 1984). According to their analyses, the annual rate of population growth experienced by frontier farmers can hardly have been greater than 1%, and ethnographically observed rates of migratory activity in similar situations never exceed 2,000 km^2 per generation. This rate carries the implication that the model will not work if observed rates for the spread of farming are greater than 2 km/year. Ammerman and Cavalli-Sforza calculated that the average rate from Greece to the British Isles was close to 1 km/year, which is within the model's expectations. Although regional variations in this overall rate were noted, their causes were thought to lie in environmental specificities (as in the Alpine area) or lack of data (as in the eastern Linear Bandkeramik area and Iberia), and it was not conceived that their order of magnitude could be such as to threaten the validity of the model.

As more results were obtained, however, it became clear that in some cases (such as the Linear Bandkeramik and Cardial spreads) these regional rates were more than just 'somewhat faster' or 'somewhat slower' (Zilhão 1993, Price et al. 1995). For instance, because the data indicated an inception of the Neolithic ≈5800 calendar (cal) B.C. in Ligúria, ≈5600 cal B.C. in Valencia, and ≈5400 cal B.C. in central Portugal, the rate of spread across the 2,000 km separating the two ends of the geographic distribution of the Cardial had to have been in the range of 5 km/year. Moreover, the style of decoration in ceramics from undated sites in Portugal implies contemporaneity with the earliest Cardial sites of eastern Spain (Zilhão 1993, Arnaud 1982) and hence an even faster spread. It was suggested, therefore, that the mechanism best explaining the west Mediterranean process was maritime pioneer colonization (Zilhão 1993, 2000), a hypothesis first entertained for southern Portugal by Arnaud (1982) and for which the occupation of the Pacific islands (Kirch 1984, Irwin 1992) provides an analogy.

Chronology of the Iberian Cardial

Evidence accumulated over the last few years in Iberia shows that eliminating disturbed sites and mixed levels is not sufficient to obtain an adequate evaluation of the rate of spread of the Cardial. There are also problems with the dating of bulk charcoal or wood samples from levels that are unquestionably in situ because of the old wood effect.

Most frequently, this effect is produced by organic material including charcoal, derived from erosion of the

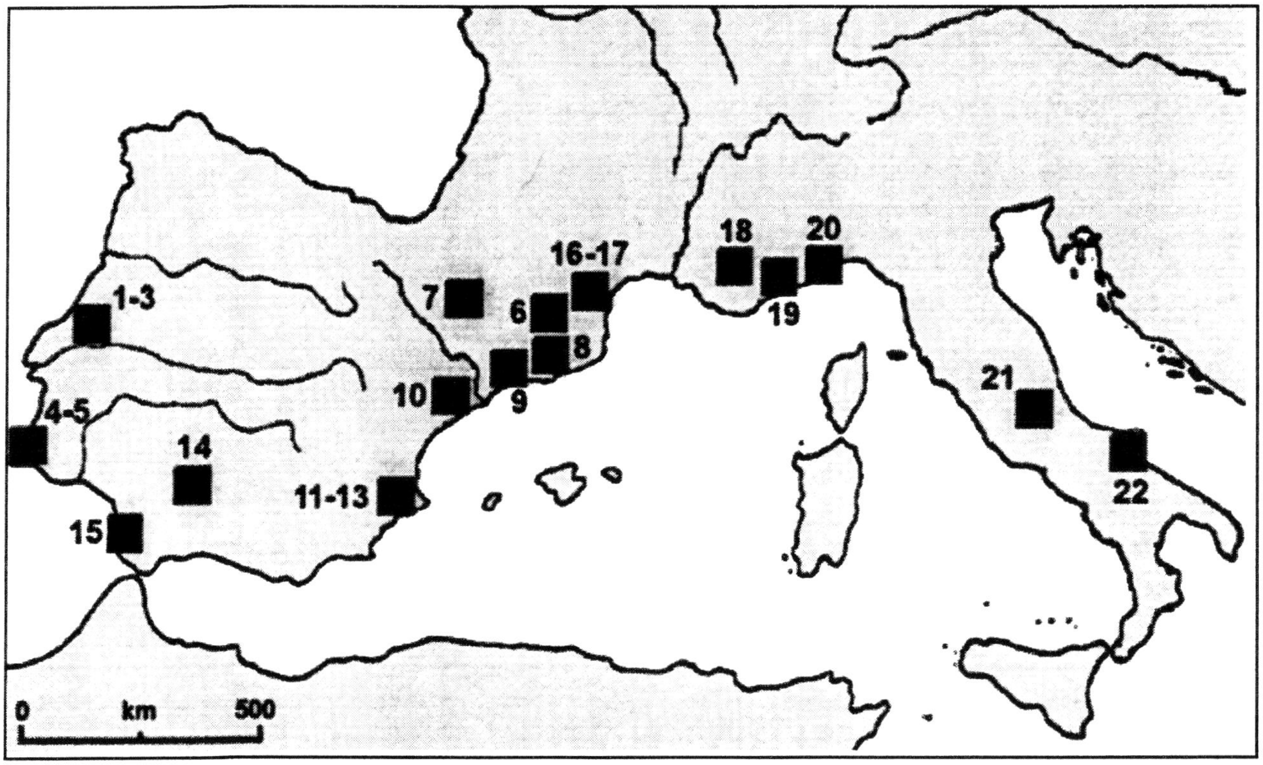

Figure 11.1. Location of sites mentioned in the text. (1-3) Caldeirão, Pena d'Água, and Cisterna; (4-5) Cabranosa and Padrão; (6) Margineda, (7) Chaves, (8) La Draga, (9) Frare, (10) Fosca; (11-13) Cendres, Or, and La Falguera; (14) La Dehesilla, (15) El Retamar; (16-17) Dourgne and Gazel; (18) Baratin, (19) Pendimoun, (20) Arene Candide, (21) San Marco, (22) Coppa Nevigata.

surrounding early Holocene soils that contributed to the formation of deposits accumulated in caves or rock shelters in early Neolithic times. In addition, vertically displaced charcoal from underlying Pleistocene levels may penetrate early Neolithic levels even when no signs of disturbance are visible macroscopically and, hence, go unnoticed during excavation. Finally, the use as timber or as fuel of the centuries-old oak trees that constituted the Mediterranean forest cleared by the first agro-pastoral groups may result in the presence in early Neolithic deposits of wood or charcoal material belonging to the inner rings of those trees, the carbon-14 content of which had begun decaying before they were felled or burned. Ages obtained from samples of such material or including such material may be significantly older than the archaeological events for which dating was sought.

On the basis of evidence from the Portuguese cave site of Caldeirão (Zilhão 1992) and the underwater lakeside dwelling of La Draga in northern Catalonia (Tarrús et al. 1994), it already had been suggested that this problem affected the dating of the earliest Neolithic in west Mediterranean Europe (Zilhão 1993). At Caldeirão, AMS dates on short lived samples of diagnostic Neolithic material (human and sheep bone) indicated an age of ≈5150 cal B.C. for the site's Cardial occupation (Table 11.1). A conventional date on wood charcoal, however, indicated an age of ≈5750 cal B.C., 600 years earlier. At La Draga, conventional dates on charred cereal seeds indicated an age of ≈4900 cal B.C., as did similar dates on wood charcoal from hearth features, but the wood of an oak pillar was dated to ≈5350 cal B.C., 450 years earlier.

A recently published dating experiment (Bernabeu 1989) demonstrated conclusively that the old wood effect was indeed a major factor to be considered when accounting for the variability in dates from early Neolithic Iberian sites. The first result to be obtained for the Cardial deposits in Cova de les Cendres (Valencia) is 7540±140 BP. The sample was made up of bulk charcoal, and the result was much earlier than the ≈6700 BP that would have been expected based on the dating of the Cardial elsewhere in Spain and France. This result prompted an investigation of the plant species represented in the charcoal assemblage and the consequent identification of *Pinus nigra* and *Juniperus* sp., the most common taxa in the underlying Pleistocene levels. Individual pieces of *P. nigra* and *Quercus* sp. found in the Cardial deposits were then submitted to AMS dating, with results of 20430±170 BP for the former and 8310±80 BP for the latter.

These results proved beyond a doubt that pieces of Pleistocene and early Holocene charcoal had made their way into the Cardial deposits. Dating a new series of bulk charcoal samples from which *P. nigra* and *Juniperus* sp. were excluded provided results closer to what might have been expected and that were accepted as good by the au-

Table 11.1. Radiocarbon dates for the Cardial Neolithic in Iberia.

Site name	Site type	Provenience	Sample	Lab number	Date BP	Cal B.C. 1σ	Cal B.C. 2σ
Caldeirão	Cave	Layer Eb	Wood charcoal	ICEN-296	6870 ± 210	5970-5570	6120-5370
		Horizon NA2	*Ovis aries* bone	OxA-1035	6330 ± 80	5348-5231	5480-5079
		Horizon NA2	*Ovis aries* bone	OxA-1034	6230 ± 80	5302-5072	5340-4940
		Horizon NA2	Human bone	OxA-1033	6130 ± 90	5226-4941	5296-4843
Pena d'Água	Rock-shelter	Layer Eb (base)	*Olea* sp. charcoal	Wk-9214	6775 ± 60	5724-5625	5766-5561
		Layer Eb (base)	Wood charcoal	ICEN-1146	6390 ± 150	5467-5215	5579-4993
Almonda	Cave	Cisterna-AMD2, level 1	Pierced *Cervus elaphus* canine	OxA-9287	OxA-92871	5473-5369	5477-5321
		Cisterna-AMD2, level 1	Bone bead	OxA-9288	6445 ± 45	5473-5369	5477-5321
Cabranosa	Open air	Neolithic hearth	*Mytilus* sp. shells	Sac-1321*	6930 ± 60	5563-5389	5579-5325
Padrão	Open air	Hearth	*Tapes decussata* shells	ICEN-873*	6920 ± 60	5521-5386	5577-5318
		Hearth	*Cerastoderma edule* shells	ICEN-645*	6800 ± 50	5432-5278	5442-5255
Margineda	Cave	Level 3b base	Wood charcoal	Ly-2839	6670 ± 120	5627-5443	5725-5336
		Level 3b	Wood charcoal	Ly-3289	6850 ± 150	5839-5583	5977-5443
		Level 3a	Wood charcoal	Ly-3288	6640 ± 160	5633-5389	5773-5263
Chaves	Cave	Level I	Wood charcoal	GRN-12685	6770 ± 70	5680-5582	5726-5523
		Level I	Wood charcoal	GRN-12683	6650 ± 80	5593-5446	5662-5433
		Level I	Wood charcoal	CSIC-378	6460 ± 70	5440-5312	5521-5262
		Level I	Wood charcoal	GRN-12686	5210 ± 340†	Unacceptable, too young	
La Draga	Open air	Hearth E-6	Wood charcoal	GAK-1523†	5710 ± 170	4772-4357	4940-4175
		Hearth E-6	Wood charcoal	UBAR-245†	5920 ± 240	5194-4498	5366-4259
		Hearth E-40	Wood charcoal	UBAR-311†	5970 ± 110	4960-4718	5208-4552
		Hearth E-50	Wood charcoal	UBAR-312†	6570 ± 460	5929-5005	6360-4466
		Garbage disposal in H-30	Animal bone	UBAR-315†	6700 ± 710	6190-4854	7042-4046
		Post E-106	*Quercus* wood	UBAR-314	6410 ± 70	5432-5269	5444-5228

		Hearth E-56	Cereal seeds	UBAR-313	6010 ± 70	4951-4806	5065-4729
		Hearth E-3	Cereal seeds	Hd-15451	6060 ± 40	5031-4914	5061-4845
Frare	Cave	T22-23, level 5c	Wood charcoal	I-13030	6380 ± 310	5579-4946	5839-4594
Cendres	Cave	H19	Single *Pinus nigra* fleck	Beta-116625†	20430 ± 170	Unacceptable, too old	
		H19a	Single *Quercus* fleck	Beta-116624†	8310 ± 80	Unacceptable, too old	
		VIe	Wood charcoal, contaminants included	Ly-4302†	7540 ± 140	Unacceptable, too old	
		VII	Wood charcoal, contaminants excluded	Beta-75220	6730 ± 80	5662-5526	5712-5446
		H18	Wood charcoal, contaminants excluded	Beta-75219	6420 ± 80	5435-5269	5470-5225
		H17, fireplace	Wood charcoal	Beta-75218	6260 ± 80	5270-5076	5370-4990
		Estrato VII	*Hordeum vulgare*	Beta-142228	6340 ± 70	5457-5262	5474-5081
		VIIa	*Ovis aries* bone	Beta-107405	6280 ± 80	5280-5086	5416-5049
Or	Cave	J4, Levels 16-17	Wood charcoal	GANOP-C13	6720 ± 380	5956-5270	6356-4837
		J4, levels 14-15	Wood charcoal	GANOP-C12	6630 ± 290	5733-5270	6009-4938
		Basal Cardial (1955-58)	Cereal seeds	KN-51	6510 ± 160	5618-5318	5722-5079
		J4, level 17	*Triticum aestivum*	OxA-10192	6310 ± 70	5359-5153	5469-5067
		Upper Cardial (1955-58)	Cereal seeds	H-1754/1208	6265 ± 75	5316-5079	5459-5036
		J4, level 14	*Triticum aestivum*	OxA-10191	6275 ± 70	5317-5083	5459-5048
La Falguera	Rock shelter	EU 2051b	*Triticum monococcum*	Beta-142289	6510 ± 70	5512-5381	5616-5321
El Retamar	Open air		Marine shells	Beta-90122*	6780 ± 80	5434-5259	5470-5143

* Calibrated after subtraction of 380 ± 30 years for correction of the reservoir effect.

† Not included in Fig. 11.1 because the result is unacceptable, or the large standard deviation makes its representation redundant or irrelevant in regard to other samples from the same site.

thors of the experiment: 6730±80 BP and 6420±80 BP. These samples included *Quercus* sp., however, and given the above-mentioned 8310±80 BP result for an individual fleck of *Quercus* sp. charcoal from these levels, in all probability these new determinations are skewed also by the old wood effect. This interpretation receives considerable support from the date for another individual fleck of *Quercus* sp. collected from the uppermost preceramic level of the site: 6670±80 BP (Bernabeu 2001).

A sample of sheep bone, a short lived diagnostic Neolithic item, was dated to 6280±80 BP. An almost identical result was obtained from charcoal collected in a hearth feature: 6260 ± 80 BP. Both are indistinguishable statistically from a new AMS date of 6340±70 BP for a sample of barley, *Hordeum vulgare* (Bernabeu 2001). Together, these three dates show that the 2σ calibrated age of the earliest Neolithic occupation of Cendres lies in the interval between 5000 and 5500 cal B.C.

The major early Neolithic site in Mediterranean Spain undoubtedly is the Cova de l'Or, with its wealth of baroquely decorated ceramic vessels, sophisticated bone tools, personal ornaments, and abundance of cereal and sheep remains (Martí et al. 1980). These assemblages have been thought to date to ≈5600 cal B.C. on the basis of conventional carbon-14 results on well provenienced bulk charcoal samples from Bernat Martí's 1970s excavations. Despite their large standard deviations, these results might be taken to suggest that the earliest agro-pastoral economies of the region were only slightly later than those of Provence and Ligúria.

Zone AMD2, a burial chamber in the multicomponent Portuguese site of Galeria da Cisterna (Almonda karstic system) excavated in 1988-89 (Zilhão et al. 1991), features baroquely decorated Cardial ceramics, stylistically similar to those from the basal levels of Or, and the same range of ornaments found throughout the latter's early Neolithic sequence. This suggests that the ceramics were deposited in the cave during funerary rituals by the earliest agro-pastoral settlers of littoral-central Portugal and that at least some of the ornaments might belong also to that early Cardial context. In fact, two particular kinds of pendants that are well represented in the Cisterna assemblage, pierced red deer canines and bone beads imitating their shape, also are absent completely from both the previous Mesolithic and later post-Cardial periods of Portuguese prehistory.

Because the stylistically early Cardial ceramics from Cisterna were part of a palimpsest accumulated over time in thin Holocene deposits that included later Epicardial, Copper Age, Bronze Age, and Iron Age pottery, no faunal or human bone material could be associated securely with the early Cardial component. Thus, the Cisterna Cardial remained undated until recent publication of the ornaments from Or (Pascual 1996, 1998) revealed that the two kinds of pendants mentioned above, pierced red deer canines and bone beads imitating them, are characteristic of the earliest part of that site's Neolithic sequence. This strengthens the hypothesis that such artifacts indeed were associated with the baroquely decorated Cardial pottery from Cisterna and that the Valencian Cardial (as at Or) represents the original cultural background of the first farming societies in littoral-central Portugal. The hypothesis was tested through AMS dating of one specimen of each ornament type, which provided the same result for both samples: 6445±45 BP, that is, ≈5400 cal B.C.

These results are some 200-radiocarbon years younger than current estimates for the age of the lowermost Cardial levels from Or. Because the latter could be skewed by an old wood effect, as suggested by the Cendres experiment, two samples from the same levels, each made up of two charred wheat seeds, were submitted for AMS dating with the following results: 6275±70 BP for level 14 (upper Cardial) and 6310±70 BP for level 17 (basal Cardial). These results are identical statistically to previously available conventional dates on poorly provenienced seed samples from the 1950s excavations. At 2σ, the dates for Cisterna fall within the lower part of the time range represented by the new Or results, which therefore represent a second positive test of the hypothesis of contemporaneity and very close cultural connection between the earliest Neolithic occupations of the two sites.

Cova de l'Or also represents a fourth demonstration, with Caldeirão, La Draga, and Cendres, that the use of bulk charcoal samples blurs what in fact is a very clear picture (Fig. 11.2). When only results obtained on short lived samples of diagnostic Neolithic items are considered (including those on marine shells collected in Neolithic hearth features excavated at open air sites), all dates for the earliest Cardial, from Valencia to Portugal, are identical statistically. This conclusion receives further support from the new AMS date of 6510±70 BP for wheat seeds from the Cardial levels of the La Falguera rock shelter (Bernabeu 2001). Such contemporaneity implies a very rapid spread, within a maximum of four or five generations, of the Iberian Cardial.

Chronology of the Earliest Neolithic in France and Italy

The Iberian pattern suggests that radiometric chronology for the earliest Neolithic in neighboring regions to the east also may be skewed significantly by the old wood effect. Unfortunately, this hypothesis is difficult to test, because there are very few direct dates on diagnostic material. Those data that are available (Pascual 1998, Binder 1995, Maggi 1997, and BANADORA database, *http://carbon14.univ-lyon1.fr/banadora.htm)*, however, indicate the same problem for those regions.

At the open air site of Baratin (Binder 1995, BANADORA database) in Provence, AMS dates on individual pieces of charcoal from three different features provided ages between 6125±80 BP (Lyon-100/OxA) and 6290±70 BP (Lyon-252/OxA), several centuries younger than the previously available conventional date of 6600±140 BP (Gif-1855). At the Arene Candide cave (Maggi 1997),

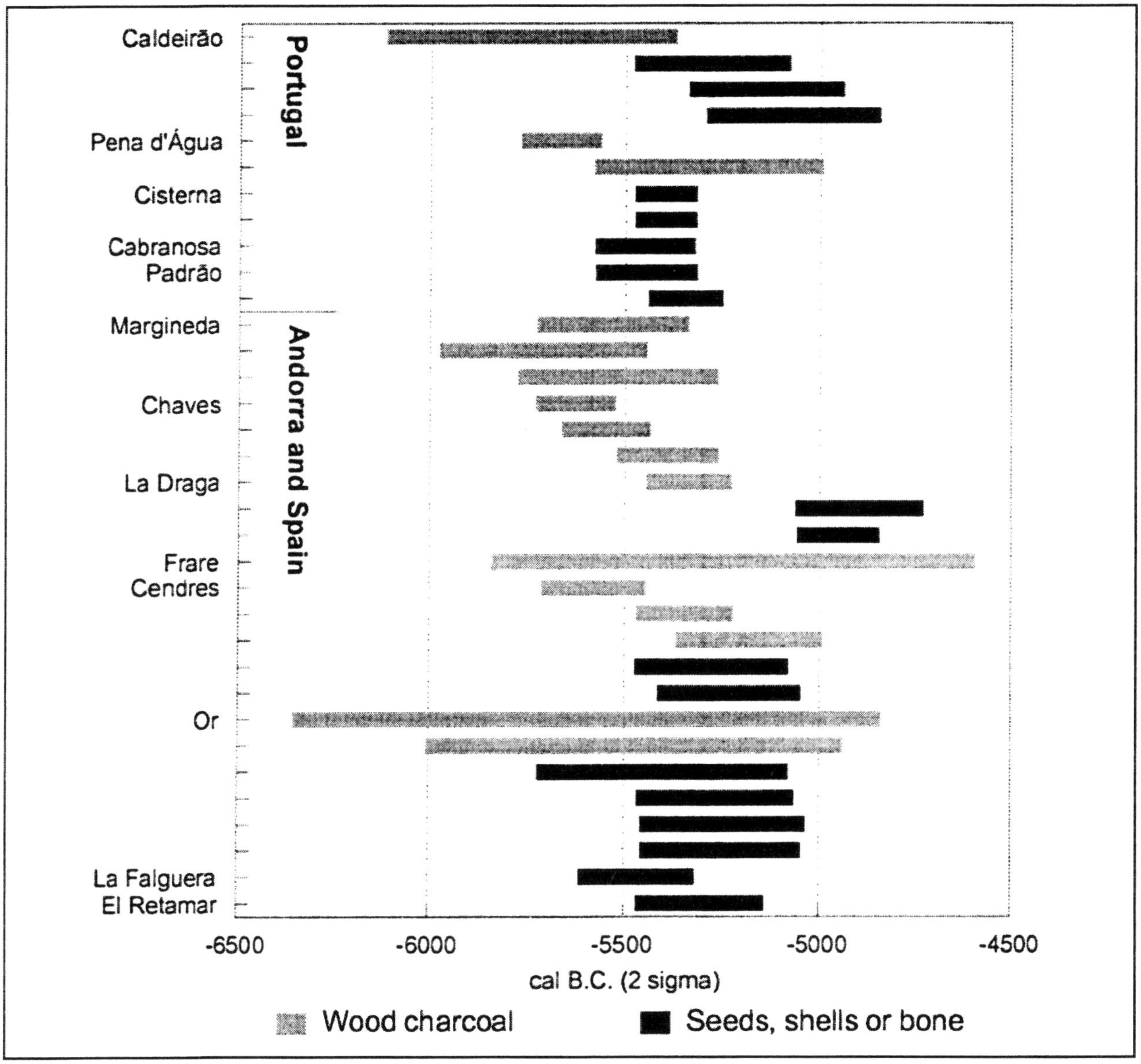

Figure 11.2. Calibrated dates for all dated Iberian Cardial sites (x ± 2σ).

AMS dating of identified charcoal showed that level 27 yielded both 6150 ± 70 BP (Beta-66552) remains of *Phillyrea* sp. and 6880±60 BP (Beta-66553) remains of *Pistacia terebintus.* At Coppa Nevigata (Hedges et al 1989, Skeates 2001), two dates on cereal seeds provided virtually identical ages of 6850±80 BP (OxA-1474) and 6880±90 BP (OxA-1475), yet the site was dated previously on conventional bulk charcoal to 7780±320 BP (BM-2557).

The only other site in these regions for which AMS direct dates on diagnostic Neolithic material are available is that of San Marco in Umbria, central Italy (Skeates 2001, Hedges et al. 1990), where three samples of *Triticum aestivum, Triticum compactum,* and *H. vulgare* seeds provided results of 6430±80 BP (OxA-1853), 6270±70 BP (OxA-1851), and 6120±90 BP (OxA-1854), respectively. Two other conventional results can be considered also as unmistakably related to the early Neolithic use of the sites in question: the 6490±75 BP (Ly-5340) date on acorns collected inside a burial feature at the Abri Pendimoun (Binder 1995, BANADORA database) and the 6255±255 BP (GX-16963G) date on human bone from the burial VII of Arene Candide (Maggi 1997).

When calibrated, these results indicate that the Neolithic begins significantly earlier in southern Italy, perhaps as early as 6000 cal B.C. but that northward and westward all available dates on short lived diagnostic samples are identical to those for Iberia (Fig. 11.3). The 2s range for the Cisterna dates (5477-5321 cal B.C.), for instance, falls inside that for the Pendimoun burial (5613-5316 cal B.C.), which is in turn virtually identical to that for La Falguera (5616-5321 cal B.C.). Therefore, the rapidity of spread mentioned above for Iberia applies to the dispersal of the Cardial and related cultures as a whole: 2,000 km from the gulf of Genoa to the estuary of the Mondego in probably no more than 100-200 years at most, that is, at a rate of at least 10-20 km/year. It should be stressed that this conclusion holds even if the French chronology based on conventional bulk charcoal samples is retained, because identical results on the same kinds of samples from in situ Cardial deposits are known in Portugal, not only at Caldeirão but also at Pena d'Água, where a sample of *Olea* charcoal recently has been dated to 6775±60 BP (Carvalho 2001). In any case, these data make it clear that the chronology of the early Neolithic in France and Italy must be revised entirely on the basis of more direct dates of domesticates or on other diagnostic short lived Neolithic materials such as bone artifacts or human bone from burials.

Conclusions

By using Ammerman and Cavalli-Sforza's equations (Ammerman & Cavalli-Sforza 1984) and an annual rate of population growth of 1%, one can calculate a rate of spread of 10 km/year, implying a rate of migratory activity of 60,000 km^2 per generation, 30 times greater than the maximum observed ethnographically. Demic diffusion therefore cannot have proceeded through a wave of advance mechanism of short distance settlement expansion wherein population growth was accommodated through gradual and slow incorporation of adjacent land.

The area covered by the 40/50-km-wide coastal strip between the Mondego River in Portugal and the cape of Nao, the southern limit of the gulf of Valencia, is precisely 60,000 km^2. In the framework of maritime pioneer colonization, the similarities in ornaments and pottery decoration observed between the two extremes of this range can be taken as evidence for a swift expansion with maintenance of cultural traditions, and the littoral placement of settlements is evidence for diffusion through sea routes. In fact, under reasonable estimates of annual population growth, the observed rate of spread across such a large area requires the operation of long distance relocation episodes. It also implies such low population densities across the whole of the settled range that large voids must be postulated between nodes of the farmers' settlement network.

This prediction is met in the archaeological record by the 'enclave' nature of early Neolithic territories in lit-

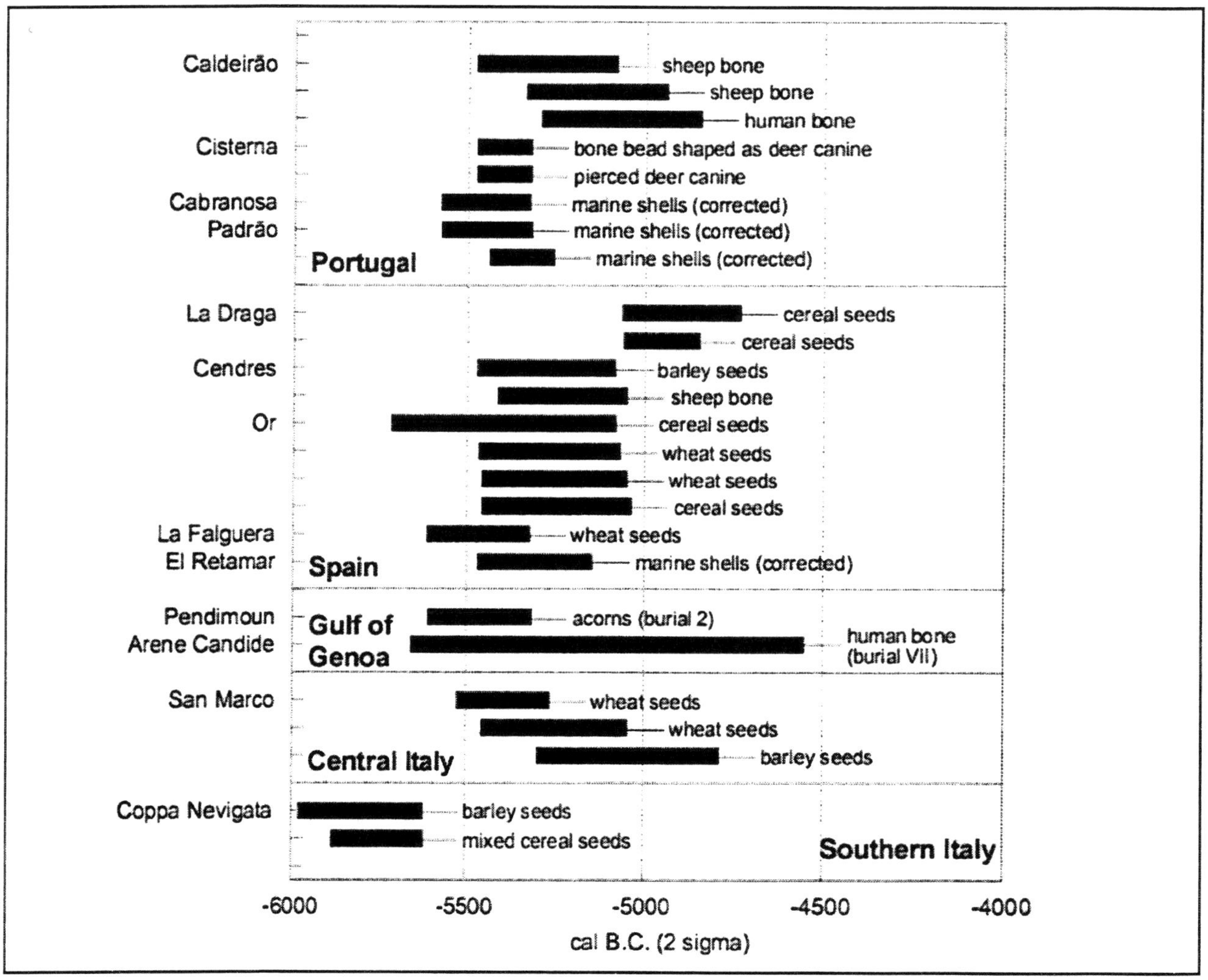

Figure 11.3. Calibrated dates on short lived diagnostic samples for the earliest Neolithic of west Mediterranean Europe ($\bar{x} \pm 2\sigma$).

toral-central Portugal, which occupy areas previously uninhabited by late Mesolithic hunter-gatherers. The latter continue to thrive in their own territories for some 500 years after initial Neolithic settlement (Zilhão 1993, 2000; Fig. 11.4), but contrasts in material culture, economy, nutrition, and mortuary behavior remain unchanged throughout this period. Neolithic sites feature pottery, polished hand axes, and heat pretreatment of flint, which are absent altogether from the Mesolithic, as are domesticates. Neolithic skeletons feature isotopic signatures of a fully terrestrial diet, whereas Mesolithic ones indicate a 50% marine component. Neolithic people are buried in special-purpose collective funerary sites, whereas Mesolithic people are buried individually in habitation sites and have never been found carrying exclusively Neolithic body ornaments (tear-shaped *Glycymeris* beads, pierced red deer canines, and bone beads imitating them).

The rapidity of the spread also indicates that long distance colonization events took place well before saturation levels were attained at the point of origin, suggesting that purely historical causes were in operation and that ultimate explanations for the phenomenon must lie within the specific features of antecedent processes triggering the expansion of agro-pastoral economies from their Middle Eastern core areas. One possibility is that after the collapse of the Levantine Pre-Pottery Neolithic B (PPNB), for which there is significant evidence of social stratification and strongly developed cult practices, the succeeding westward-spreading Neolithic societies, which lack any archaeological evidence of specially built temples and of ranking in settlement or in burial, fissioned before groups became too large and severe conflict or social inequality developed (Zilhão 2000, Özdogan 1997, Zilhão 2001). Along the north Mediterranean shores, this tendency to fission and move on would have been reinforced further because opportunities for settlement and expansion around initial enclaves were limited by physical geography and the presence of local hunter-gatherer groups.

Acknowledgements

A. Ammerman, W. K. Barnett, J. Bernabeu, B. Martí, A. Gilman, A. M. Soares, E. Trinkaus, P. J. Watson, and two anonymous reviewers contributed helpful comments on previous versions. The AMS dates for Galeria da Cisterna and Cova de l'Or were funded by the Instituto Português de Arqueologia.

Abbreviations

AMS, accelerator mass spectrometer; BP, before present; Cal, calendar.

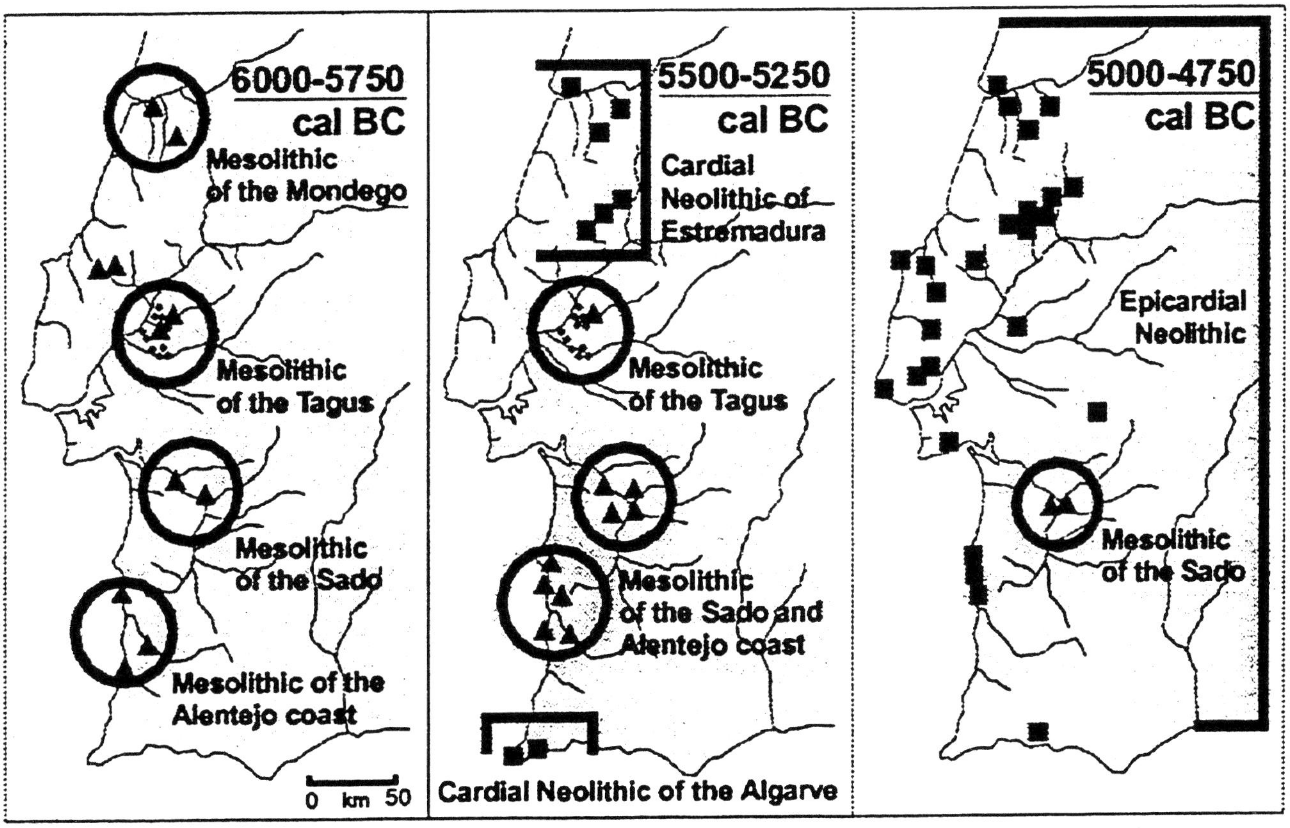

Figure 11.4. Geographic distribution of late Mesolithic and early Neolithic settlements in south-central Portugal between 6000 and 4750 cal B.C.

References

Acosta, P. & M. Pellicer
1990 *La Cueva de la Dehesilla (Jerez de la Fontera): las Primeras Civilizaciones Productoras en Andalucía Occidental.* Jerez: Consejo Superior de Investigaciones Científicas.

Ammerman, A.
2003 Looking back. In *The Widening Harvest* (A.J. Ammerman & P. Biagi, eds.), pp 3-23. Boston: Archaeological Institute of America.

Ammerman, A. J. & L.L Cavalli-Sforza
1973 A population model for the diffusion of early farming in Europe. In *The Explanation of Culture Change: Models in Prehistory* (C. Renfrew, ed.), pp. 343- 357. London: Duckworth.
1984 *The Neolithic Transition and the Genetics of Population in Europe.* Princeton: Princeton University Press.

Arnaud, J. M.
1982 *Le Néolithique Ancien Méditerranéen*, pp. 29-48. Montpellier, France: Fédération Archéologique de l'Hérault.

Baldellou, V. & A. Castán
1983 Excavaciones en la Cueva de Chaves de Bástaras (Casbas-Huesca). *Bolskan* 1: 9-37.

Baldellou, V. & P. Utrilla
1985 Nuevas dataciones de radiocarbono de la prehistoria Oscense. *Trabajos de Prehistoria* 42: 83-95.

Barnett, W. K.
2000 Cardial pottery and the agricultural transition in Mediterranean Europe. In *Europe's First Farmers* (T.D. Price, ed.), pp. 93-116. Cambridge: Cambridge University Press.

Bernabeu, J.
1989 *La Tradición Cultural de las Cerámicas Impresas en la Zona Oriental de la Península Ibérica.* Valencia: Servicio de Investigación Prehistórica.
2001 *Neolithic Landscapes of the Mediterranean.* València: Universitat de València. In press.

Bernabeu, J., M. Pérez & R. Martínez
1999 Ituesos, neolitización y contextos arqueológicos aparentes. In *Actes del II Congrés del Neolític a la Península Ibèrica* (J. Bernabeu & T. Orozco, eds.), pp. 589-596. València: Universitat de València.

Binder, D.
1995 Eléments pour la chronologie du Néolithique ancien á cèramique imprimèe dans le Midi. In *Chronologies Néolithiques de 6000 à 2000 avant Notre Ère dans le Bassin Rhodanien* (J.-L. Voruz, ed.), pp. 55-65. Ambérieu-en-Bagey: Société Préhistorique Rhodanienne.

Cardoso, J. L., J.R. Carreira & O.V. Ferreira
1996 Novos elementos para o estudo do Neolîtico antigo da região de Lisboa. *Estudos Arqueológicos de Oeiras* 6: 9-26.

Carvalho, A. F.
2001 Abrigo da Pena d'Água. *Relatório das Escavações de 2000.* Lisboa: IPA Archive.

Estévez, J.
1988 In *Cova Fosca. Un Asentamiento Meso-neolítico de Cazadores y Pastores en la Serranía del Alto Maestrazgo* (C. Olaria, ed.), pp. 281-337. Castellón: Servicio de Publicaciones de la Diputación de Castellón.

Fritz, G. J.
1994 Are the first American farmers getting younger? *Current Anthropology* 35: 305-309.

Geddes, D.
1983 Neolithic transhumance in the Mediterranean Pyrenees. *World Archaeology* 15: 51-66.

Guilaine, J., M. Barbaza, J. Gasco, D. Geddes, J. Coularou, J. Vaquer, J.-E. Brochier, F. Briois, J. André, G. Jalut & J.-L. Vernet
1993 *Dourgne. Derniers Chasseurs-collecteurs et Premiers Éleveurs de la Haute-Vallée de l'Aude.* Toulouse/Carcassonne: Centre d'Anthropologie des Sociétés Rurales/Archéologie en Terre d'Aude.

Guilaine, J. & M. Martzluff
1995 *Les Excavacions a la Balma de la Margineda (1979-1991).* Andorra: Govern d'Andorra.

Hedges, R. E., R.A. Housley, I.A. Law & C. Bronk Ramsey
1989 Radiocarbon-dates from the Oxford AMS system – archaeometry date list 9. *Archaeometry* 31: 207-234.
1990 Radiocarbon-dates from the Oxford AMS system – archaeometry date list 10. *Archaeometry* 32: 101-108.

Irwin, G.
1992 *The Prehistoric Exploration and Colonisation of the Pacific.* Cambridge: Cambridge University Press.

Kirch, P. V.
1984 *The Evolution of the Polynesian Chiefdoms.* Cambridge: Cambridge University Press.

Lazarich, M., J. Ramos, V. Castañeda, M. Pérez, N. Herrero, J.M. Lozano, E. García, S. Aguilar, M. Montañés & C. Blanes

1997 El Retamar (Puerto Real, Cádiz): Un asentamiento Neolítico especializado en la pesca y el marisqueo. In *II Congreso de Arqueología Peninsular. Tomo I-Paleolítico y Epipaleolítico* (R. Balbín & P. Bueno, eds.), pp. 49-58. Zamora: Fundación Rei Afonso Henriques.

Maggi, R.

1997 *Arene Candide: a Functional and Environmental Assessment of the Holocene Sequence (Excavations Bernabo' Brea-Cardini 1940-50).* Roma: Istituto Italiano di Paleontologia Umana.

Martí, B., V. Pascual, M.D. Gallart, P. López, M. Pérez, J.D. Acuña & F. Robles

1980 *Cova de l'Or (Beniarres, Alicante).* Valencia: Servicio de Investigación Prehistórica, Monografía No. 2.

Martin, A.

1986-89 Reflexión sobre el estado de la investigación del neolítico en Cataluña y su reflejo en la cronología radiomètrica. *Empúries* 48-50: 84-102.

1990 In *Autour de Jean Arnal* (J. Guilaine & X. Gutherz, eds.), pp. 37-54. Montpellier: Université des Sciences et Techniques du Languedoc.

Monge Soares, A.

1993 *Isotope Techniques in the Study of Past and Current Environmental Changes in the Hydrosphere and the Atmosphere*, pp. 471-485. Vienna: International Atomic Energy Agency.

Olaria, C.

1988 *Cova Fosca. Un asentamiento Meso-neolítico de Cazadores y Pastores en la Serranía del Alto Maestrazgo.* Castellón: Servicio de Publicaciones Diputación de Castellón, Monografies de Prehistoria i Arqueologia Castellonenques No. 3.

Özdogan, M.

1997 The beginning of Neolithic economies in southeastern Europe: an Anatolian perspective. *Journal of Mediterranean Archaeology* 5: 1-33.

Pascual, J. L.

1996 *Recerques del Museu d'Alcoi* 5: 17-52.

1998 *Utillaje Óseo, Adornos e Ídolos Neolíticos Valencianos.* Valencia: Servicio de Investigación Prehistórica.

Price, T. D., A. Gebauer & L. Keeley

1995 The spread of farming into Europe north of the Alps. In *Last Hunters, First Farmers: New Perspectives on the Prehistoric Transition to Agriculture* (T. D. Price & A. Gebauer, eds.), pp. 95-126. Santa Fe: School for American Research Press.

Rowley-Conwy, P.

1995 Making first farmers younger: the west European evidence. *Current Anthropology* 36: 346-353.

Skeates, R.

2003 Radiocarbon dating and interpretations of the Mesolithic-Neolithic transition in Italy. In *The Widening Harvest* (A.J. Ammerman & P. Biagi, eds.), pp 157-187. Boston: Archaeological Institute of America.

Spriggs, M.

1989 The dating of the island Southeast Asian Neolithic – an attempt at chronometric hygiene and linguistic correlation. *Antiquity* 63: 587-613.

Stuiver, M., P.J. Reimer, E. Bard, J.W. Beck, G.S. Burr, K.A. Hughen, B. Kromer, F.G. McCormac, J. van der Plicht & M. Spurk

1998 INTCAL98 radiocarbon age calibration, 24,000-0 cal BP. *Radiocarbon* 40: 1041-1083.

Tarrús, J., J. Chinchilla & A. Bosch

1994 La Draga (Banyoles): Un site lacustre du Néolithique ancien cardial en Catalogne. *Bulletin de la Société Préhistorique Française* 91: 449-456.

Whittle, A.

1990 Radiocarbon dating of the linear pottery culture – the contribution of cereal and bone samples. *Antiquity* 64: 297-302.

1996 *Europe in the Neolithic: the Creation of New Worlds.* Cambridge: Cambridge University Press.

Zilhão, J.

1992 *Gruta do Caldeirão. O Neolítico Antigo.* Lisboa: Instituto Português do Património Arquitectónico e Arqueológico.

1993 The spread of agro-pastoral economies across Mediterranean Europe: a view from the Far West. *Journal of Mediterranean Archaeology* 6: 5-63.

2000 From the Mesolithic to the Neolithic in the Iberian Peninsula. In *Europe's First Farmers* (T. D. Price, ed.), pp. 144-182. Cambridge: Cambridge University Press.

2003 The Neolithic transition in Portugal and the role of demic diffusion in the spread of agriculture across west Mediterranean Europe. In *The Widening Harvest* (A.J. Ammerman & P. Biagi, eds.), pp 207-223. Boston: Archaeological Institute of America.

Zilhão, J. & A. F. Carvalho
1996 O Neolítico do Maciço Calcário Estremenjo. Crono-estratigrafia e povoamento. In *Actes de I Congrés del Neolític a la Península Ibérica, Vol. 2,* pp. 659-671. Gavà: Museu de Gavà.

Zilhão, J. , J. Maurício & P. Souto, P.
1991 A arqueologia da gruta do Almonda. Resultados das escavações de 1988-1989. In *Actas das IV Jornadas Arqueológicas. (Lisboa 1990),* pp. 161-171. Lisboa: Associação dos Arqueólogos Portugueses.

Zvelebil, M. (editor)
1986 *Hunters in Transition: Mesolithic Societies of Temperate Eurasia and their Transition to Farming.* Cambridge: Cambridge University Press.

12

Subsistence Diversification and Mesolithic Technology: a Study of the Portuguese Mesolithic

Bradley J. Vierra
University of Texas, San Antonio

Abstract

Why do foraging societies use a variety of technologies to produce subsistence resources? How do we explain the current and past diversity in foraging technology, as well as the long-term process of technological change which led to this diversity? Given that stone tool technology is one of the most ubiquitous remnants of past human behavior, these are indeed important questions to ask. They are also difficult questions to answer, but they are ones that hunter-gatherer archeologists are beginning to address. Yet despite the importance of technology to studies of human evolution, most research has focused on methodological and not theoretical issues concerning prehistoric stone tool technology. One possible explanation for these long-term changes in technology is that of subsistence diversification. Changes in the number and types of resources being procured, can have an important effect on the organization of labor and foraging technology. These changes may in part account for the shift from non-geometric to geometric based technologies during the Portuguese Mesolithic.

Introduction

The current empirical evidence in southwestern Europe indicates that hunting and gathering groups began to diversify the range of resources being exploited about 20-14,000 years ago (Straus et at. 1981, Straus & Clark 1986). Concurrent with this shift in subsistence economy, was the transition from bifacial to microlithic technologies. This process of increasing the number of different food items being consumed (i.e., subsistence diversification) and the use of microlithic technologies continued and became most pronounced during the late Mesolithic (ca. 8000-6000 BP). Given that the use of microlithic technologies appears to correspond to a long-term trend in subsistence diversification that cross-cuts the Pleistocene-Holocene boundary, these changes may be a response to overall subsistence diversification and to an increasing reliance on aquatic resource use. This paper will evaluate subsistence diversification as a possible explanation for changes in technology. A cross-cultural study will first be used to evaluate the relationship between subsistence diversification and technology. Given the implications of this study, a set of inductively derived predictions will be generated and subsequently tested on a sample of Portuguese Mesolithic sites.

Subsistence Diversification and Technological Specialization

The Mesolithic of western Europe is generally characterized by increasing subsistence diversification, a growing reliance on aquatic resources, storage, decreasing residential mobility and the use of composite geometric-based tools (Bonsall 1985, Vermeersch & Van Peer 1990). What is the relationship between prehistoric foraging strategies, foraging tactics and technology? *Foraging strategies* refers to what foods were being consumed, *foraging tactics* to how these foods were being procured, and in this study, *technology* refers to the tools that were primarily being used to procure these foods. This section will empirically evaluate the assumptions underlying subsistence diversification as a possible explanation for changes in foraging technology. Binford (1980) and Kelly (1983) have discussed the relationship between foraging tactics and resource structure. Their studies predict the increasing use of logistical mobility in heterogeneous environments characterized by shortened growing seasons, greater spatial and temporal incongruity in the distribution of resources and a reliance on storage. In respect to technology, Torrence (1983) has argued that we can also expect greater technological specialization in heterogeneous environments marked by greater seasonality, shorter periods of resource availability and less accessible resources. Ethnographic research shows that these conditions generally occur in mid to high latitudes above 40° latitude (i.e., with growing seasons of less than about 200 days, also see Vierra [1989]).

It was Torrence (1983: 14) who introduced the concept of 'time-stress', that can be measured by the length of time that a resource is available and the type of resource being exploited. The latter includes plants that are most accessible, terrestrial animals that are less accessible and aquatic resources that are the least accessible. Although not implied in her original use of the concept, time-stress can be defined as *temporal constants that limit an individual's ability to procure subsistence resources and thereby inhibit their ability to survive and reproduce.* In an evolutionary sense, time-stress can provide a selective pressure, whereby tool efficiency varies as a function of resource availability and accessibility and does not simply imply a progressive notion of evolution (e.g., as with the vitalistic notion of increasing bow-and-arrow efficiency during the Mesolithic [Fisher 1985: 38]).

Nonetheless, efficiency is a relative term. In this study it is defined as *a tool's ability to reduce the amount of time needed to complete a task and, therefore, to increase foraging returns per unit of time.* Therefore, given conditions when time is a limiting factor, selection should favor the use of more efficient tools. As suggested by Torrence, it is assumed that 'highly specialized tools (i.e., artefacts designed to be used in a small number of function) are more efficient than general-purpose equipment' (Torrence 1983: 13).

Although time-stress may be related to variations in environment and resource structure, it could also be a function of increasing subsistence diversification and scheduling conflicts, it has been argued that increasing diet-breadth may be the result of regional population growth, regional packing and declining foraging returns (Binford 1983: 211; Cohen 1977, Layton et al. 1991, Clark & Yi 1983, Clark 1987, Price & Brown 1985: 8, Straus & Clark 1986). Given a density-dependent situation, foods appear to be added to the decreasing order of profitability; as more profitable items become less abundant, more costly foods in terms of greater acquisition costs are eventually added to the diet (cf. the diet-breadth model in optimal foraging theory). These include smaller food package sizes and, in some settings, a greater reliance on aquatic resources. It is these latter resources which require very complex technologies for procurement and processing (Oswalt 1976). In conjunction with this, increasing diet-breadth may eventually lead to sequential or simultaneous resource acquisition, and therefore to scheduling conflicts, whereby specific task groups may be formed with specialized technologies geared toward the efficient procurement of a particular resource. Although sequential or simultaneous resource acquisition always occurs to some degree, the difference is in how critical the procurement of a resource is for immediate consumption or storage for later consumption due to its limited availability, and how this affects a shift in the organization of labor and technology in order to increase or ensure greater success in resource acquisition rates. Therefore, greater technological specialization is also expected with increasing subsistence diversification given conditions when time is a limiting factor.

Archeologists often perceive generalist foraging strategies as being more secure than specialist strategies (e.g., Torrence 1983: 13). This assumes that groups dependent on a few species are more prone to starvation when a particular resource fails, whereas groups dependent on a greater number of species are less prone to starvation because of the lessened effect of individual resource failure on the overall subsistence economy. This perspective, however, contrasts with the predictions of the diet-breadth model in optimal foraging theory. This model uses a cost-benefit analysis to rank species with respect to energetic costs of search and handling times vs. the energetic content of a particular prey species. It predicts that lower-ranking resources should only be added to the diet as the result of the decreased abundance and increased foraging costs of higher-ranked resources (MacArthur & Pianka 1966, Emlen 1966, Stephens & Krebs 1986: 17-24). Therefore, foragers residing in food abundant environments should exploit a smaller number of preferred foods (i.e., be specialists), whereas those residing in more unproductive environments should exploit a wider variety of foods (i.e., be generalists).

Torrence (1983: 18-19) argues that technological specialization should decrease as a function of the increasing number of foods added to the diet. This assumes that tool efficiency is a more critical factor for groups who must procure a few species in a limited amount of time. This perspective holds that generalist foraging strategies are an opportunistic response to resource abundance. In contrast, I would argue that technological specialization should increase as a function of the increasing number of foods added to the diet. Following the diet-breadth model, it is assumed that exploiting a large number of species would be more time-stressful than exploiting a few species, since increasing subsistence diversification is not an opportunistic response to resource abundance, but rather to the declining foraging returns of higher ranked species.

The Cross-Cultural Sample

A cross-cultural sample of 53 hunting and gathering societies was selected to test the predicted relationships between subsistence diversification and technology. The ethnographic information used in this chapter represents only part of a larger study presented elsewhere (Vierra 1992, 1995). Data were collected from the original ethnographies on number of subsistence foods, number of subsistants and number of technounits. The number of subsistence foods was recorded for each group and will be used as a measure of subsistence diversification (i.e., diet breadth). A subsistant is a single tool or facility used to obtain food (Oswalt 1976: 38), and a technounit is the number of component parts that make up the tool or facility (Oswalt 1976: 46). The number of subsistants will be used as an indication of tool diversity, and technounits of tool complexity. Together they provide an indication of technological specialization as it relates to procurement technology. Although not included here, a complementary study of processing technology should also be done.

Torrence's 'resource abundance' hypothesis predicts decreasing tool specialization with increasing subsistence diversification, that is, a negative linear relationship between the number of subsistence foods and the number of subsistants and technounits. In contrast, the 'subsistence diversification' hypothesis predicts an increase in tool specialization with increasing subsistence diversification, that is, a positive relationship between the number of subsistence foods and number of subsistants and technounits.

These two alternative hypotheses can easily be tested by doing a regression analysis of these variables. The results show that there is a significant positive correlation between the number of subsistence foods and number of subsistants (r = 0.45, p = <0.0001) and technounits (r = 0.45, p = <0.0001), thus supporting the 'subsistence diversification' hypothesis. That is, groups exploiting a wider variety of resources are more time-stressed and, therefore, at greater risk than those groups exploiting a few resources.

Figure 12.1, however, shows that there are two linear distributions represented. Separate regression lines have been plotted to illustrate these distributions. The question is, what specific resources are conditioning the increase in tool specialization exhibited by the distribution with the higher intercept? In order to answer this question, the two distributions were separated into three parts: (a) groups in the upper distribution (n = 6), (b) groups above the regression line in the lower distribution (n = 23) and (c) groups below the regression line in the lower distribution (n = 23).

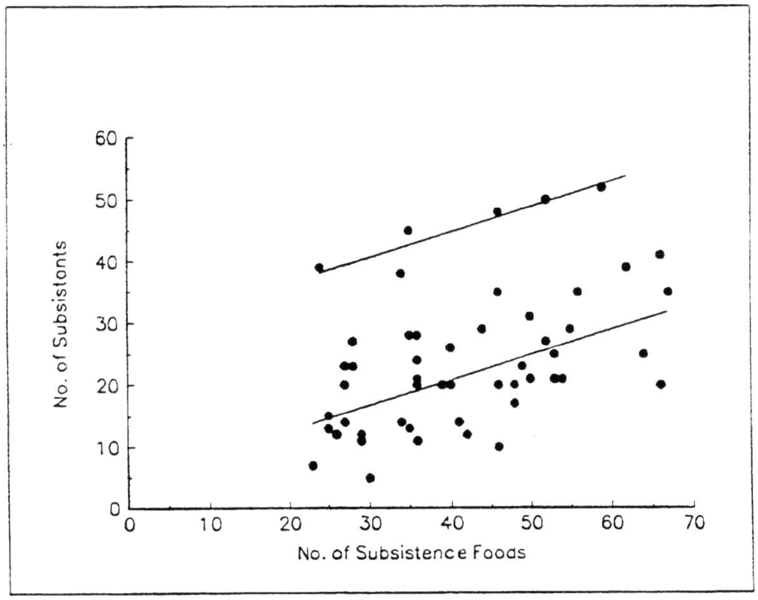

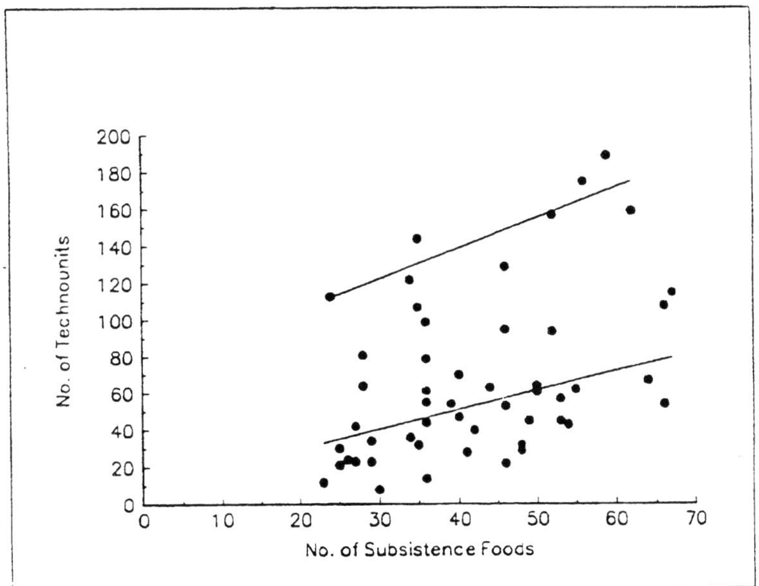

Figure 12.1 Regression analysis of number of subsistants (top) and number of technounits (bottom) on number of subsistence foods.

Information on subsistence economy was obtained from Murdock (1981), and compared for the three groups.

The hunter-gatherer groups in the upper distribution appear to be mostly dependent on aquatic and terrestrial animal resources. That is, their subsistence economies depend on a mean of 51.6% fishing (sd = 10.6), 36.6% hunting (sd = 11.0) and only 11.6% gathering (sd = 10.6). Much of the fishing includes anadromous fish or sea mammals. In contrast, the hunter-gatherer groups which are situated above the regression line in the lower distribution are characterized by economies which are more evenly distributed between fishing, hunting and gathering. They exhibit a mean dependence on 39.5% fishing (sd = 17.9), 33.0% hunting (sd = 11.9) and 26.9% gathering (sd = 14.2). Lastly, the hunter-gatherer groups which are situated below the regression line in the lower distribution are primarily dependent on gathering and hunting. These groups exhibit a mean of 45.2% gathering (sd = 18.6), 30.8% hunting (sd = 10.1) and 22.1% fishing (sd = 18.4). The differences between these three groups reflect the increased dependence on aquatic resources at mid to high latitudes. For example, the groups in the upper distribution are situated at a mean latitude of 59.1° (sd = 9.1) and those below the regression line in the lower distribution at a mean latitude of 34.3° (sd = 18.4) (see also Keeley 1988, Lee 1968, Pálsson 1988, Vierra 1989).

As suggested by Torrence (1983) technological specialization is a function of resource availability (i.e., seasonality) and accessibility (i.e., resource type). In respect to resource availability, the potential for prey escape is greater, and means of pursuit more limited, for aquatic resources, less so for terrestrial animal resources and the least for terrestrial plant resources. I would, therefore, expect a strong positive relationship between technology and dependence on fishing, a weaker positive relationship with dependence on hunting and a strong negative relationship with dependence on gathering. Spearman's rank order correlation coefficients indicate that there is a significant positive correlation between technology and fishing (subsistants: $r_s = 0.63$, p = <0.0001); technounits: $r_s = 0.73$, p = <0.0001), a significant negative correlation with gathering (subsistants: $r_s = -0.60$, p = <0.0001; technounits: $r_s = -0.67$, p=<0.0001) and a weak nonsignificant correlation with hunting (subsistants: $r_s = 0.01$, p = 0.8878; technounits: $r_s = 0.08$, p = 0.5261). Therefore, dependence on aquatic resource use is also an important factor conditioning technological specialization (also see Oswalt 1976, Torrence 1983; Vierra 1989, 1992, 1995).

Aquatic resources are often stored at higher latitudes to make up for declining terrestrial productivity (Vierra 1989, 1992, 1995). In temperate settings these resources may be used as an alternative source of protein during times of terrestrial shortfalls (Erlandson 1980), or in the case of the arctic, they may act as an important source of carbohydrates in the absence of plants (Yesner 1987). The number of stored foods is both an indication of seasonality and of subsistence diversification; that is, storage aids in the exploitation of a greater number of foods, since they can be stored for later consumption. Information on the number of stored foods was obtained from Panowski (1985) and from the original ethnographies (Vierra 1992, 1995). In respect to the previous scatter plot, groups in the upper distribution stored a mean of 16.8 foods (sd = 3.5), groups in the lower distribution above the regression line a mean of 15.1 foods (sd = 5.0), and groups in the lower distribution below the regression line a mean of 6.0 foods (sd = 6.3). Indeed, there is a significant correlation between number of stored foods and number of subsistants ($r = 0.73$, p = <0.001) and technounits ($r = 0.68$, p = <0.0001). That is, technological specialization and tool efficiency are critical given conditions when resources must be procured in bulk within a limited amount of time.

Although Binford (1980) and Kelly (1983) suggest that residential mobility is inhibited by a dependence on stored foods, this may also reflect the increasing use of logistical mobility (i.e., task groups) for the bulk acquisition of these foods. Given the inverse relationship between residential and logistical mobility, I would expect not only a negative correlation between number of residential moves with number of stored foods, but also with number of subsistence foods. Data on the number of annual residential moves was collected from the ethnographies (Vierra 1992, 1995). A regression analysis shows that there is a significant negative correlation between number of residential moves with number of stored foods ($r = -0.53$, p = 0.0006), and number of subsistence foods ($r = -0.50$, p = 0.0007). This indicates that a reorganization of labor also occurs with an increasing dependence on a variety of subsistence resources.

The implications of this study are that groups dependent on an increasing number of different terrestrial and aquatic resources will be affected by the fact that these foods may only be seasonally available, and that they will need to be acquired either sequentially or simultaneously. These conditions would increase the importance of the bulk acquisition, processing and storage of foods for future consumption that would take place during periods of low productivity (e.g., in winter) or because of resource failure. The result would be the reorganization of labor into task groups that are equipped with specialized technologies (i.e., including more diverse and complex tools) for the efficient exploitation of particular target species. It is the need for storage in conjunction with the rescheduling of subsistence pursuits, and possibly regional packing, that reduce residential mobility and increases logistical mobility.

In respect to the Mesolithic of western Europe, we can expect increasing technological specialization with increasing subsistence diversification. Second, we can also expect increasing technological specialization with an increasing dependence on storage and aquatic resource use; most notably above 40° latitude (i.e., above northern Portugal and Spain). Third, the combined effects of both seasonality and subsistence diversification will be most profound in northern Europe. Fourth, we can expect decreasing residential and increasing logistical mobility with greater subsistence diversification and storage.

The Archeological Implications

It has just been suggested that tool specialization is a response to conditions of limited resource availability and accessibility associated with increasing subsistence diversification. Therefore, composite microlithic tools composed of a series of geometric components may also reflect the increased demands for tool efficiency due to subsistence diversification and time constraints. It was Clarke who first described Mesolithic technologies as composite tools with"small, rapidly replaceable and interchangeable, standardized and mass-produced units, manufactured in advance and kept in readiness for inevitable wear and tear – a pull-out and plug-in construction" (1976: 457). As such, they may represent a response to the conflicting demands of increasing tool diversity, while reducing tool production/maintenance costs.

Composite microlithic tools probably represent a reliable technology designed for the exploitation of specific resources (*sensu* Bleed 1986). At least certain aspects of this technology were produced and maintained during intervals of 'downtime' prior to anticipated windows of resource acquisition. This would be especially true for the production and maintenance of hafts, which were used for a variety of tasks (e.g., arrow foreshaft, leister, or knife; see Clarke 1976: 454-455). In contrast, the microliths themselves appear to reflect a more maintainable and expedient aspect of the technology, being interchangeable between various functionally differentiated tools and having a shorter use-life. Although these microliths may reflect a "pull-out and plug-in construction" (Clarke 1976: 457), the tool itself would not have been easily modified, since the hafts would have been constructed for particular tool functions. Therefore, 'rehafting' (i.e., haft replacement) would have occurred more infrequently than 'retooling' (i.e., microlith replacement – Keeley 1982).

One important aspect of these composite microlithic tools is the high component replacement rate for the individual microliths. Kuhn (1989) has argued that technologies that exhibit continuous use/maintenance schedules should be characterized by low tool replacement rates because of the use of a 'replacement when exhausted' strategy. This strategy is most suitable for generalized activities with more unpredictable use schedules. These technologies are often used by residentially mobile groups, whereby small foraging parties procure resources on a daily encounter basis. On the other hand, technologies that are produced during periods of downtime in anticipation of predictable periods of heavy use should be characterized by a 'replacement based on probability of failure' strategy. This strategy is most suitable for highly specialized and repetitive activities with predicable use schedules. These technologies are often used by more logistically organized groups that target the exploitation of specific resources which are often procured in bulk, stored, and later consumed at the residential site (see also Bleed 1986). As pointed out by Kuhn, logistical hunter-gatherers tend to invest more in the manufacture of complex, sophisticated technologies, paying the price of high manufacture costs in return for high reliability. An alternative strategy for insuring reliability is to replace implements before they show signs of advanced wear (1989: 16).

It is therefore suggested that composite microlithic tools are designed for greater reliability and with a higher microlith component replacement rate to ensure tool dependability.

What are the archeological implications of long-term changes in foraging strategies and tactics on the organization of technology? Based on the assumptions underlying the subsistence diversification hypothesis, and the use of an organizational perspective, we can now begin to make some predictions concerning shifts in the organization of technology with respect to tool design, raw material procurement, tool production, tool use and tool discard patterns (Binford 1979, Nelson 1991). How might these various aspects of tool organization affect the structure of the archeological record? The following implications are outlined in Table 12.1. For analytical purposes the comparison is made between the Early and Late Mesolithic.

With respect to overall *tool design*, a shift towards increasing tool specialization, diversity and complexity is expected as a means of reducing resource acquisition and processing costs with increasing subsistence diversification during the Late Mesolithic.

Lithic material procurement strategies would be characterized by the increasing use of local materials during the Late Mesolithic due to decreasing residential and increasing logistical mobility. That is, assuming that lithic materials were collected using an embedded procurement tactic (*sensu* Binford 1979).

Standardization would act as a means of reducing *tool production* costs, something especially important given such high microlith replacement rates. This would be reflected in the decreasing variation in blank morphology from Early to Late Mesolithic, resulting from an increasing investment in core preparation as a means of producing a standardized tool blank.

In respect to *tool use*, I would expect an increasing number of different tool functions from Early to Late Mesolithic as a result of subsistence diversification. In particular, this would include greater variation in use-wear, utilized edge angles, and tool hafting techniques.

Tool discard patterns would be affected in two ways. First, I would expect a greater number of broken microlithic tools in Early Mesolithic assemblages and a greater number of unbroken microlithic tools in Late Mesolithic assemblages. This prediction is based on the assumption that Early Mesolithic individuals may have emphasized a 're-

Table 12.1. Early vs Late Mesolithic predictions.

Technological Organization	Early Mesolithic	Late Mesolithic
Tool Design	< tool specialization	> tool specialization
Raw Material Procurement	< local material	> local material
Tool Production	< standardization	> standardization
Tool Use	< tool diversity	> tool diversity
Tool Discard	> exhausted tools < spatial segregation	< exhausted tools > spatial segregation

placement when exhausted' strategy. Tools were primarily used until they were broken and no longer useable (i.e., low tool replacement rates). This would be expected given a greater emphasis on hunting and residential mobility. In contrast, Late Mesolithic individuals might have emphasized a 'replacement based on probability of failure' strategy, whereby tools were replaced prior to breakage when simply worn or slightly damaged. These high tool (or component) replacement rates reflect the increased requirements for tool efficiency, and reliability. This would be expected given an increasing emphasis on a variety of subsistence resources, and logistic mobility. Second, I would also expect greater spatial segregation of tool production maintenance and discard areas in Late Mesolithic sites. This spatial segregation would be the result of specialized task groups producing and maintaining their own gear (cf. Flenniken 1981, Murdock & Provost 1973), and because of increased residential site occupations and the maintenance of activity areas.

The Archeological Record: Subsistence Diversification and the Portuguese Mesolithic

Approximately 4100 stone tools and the by-products of manufacturing stone tools were analyzed from five Portuguese Mesolithic sites in order to test the implications of the subsistence diversification hypothesis. This consisted of the Early Mesolithic sites of Ponta da Vigia, Cabeço do Curral Velho and Palheirões do Alegra and the Late Mesolithic sites of Vidigal and Fiais (Arnaud 1986, Araújo 1991, González Morales & Arnaud 1990, Raposo et al. 1989, Straus et al. 1990, Straus & Vierra 1989, Vierra 1992, 1995, Zilhão et al. 1987). Ponta da Vigia and Cabeço do Curral Velho are situated in the Estremedura region northwest of Lisbon, while Palheirões do Alegra, Vidigal and Fiais are located in the Alentejo region of southwestern Portugal (Fig. 12.2). Caution should, however, be taken when making generalizations about the nature of Portuguese Mesolithic technology from such a small sample of sites. Nonetheless, the comparison between Early and Late Mesolithic samples is based on the shift from nongeometric to geometric-based technologies. It is necessary to dichotomize the data set, in order to make a temporal comparison with this small sample. It will, therefore, provide an initial step toward describing and understanding possible long-term changes in microlithic technology. Table 12.2 summarizes the lithic artifact data for the Early and Late Mesolithic sites (see Vierra 1992, 1995 for a detailed discussion of these sites).

Early Mesolithic (Epipalaeolithic) sites in central and southern Portugal date between approximately 10,000 to 8000 B.P. (Table 12.3). They consist of surface lithic scatters and shell middens (*concheiros*), with the latter containing mainly shellfish and a few terrestrial animal remains (Bicho 1992, 1993; Vierra 1992, 1995). For example, Ponta da Vigia and Palheirões do Alegra are lithic scatters, and Cabeço do Curral Velho is a shell midden. Late Mesolithic sites date between circa 7500 to 5000 B.P. (Table 12.3). They range from small shell middens like at Vidigal and Fiais, to the very large middens associated with the Muge and Sado River valley site complexes (Arnaud 1985, 1987, González Morales & Arnaud 1990, Roche 1972a,b; Vierra 1992, 1995, Zilhão 1993).

Although it is difficult to quantify, there appears to be an increase in site size, artifact density and number of exploited species through time. The Early Mesolithic sites are generally about 100 m^2 in size, the smaller Late Mesolithic sites cover about 1000 m^2, and the larger Late Mesolithic sites cover approximately 4000 m^2. A comparison of artifact densities at two nearby Early and the Late Mesolithic shell middens indicates that the early midden contains an estimated 3.8 artifacts per m^3, and the late midden 13.0 artifacts per m^3. In addition, the larger Late Mesolithic sites also contain evidence of structures, storage features and multiple burials; all of which is indicative of increased length of site occupation during the Late Mesolithic. The number of exploited species also drastically changes from Early to Late Mesolithic. The early shell middens contain a mean of eight species, in contrast to the late shell middens which contain a mean of 30 species. Analyses have shown that this pattern is not solely the result of differing sample sizes, but does reflect changes in foraging strategies. Lastly, studies show that the Early Mesolithic assemblages are characterized by a steeper regression slope when comparing assemblage size and tool

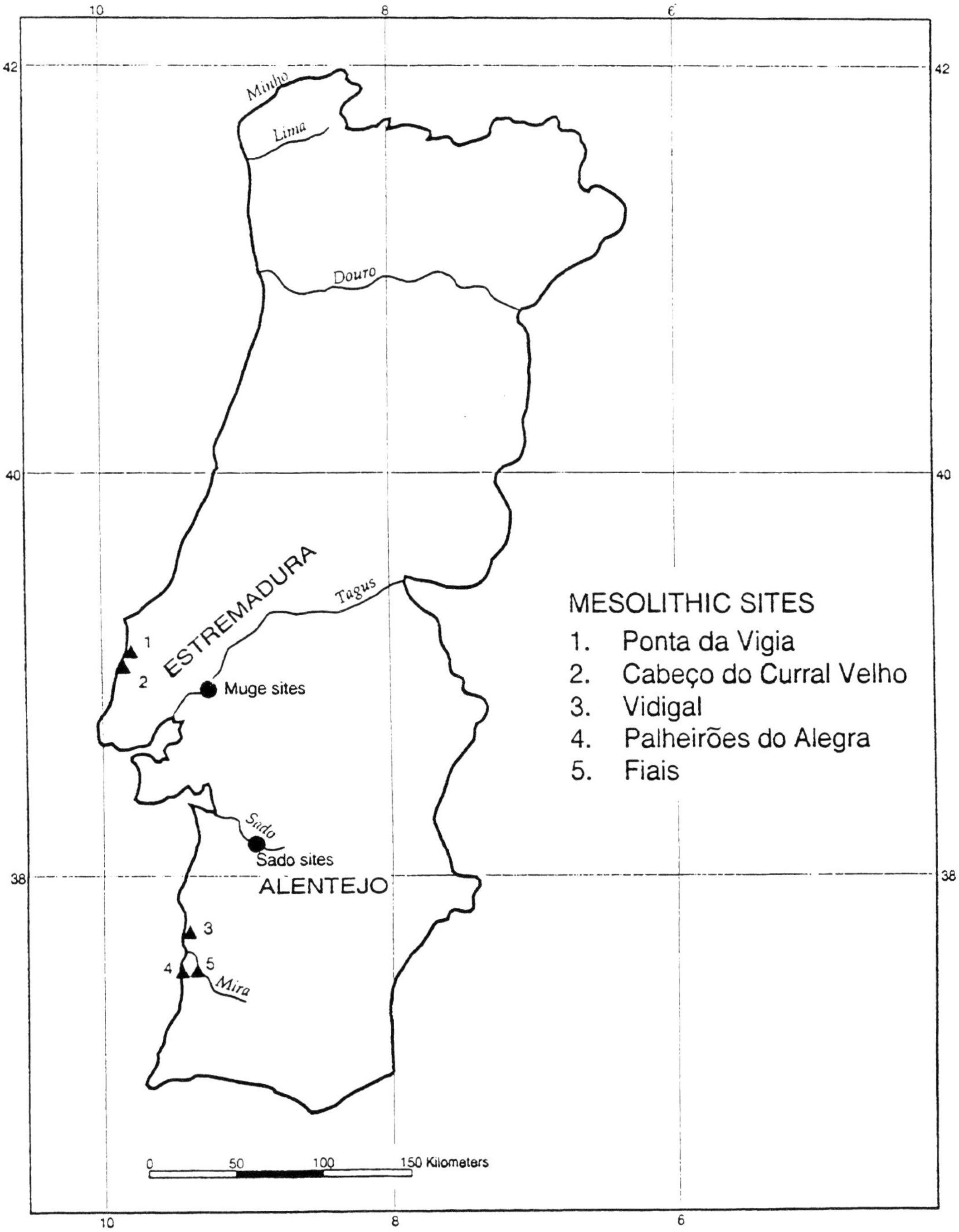

Figure 12.2 Map of Portugal showing Early Mesolithic sites (Ponta da Vigia, Cabeço do Curral Velho, Palheirões do Alegra) and Late Mesolithic sites (Vidigal, Fiais) analyzed here.

class richness, vs. a lower slope for the Late Mesolithic sites (Vierra 1992, 1995). These Portuguese data, therefore, reflect a long-term trend of decreasing residential mobility and increasing subsistence diversification.

How does the Portuguese Mesolithic data compare with the predictions made involving the various aspects of technological organization and the structure of the archeological record? There are major changes in tool design between the Early and Late Mesolithic (Fig. 12.3). Early Mesolithic assemblages are dominated by scrapers, with fewer backed bladelets, burins, and other tools; whereas, Late Mesolithic assemblages are dominated by geometric microliths, with fewer notches, retouched pieces, and other tools. This pattern can also be seen in a comparison of the assemblages from the Early Mesolithic site of Areeiro III (Bicho 1991, 1992) and the Late Mesolithic site of Moita do Sebastião (Roche 1972a) (Fig. 12.4). The shift to a composite geometric-based technology coincides with the increased exploitation of a variety of resources, including a marked emphasis on estuarine and littoral fish. This evi-

Table 12.2. Early and Late Mesolithic assemblages—Summary of lithic data.

Artifact Type		Early Mesolithic	Late Mesolithic
	Flake cores	22	39
	Bladelet cores	6	111
	Mixed cores	1	1
Cores and Heavy	Bipolar cores	1	0
Duty Tools	Core fragments	6	13
	Cobble unifaces	38	1
	Cobble bifaces	0	1
	Tested cobbles	3	3
	Hammerstones	3	0
	Subtotal	80	169
	Flakes	1345	931
	Angular debris	205	171
	Bladelets	139	842
	Blades	6	0
Debitage	Microburins	0	61
	Krukowski microburins	0	2
	Piecès esquilleés	2	2
	Burin spalls	12	5
	Crested ridge bladelets	4	5
	Core trimming flakes	12	23
	Subtotal	1725	2042
	Endscrapers	32	4
	Nosed scrapers	0	2
	Sidescrapers	2	0
	Perforators	2	3
	Burins	8	5
	Retouched pieces	10	23
Retouched Tools	Pointed bladelets	3	8
	Backed bladelets	12	12
	Truncations	8	11
	Notches	9	25
	Denticulates	4	9
	Trapezes	2	23
	Triangles	1	30
	Segments	4	40
	Combination tools	1	3
	Various tools	12	1
	Subtotal	110	199
Total		1915	2410

dence would seem to support the expectations of the subsistence diversification hypothesis, with the coincident switch to increasing tool specialization with aquatic resource use. Although bone tools are generally lacking from the Late Mesolithic sites of Vidigal and Fiais, the larger collections from the Muge sites do contain some. These tools include awls, perforators, axes, daggers (?), buttons (?), hafts, handles, polished ribs, and worked bone (Roche 1951: 142-145, 1967: 233-240 1972a: 83-85, 1980: 16); however, there does not appear to be any evidence for the use of bone in the manufacturing of specialized fishing gear. Although the absence of specialized bone fishing gear contrasts with the situation in northwestern Europe, this is not unexpected. The Mediterranean mixed oak woodlands provided a range of nuts and seeds suitable for storage and consumption during the winter months in Portugal; however, the paucity of these resources in the northern latitudes increased the reliance on seasonally available aquatic resources for storage by coastal Mesolithic groups. This included anadromous fish and sea mammals, both of which

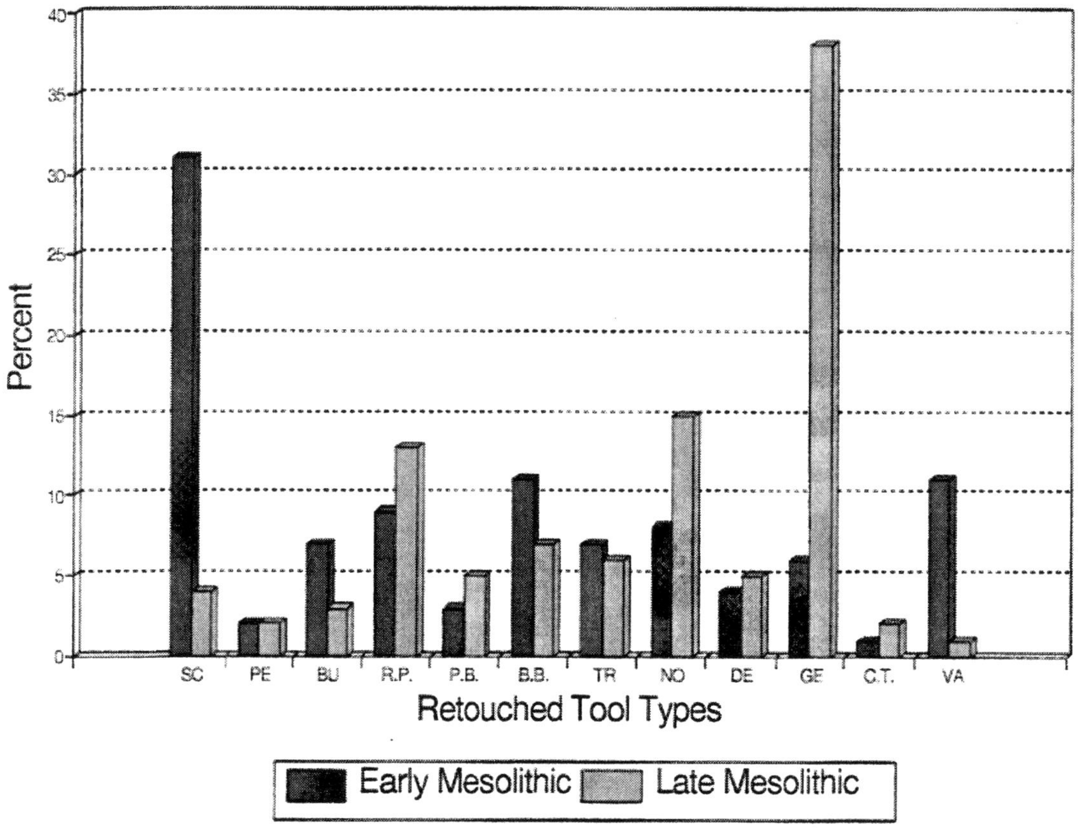

Figure 12.3 Comparison of major retouched tool groups for the Early and Late Mesolithic.

are absent in Portugal. It appears that these two aspects of technology were differentially affected, with bone tool specialization being related to increased aquatic resource use and greater time stress in these northern latitudes (e.g., Larsson 1990, Price 1985, Rowley-Conwy 1983).

The second prediction, involving changes in lithic procurement strategies, could not be evaluated. The Early Mesolithic site of Palheirões do Alegra was to be compared to the nearby Late Mesolithic sites of Vidigal and Fiais. Analyses showed that the inhabitants of the early site were obtaining cherts and rhyolites from secondary beach gravels, whereas the cherts and rhyolites on the later sites were obtained from primary sources. A survey of the present beach gravels revealed that these lithic materials are extremely rare. It therefore appears that the shift from using secondary to primary raw materials may actually reflect changes in raw material availability and not in foraging tactics. Similar changes in coastline configuration and raw material availability have been reported in other areas of Europe (e.g., Pitts & Jacobi 1979: 174-175).

The predictions for tool production techniques involved a shift towards greater standardization during the Late Mesolithic. The Early Mesolithic is characterized by a greater diversity of tool blank types including chert flakes, bladelets and blades. In contrast, Late Mesolithic tools were primarily made on chert bladelets. Although Early Mesolithic chert flakes and bladelets tend to be larger than their Late Mesolithic counterparts, the metrical attributes of the flakes exhibit similar coefficients of variation. For example, Early Mesolithic chert flakes have a mean length of 19.0 mm (sd = 8.4, cv = 0.44) and Late Mesolithic flakes a mean length of 16.3 mm (sd = 6.8, cv = 0.42). There is a significant difference between these means lengths (t = 3.99, df = 588, p = <0.001). Late Mesolithic chert bladelets are also shorter than their Early Mesolithic counterparts with a mean length of 18.2 mm (sd = 6.1, cv = 0.33) *vs* a mean length of 21.3 mm (sd = 2.7, cv = 0.39), respectively. There again is a significant difference between these mean lengths (t = 3.41, df = 362, p = <0.001). The Late Mesolithic chert bladelets do, however, exhibit slightly smaller coefficients of variation than the Early Mesolithic bladelets. Overall, Early Mesolithic scrapers and microliths were made on larger flakes and bladelets in contrast to Late Mesolithic geometrics that were made from smaller bladelet blanks using a microburin technique. The use of this technique is a qualitative indication of increased standardization, since it was used to produce bladelet midsections that served as blanks for geometrics. Evidence for the use of the microburin technique is lacking from the Early Mesolithic assemblages in this study, but does rarely occur in Portuguese Magdalenian and Early Mesolithic (Epipalaeolithic) assemblages (Bicho 1992, Marks et al. 1994).

Besides the variation in blank morphology, we can also look at the investment in chert core platform preparation as a means of measuring standardization. Core platform type was recorded as cortical, unfaceted, finely fac-

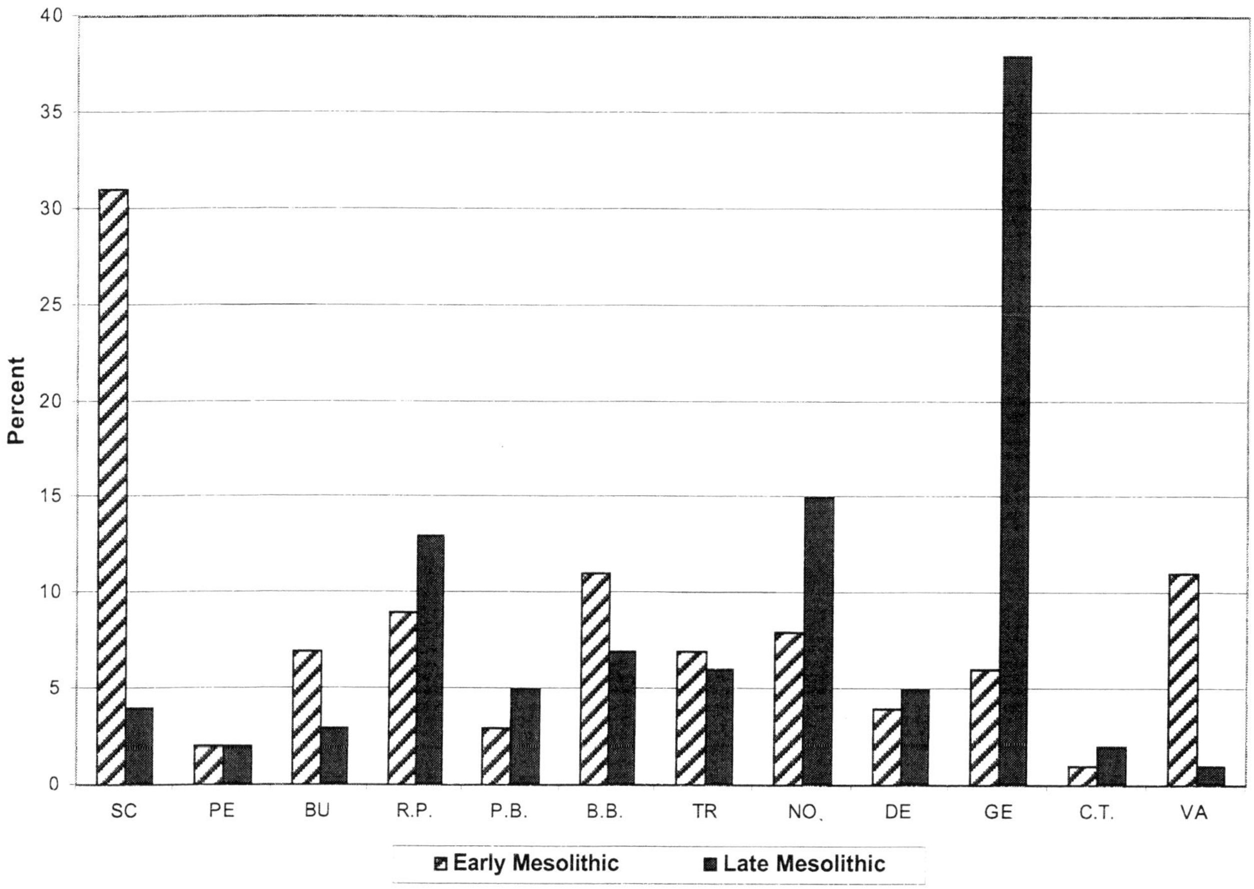

Figure 12.4. Comparison of major retouched tool groups for the Early Mesolithic site of Areëiro III and the Late Mesolithic site of Moita do Sebastião.

eted, faceted or cortical and unfaceted. A chi-square test indicates that there is a significant difference in chert core platform types between the Early and Late Mesolithic assemblages (χ^2= 16.512, df = 2, p = <0.001). Adjusted residuals (d) were calculated to determine which of the cells was contributing to the significant results. Adjusted residuals greater than 1.96 or -1.96 are significant at the .05 level (Everitt 1977). It appears that the Early Mesolithic chert cores exhibit more unfaceted platforms (d = 3.4) and Late Mesolithic chert cores more finely faceted platforms (d = 4.0) than expected. These differences are also present on the chert debitage. Debitage platform types were recorded as cortical, unfaceted, finely faceted, faceted, punctiform, crushed, battered, and collapsed. There is a significant difference in flake (χ^2= 35.9, df = 5, p = <0.001) and bladelet (χ^2= 126.1, df = 5, p = <0.001) platform types between Early and Late Mesolithic assemblages. Early Mesolithic chert flakes exhibit more crushed platforms than expected (d = 4.7) and Late Mesolithic flakes more unfaceted platforms than expected (d = 3.8); whereas, the Early Mesolithic chert bladelets exhibit more cortical (d = 2.3), crushed (d = 8.3) and collapsed (d = 5.5) platforms, and the Late Mesolithic bladelets more finely faceted platforms (d = 6.0) than expected. These finely faceted platforms may have acted as seats for an indirect punch technique during the Late Mesolithic, whereas a direct percussion removal technique could have been used with the unfaceted and crushed platforms during the Early Mesolithic.

The study of tool use patterns shows that there is an increase in the percentage of utilized debitage from Early to Late Mesolithic. Two percent of the Early Mesolithic flakes and bladelets exhibit evidence of macroscopic edge damage (as visible with a 10x hand lens), as compared to 4.6% of the Late Mesolithic debitage. The problem is that two of the three Early Mesolithic sites are located in dunes, and the patina present on some of this material may be obscuring the presence of edge damage. Nonetheless, the location of this edge damage mirrors the variation exhibited by the retouched tool assemblages. That is, most of the Early Mesolithic use wear is on flakes (76.6%) with some bladelets (23.4%) vs. mainly bladelets (90.6%) with fewer flakes (9.4%) for the Late Mesolithic. Although this use-wear is primarily situated along the lateral edges of these flakes and bladelets during both the Early (70.0%) and Late Mesolithic (90.6%), 30.0% of it was present on the ends of Early Mesolithic debitage, vs. only 9.4% on Late Mesolithic debitage. The Early Mesolithic damaged ends have a mean edge angle of 60° (sd = 20.2), and sides of 46.1° (sd = 14.3); whereas, the Late Mesolithic damaged ends have a mean angle 51.0° (sd = 18.5) and sides of

34.2° (sd = 11.4). Overall, the Early Mesolithic edge angles are steeper than the Late Mesolithic ones. There is a significant difference in the mean end (t = 0.79, df = 10, p = 0.45), and lateral edge angles (t = 4.04, df = 10; p = <0.001) between the Early and Late Mesolithic assemblages. The shift from using steeper to more acute edge angles may be related to the switch from an emphasis on the processing of terrestrial game to processing both terrestrial fauna and fish (e.g., see Flenniken [1981] for a discussion of hafted microlith use for processing fish). This switch may be reflected in the importance of scrapers during the Early Mesolithic, and possibly knives during the Late Mesolithic.

In respect to tool discard patterns, the relationship between whole and broken tools is similar for both periods. That is, backed bladelets tend to be broken and geometrics tend to be whole; eight (66.6%) of the 12 Early Mesolithic backed bladelets are broken; so are nine (81.8%) of the 11 Late Mesolithic backed bladelets. This contrasts with the pattern for geometrics. Four (57.1%) of the seven Early Mesolithic geometrics are whole, as are 48 (71.6%) of the 67 Late Mesolithic geometrics. The sample sizes are very small, but the overall breakage patterns appear to be similar. What is important is that the emphasis in procurement technology probably changed from relatively more backed bladelets in the early assemblages to more geometrics in the late assemblages. It is noteworthy that 18 of the 22 Late Mesolithic geometrics with edge damage were whole, indicating that they were worn, but replaced prior to breaking. This presumably reflects a shift from the use of backed bladelets as mostly points, to geometrics as both points and lateral components on arrows, as well as components on other various tools. For example, these composite tools could have been used to increase the 'killing power' of arrows for hunting terrestrial game (Myers 1989: 81), or as points and barbs on arrows or spears for shooting larger fish (e.g., Akazawa 1986: 157).

The predictions concerning the intrasite spatial aspect of tool discard patterns could not be evaluated, since sufficient spatial analysis information was not available for the Early Mesolithic sites. Differing intrasite activity areas could, however, be identified at the Late Mesolithic sites of Vidigal and Fiais. For example, at Vidigal there is a central shell midden dump that contains expedient quartzite flake tools and microliths *vs* the surrounding periphery of the site where the by-products of manufacturing microlithic tools are present (Straus et al. 1990, Straus & Vierra 1981; Vierra 1992, 1995). Given the paucity of quartzite cores at the site, flakes made of this material were presumably produced from beach gravels, and then brought back to the site for use and eventual discard (see Donnan & Moseley [1968] for a similar example). The three block excavation areas at Fiais can be differentiated into a tool production/maintenance area with midden and possible structure, a bone dump containing mostly geometric microliths and a midden with relatively more retouched pieces and ground stone implements (Vierra 1992, 1995). Activity areas were maintained at both sites, with the trash being dumped over older occupation surfaces.

Conclusions

This paper has attempted to evaluate subsistence diversification as a possible explanation for the shift to geometric based tool systems during the Late Mesolithic in Portugal. It was suggested that subsistence diversification was possibly due to regional packing, and the result of the declining foraging returns of higher ranked species. Given the assumptions of the diet-breadth model, foraging societies exploiting a variety of resources are probably at a greater risk than those exploiting a few resources. Therefore, technological specialization may be the result of a generalist foraging strategy and the increased use of logistical mobility (i.e., task groups) as a means of ensuring greater success in resource acquisition rates. Given conditions of limited resource availability and accessibility, composite tools could have provided a selective advantage by meeting the conflicting demands of increased tool diversity, while reducing tool production costs, and increasing tool efficiency.

The preliminary archeological evidence from Portugal does show some support for the subsistence diversification hypothesis in respect to changes in tool design, tool production and tool use patterns. The switch to composite geometric-based tool systems did coincide with a shift in foraging strategy, and a marked increase in the use of aquatic resources. In addition, Late Mesolithic technologies were characterized by greater standardization, possible changes in the intensity of tool use, and maybe even high microlith replacement rates prior to breakage. These, however, are all tentative observations in need of further empirical evaluation.

Nonetheless, one does not have to venture very far to see that the relationship between subsistence diversification, and composite geometric technologies may be a complex one. For example, in northern Spain the sequence for increasing subsistence diversification has been well documented and the shift to geometric-based tools occurs at about the same time as that in Portugal. Yet few, if any, geometrics have been found in the Mesolithic Asturian shell middens; whereas, geometrics produced using a microburin technique are present at inland sites the Basque region (Barandiarán 1979, Barandiarán and Cava 1985, Barandiarán 1990, Cava 1989, Clark 1983, Clark & Straus 1983, González Morales 1982, González Morales & Arnaud 1990, González Morales & Márquez 1978; Straus 1991, 1992; Straus & Clark 1986). The shift to a geometric-based technology in northern Spain appears to have occurred in a very different context than that exhibited by the Portuguese Mesolithic. It seems likely that these composite tools were differentially integrated into existing technologies, depending on local ecological conditions. This only highlights the variation that might be expected when studying the Mesolithic of western Europe. Explaining why geometric microliths are so common in the Late Mesolithic Portuguese shell middens, and yet seem to be absent from the Mesolithic Asturian middens, is an important problem that future research needs to address.

Acknowledgements

My work in Portugal was made possible by grants from the Office of Graduate Studies, Department of Anthropology, Graduate Student Research Committee and a Regents Endowed Fellowship from the University of New Mexico. Excavations at Vidigal were supported by grants from the National Geographic Society and LSB Leakey Foundation to Lawrence Straus. I am especially indebted to José Morais Arnaud for allowing me to participate in the excavations at Fiais and Vidigal, as well as study the material from these sites. The analysis of collections are from Palheirõa do Alegra, Ponta da Vigia and Cabeço do Curral Velho were made possible through the generous permission of Luis Raposo, João Zilhão, and Ana Christina Araújo. Franciso Alves, director of the Museu Nacional de Arqueologia e Etnologia, graciously provided me with space in the museum for my analysis.

References

Akazawa, T.

1986 Hunter-Gatherer Adaptations and the Transition to Food Production in Japan. In *Hunters in Transition* (M. Zvelebil, ed.) pp. 151-166. Cambridge: Cambridge University Press.

Araújo, A. C.

1991 O Concheiro, Epipaleolitico do Cabeço do Curral Velho (Cambelas, Torres Vedras). *O Arqueólogo Portguês*. In press.

1993 A Estaço Mesolítico do Forna da Telha (Rio Maior). *Trabalhos de Arqueologia e Etnologia* 23: 15-44.

Araújo, A. C.

1985 The Mesolithic Communities of the Sado Valley, Portugal, in their Ecological Setting. In *The Mesolithic in Europe* (C. Bonsall, ed.), pp. 614-631. Edinburgh: John Donald Publishers.

1986 Post-Glacial Adaptations in Southern Portugal. Paper presented at the World Archaeological Congress, Southampton.

1987 Os concheiros Mesolíticos dos Vales do Tejo e Sado: semelhanças e diferenças. *Arqueologia* 15: 53-64.

Arnaud, J.M. & J.D.B. Alves

1988 Charaterizaço da ocupaço prehistórica da Gruta da Casal Papagaio (Fátima, Vila Nova de Ourém). *Buletim da Sociedade Portuguesa de Espeleologia* 2: 27-34.

Barandiarán, I.

1979 Azilien et Post-Azilien dans le Pays Basque Mérdional. In *La Fin des Temps Glaciaires en Europe* (D. de Sonneville-Bordes, ed.), pp. 721-732. Paris: Centre National de la Recherche Scientifique.

Barandiarán, I., & A. Cava

1985 The Evolution of the Mesolithic in the North East of the Iberian Peninsula. In *The Mesolithic of Europe* (C. Bonsall, ed.), pp. 572-581. Edinburgh: John Donald Publishers.

Berganza, E.

1990 El Epipaleolítico en el País Vasco. *Munibe* 42: 81-89.

Bicho, N. F.

1991 Areëiro III: an open air site dated to 8,850 BP (Rio Maior, Portugal). *Mesolithic Miscellany* 12: 1-10

1992 *Technological Change in the Final Upper Paleolithic of Rio Maior, Portuguese Estremadura.* Unpublished Ph.D. Dissertation, Department of Anthropology, Southern Methodist University, Dallas.

1993 Late Glacial Prehistory of Central and Southern Portugal. *Antiquity* 67: 761-775.

Binford, L.R.

1979 Organization and formation processes: looking at curated technologies. *Journal of Anthropological Research* 35: 255-273.

1980 Willow smoke and dogs' tails: hunter-gatherer settlement systems and archeological site formation. *American Antiquity* 45: 4-20.

1983 *In Pursuit of the Past.* London: Thames and Hudson.

Bleed, P.

1986 The optimal design of hunting weapons: maintainability or reliability? *American Antiquity* 51: 4-20.

Bonsall, C. (editor)

1985 *The Mesolithic in Europe.* Edinburgh: John Donald Publishers.

Cava, A.

1989 La industria lítica: los utensilios. *Trabajos de Arqueologia Navarra* 8: 37-136.

Clark, G.A.

1983 *The Asturian of Cantabria: Early Holocene Hunter-Gatherers in Northern Spain.* Tucson: University of Arizona Press, Anthropological Papers of the University of Arizona No. 41.

1987 From the Mousterian to the Metal Ages: long term change in the diet of Cantabrian Spain. In *The Pleistocene Old World: Regional Perspectives* (O. Soffer, ed.), pp. 293-316. New York: Plenum

Clark, G.A. & S. Yi

1983 Niche-width variation in Cantabrian archaeofaunas: a diachronic study. In *Animals and Archeology I: Hunters and Their Prey* (J. Clutton-Brock & C. Grigson, eds.), pp. 183-208. Oxford: BAR International Series No. 163.

Clark, G.A. & L.G. Straus

1983 Late Pleistocene hunter-gatherer adaptations in Cantabrian Spain. In *Hunter-Gatherer Economy in Prehistory* (G. Bailey, ed.), pp. 131-148. Cambridge: Cambridge University Press.

Clarke, D.

1976 Mesolithic Europe: the economic basis. In *Problems in Economic and Social Archaeology* (G. Sieveking & I. Longworth, eds.), pp. 449-481. London: Duckworth.

Cohen, M.N.

1977 *The Food Crisis in Prehistory.* New Haven: Yale University Press.

Daveau, S., A.R. Pereira & G. Zbyszewski
1982 Datation au ^{14}C du site archéologique de la Plage de Magoito (Portugal). Scellé par une dune consolidée. *Clio* 4: 133-137.

Delibrias, G. & J. Roche
1965 Chronologie absolue des amas coquilliers mésolithique de Muge (Portugal). *Comptes Rendues des Séances de L'Académie des Sciences de Paris* 260: 2005-2006.

Donnan, C.B. & M.E. Moseley
1968 The utilization of flakes for cleaning fish. *American Antiquity* 33: 502-503.

Emlen, J.M.
1966 The role of time and energy in food preference. *American Naturalist* 100: 611-617.

Erlandson, J.M.
1988 The role of shellfish in prehistoric economies: a protein perspective. *American Antiquity* 53: 102-109.

Everitt, B.S.
1977 *The Analysis of Contingency Tables.* London: Chapman and Hall.

Fischer, A.
1985 Hunting with flint-tipped arrows: results and experiences from practical experiments. In *The Mesolithic in Europe* (C. Bonsall, ed.), pp. 29-39. Edinburgh: John Donald Publishers.

Flenniken, J. J.
1981 *Replicative Systems Analysis: a Model applied to the Vein Quartz Artifacts from the Hoko River Site.* Pullman: Washington State University, Laboratory of Anthropology Reports of Investigation No. 59.

González Morales, M.R.
1982 *El Asturiense y Otras Culturas Locales.* Santander: Centro de Investigación y Museo de Altamira Monografías No. 7.

González Morales, M.R., & J.E.M. Arnaud
1990 Recent Research on the Mesolithic in the Iberian Peninsula: Problems and Perspectives. In *Contributions to the Mesolithic in Europe* (Vermeersch P. & P. Van Peer, eds.), pp. 451-461. Leuven: Leuven University Press.

González Morales, M. & M.C. Márquez Uría
1978 The Asturian shell midden of Cueva de Mazaculos II (La Franca, Asturias, Spain). *Current Anthropology* 19: 614-615.

Keeley, L. H.
1982 Hafting and retooling: effects on the archaeological record. *American Antiquity* 47: 798-809.

Keeley, L. H.
1988 Hunter-gatherer economic complexity and 'population pressure': a cross-cultural analysis. *Journal of Anthropological Archaeology* 7: 373-411.

Kelly, R.L.
1983 Hunter-gatherer mobility strategies. *Journal of Anthropological Research* 39: 277-306.

Kuhn, S. L.
1989 Hunter-gatherer foraging organization and strategies of artifact replacement and discard. In *Experiments in Lithic Technology* (D. S. Amick & R. P. Mauldin, eds.), pp. 33-47. Oxford: BAR International Series No. 528.

Layton, R., R. Foley, & E. Williams
1991 The transition between hunting and gathering and the specialized husbandry of resources: a socio-ecological approach. *Current Anthropology* 32: 255-274.

Larsson, L.
1990 The Mesolithic of southern Scandinavia. *Journal of World Prehistory* 4: 257-309.

Lee, R. B.
1968 What hunters do for a living, or, how to make out on scarce resources. In *Man the Hunter* (R. Lee & I. DeVore, eds.), pp. 30-48. New York: Aldine Publishing.

Lubell, D. & M.K. Jackes
1988 Portuguese Mesolithic-Neolithic subsistence and settlement. *Rivista di Antropologia, Supplemento* 46: 231-248.

MacArthur, R.H. & E.R. Pianka
1966 On optimal use of a patchy environment. *American Naturalist* 100: 603-609.

Marks, A.E., N. Bicho, J. Zilhão & C.R. Ferring
1994 Upper Pleistocene prehistory in Portuguese Estremedura: results of preliminary research. *Journal of Field Archaeology* 21: 53-68.

Meiklejohn, C., M.K. Jackes & D. Lubell
1986 Radiocarbon dating of human skeletal material from two sites in Portugal. *Mesolithic Miscellany* 7(2): 4-6.

Murdock, G. P.
1981 *Atlas of World Cultures.* Pittsburgh: University of Pittsburgh.

Murdock, G. P. & C. Provost
1973 Measurement of Cultural Complexity. *Ethnology* 12: 379-392.

Myers, A.
1989 Reliable and maintainable technological strategies in the Mesolithic of mainland Britain. In

Time, Energy, and Stone Tools (R. Torrence, ed.), pp. 78-91. Cambridge: Cambridge University Press.

Nelson, M. C.
1991 The study of technological organization. In *Archaeological Method and Theory Vol.* 3 (M. Schiffer, ed.), pp. 57-100. Tucson: University of Arizona Press.

Oswalt, W.H.
1976 *An Anthropological Analysis of Food-Getting Technology*. New York: John Wiley and Sons.

Pálsson, G.
1988 Hunters and gatherers of the sea. In *Hunters and Gatherers: History, Evolution, and Social Change* (T. Ingold, D. Riches & J. Woodburn, eds.), pp. 189-204. Oxford: Berg.

Panowski, E.T.
1985 *Analyzing Hunter-Gatherers: Population Pressure, Subsistence, Social Structure, Northwest Coast Societies and Slavery.* Unpublished Ph.D dissertation, Department of Anthropology, University of New Mexico.

Pitts, M. & R. Jacobi
1979 Some aspects of change in flaked stone industries of the Mesolithic and Neolithic in southern Britain. *Journal of Archaeological Science* 6: 163-178.

Price, T.D.
1985 Affluent Foragers of Mesolithic Scandinavia. In *Prehistoric Hunter- Gatherers: The Emergence of Cultural Complexity* (T. D. Price & J. A. Brown, eds.), pp. 341-364. New York: Academic Press.

Price, T. D., and J. A. Brown
1985 Aspects of Hunter-Gatherer Complexity. In *Prehistoric Hunter-Gatherers: The Emergence of Cultural Complexity* (T. D. Price & J. A. Brown, eds.), pp. 3-20. New York: Academic Press.

Raposo, L., C. Peñalva & J. P. Pereira
1989 Noticia da Descoverta da Estaço Mirense de Palheirões do Alegra, Cabo Sardo (Odemira, Portugal). Paper presented at the II Encontro do Quaternário lbérico. Madrid.

Renouf, M.
1984 Northern hunter-fishers: an archeological model. *World Archaeology* 16: 18-27.

Roche, J.
1951 *L'industrie Préhistorique du Cabeço d'Amoreira (Muge).* Porto: Instituto para a Alta Cultura.

1957 Premiére datation du Mésolithique Portugais par la méthode do carbone 14. *Boletim da Academia da Ciéncias de Lisboa* 29: 294-296.

1967 Les collections Mésolithiques du Museu des Services Géologiques du Portugal: Cabeço da Arruda, Muge. *Comunicações dos Serviços Geologicos de Portugal* 51: 221-241

1972a *Le Gisement Mésolithique de Moita do Sebastião (Muge, Portugal).* Lisboa: Instituto para Alta Cultura.

1972b Les amas coquilliers (concheiros) Mésolithiques de Muge (Portugal). *Fundamenta* 3: 72-107.

1980 Algumas caracteristicas da indústria do Mesolítico Antigo de Muge (Portugal). *Arqueologia* 2: 12-18.

Rowley-Conwy, P.
1983 Sedentary hunters: the Ertebølle example. In *Hunter-Gatherer Economy in Prehistory* (G. Bailey, ed.), pp. 111-126. Cambridge: Cambridge University Press.

Soares, A.M. & J.M.P. Cabral
1984 Datas convencionais de radiocarbono para estaçoes arqueológicas Portuguesas e a sua calibraço: reviso crítica. *O Arqueólogo Português* 2: 167-214.

Stephens, D. W. & J. R. Stephens
1986 *Foraging Theory.* Princeton: Princeton University Press,

Straus, L. G.
1991 Epipaleolithic and Mesolithic adaptations in Cantabrian Spain and Pyrenean France. *Journal of World Prehistory* 5: 83-104.

1992 *Iberia Before the Iberians.* Albuquerque: University of New Mexico Press.

Straus, L. G., J. Altuna, G. Clark, M. González Morales, H. Laville,. A. Leroi-Gourhan, M. Menéndez Hoz & J. Ortea
1981 Paleoecology at La Riera (Asturias, Spain). *Current Anthropology* 22: 655-682.

Straus, L. G., J. Altuna & B. Vierra
1990 The concheiro at Vidigal: a contribution to the late Mesolithic of southern Portugal. In *Contributions to the Mesolithic in Europe* (P. M. Vermeersch & P. Van Peer, eds.), pp. 463-474. Leuven: Leuven University Press.

Straus, L. G. & G. A. Clark
1986 *Stone Age Hunter-Gatherer Adaptations in Northern Spain.* Tempe: Arizona State University Anthropological Research Papers No. 36.

Straus, L. G. & B. J. Vierra
1989 Preliminary investigation of the *concheiro* at Vidigal (Alentejo, Portugal). *Mesolithic Miscellany* 10: 2-11.

Torrence, R.
1983 Time budgeting and hunter-gatherer technology. In *Hunter-Gatherer Economy in Prehistory* (G. Bailey, ed.), pp. 11-22. Cambridge: Cambridge University Press.

Vermeersch P.M. & P. Van Peer (editors)
1990 *Contributions to the Mesolithic in Europe*. Leuven: Leuven University Press.

Vierra, B. J.
1989 Post-Pleistocene Adaptations Revisited: Aquatic Resource Use and Hunter-Gatherer Complexity. Paper presented at the Conference on 'The Organization of Land and Space Use, Technology and Activities in Past and Present Societies'. Albuquerque: University of New Mexico.

1992 *Subsistence Diversification and the Evolution of Microlithic Technologies: A Study of the Portuguese Mesolithic*. Ph.D Dissertation, Department of Anthropology, University of New Mexico. Ann Arbor: University Microfilms.

1995 *Subsistence and Stone Tool Technology: an Old World Perspective*. Tempe: Arizona State University Anthropological Research Papers No. 47.

Zilhão, J.
1993 The spread of agro-pastoral economies across Mediterranean Europe: a view from the Far West. *Journal of Mediterranean Archaeology* 6: 5-63.

Zilhão, J., E. Carvalho & A. C. Araújo
1987 A estação Epipaleolitica da Ponta da Vigia (Torres Vedras). *Arqueologia* 16: 8-18.

13

The Azilian in the Cantabrian Region

Juan Antonio Fernández-Tresguerres Velasco
University of Oviedo

Abstract

The Azilian is the latest derivative of the cultures of the late Upper Paleolithic. During the last oscillations of the Würm glaciation and the earliest part of the Holocene, a culture developed in southern France and northern Spain that has its origins in the latest Magdalenian. One of the principle aspects of this change is the mental transformation expressed in the differences in artistic expression between the Azilian and the Magdalenian.

The Azilian

The Azilian was found in the earliest excavations carried out on the Cantabrian Coast at the start of the 20th century. It was first thought to be a relatively simple industry in its development and structure, but was later seen to be rather more complex, and the fact is that we still do not know the industry fully in all its dimensions. When the Azilian was first discovered it seemed to be a remnant, a 'degeneration' of the Magdalenian, but this preconception was based on insufficient knowledge of the Azilian, as well as the inevitable comparison with the spectacular Magdalenian from which it is derived.

The Azilian succeeds the Magdalenian stratigraphically, although it was not unusual to find flowstone deposits separating this new culture from its predecessor. The Azilian thus was thought to represent the prolongation of Upper Paleolithic cultures in the new climatic conditions which became established after the retreat of the glaciers. At times there could be difficulties in its identification. In Cueto de la Mina rockshelter, in Asturias, the Conde de la Vega del Sella could not differentiate the supposed Azilian layer(s), observing that the characteristics of this industry, in the absence of archaeological 'index fossils', were not very different from those of the Late Magdalenian. The Azilian, he concluded, was an 'impoverished version' of the Magdalenian (1916).

The known Azilian levels gave little data – the collections were not very abundant – and, when they were placed on a time scale, they could not be analyzed as an evolutionary phenomenon. It was a period which pretty clearly emerged from a Magdalenian base, yet no clear explanation for this transformation was proposed. More recent excavations allow us to go a step further, and examine at least partially, the Azilian phenomenon in its development through time. Changes in behavioral forms can be seen which, even if they do not acquire the important dimensions observed in the Magdalenian, are nevertheless significant – they link the Azilian more firmly with the late Upper Paleolithic in a process during which profound and significant changes occurred, especially in respect of symbolism.

The Geography of the Cantabrian Azilian

The Azilian is found from the center of the Cantabrian Coast – with a rare exception on the south slopes of the Cantabrian Mountains, in the region of León (Cueva de la Uña – Bernaldo de Quirós & Neira Campos [1993a,b; 1994]) – to the center of France (and even to the north-east in the area of Doubs). The number of known sites with Azilian levels has increased notably in recent years. In Asturias we have the sites of La Lluera (Rodríguez Asensio, 1990), Cueva Oscura de Ania (Gómez Tabanera et al. 1975), Cueva Oscura de Perán (Fernández Rapado & Mallo Viesca 1965), Los Azules (Fernández-Tresguerres 1981), to add to the earliest 'classic' sites which had produced relatively abundant remains (e.g., Cueva de la Riera, re-excavated by Straus and Clark [1986], Cueto de la Mina [Vega del Sella 1916], and Cueva de Balmori [Vega del Sella 1930]). To the east, in Cantabria, there are some new sites which have only been tested, such as Cueva de San Juan (Molinero Arroyarbe & Arozamena 1984), but others are certainly important, such as the El Piélago group (García Guinea 1985), with a relevant stratigraphy, the restudy of El Pendo (González Echegaray & Freeman 1980), excavations in Rascaño (González Echegaray & Barandiarán 1981), La Pila (Bernaldo de Quirós et al. 1992) and Cueva del Perro, dug recently by González Morales (González Morales & Díaz Casado 1991-1992).

On the other hand, the Azilian has been known since the beginning of the century in Cueva del Castillo (Cabrera Valdés 1984), in Cueva del Valle (Breuil & Obermaier 1912a,b; Cheynier & González Echegaray 1964), in Salitre (Obermaier 1925) as well as Morín (re-excavated by Freeman and González Echegaray in the 1960s (Freeman & González Echegaray et al. 1973) and the upper levels of La Chora (González Echegaray et al. 1963).

Sites with Azilian levels are also abundant in the Basque Country and Navarra. Recently work has been done in caves such as Ekain (Altuna et al. 1984) or Berroberría (excavated by the Marqués de Loriana and later by Maluquer de Motes and, again, by Barandiarán 1990). These sites together with Urtiaga document the presence of this industry in the eastern sector of the Cantabrian Coast, in contact with the Pyrenees. New excavations, such as those at Ekain mentioned above, or in Zatoya (Barandiarán et al. 1989) and Abauntz (Utrilla 1982) have greatly broadened our horizons in respect of the development of the Epipaleolithic. To these relatively well-reported sites, others (still not well known) may be added, such as Santa Catalina (Berganza 1991), Urratxa and Anton Koba (Armendáriz 1993). All these caves provide a large amount of data, allowing us to understand better the Azilian in all of its complexity.

In the western zone of the Cantabrian paleolithic, recent excavations have identified certain levels which doubtless represent the most primitive period of the Azilian. This is called the *Early Azilian*, and has characteristics to be described later which distinguish it from the *Recent* or *Classic Azilian*, but which shows tendencies leading towards this second phase. Level 5 in Cueva de Los Azules (Cangas de Onis) and level II of Cueva de la Lluera I (San Juan de Priorio, Oviedo) certainly belong to this period. We are less sure, from the published data, about the Early Azilian in level 2 of Cueva Oscura de Ania (Las Regueras) (Gómez Tabanera et al. 1975). Outside Asturias this early phase might be documented in level V of Ekain, although there is so little industry there that its assignation is rather doubtful. This aspect is quite important, as we may be observing a step towards a more regionalized Azilian.

In any case, this regionalization eventually became evident, even though the data required to define it properly are often spotty. The differences need not have a regional 'cultural' meaning, as they could also be due to a diversification of activities. But this does not seem to be the case. It is clear that at the end of the Azilian the discordance between parts of the Cantabrian Coast becomes much more pronounced, possibly because of greater isolation between the western and eastern sectors, with the boundary between those in the west of Cantabria (Fernández-Tresguerres 1983). We can note that, throughout its existence, the Azilian undergoes transformations and breaks, both in its geographic and chronological development and in the apparent homogeneity which once seemed to characterize it.

Character of the Azilian

From the beginning, the Azilian was defined in comparison with the Magdalenian, as a time when foragers abandoned a certain way of manufacturing tools, marked until then by the high quality of Magdalenian worked bone objects. Because of this, the Azilian was usually described as 'coarse' when compared to the Magdalenian. Its technical and artistic aspects differed precisely in the more careless way of manufacturing technical forms, which were nevertheless derived from Magdalenian precursors. This viewpoint seems unsatisfactory to us today.

But what exactly defines the Azilian *vis à vis* the culture which preceded it? The break between the two is obviously great enough to have allowed the identification of the Azilian as 'something different.' This break can be observed, above all, in the three aspects which mark the difference.

The first, the clearest and perhaps the most dramatic, is the disappearance of the traditional idioms of artistic expression. Secondly, there is a certain discontinuity in the forms of the bone and lithic industries, especially in the greater or smaller proportion of certain types (e.g., endscrapers and burins). Third, there are technological changes, notably a new method of manufacture, noticeable above all in the form of the harpoons. The manufacturing techniques appear simpler, but at the same time more efficient, producing tools where the technical and practical aspects tend to predominate over their appearance. It is, however, in the field of technology – and we must insist on this – where we can see quite clearly a relative continuity between the two periods.

Other perceptible differences between the end of the paleolithic and beginning of the Azilian in France (although they are not so dramatic even there), and which were long accepted as being quite evident, are rather more problematic in the context of the early Cantabrian Azilian. This can be seen in the changed economic circumstances of the earliest Epipaleolithic populations. The means of subsistence must have been transformed by the climatic changes at the end of the last glaciation and with the onset of the Holocene, bringing with it more or less dramatic changes in the flora and fauna. But this economic transformation used to be explained precisely by the disappearance of the reindeer, which was the dietary mainstay (and the basis for the symbolic world) in the late paleolithic of France, to a much greater extent than it ever was in Cantabria. Pretty clearly, faunal change was more pronounced in France. Around 12,000 BP the cold adapted species were replaced by a more temperate fauna, although present data suggest that reindeer persisted in small herds during the initial phases of the Azilian in mountainous areas (Depelch 1989) or in Les Landes, as can be observed at the site of Duruthy (Arambourou et al. 1978). Although this is true, it is clear that reindeer no longer had the economic importance they once held.

The situation is quite different along the Cantabrian Coast, where it is not so easy to see drastic environmental change. Instead it seems that climatic conditions at the end of the Magdalenian were not so different from those that accompanied the earliest phases of the Cantabrian Azilian. In fact, the start of the Azilian fits perfectly in the climatic sequence of the late glacial period and is in no way the result of dramatic environmental changes at the start of the Holocene. Indeed, it may be said that it is the Azilian which is profoundly transformed, although the chronology is not always clear in respect of the sequence of events or processes.

Chronology

Chronological data are not always helpful in understanding culture process questions. The changes can happen too quickly for us to trust absolutely in the ^{14}C dates, which doubtless are significant with all the reservations that we might have about them. On the other hand, there are synchronous events in the evolution of the culture along the Cantabrian Coast, especially in its later moments. This can be observed in the fact that the Classic Azilian continues in the west, whereas geometric microliths gradually appear in the eastern sector, nearest the Pyrenees. This happens in the Holocene, when the increasing density of the forests, combined with the complex Cantabrian relief, probably contributed to the isolation of those Epipaleolithic populations living in the Basque country.

The first changes at the end of the Magdalenian and the earliest appearance of the Azilian occurred during the Alleröd. The first radiocarbon dates are from shortly after 12,000 BP, coinciding with the end of the Magdalenian. The dates overlap, and within the limits of variation it is difficult to establish even approximately the moment when the first signs of the Azilian appear (see also Clark 1989). There have been few dates corresponding to an Azilian older than c. 11,000 BP (i.e., Azules 3e base, two from Zatoya II [final Magdalenian-Azilian], and another from Zatoya B3 [also final Magdalenian-Azilian]; dates from Anton Koba and La Pila). An important block of five dates is concentrated between 10,500 and 10,940 BP (Rascaño I, Azules 3e and 3f, Piélago and Zatoya [final Magdalenian-Azilian]). Another eight dates correspond to the end of that millenium and the start of the following one (i.e., between 10,400 and 9,430 BP: Rascaño I, Azules upper layers of 3e and 3, Portugaín, Piélago and Urtiaga D). Dates from Ekain, levels II and IV, Abauntz level D, Azules upper layer 3 (although there is a difference between this date and the ones obtained later from the British Museum) all post-date 10,000 BP. Urtiaga C and Berrobería C2 are dated in the late ninth millennium BP.

In some cases, the dates for Early Azilian levels are older than those which correspond to Recent or Classic Azilian in the same caves. This is the case at La Lluera and Los Azules. In the latter site level 3e was dated to 11,190 BP. It is therefore logical to suppose the Early Azilian had its origin at some time prior to this date.

Paleoclimate

Climatic and palynological data indicate that the Azilian developed during the Tardiglacial and the onset of the Holocene. It is dated in Cantabria to the Alleröd (Würm IV - Cantabrian VIII), that is to say, during a relatively mild episode. It developed fully during Dryas III (Würm IV – Cantabrian IX) and the Pre-Boreal (González Sainz 1994, Hoyos Gomez 1995)

There is nothing in the development of the Magdalenian that would indicate that they were not responsible for the great cultural transformation that signals the start of the Epipaleolithic. The first Azilian populations lived in changing climatic and environmental conditions which were similar, if not identical to, those that characterized the Magdalenian sequence. This means that initially, although not later, Azilian paleoeconomies did not differ substantially from those of the Magdalenian, at least on the Cantabrian Coast.

Azilian Novelties and the Continuation of the Magdalenian Tradition

We have already mentioned these two aspects, emphasized in the literature on the Azilian. There was certainly a break in graphic symbolism, and continuity in the presence of certain elements of Magdalenian technology. Nowadays it seems that the break was less abrupt than once thought, especially if we take into account only the objects found in Classic or Recent Azilian levels. Although certain forms are maintained, there is a complete break in the ideological background of these people compared with those of their Magdalenian ancestors. We shall therefore start with the art, which is possibly the most significant aspect.

Azilian Art and Symbolism

Comparing the two worlds, Magdalenian art and that of the fully developed Azilian, and momentarily leaving other considerations aside, (e.g., possible relations between the two), we can see at once that these two symbolic universes are completely different (Fernández-Tresguerres 1994) (Figs. 13.1, 13.2). The disappearance of cave art represented the destruction of a 'language' that transmitted a determinate and well-defined representation of the world. In the Azilian, a generalized ignorance of the animal figure immediately becomes apparent, along with the disappearance of the 'container' and center of the art: the cave. Former sanctuaries or shrines are abandoned, and with it one more symbol in the group of signs of the paleolithic language. The universe of animals transformed in symbols, transmitting primitive myths, so important in Magdalenian expression, disappears quickly (after a brief reappearance in the Early Azilian). This was the consequence of a radical change in the Azilian conception of the world, although it is difficult for us to determine today the causes for this transformation and the specific content of their new conceptions. Considering the signs which form the vocabulary of the new language, we find a transformation of those characteristic of paleolithic art, as they become replaced by others. Furthermore, despite the limitations of mobiliary art, there is even a complete change in the kinds of raw material used, as plaquettes are substituted for cobbles, rarely used in the Upper Paleolithic. And bone, widely used in earlier periods as a medium for engraving, is now utilized much less frequently.

This analysis, which emphasizes the full-blown, 'Classic' Azilian, should be complemented with evidence from the Early Azilian. Although in Cantabria the absence of a good sample of indisputably Classic Azilian art does not alter the abruptness of the break between the two worlds, it does make the manner of the transition more comprehensible. This intermediate episode between the Magdalenian and the Classic Azilian is more clearly represented in France. The presence in the Magdalo-Azilian or Early Azilian of plaquettes decorated with animals in Pégourié, Abri Murat, Pont-d'Ambon, Abri Morin and La Borie del Rey are clear indications of the persistence of art with animal themes, so deeply rooted in the mythology of Upper Paleolithic populations (Llorbranchet 1989). But this survival is, nevertheless, limited. Caves disappear from the Azilian mental universe, but for a short time plaquettes continue to be used as a form of portable shrine. Was this due to changes in the mobility of Epipaleolithic foragers, perhaps corresponding to differences in isolation of the French and Cantabrian populations? The manner of territorial occupation may have changed so much that the relation of a group with a particular sanctuary site could have been lost. But, although territories probably became restricted during the Azilian – provoking a greater fragmentation of cultural formulae – it is unlikely that this happened in its earlier phases, unless the spiritual and social unity of western European populations began to break up at the end of the Magdalenian. This in turn would have provoked the disappearance of cave art if its *raison d'être* no longer existed (cf. Barton et al. 1994, Clark et al. 1996).

In Cantabria, no remains of animal art have been found which clearly post-date the Magdalenian, although some are known from Cataluña. Two plaquettes with animal engravings from the cave of Sant Gregori could be included in the early period, at least stylistically. On the one found by S. Vilaseca in level 2 of the cave (which, according to Fortea Pérez [1973] corresponds with an Azilian-type Epipaleolithic industry) we can see the image of a stylized hind whose body is covered with small marks, thought to represent fur, and which are also seen on the plaquettes of Pont-d'Ambon, corresponding to the Magdalo-Azilian. The second, published by Fullola, Viñas and García-Argüelles (1990) was found in what was probably Vilaseca's spoil heap. In this case we can see a bovid, a horse's head (?) and a hind, together with a zig-zag line motif. As on the first plaquette these are strongly stylized. Although with so few data conclusions can only be very tentative, the association of animals and signs, so characteristic of Upper Paleolithic art, appears to survive. These plaquettes, together with those found in French caves, may be the last manifestations of an art, symbol system and a mentality which refused to die, but which had inevitably to disappear. The signs and animal figures formed a language with a precise syntax, which could not accept too many alterations in its essential qualities, since – taking into account regional variations, or 'dialects' – it appears to have maintained a powerful unity as well as being perfectly structured in its elements. This apparent inflexibility might have been the cause of its disappearance, when it was unable to adapt easily to the new circumstances resultant from the changed environment.

In Cantabria, the animal figure disappears from art at the end of the Magdalenian. The schematic bovid on a pebble-pendant from Balmori, or the more naturalistic one engraved on a chelonid plaquette from the same cave are of too doubtful provenience stratigraphically to be of much use. The possible hindquarters of an animal engraved on a bone fragment found in Atxeta require further data about the Azilian in that cave (Barandiarán 1973).

In the Early Azilian of Asturias, art is reduced to some decorated harpoons found in the caves of La Lluera and Los Azules. Similar decorations are found on the same kind of instrument at two sites 70 kilometres apart. It is surprising because decoration on bone tools – in this case harpoons – disappears during the Azilian. This Upper Paleolithic tendency, well-developed in the Magdalenian, is preserved for a short time at the start of the Azilian. Possibly decoration of this kind of tool is one of the few links with the symbol system of the Magdalenian.

The advance of the new Azilian culture, and the definitive establishment of its characteristics resulted in the end of the Magdalenian cave art, replaced by another understood in a very different way from the relative and peculiar Magdalenian realism. Abstract signs, without any reference to natural forms, become the exclusive form of graphic communication. There is also a strong tendency towards the most absolute simplification. But we would be mistaken if we were to think that this apparent limitation means a loss in complexity. The signs painted or engraved on cobbles have a definite capacity for combination, perhaps not so intricate as the syntax of the symbols of cave art, but which doubtless was organized as a language to express another universe of different mythical content. All this was suspected some years ago, mainly at the level of intuition. These signs, it was thought, were pieces of a primitive alphabet, or simple elements of accountancy, or in some cases part of funeral rites, symbolizing the deceased.

Couraud has the distinction of discovering in the simple Azilian cobble art the structure and the syntax of a complex language. By establishing a chart of the distribution of the signs it was seen that Azilian art has a capacity for combinations, associations and meaning, depending on the number of symbols. The color and technique (painting

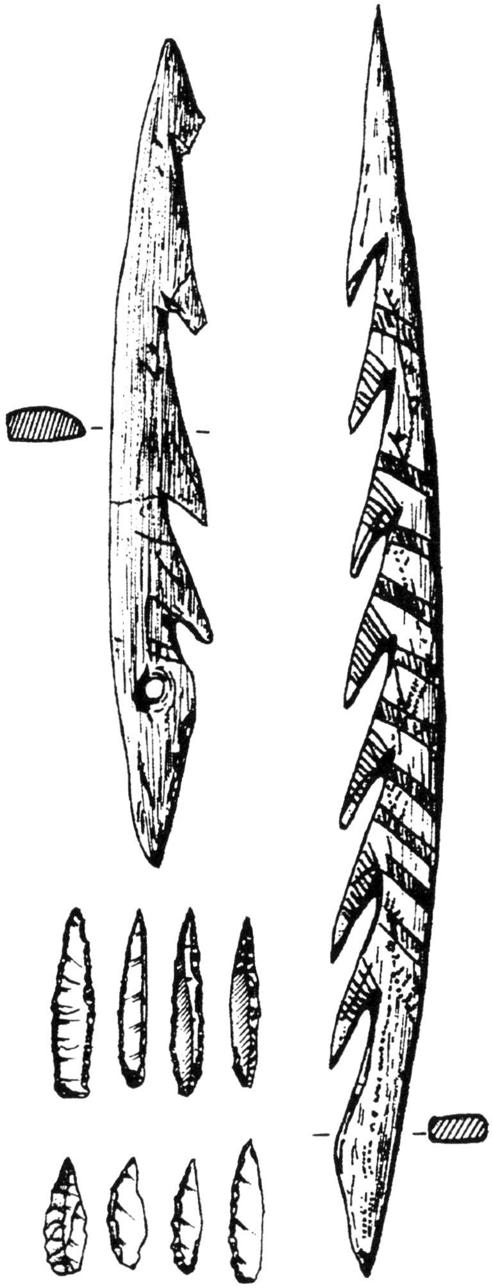

Figure 13.1. Cueva de los Azules – uniserial harpoons, backed and pointed bladelets typical of the Early Azilian.

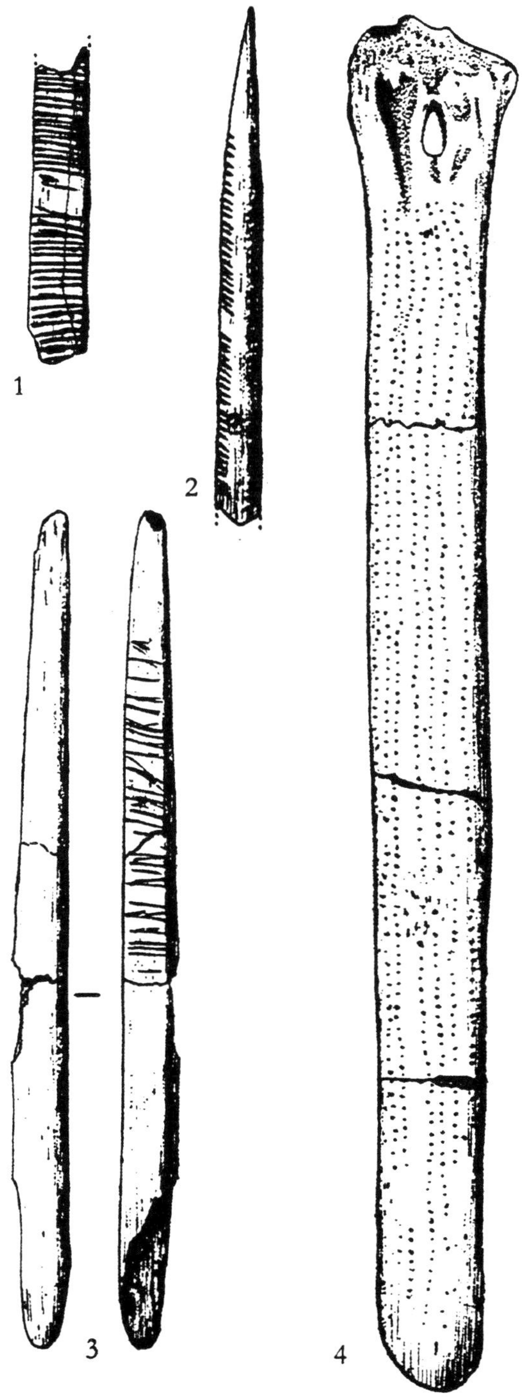

Figure 13.2. Cueva de los Azules – spatulate pieces and a decorated awl from Level 3.

or engraving) also have meaning, although it is quite simplified compared with the complex Magdalenian syntax (Couraud 1985). Later, d'Errico reached similar conclusions in a broad analysis of engraved Azilian art (1994).

The difficulties in understanding Azilian art are increased by the scarcity of decorated objects. The case of Mas d'Azil, with more than a thousand painted and engraved cobbles, is virtually unique. Few cobbles have been discovered in the Cantabrian region, and the few that have been found show a clear regionalization. This certainly goes beyond what is observed in paleolithic art, where an admirable concept of unity predominates over the variations found in different zones. It is true that in the Classic or Recent Azilian we always see the same tendency toward abstraction and, leaving aside art on bone tools, the generalized use of cobbles. But beyond that, each group seems to have used a limited number of signs, within a very small area. In Cantabria these do not go much beyond the confines of a single site.

In the developed Cantabrian Azilian, few painted or engraved cobbles have been found (e.g., in El Pindal cave). Adding to the confusion, many of them have been lost, for the most part as a consequence of the Civil War and its aftermath. Among the ones we have – and taking into account one found outside the region, in level 4 of Cueva del Filador (Fullola & Couraud 1988) – we notice great differences in conception. In Los Azules nearly all have dots painted in black. In Pindal they are dominated by the cobble; in Filador, thick red parallel lines (Fullola & Couraud 1988). In the case of the lost cobbles we have descriptions of indeterminate marks, or – often – no description at all (Obermaier 1925).

As regards Cueva de Los Azules, where there were a relatively large number of cobbles (29), 19 of them were related to a burial and were concentrated in the upper levels that contained it (less frequently in levels 3f, 3g). So far, no cobbles have been found in the Early Azilian of this cave.

In mobiliary art on bone objects we find a certain change from Magdalenian stylistic attributes. In contrast with what the painted cobbles seem to suggest, there is a greater uniformity in the Classic Azilian between western and eastern Cantabria, a greater unity of conception. But this requires comment. The decorated objects found in recent excavations resemble each other more than the ones found in old excavations. This could be due to problems the excavators had in ascribing the objects to their correct stratigraphic proveniences. It is surprising, for instance, to note the scarcity of decorated *sagaies* (spear points) in the recent discoveries with the relatively large number found in the early excavations (e.g. in Asturias, in La Paloma six examples, one in La Riera). *Sagaies* are not very common in the Azilian, and decorated ones are rarer still, which makes one suspect that some of these objects, whose decoration looks Azilian in its simplicity, really belong to the end of the Magdalenian. What is more, tools with a short 'use life' were not much decorated in the Azilian.

Decoration is more common on objects used over a longer period. We find it on spatulas, whether these are well manufactured (e.g., in Los Azules) or simple ribs (in Los Azules, Arenaza – Fernández-Tresguerres 1994, Apellániz, 1982). Other tools made from ribs, but discovered in fragments, have oblique parallel incised lines grouped in series with empty triangular spaces between them. This theme is seen from Los Azules to the Basque Country (Arenaza, Anton Koba, and possibly Aitzbitarte), including Cantabria (La Cubera [Chaline 1965], and in the possible late Magdalenian site of La Chora [González Echegaray et al. 1963]). Another more or less simple form of decoration is found on the awls with series of deep oblique parallel lines, found in Los Azules and Anton Koba. We can, therefore, find motifs which extend throughout the Cantabrian region, while others are extremely localized. This is the case with the pendants found in Cantabria at El Piélago, and at some other sites in its vicinity (Morín, San Juan, La Chora) (González Sainz 1982), but in archaeological contexts of which are not always clear.

This brief summary of some aspects of Azilian art seeks to demonstrate the radical change in the language of the populations of the late Magdalenian and the Azilian. It marks the end of a world which sees the loss of its universe of symbolic images and its replacement by another, while its other cultural forms do not seem to suffer such drastic alteration. This indicates that the great revolution which took place between the end of the Magdalenian and the Azilian is fundamentally symbolic – a break or transformation in a vision of the world. And as the Azilian populations were not so different physically or in terms of adaptation from the Magdalenian, there can be no other reason for the disappearance of this language than the loss of its functionality. Why this happened is less clear, but it has already been mentioned that it may have been connected with the structure of society and its dispersion over territories formerly so inhospitable that they could not have been occupied (Clark et al. 1996). The causes seem to be much more complex than the simple alteration provoked by climatic change. It is evident that art is the clearest indicator, within its impregnable ambiguity, of that great transformation.

Technology

Technology obviously also underwent modifications, as can be seen from the simple comparison of Magdalenian and Azilian worked bone and stone objects. Nevertheless, these express continuity. It is enough to see how the forms of the late Magdalenian start to take on an Azilian appearance. This is most clearly perceptible in bone tools, where the continuity of form in the harpoons is definitely inherited from the Magdalenian.

The Early Azilian –Only two sites in Asturias are known so far which contain levels that are both 'post-Magdalenian' and clearly earlier than the Classic Azilian. So the present scheme is only valid for this zone. It is possible that this is another symptom of the regionalization of the culture, although it is in the later Azilian that the differences between western and eastern Cantabria become most apparent.

In 1983, ongoing excavations showed that in Cueva de Los Azules, below level 3 (an absolutely Classic Azilian, rich in materials and accumulated over a short time), there was a sterile layer and in one sector of the cave vestibule, beneath this layer, a Magdalenian level with harpoons (level 6) was found. But in another sector of the same vestibule, which had been badly eroded at the end of the Magdalenian, the industry had changed, being similar to the Azilian in some aspects and different in others (level 5). This shows that there is a stratigraphic sequence within the Azilian: (1) an Early Azilian with perfectly identifiable characteristics, and (2) a more recent period, well known as the traditional Azilian industry. Level 5 was divided into three layers. The oldest was defined by its very small, wide and thick backed bladelets. But this early period is still not well understood. Immediately above this layer was another, defined by its long and narrow double-backed points, similar to Sauveterrian points, and also by short, more or less circular scrapers, characteristic of both the late Magdalenian and the Azilian (Fig. 13.3).

Bone-working is well defined by the harpoons, which are present from the start of the Azilian. Although they have a flattened cross-section, they are not exactly like the classic Azilian harpoon. One has a practically rectangular cross-section, and another is plano-convex, similar to the harpoon with a double row of barbs from the site of La Chora (González Echegaray et al. 1963). One lacks a basal perforation, and in another this is circular. The tendency for the base to have a small convex protuberance is not generally characteristic of Azilian harpoons. Finally, most of the fragments and one complete harpoon are decorated, which clearly differentiates them from the typical style of Epipaleolithic harpoons (Fernández-Tresguerres & Junceda 1994).

The level in La Lluera belongs to the same period. A similar harpoon was found here, although this was decorated on both faces. It was associated with points of the same style described above – long and double-backed. The publication of Cueva Oscura de Ania (Gómez Tabanera et al. 1975) refers to very similar backed points, but these were associated with small harpoons, which lacked basal perforations. Stratigraphically and culturally these levels define an Early Azilian, which seems to have been quickly overtaken by the Classic period.

Levels 25 and 26 in La Riera are less easy to define – which may be significant – and they could be an intermediary stage between the Magdalenian and the Azilian. However the limited industry makes it difficult to identify them more precisely (Straus & Clark 1986).

In the Basque Country, level V in Ekain, dated to 11,400 BC, could correspond to this moment of transition. But unfortunately the industry is not too expressive. Nevertheless, in its limited sample, it can be seen that the percentages of tools made on bladelets, as well as Azilian points, is quite low, while both increase in overlying levels IV and III. This tendency is also observed in Los Azules. There is also a noticeable increase in the number of burins, which are better finished than in Azilian levels. The endscrapers found are not typical of the Azilian, but the collection is poor and is not characteristic.

In Piélago II levels 5 (a and b), 6 (Proto-Azilian) and 4 (Azilian 1) have a lithic assemblages that are distinct from the upper levels, especially in the rarity or absence of geometric forms. But in level 4 the harpoon type corresponds to the Classic Azilian and level 6 is too poor to be diagnostic.

Evidently, then, there are too few sites with levels belonging to the Early Azilian to offer much certainty to the observations made here – they are only indicative. Besides, the small size of the collections means that a full description of the industry or any kind of statistical calculation is not feasible. Nevertheless, we can see in it the first stage of development towards a full Azilian character. It still has inherited tendencies, but it has lost some of the perfectionist style so typical of the Magdalenian.

Classic or Recent Azilian – The classic period is found throughout the Cantabrian region, with numerous levels attesting to its widespread expression. From the point of view of stone-working, Azilian assemblages essentially consist of small endscrapers made on more or less circular flakes; few, generally badly-made burins, abundant backed bladelets, and moderate numbers of Azilian points – in some cases, (e.g., the recent excavations in Morín) these are microgravettes – and a relatively large presence of other tools. They usually have a coarse, careless aspect, which differentiates them visually from the Magdalenian (Fernández-Tresguerres 1980) (Fig. 13.4).

The percentages of endscrapers are almost always higher than those of burins, which is the opposite of the Magdalenian tendency. For instance, in Los Azules, Piélago II, Abauntz and Zatoya, the endscraper percentages dominate those of burins. But it is not always so. In levels III and IV of Ekain this pattern is reversed: there are many more burins than scrapers. This is also the case at Urtiaga.

The massive presence of microliths is also noticeable in most sites. Sometimes, as in Los Azules or Ekain III-IV, the percentages can be >50% of the retouched pieces. In other caves (e.g. Abauntz, Piélago, La Riera) they are lower but still highly significant.

Azilian lithic assemblages are generally quite different from those of the typical Magdalenian. Nevertheless, the tendencies began to appear in the final phases of the latter culture: small scrapers became more common, Azilian points appear, the number of bladelets is quite high, and there are fewer burins. Thus, it is not unexpected that sometimes, in the absence of the 'index fossil' tool types, it is not easy to differentiate Late Magdalenian levels from Early Azilian ones.

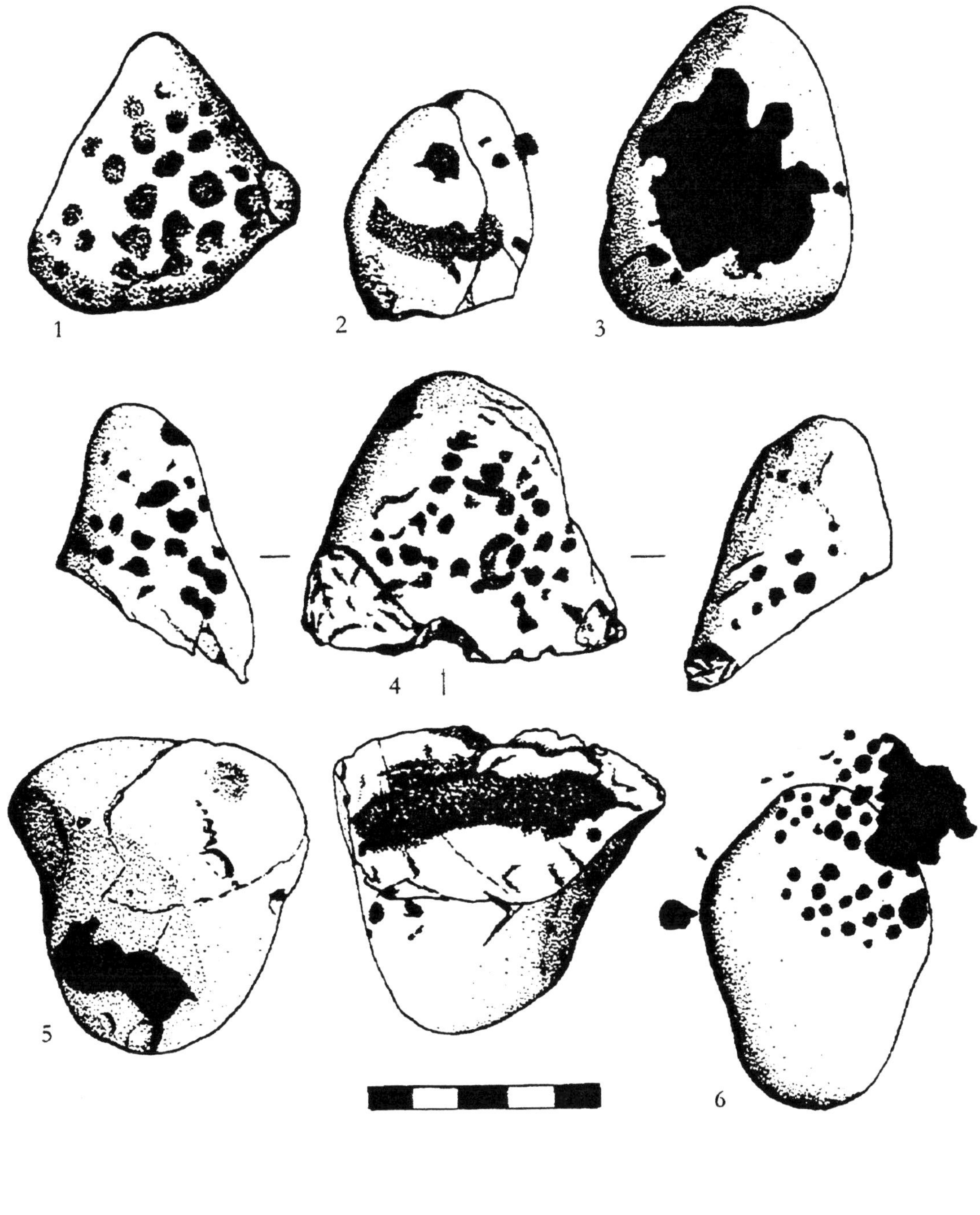

Figure 13.3. Cueva de los Azules – painted pebbles.

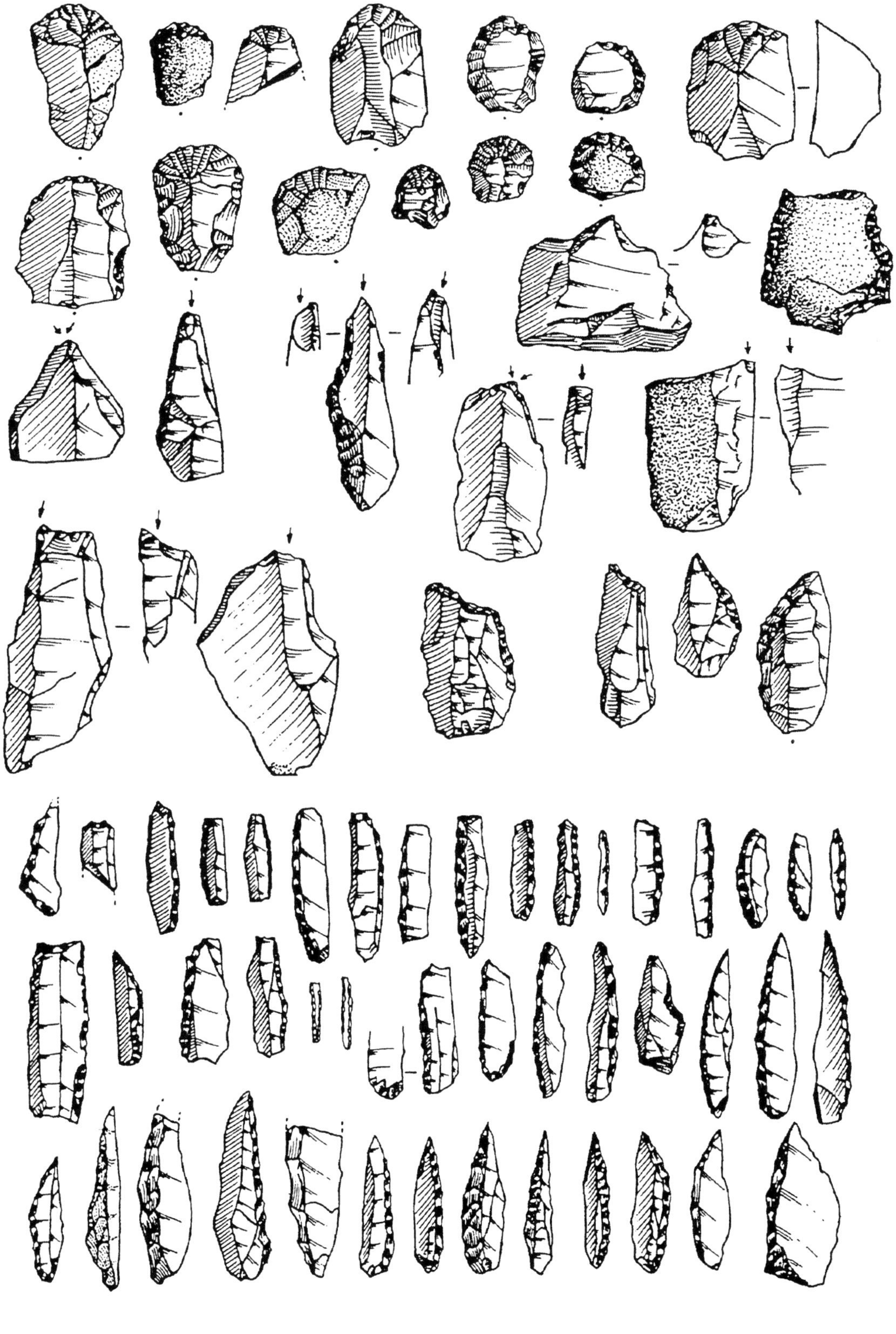

Figure 13.4. Cueva de los Azules – lithic industry from the Classic Azilian.

Lithic Raw Materials in Early and Classic Azilian

The comparison of certain aspects of level 5 with those of level 3 in Cueva de Los Azules reveals tendencies which appeared in the early phase and which became more pronounced over time, as the Classic Azilian developed. One of them is the progressive tendency towards a certain indifference as regards the final appearance of tools. Azilian populations were more concerned with the practical aspects of technology than with its appearance. This had a clear influence on the choice of raw material. The greater incidence of flint than quartzite among the stone-working debris of level 5 is significant:

Level	Flint	Quartzite	Quartz
3	50.3%	48.2%	1.5%
5	75.2%	24.2%	0.5%

If we consider the relationship between raw material and the tool types, this tendency in the Classic Azilian to take less care in the aspect of the tools becomes clearer in the persistent choice of quartzite in level 3:

Level 3	Flint	Quartzite	Quartz
Various types	55.0%	40.4%	2.0%
Endscrapers	91.2%	7.1%	1.5%
Bladelet tools	98.1%	1.8%	—
Level 5			
Various types	85.4%	14.6%	—
Endscrapers	97.1%	1.4%	1.4%
Bladelet tools	98.8%	1.2%	—

The number of unretouched bladelets declines after the Late Magdalenian in the cave. This indicates a change in flint-working, which becomes more careless. In the Magdalenian (level 6) they worked to a higher standard, as can clearly be seen in the larger number of blade cores, whereas in the Azilian they tended to use any fragment which had a suitable size and shape.

Unretouched bladelets:

Level	Flint	Quartzite
3	3.7%	2.0%
5	6.95%	4.3%
6	9.45%	2.1%

Azilian loss of interest in form doubtless influenced the choice of raw material (or *vice versa).* As the Azilian advanced, poor quality flint apparently became more common in the landscape, and they often chose radiolarite which was full of fractures, impeding proper knapping, whereas in levels 5 and 6 better quality nodules predominate (Fernández-Tresguerres 1989). It is true that this could reflect an attitude of the population, as already mentioned, but it could also be the result of greater difficulties in accessing the sources of higher quality raw materials available earlier, possibly due to the increasingly difficult movement caused by forest expansion and/or changes in the hydraulic regimens of the streams and rivers in the region.

Bone-working

There is also a noticeable contrast in quality with Magdalenian bone-working – in the Azilian, it becomes much poorer. There are fewer types in Azilian levels, and with little variation. Roughly made flat harpoons with one or two rows of barbs form the basic groups of bone technology. There are few *sagaies* or spatulas. It is sometimes said that the bone tools are equally carelessly made, although this is not always so. This opinion is probably influenced by the high quality of Magdalenian bone tools. But the fact is that the utilitarian approach can also be seen here. This does not mean that the bone tools are always badly manufactured, as we can find some Azilian tools which are perfectly worked. Nevertheless – to insist on this – the lack of decoration, and in most cases a greater simplicity in the forms, makes them less remarkable.

In these aspects the break with the Magdalenian seems to be more clearly defined, yet continuity is also evident, especially in the most characteristic Azilian bone tool: the harpoon. The largest collection of harpoons so far discovered in Cantabria was in Cueva de Los Azules (Fernández-Tresguerres & Junceda 1994) (Fig. 13.5). The study of nearly a hundred harpoons from this site, together with their stratigraphy, allows us to deduce certain conclusions about the possible evolution of this tool during the Azilian.

The harpoon appears in Los Azules in its distinctive, novel form in level 5. The Magdalenian type has been transformed into a flattened instrument, with angular barbs and, in the base, either a simple protuberance or a circular hole. As they appear here, they have their roots in Magdalenian examples (one found in level 6 is a possible predecessor). At once we see the process of simplification in the manufacturing technique: a simple piece of deer antler in which barbs are cut after quickly scraping off the surface.

At the base of level 3 the Azilian harpoon is absolutely typical. All those found in levels 3 and 2 have a single row of barbs. What is more, the examples found in Asturian caves are all uniserial. In the Basque Country most are of this kind (the exceptions are an example from Cueva de Agarre and another from Pikandita, discovered out of stratigraphic context). In Cantabria, however, biserial types are relatively abundant (Piélago, Morín, Meaza, Castillo). Together with the decorated pendants mentioned earlier, this suggests a degree of possible Azilian regionalization. We should proceed with caution, however, since there is a definite stratigraphic series only at Piélago II. At this site the type of decoration only appears in levels 3 and 2 – one without, and one with only a few geometric microliths. Thus it will be necessary for Azilian studies to be more concentrated geographically in order to define more clearly any local or regional variations which might exist.

In more recent layers of Los Azules (2, upper layers of 3) the harpoons have larger, better defined barbs, with a more rectangular shape. In contrast, the harpoons from layers underlying level 3 have predominately angular barbs,

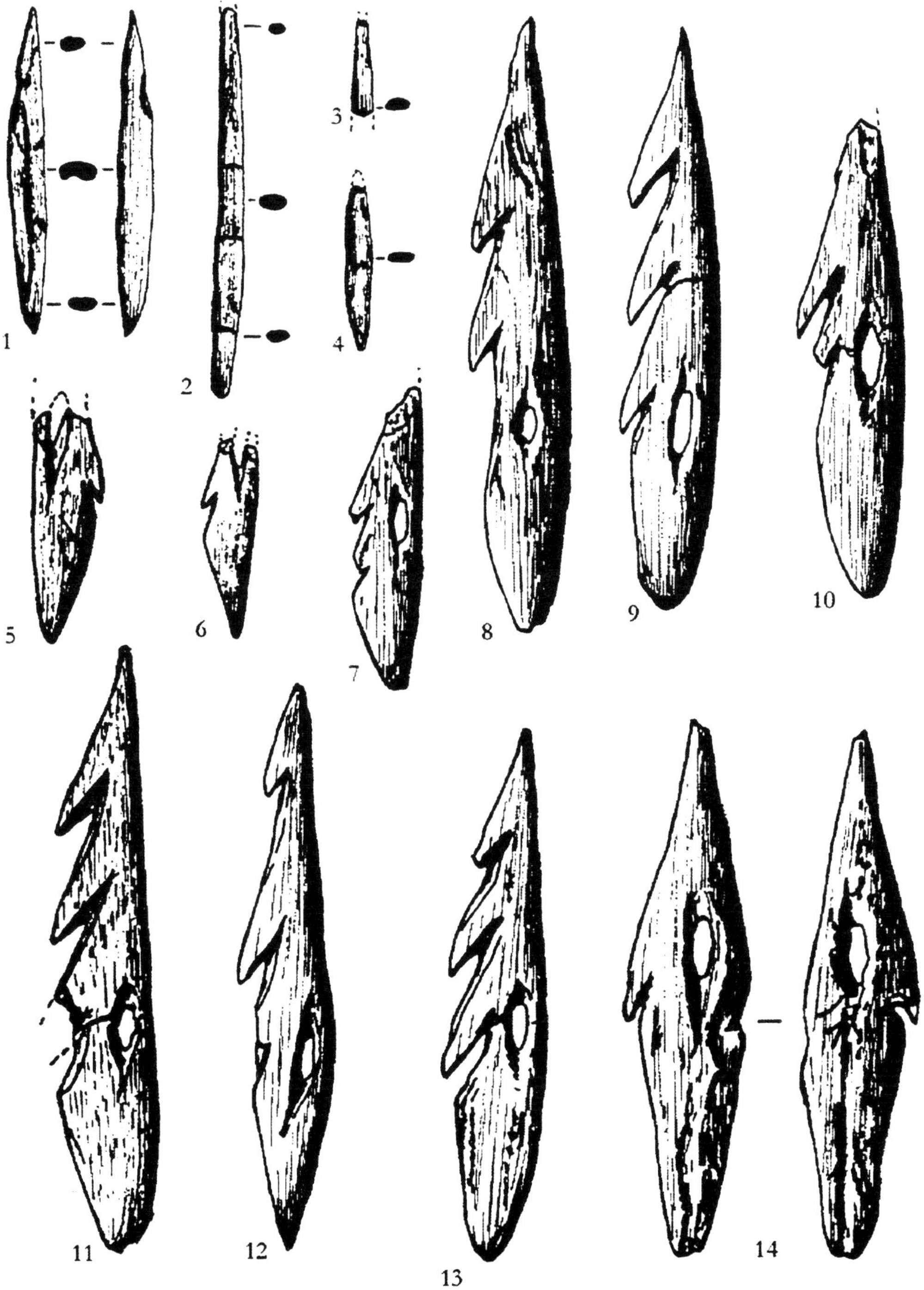

Figure 13.5. Cueva de los Azules – harpoons from the Classic Azilian.

the two sides are straight while the upper surface is curved, as seen in level 5. It is not unusual for the barb nearest the point to be largest. The base is also longer in the harpoons from the lower levels (3e-g) – that is, excepting the single-barbed examples where the base is logically much longer. In these levels, the base can comprise more than a third of the length of the instrument, sometimes reaching a half. In the higher levels, the base makes up less than a third of the length of the harpoon. The oval, eye-shaped hole in the base is characteristic of the Classic Azilian. The position of the center of the hole, which would be the place of greatest tension once the harpoon was stuck in the prey's body, can, however, vary. The example from level 5 had a circular hole, which was located in the very base. It is also in the base on the only complete harpoon found in level 3h. But it starts to move upwards in level 3g, as far as the top of the base, but never higher than the lowest barb. It stays in this position in some examples from 3f, but in others it moves to the center of the harpoon (in one case clearly above the middle), above the base and the lowest barb. In level 3e some instruments still show this tendency, but the eye-hole returns to its position in the base in the upper layers of level 3 and in level 2. At the moment it is difficult to assess the meaning of these modifications in harpoon manufacture. They may be local variations, but we cannot be sure until we have a greater number of examples from sites over a wider area.

Other aspects of Azilian bone-working are less significant. Awls and spatulas, except for their decoration, have traditional forms, generally highly simplified. Awls are normally manufactured from an epiphysis, keeping the articulation as the handle of the tool. They are, therefore, sharpened bone splinters, with a circular cross-section, well-polished and sometimes have a more or less complex decoration (Los Azules, Anton Koba, Valle, Piélago II). The spatulas are manufactured either from the epiphysis of a cervid long bone (unusual in the Azilian) or from a rib, with an end which has been cut, rounded and polished with use. The may also be decorated, usually more schematically than the awls, except for the example from Los Azules, which even so was not so sophisticated. The decoration on the ribs is reduced to superficial rough incisions on the concave face. Other ribs with a more complex decoration, broken into fragments, may not be true spatulas. There are few *sagaies,* taking into account their abundance in earlier periods. We already mentioned the discrepancy between the discoveries in old and modern excavations. Other instruments which appear in Los Azules are small bone hooks manufactured from small, straight, double-pointed splinters, with a flat cross-section.

Azilian bone-working is clearly simpler and more impoverished than that of the Magdalenian, as if only the most basic elements were retained, simplified and without anything which did not contribute to technical efficiency. It is possible that the increase in the percentages of microliths is the cause of this regression in bone-working, as bone tools were substituted by complex multicomponent instruments which also had their origins in the Magdalenian (e.g., *sagaies,* where the bone support can easily be substituted by wood).

Funeral Rites

We only have one example from the Cantabrian Azilian to give us information about funeral rites in this period: the burial of an adult male in Cueva de Los Azules. It was found between layers 3a and 3d, near the end of the occupation of the cave. The body was laid in a pit near the entrance, where the floor had to be lowered as it nearly reached the roof. The head of the pit ended in a large fallen block, the cave wall formed the western side of the pit, and small blocks lined the eastern side of the body. Lying on his back, he was buried on a floor dyed red; one arm lay next to the body, while the other was folded with the hand over the pelvis. Much of the skull had disappeared due to the erosion of a small stream. The grave goods formed four groups consisting of raw materials (flint, quartzite, antler) and some tools, painted cobbles and a large group of mussel (*Modiolus* sp.) shells. A tumulus of stones and earth covered everything, on top of which four flagstones were placed in the area covering the legs.

The skeleton was that of an adult male about 50 years old, with a powerful jaw and heavily worn teeth. Bones which have allowed measurement, such as clavicles, humeri and femora, are robust, with well-marked muscle insertion points, and indicate a stature of 169.5 cm (Garralda 1980). Regarding palaeopathology, some bones contain osteophytes and the phalanges are deformed (by osteoarthritis?). Like other skeletons discovered in Europe, the Los Azules man seems to be an evolved form of Upper Paleolithic human, but we are unable to see any clear variation due to the lack of skull. And this single example cannot offer many conclusions, since we have no examples from the Magdalenian with which to compare it (Garralda 1980).

The Late Azilian

Until the 1970s, what we knew of the Azilian was based on a few excavations which gave a very limited conception of the period. Since then, a series of excavations have provided a new, much more complex perspective, in which we see an early phase defined in Asturias, the classic period found throughout Cantabria, and a final period, still known only from a few stratigraphic sequences, but with clear differences between the west and the east of the Cantabrian Coast. This last is observed in a few sites with a full stratigraphy of advanced Epipaleolithic levels.

In the east and center, geometric elements slowly appear in Azilian levels, coexisting with the other characteristic lithic elements which also evolve. In the center of the Cantabrian region we have the long-known example of Cueva del Valle (e.g., Cheynier & González Echegaray

1964). In recent excavations we have seen this pattern in Piélago, where geometrics are not found in lower levels but become more frequent over time. There are a few in level 3 (a triangle and two or three circle segments), they increase a little in level 2, and become abundant in level 1 (García Guinea et al. 1985). In Ekain there are a few triangles in level II (Azilian with hints of Sauveterrian), but they are not found in the lower Azilian levels. In Zatoya the geometric forms (trapezes, triangles and a few segments) also become more common in the upper part of the stratigraphy, substituting for backed bladelets (Barandiarán & Cava 1989). They increase from very small percentages (2.2% in IIb) to much more significant ones (26.8% in I).

The situation in the west looks very different. Geometric microliths are not found in any sites in Asturias. In Los Azules, where we probably have the complete sequence from its beginnings to a point in time very near to its end, and with a large collection of stone tools (c. 300,000), no geometric microliths are known.

It is true that the percentage of backed bladelets increases, but the general aspect of lithic technology is still traditional, with a few modifications in the details, but little more. The presence of the burial could distort some of our perceptions. For example there are more painted cobbles (rare in the lower levels), but this is because of the grave goods. At the moment it seems that mobiliary art on bone becomes less common, but this has to be checked against the results of future excavations. We need to draw a line between the two phases, if it exists, and determine the causes for the changes.

In the eastern part of the region, stone and bone-working were influenced by developments in the Pyrenees and France, while in the west the Azilian evolves into the Asturian, and was accompanied by important changes in population distribution, which became concentrated along the coastal fringe. Inland areas were apparently abandoned (although cf Diez Castillo 1998). The chronology of the end of the Azilian may be very close to the start of the Asturian (Clark 1989, González Morales 1995). But the new Asturian technology, often dominated by crude quartzite heavy duty tools, has few similarities with the more complex industry of its predecessor.

Azilian Economy

The Azilian population followed the same economic pattern as their ancestors, at least at first. They mainly exploited the more abundant medium-sized herbivores to be found in the immediate environment. As a general rule, red *(Cervus elaphus)* and roe deer *(Capreolus capreolus)* were the most commonly hunted species. At the same time, at some sites, there was an increase in wild boar *(Sus scropha)* and chamois *(Rupicapra rupicapra)* (La Riera), or roe deer (Ekain). However, the incidence of larger species gradually declined until it reached low levels. Even so, horse *(Equus caballus)* still attracted Azilian populations in some zones (e.g., at Morín, Aitzbitarte), but its presence is hardly significant regionally (Altuna 1979, 1995).

The situation seen in La Riera is also true of Los Azules, La Paloma and Ekain. All have archaeofaunas with a high proportion of deer, while wild boar appears (more important than in the late Magdalenian), with a few roe deer (although this increases slightly in Ekain) and ibex *(Capra pyrenaica).* In Zatoya the deer, abundant in IIb, is replaced by wild boar. This variability clearly depends on the immediate environment in which a particular site is found. Thus ibex *(Capra pyrenaica)* is more common than deer in the rocky environs of Piélago. In my opinion, this is a clear indication of the conservative attitude of the Azilian population in certain aspects of their economy. It is evident that large species disappear from the spectrum of fauna found in caves, but the exploitation of medium-sized species, especially deer, is still quite intense. It seems that they hunted the most abundant species in their surroundings, which suggests that there was no particular specialization, nor specific cultural inclination.

We must look at other aspects of Azilian economy to see if there were any new behavioral patterns, or an economic inclination which did not have its origin in earlier periods. In a cave like Los Azules, with a full stratigraphy, we can see two aspects which seem to show a change in behavior during the Azilian. One of them is the slow increase in remains of freshwater and anadromous fish (trout, salmon) as the Azilian progresses. The Sella and Güeña rivers are very near the cave, so it is logical, even if not absolutely necessary, that their resources would have been exploited. The exploitation was intense in the upper levels (which could be a reason for the high number of harpoons there), and it was less intense in the lower levels. Consequently, the impression is that the exploitation of riverine sources intensified during the Azilian. We also have a higher incidence of marine and estuarine molluscs in the upper levels, especially limpets, and *Trivia* (used for adornment). This is significant if we consider that the cave is found inland, not too far from the coast as the crow flies, but separated from it by the rugged coastal mountains. The presence of mussels *(Modiolus modiolus barbatus)* in the burial suggests that the connection with the coast was relatively important and stable.

La Riera, which is nearer the coast-line, has a much higher incidence of molluscs in Azilian level 27 than in lower levels, with an absolute predominance of *Patella vulgata*, indicating a preference for gathering in the sheltered zones of a rocky estuary. From level 25 there is also an increase in the capture of birds. Vega del Sella (1930) noted the presence of salmonids in the Azilian levels, which was confirmed by more recent excavations. Although these were not too abundant, *Salmo trutta trutta* (trout) and *Salmo trutta fario* (sea trout, or *reo*) were identified. The incidence of marine fish increased in level 26 and above. All this shows that the exploitation of the area's resources increased in scope and intensity over time, with a greater interest in salt and brackish water resources than previously.

A similar situation can be observed in Ekain, nearer the present coast-line than Los Azules. In levels II and III there is a relative increase in shellfish (limpets and topshells [*Monodonta lineata*]). At this site, this was accompanied by a small increase in the numbers of small passeriform birds. The seasonality of site occupation can be established through the analysis of the fauna found there. This was possible in Ekain V-III, where the age at death of young ungulates could be determined from their dentitions. The same is true of two sites in Navarra: Abauntz and Zatoya.

Azilian populations had a territorial mobility which was, at first, probably not very different from that of their Magdalenian ancestors. However this raises several problems, the first of which is the very tricky definition and limits of their territories. It is possible that geographically territorial definitions and limits coincided. At the beginning of this chapter, I mentioned a social entity which included numerous groups with common characteristics and which possibly shared the same world view, despite differences in their symbolic repertoires. This coherence is clearly expressed in the art of the Cantabrian-Pyrenees-Aquitaine region. But it becomes more complex if we examine the economy of these regions. Ethnology clearly shows that territories can be shared, that there may be aggregations of several groups at certain times of the year for various purposes, and that at other times these same groups behaved as independent units. As is well-known from the work of Bordes, the occupation of a territory by different groups can be seen in prehistory in the variations among the different Mousterian facies (e.g., 1968, 1972). In this way, we can imagine that the unity represented by Upper Paleolithic art was lost at the start of the Epipaleolithic, taking with it the whole social structure. At least there are no remains of important coherent artistic traditions to convey the impression of large human groups which shared the same view of reality. At the start of the Azilian, it seems that the common conscience which human groups once possessed had shrunk, and the clear fragmentation of signs possibly indicates more limited social (and possibly economic) territories. It is interesting to compare French animalistic art – and its reflection in San Gregori – with what is found in the Asturian region. The appearance here of signs which are known only in this geographical zone may be significant. What is more, they are found in sites that are relatively close – within 70 km of each other – and which have good communications along the narrow coastal plain. Another case of signs limited to a certain part of Cantabria are the signs engraved on pendants from Piélago, Cueva San Juan, etc. The regional character of these signs is not accompanied by such a clear common background as happens during the Upper Paleolithic. Perhaps the Cantabrian Epipaleolithic reproduces at a smaller scale the greater and more widespread unity of the Magdalenian world.

Finally, there is another aspect of territoriality that bears on this topic: the specific area of exploitation of a particular local group. As we saw, populations apparently moved over ill-defined territories to exploit resources, but so far it is impossible to identify precisely the range of movement. At a site like Los Azules, over time, the territory was expanded in order to exploit new resources. The nearby rivers were fished for trout and salmon, but the economic territory also extended to the coast (to gather molluscs, which may even have become a part of their symbolic universe – as in the example of the mussel shells found with the burial). The only mobility that can be clearly demonstrated is the relationship to the coast. The signs engraved on the harpoon from level 5 are practically identical to those found in La Lluera, which could indicate another, wider territory of a larger social aggregate of some kind.

The fact that the resources of a territory are being more intensely exploited over time (we have no information about plant gathering) could be evidence of more restricted movements, hence social circumscription. There is a wider range of consumable resources easily obtainable without any great local alterations, at least for the group as a whole. The greater exploitation of the rivers could be an example of this. And if we take into account the diversity of microenvironments found in relative juxtaposition to one another within small areas of Cantabria, movement could have been reduced even more. If Epipaleolithic communities began to act this way, the tendency towards smaller groups in more restricted areas would make the great Paleolithic art unnecessary, as it had lost its meaning and function. The later Asturian culture became even more isolated as, from the data we have, it was limited to the coastal zone or adjoining mountain areas. Meanwhile, in eastern Cantabria, contacts with the Pyrenees were the cause of a more richly varied evolution.

Addendum (July, 2003)

The site of Cueva Oscura de Ania (Adán et al. 1999) has produced harpoons that correspond to an early phase of the Azilian similar to that which was detected and dated at Cueva de los Azules. For now, this early Azilian appears to be confined to the western sector of paleolithic Cantabria.

Regarding Azilian art, the western Cantabrian art appears to express a stylistic reality quite distinct from that observed in the Pyrenees and the south of France, where some examples of parietal art and animal depictions on portable art still persist. The discovery of painted cobbles in the cave of Estebanvela (Segovia), on the *mesetas* of Old Castile, with abstract motifs similar to those found in French caves and in areas outside the Azilian 'heartland' (e.g, Italy) reflects the separation in 'worldview' between the Cantabrian west and the rest of the Epipaleolithic world after this point in time.

References

Adán Álvarez, G., E. García Sánchez, J. M. Quesada López

1999 El Aziliense de la Cueva Oscura de Ania (Las Regueras, Asturias): primera aproximación y su contexto en la cuenca del Nalón. *Espacio, Tiempo y Forma, Serie I – Prehistoria y Arqueología* 12: 215-267.

Altuna, J.

1979 La faune des ongulés du Tardiglaciaire en Pays Basque et dans le reste de la région Cantabrique. In *La Fin des Temps Glaciaires en Europe* (D. de Sonneville-Bordes, ed.), pp. 85-95. Paris: Éditions du CNRS.

1995 Faunas de mamíferos y cambios ambientales durante el Tardiglacial cantábrico. In *El Final del Paleolítico Cantábrico. Transformaciones Ambientales y Culturales durante el Tardiglacial y Comienzos del Holoceno en la Región Cantábrica* (A. Moure & C. González Sainz, eds.), pp. 77-117. Santander: Universidad de Cantabria.

Altuna, J. et al.

1984 *El Yacimiento Prehistórico de la Cueva de Ekain (Deba, Guipúzcoa)*. Bilbao: Sociedad de Estudios Vascos.

Apellániz, J.M.

1982 *El Arte Prehistórico del País Vasco y sus Vecinos*. Bilbao: Universidad de Deusto.

Arambourou, R. et al.

1978 *Le gisement préhistorique de Duruthy à Sorde-l'Abbaye (Landes. Bilan des recherches de 1958 à 1975)*. Paris: Mémoires de la Société Préhistorique Française No. 13.

Armendariz, A.

1993 Anton Koba (Oñati), *Arkeoikuska* 92: 190-193.

Barandiarán Maestu, I.

1973 *Arte Mueble del Paleolítico Cantábrico*. Zaragoza: Portico Librerias.

1990 Revisión estratigráfica de Berroberría. Datos en 1990. *Veleia* 7: 7-33.

Barandiarán Maestu, I. et al.

1989 *El Yacimiento Prehistórico de Zatoya (Navarra)*. Pamplona: Trabajos de Arqueología Navarra No. 8.

Barton, C.M., G.A. Clark & A. Cohen

1994 Art as information: explaining Upper Paleolithic art in western Europe. *World Archaeology* 26: 185-207.

Berganza, E.

1991 Cueva de Santa Catalina (Lekeitio). *Arkeoikuska* 89: 58-60.

1992 Nouvelles données sur la transition Magdalénien Superieur-Azilien. La grotte de 'La Pila' (Cantabria, Espagne). In *Le Peuplement Magdalénien – Paléogéographie Physique et Humaine*, pp. 259-269. Paris: Éditions du CNRS.

Bernaldo de Quiros, F. & A. Neira Campos

1993a Paleolítico superior final de alta montaña en la Cordillera Cantábrica (noreste de León). *Pyrenae* 24: 17-22.

1993b Occupations de haute montagne dans la Région Cantabrique espagnole (nord-est de la Province de León). In *Actes du Congrès National de Sociétés Historiques et Scientifiques* 118: 193-203.

1994 Mountain occupation sites in the Cantabrian Range (Spain). *Preistoria Alpina-Museo Tridentino di Scienze Naturali* 28: 49-58.

Breuil, H. & H. Obermaier

1912a Les premiers travaux de l'Institut de Paléontologie Humaine. *L'Anthropologie* 23: 1-16.

1912b Institut de Paléontologie Humaine.Travaux exécutés en 1912. *L'Anthropologie* 24: 1-27.

Cabrera Valdés, V.

1984 *El Yacimiento de la cueva de 'El Castillo', Puente Viesgo, Santander*. Madrid: Bibliotheca Præhistorica Hispana Vol. 22.

Chaline, J.

1965 Observaciones préliminares sobre los terrenos cuaternarios de los alrededores de Arredondo (Provincia de Santander). *Cuadernos de Espeleología* 1: 3-17.

Cheynier, A. & J. González Echegaray

1964 La Grotte del Valle. In *Miscelánea en Homenaje al Abate Breuil I,* pp. 327-345. Barcelona: Universidad de Barcelona.

Clark, G.A.

1989 Site functional complementarity in the Mesolithic of northern Spain. In *The Mesolithic in Europe* (C. Bonsall, ed.), pp. 589-603. Edinburgh: John Donald Publishers.

Clark, G.A., C.M. Barton & A. Cohen

1996 Explaining art in the Franco-Cantabrian refugium: an information exchange model. In *Debating Complexity* (D. Meyer et al., eds.), p. 241-253. Calgary: AAUC.

Courad, Cl.

1985 *L'Art Azilien. Origine-Survivence*. Paris: Éditions du CNRS.

Delpech, F.
1989 L'environnement animal des Magdaleniens. In *Le Magdalenien en Europe. La structuration du Magdalenien* (J.-Ph. Rigaud, ed.), pp. 5-30. Liège: Études et Recherches Archéologiques de l'Université de Liège No. 38.

d'Errico, F.
1994 *L'Art Gravé Azilien – De la Technique à la Signification*, XXXIe Supplément à *Gallia Préhistoire*. Paris: Éditions du CNRS.

Fernández Rapado, R. & M. Mallo Viesca
1965 Primera cata de sondeo en Cueva Oscura de Perán. *Boletín del Instituto de Estudios Asturianos* 54: 65-72.

Fernández-Tresguerres Velasco, J.A.
1981 *El Aziliense en las Provincias de Asturias y Santander.* Santander: Monografías del Centro de Investigación y Museo de Altamira No. 2.
1983 Visión general del Epipaleolítico cantábrico. In *Homenaje a Martín Almagro* (A. Balil et al., eds.), pp. 231-237. Madrid: Ministerio de Cultura.
1989 Thoughts on the transition from the Magdalenian to the Azilian in Cantabria: evidence from the Cueva de Los Azules, Asturias. In *The Mesolithic in Europe* (C. Bonsall, ed.), pp. 582-588. Edinburgh: John Donald, Publishers.
1994 El arte aziliense, *Complutum* 5: 81-95.

Fernández-Tresguerres Velasco, J.A. & F. Junceda Quintana
1994 Los arpones azilienses de la cueva de Los Azules (Cangas de Onís, Asturias). In *Homenaje al Dr. Joaquín González Echegaray* (J.A. Lasheras. ed.), pp. 87-95. Madrid: Monografías del Centro de Investigación y Museo de Altamira No. 17.

Fortea Pérez, J.
1973 *Los Complejos Microlaminares y Geométricos del Epipaleolítico Mediterráneo Español.* Salamanca: Memorias del Seminario de Prehistoria y Arqueología.

Fullola i Pericot, J.M. & C. Couraud
1988 Le galet peinté de l'Abri Filador (Catalogne, Espagne). *L'Anthropologie* 88: 119-123.

Fullola i Pericot, J.M., R. Viñas, R. Vallverdu & P. García Argüelles i Andreu
1990 La nouvelle plaquette gravée de Sant Gregori (Catalogne, Espagne). In *L'art des objets au Paléolithique I: l'art mobilier et son contexte* (J. Clottes, ed.), pp. 279-285. Foix: Colloque International de Foix – Le Mas d'Azil.

García Guinea, M.A. et al.
1985 Las cuevas Azilienses de El Piélago (Mirones, Cantabria). *Sautuola* 4: 11-154.

Garralda, M.D.
1980 El esqueleto aziliense de la Cueva de Los Azules I (Cangas de Onís, Oviedo). In *Actas del 1° Congreso Español de Antropología*, pp. 573-580. Barcelona: Asociación Española de Antropología.

Gómez Tabanera, J.M., M. Pérez Pérez & J. Cano Díaz
1975 Premiére prospection de 'Cueva Oscura de Ania' dans le bassin du Nalón (Las Regueras, Oviedo) et connaissance de ses vestiges d'art rupestre. *Bulletin de la Société Préhistorique de l'Ariège* 30: 59-69.

González Echegaray, J. & L.G. Freeman
1973 *Cueva Morín – Excavaciones 1969.* Santander: Patronato de Cuevas Prehistóricas, Diputación Provincial de Santander.
1980 *El Yacimiento de 'El Pendo' (Excavaciones 1953-57).* Madrid: Bibliotheca Praehistorica Hispana Vol. 17.

González Echegaray, J., M.A. García Guinea & A. Begínes Ramirez
1963 *Cueva de La Chora (Santander).* Madrid: Excavaciones Arqueológicas en España No. 26.

González Echegaray, J. & I. Barandiarán Maestu
1981 *El Paleolítico superior de la Cueva del Rascaño (Santander).* Madrid: Monografías del Centro de Investigación y Museo de Altamira No. 3.

González Morales, M.
1995 La transición al Holoceno en la región Cantábrica: el contraste con el modelo Mediterráneo Español. In *Los Últimos Cazadores – Transformaciones Culturales y Económicas durante el Tardiglaciar y el Inicio del Holoceno en el Ámbito Mediterráneo.* V. Villaverde Bonilla, ed., pp. 63-78. Alicante: Diputación de Alicante e Instituto de Cultura 'Juan Gil-Albert.'

González Morales, M. & Y. Díaz Casado
1991-1992 Excavaciones en los abrigos de la Peña del Perro (Santoña, Cantabria). Estratigrafía, cronología y comentario préliminar de sus industrias. *Veleia* 8-9: 43-64.

González Sainz, C.
1982 Un colgante decorado de Cueva Morín (Santander). Reflexiones sobre un tema decorativo de finales del Paleolítico superior. *Ars Praehistorica* 1: 151-159.

1989 *El Magdaleniense Superior-final de la región Cantábrica*. Santander: Prensa Universitaria de la Universidad de Cantabria.

1994 Sobre la cronoestratigrafía del Magdaleniense y Aziliense en la región Cantábrica. *Munibe* 46: 53-68.

Hoyos Gómez, M.

1995 Paleoclimatología del Tardiglacial en la Cornisa Cantábrica basada en los resultados sedimentológicos de yacimientos aruqeológicos kársticos. In *El final del Paleolítico Cantábrico – Transformaciones ambientales y culturales durante el Tardiglacial y comienzos del Holoceno en la Región Cantábrica* (A. Moure & C. González Sainz, eds.), pp. 15-75. Santander: Prensa Universitaria de la Universidad de Cantabria.

Lorbranchet, M.

1989 De l'art naturaliste des chasseurs de rennes à l'art géométrique du Mésolithique dans le sud de la France. *Almasor* 7: 95-124.

Molinero Arroyabe, J.T. & J.F. Arozamena

1984 Reseña arqueológica del karst de Helguera. *Boletín Cántabro de Espeleología* 5: 29-35.

Moure Romanillo, A. & C. González Sainz (editors)

1995 *El final del Paleolítico Cantábrico – Transformaciones ambientales y culturales durante el Tardiglacial y comienzos del Holoceno en la Región Cantábrica.* Santander: Prensa Universitaria de la Universidad de Cantabria.

Obermaier, H.

1925 *El Hombre Fósil.* Madrid: Memoria No. 9 de la Comisión de Investigaciones Paleontológicas y Prehistóricas.

Rodríguez Asensio, A.

1990 Excavaciones arqueológicas realizadas en el cueva de 'La Lluera' (San Juan de Priorio, Oviedo). In *Excavaciones Arqueológicas en Asturias (1983-1986),* pp. 15-17. Oviedo: Diputación Provincial.

Straus, L.G. & G.A. Clark

1986 *La Riera Cave – Stone Age Hunter-Gatherer Adaptations in Northern Spain.* Tempe: Arizona State University Anthropological Research Papers No. 36.

Utrilla Miranda, P.

1982 El yacimiento de la cueva de Abauntz (Arraiz-Navarra). *Trabajos de Arqueología Navarra* 3: 203-345.

Vega del Sella, Conde de la

1916 *Paleolítico de Cueto de la Mina (Asturias).* Madrid: Comisión de Investigaciones Paleontológicas y Prehistóricas, Memoria No. 13.

1930 *Las Cuevas de la Riera y Balmori.* Madrid: Comisión de Investigaciones Paleontológicas y Prehistóricas, Memoria No. 38.

14

Nine Decades of Research on the 'Asturian' of Cantabria

M. Á. Fano Martínez
University of Deusto

M. R. González Morales
University of Cantabria

Abstract

The term 'Asturian' is commonly used to refer to the Mesolithic of the western Cantabrian coastal plain.[1] *The archaeological record there is characterized by numerous shell middens which are almost always found in cave and rock shelter contexts. From the 9th through the 6th millennium cal BC, these are the only archaeological sites found along the west Cantabrian coast. In contrast, the east Cantabrian coast has yielded far fewer shell middens than found in western Cantabria and in Asturias province, to the west. Among the most prominent early Cantabrian prehistorians were the Count of Vega del Sella and Hugo Obermaier. In fact, it was this German author who first brought Peninsular prehistory to the forefront with his book El Hombre Fósil (1916; 2nd edition, revised 1924, English translation 1925), which included all Cantabrian Mesolithic research up to the onset of World War I. After several decades of inactivity due in part to the revolution in Asturias (1931), the Spanish Civil War (1936-9) and its aftermath, Francisco Jordá and his collaborators presented a series of proposals that linked the Mesolithic to a normative vision of culture, questioned in the subsequent decades by G.A. Clark and L.G. Straus, who introduced anglophone 'New Archaeology' to Cantabrian Mesolithic research. Research during the last two decades has attempted to mitigate the estrangement of the discipline of History from the processual approach. However, key aspects of the processual approach have continued to prevail, such as the notion of hunter-gatherer environmental 'adaptation', and the method of initiating research through the formulation of explicit hypotheses.*

Mesolithic Research from 1900 until the 1960s

Mesolithic research began with Ricardo Duque de Estrada, the eighth Count of Vega del Sella, in the years bracketing World War I. This aristocrat was one of the pioneers of Cantabrian prehistory, and the man responsible for identifying its 'signature' Mesolithic expression, the Asturian (Fig. 14.1). In the excavations at El Penicial cave, in eastern Asturias, Vega del Sella discovered a lithic industry without precedent in the region (Vega del Sella 1914). The most distinctive feature of the Penicial lithic assemblage was the presence of a unifacial, pointed artifact made on a quartzite cobble and later named the 'Asturian pick'.

During these early years, and decades before the development of radiometric dating techniques, the most basic question addressed by the archaeologists of the time was how to establish a chronology for the finds from El Penicial. This debate can be followed in the correspondence of that era (Márquez Uría 1988: 487). Certainly, the limited stratigraphy of El Penicial did not permit secure chronological conclusions to be drawn, so Vega del Sella had to rely upon typological criteria in order to present an assessment of the age of the materials and this initially led him astray. In his monograph on the site (Vega del Sella 1914) he attributed the finds to the Lower Paleolithic (or, more precisely, to a period of transition between the Acheulean and the Mousterian). Shortly afterwards, however, work he directed in the nearby caves of Fonfría and Mazaculos demonstrated unequivocally that the Penicial lithic assemblage was very similar to those recovered from cave and rock shelter sites containing shell middens (Vega del Sella 1916: 63).

One year later, Hugo Obermaier officially named the period characterized by shell middens and quartzite-dominated tool assemblages with the diagnostic picks as the 'Asturian'. This was the term Obermaier used in the first edition of *El Hombre Fósil*, already convinced that the Asturian was a post-Paleolithic culture (Obermaier 1916: 334). It was evident that while Obermaier was writing *El Hombre Fósil*, he was fully informed of the work of Vega del Sella at Cueto de la Mina (Fano n.d.). This was a cave with a long stratigraphy that strongly suggested a post-Paleolithic age for the Asturian deposits (Vega del Sella 1916). But it was another Asturian cave, nearby La Riera (also excavated by the Count), that provided the least ambigu-

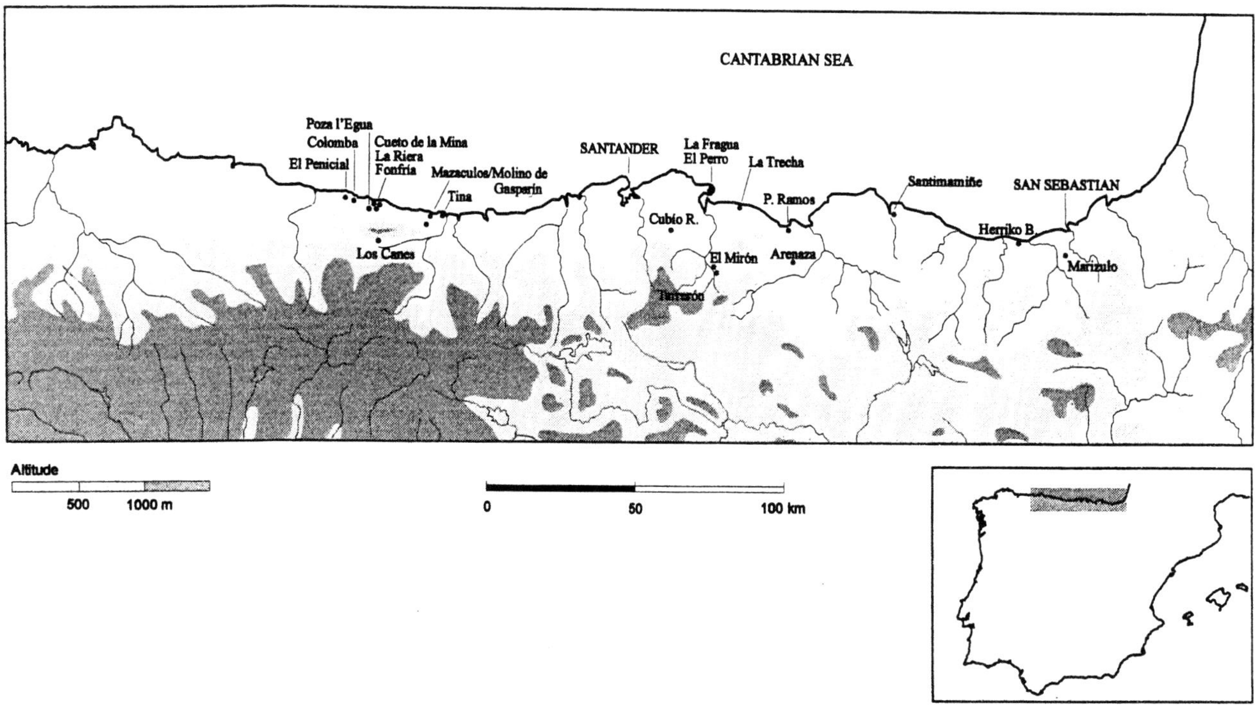

Figure 14.1. Location in the Cantabrian region of the archaeological sites mentioned in the text.

ous and most convincing stratigraphy to fix the relative age of the Asturian. La Riera had a clear succession of archaeological deposits sealed under a Holocene-age stalagmitic crust that directly corresponded to the Upper Paleolithic (Aurignacian, Solutrean, Magdalenian), the Epipaleolithic Azilian, and the Mesolithic Asturian, here represented by a classic Asturian *conchero* (Vega del Sella 1923: 47; 1930: 9).

In summary, the archaeological work carried out by Vega del Sella subsequent to the excavation of El Penicial caused him to reject his initial hypothesis that the 'Asturian' was part of the Lower Paleolithic. However, as indicated below, the 'coarseness' of the Asturian lithic assemblage continued to be a problem in the following decades. This confusion was caused entirely by the apparent contradiction between the formal and technical aspects of the industry, and its stratigraphic position.

Although it withstood critical scrutiny for almost 40 years, Vega del Sella's assessment of a Mesolithic age for the Asturian came under fire during the 1950s and 1960s. Francisco Jordá Cerdá (1959) and Noël Llopis Lladó (1953, 1970: 177) proposed different interpretations of the Mesolithic record based primarily on archaeological and geological arguments (Fano 1998a). Llopis' geological observations in caves with Asturian shell middens led him to believe that the formation of these deposits were *earlier* than the formation of the archaeological levels corresponding to the Upper Paleolithic. Perhaps motivated by these observations, the archaeologist Jordá decided to resurrect the 'primitive', Acheulean-like character of the picks and the generally impoverished Asturian lithic inventory as the basis for a chronological argument that the Asturian was of Lower Paleolithic age (Jordá 1959: 66). This view was eventually questioned by Clark as part of his Ph.D research (e.g., Clark 1971).

These efforts to revise the chronological placement of the Asturian during the 1950s and 1960s indicate the extent to which Spanish prehistory continued to be understood in terms of typological systematics that were strongly linked to a 'normative' view of culture (e.g., Binford 1965). In this particular instance, tool typologies as cultural indicators are emphasized and Jordá, at least, tied it to a unilinear model of cultural evolution.

The Theoretical Impact of Processual Archaeology

In the late 1960s and early 1970s, and guided by their mentor, Leslie G. Freeman, North American archaeologists G. A. Clark and L. G. Straus sought to apply New Archaeology's conceptual framework to research on the prehistory of Cantabrian Spain. One key contribution was the incorporation of radiocarbon dating to the study of the Cantabrian Mesolithic. By chronometrically dating Mesolithic levels for the first time, archaeologists were able to confirm the post-Paleolithic age of the Asturian (Clark 1972). But aside from the use of specific techniques, what was truly important was the development of systematic survey and excavation projects that set out to assess the credibility of previously formulated hypotheses in an archaeology that was previously almost entirely inductive. The generation of test implications from specific hypotheses developed prior to excavation was part of an effort to make archaeology a more rigorous enterprise, and one squarely within the framework of a positivist concept of science (e.g., Watson, LeBlanc & Redman 1971). The concept of culture as an adaptive system had now been transformed into an interest in the environment. Questions were focused on the subsistence and population patterns of Asturian foragers. A good example of this new methodological application is Geoff Clark's doctoral dissertation (Clark 1976, 1983a).

The crude-looking, quartzite heavy duty tools found in Asturian middens were very different from the microlithic industries made on flint and chert associated with the Azilian.[2] Initially, Clark attributed the disparity between the Azilian and Asturian to a problem of sampling error and tried to evaluate this hypothesis, without success, through archaeological test pits in the entrances of several caves. His intention was to try to identify a more 'conventional' Mesolithic industry that consisted of occupation levels that differed from the shell middens, taken to be garbage heaps. In Clark's approach, the possibility of approaching this problem from a functional perspective began to emerge, but it was Straus who proposed that the Azilian and Asturian might be two functional facies (interior and coastal, respectively) produced by the same human groups (Straus 1979). Key arguments in the formulation of this hypothesis were the apparent chronological overlap of some Azilian and Asturian sites, as well as the location of the two types of sites with respect to the contemporary coastline. The evolution of this hypothesis can be followed in subsequent works by both workers (Straus 1981, 1985, 1986, 1992, 1995; Straus et al. 1983; Straus & Clark 1986; Clark 1983b, 1989, 1995, 2000).

Curiously, Straus did not hesitate to use arguments of a normative character to reinforce his hypothesis, such as the presence of parallels in art and funerary contexts (Straus 1992: 228). Regarding the 'Azilian' painted cobbles occasionally recovered in Asturian contexts, he applies the normative concept of culture by interpreting these objects as an expression of a cultural norm characteristic of a specific human group. The same thing occurs when Straus touches upon the similarities observed between the Azilian human burial found at the Los Azules rockshelter near Cangas de Onís (Fernández Tresguerres 1976), and the Asturian burial at Molino de Gasparín (Colombres) (Carballo 1960). Here the cultural norm invoked refers to a concrete way of burying the dead.

Finally, it is convenient to cite a typical characteristic of the processual focus, what might be called 'the remoteness of History'. In fact, the New Archaeologists who worked in northern Spain were more interested in knowing how Mesolithic foragers adapted to their changing environments than they were about the genesis of Mesolithic historic processes (e.g., Clark's lack of interest in the Mesolithic/Neolithic transition [although cf. Clark 1986, 1990]).

Research in the 1980s and 1990s

The last 20-25 years have seen the development of research projects oriented toward the evaluation of hypotheses about specific aspects of Asturian adaptations, and by the integration of Asturian research into the wider perspectives of Iberian and European Mesolithic studies.

One of the main results of modern-era Mesolithic research has been marked improvement in our knowledge of the lithic and bone industries associated with Asturian shell middens. A case in point is the work of M. R. González Morales at Mazaculos Cave between 1976 and 1983 (González Morales 1978, 1995; González Morales & Márquez Uría 1978; González Morales *et al.* 1980), and the systematic re-analysis of Asturian museum collections by González Morales (1982) and P. Arias (1991), and the re-examination of non-Asturian Mesolithic sites in the Basque Country. As noted above, much of the information on Asturian industries before the end of the 1970s originated with Vega del Sella's excavations of shell middens of eastern Asturias prior to 1920, while the evidence for the Mesolithic of the Basque Country was based mainly on the results of the excavations at Santimamiñe Cave (Aranzadi, Barandiarán & Eguren 1931).

The Mazaculos Asturian shell midden was originally excavated in 1915 by Vega del Sella (1923). Its re-excavation offered the unique opportunity to expose intact Mesolithic deposits using modern techniques, including fine mesh screening of all sediments. This allowed us to investigate one of the most debated questions regarding Asturian lithic industries: whether or not the apparent absence of a microlithic component in Asturian collections was due to inadequate excavation procedures. Tiny, fragile, fragmented bone and fish remains (fish bones and teeth, sea urchin spines, micromammals , etc.) were recovered in quantity, but very few microliths. This supported the hypothesis that the absence of microliths is a substantive aspect of 'classic' Asturian shell middens in caves. The recovery of a number of straight, bi-pointed 'fish hooks' or gorges also confirmed observations made on the Asturian bone industries by Vega del Sella some six decades earlier (1915, 1923).

The reanalysis of museum collections with an eye towards a critical review of both Vega del Sella's publications and Clark's test pits pointed in the same direction, especially with regards to the supposedly Asturian open air site of Liencres (better known as Rostrío de Ciriego at the beginning of the century) (Clark 1975, 1979). Liencres was no longer accepted as a single-component Asturian site, but rather as a flaking area linked to the presence of local flint outcrops (very scarce in the region) where most of the production seems to technotypologically 'later', perhaps of Chalcolithic- age. This was indicated by a typical Chalcolithic arrowhead (Clark 1975: 23), and by what we regard as lack of conclusive evidence for the association of the quartzite Asturian picks and flint *dèbitage* (however, cf. Papalas et al. 2002 for a statistical argument to the contrary). Thus, invoking Liencres as evidence to substantiate a claim for the synchronic occurrence of Azilian and Asturian industries became difficult to accept.[3]

The determination of occupational seasonality using $^{16/18}O$ analysis on shell and renewed faunal studies of the dentitions of juvenile cervids were applied to test hypotheses of functional similarity and difference by González Morales (1992) at Mazaculos, and by Clark and Straus at La Riera (Deith & Shackleton 1986). The results at Mazaculos showed that the site was being used during different seasons of the year. This was evident from the remains of animals hunted from late spring to fall, and from the collection of *Monodonta* sp. (topshell) during the late fall and winter. Year-round site utilization, but with seasonal differences in resource exploitation, was also documented at La Riera (Straus & Clark 1986). These results were highly relevant to the notion of Asturian shell middens as temporary, seasonal components of larger Mesolithic settlement-subsistence systems that included central base camps located in the montane interior, as proposed by Clark and Straus (1983).

The isotopic and faunal data from Mazaculos showed a different situation. Analyses conducted by Deith on samples from other sites in eastern Asturias gave results consistent with a marked seasonal peak for *Monodonta* collection during the fall and winter (González Morales 1992). But in the case of Mazaculos, the faunal remains, dominated by red deer, included a sizable percentage of very young individuals, suggesting a late spring and summer presence at the site. Other evidence for year-round occupation of Mazaculos is reinforced by the scarcity of the Asturian inland base camps proposed by Clark (1983). For some researchers, this implies the systematic abandonment of the Cantabrian highlands during the Mesolithic (see discussion in Fano 1996, González Morales 1999, Straus & González Morales 2002, Arias & Fano n.d.a). It is hoped that paleodietary studies currently underway will demonstrate whether or not there were two different economic strategies being practiced: one corresponding to coastal dwelling foragers, and the other by those who inhabited the inland areas of the region (Arias & Fano n.d.b).

A second important issue is the analytical significance of the shell midden deposits. If they could be shown to be year-round residential base camps, and not temporarily occupied seasonal sites, the idea of Asturian sites as mere garbage dumps (e.g., Straus 1979) could be challenged. The identification of habitation surfaces by González Morales at Mazaculos (1982), including bone and lithic scatters and hearths at the base of the shell midden, represented a new direction in understanding Asturian occupations and revealed – for the first time – a Mesolithic occupation that contained living areas with conventional finds, including typical Asturian lithics (i.e., unifacial picks, choppers, quartzite and flint flakes), very few retouched implements, a substantial fauna dominated by red deer *(Cervus elaphus)* but with aurochs *(Bos primigenius),* wild boar *(Sus scropha)* and other ungulates in lower numbers); small bone points, and constructed hearths with burnt and fire cracked rock (Fig. 14.2). It was only after the excavation that the origin of these deposits became clear. Everything had been progressively covered by the mound of debris growing from the rear wall of the shelter, where the initial accumulation of shell seemed to take place. In most Mesolithic cave sites, because of the erosion of unconsolidated shell deposits, only the concreted cornices of the middens are left intact. However, uniquely at Mazaculos, Mesolithic life-ways could be clearly reconstructed because the former living surfaces were literally engulfed by the ongoing deposition of shells. The human occupation probably spread outward from under the shelter overhang; traces of new living spaces appeared on what were the peripheries of the shell mound.

A third area of recent research which has offered new lines of evidence for the interpretation of Cantabrian Mesolithic adaptations is the analysis of what might be called 'habitability variables' – things that shed light on why Mesolithic foragers chose to live where they did. Especially important in this highly seasonal environment is research by M. A. Fano on solar exposures as they affected site locations (Fig. 14.3). Fano (1998b, 2001) assumed that there was intentionality behind the selection of a precise settlement location by Mesolithic groups and that there was (or should be) a fairly direct relationship between the characteristics of the settlement and the activities carried out in and/or from that place. Ultimately, Fano considered the possibility of finding a link between the characteristics of the habitat and the remains derived from the human group activities.

Analyzing the factors determining the conditions of habitability of a site is especially interesting because it allows us to control for settlement characteristics and thereby gain some insight into the decisions groups made regarding the selection of one site location over another. The location and physical characteristics of where people decided to settle can be analyzed scientifically to yield information that can be correlated with data from the archaeological record *sensu stricto* (Fano 2001). Once the settlement location is chosen (according to economic or other criteria), it is reasonable to suppose that prehistoric people tried – as people do today – to protect themselves from the natural elements in an efficient way. However, this is something of a balancing act. Sometimes the whole group (or a part thereof) would opt for a less favorable location, in terms of habitat, but more strategic one, based on the principal activities to be conducted there. So, the study of the immediate environs of a site – the place where humans establish their residence in a more or less stable way – must be taken into account as part of the overall research strategy when posing questions about the roles of different settlements. As in the study of tool function, the goal is to determine whether or not there are significant differences among settlement characteristics.

To apply this approach to determine why shell middens are located where they are requires a reliable sample of Asturian sites in relation to the following criteria for reliability: (1) number of recorded sites, (2) their precise geographic locations, and (3) confidence that the sites studied do, in fact, date to the Mesolithic. To do this, Fano (1998) first had to evaluate the impact of the Flandrian Transgression on the archaeological record of the north Spanish coastal plain. He also considered the relationship between the geological substrate and site preservation, site location and the intensity of survey in the different parts of the Cantabrian region, and site chronology as determined by radiocarbon and by the malacological composition of the deposits. This resulted in a sample of 81 Mesolithic sites. Using a GIS program, models of solar exposure were constructed that estimated the potential solar exposure at each site location for different seasons of the year. Combining the GIS data with the results of seasonality studies based on oxygen isotope analysis of shell, Fano concluded that, in general, Mesolithic groups did not settle in those parts of the region that were perceptibly affected by topographic shading (poor exposure to the sun) during the coldest periods of the year (1998).

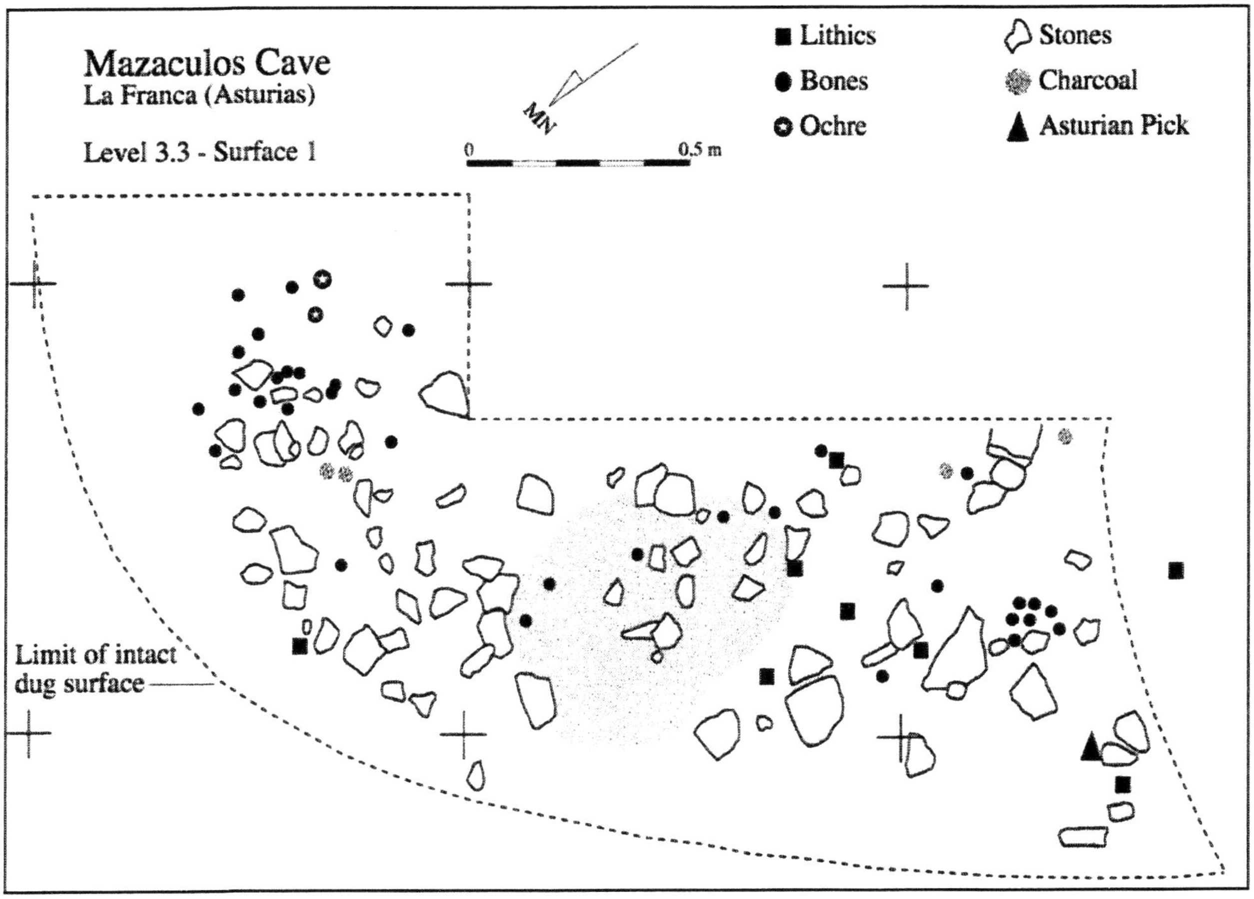

Figure 14.2. Occupation surface in Level 3.3, Mazaculos Cave (Asturias), showing the distribution of animal bones, lithics, charcoal accumulations and burnt stones.

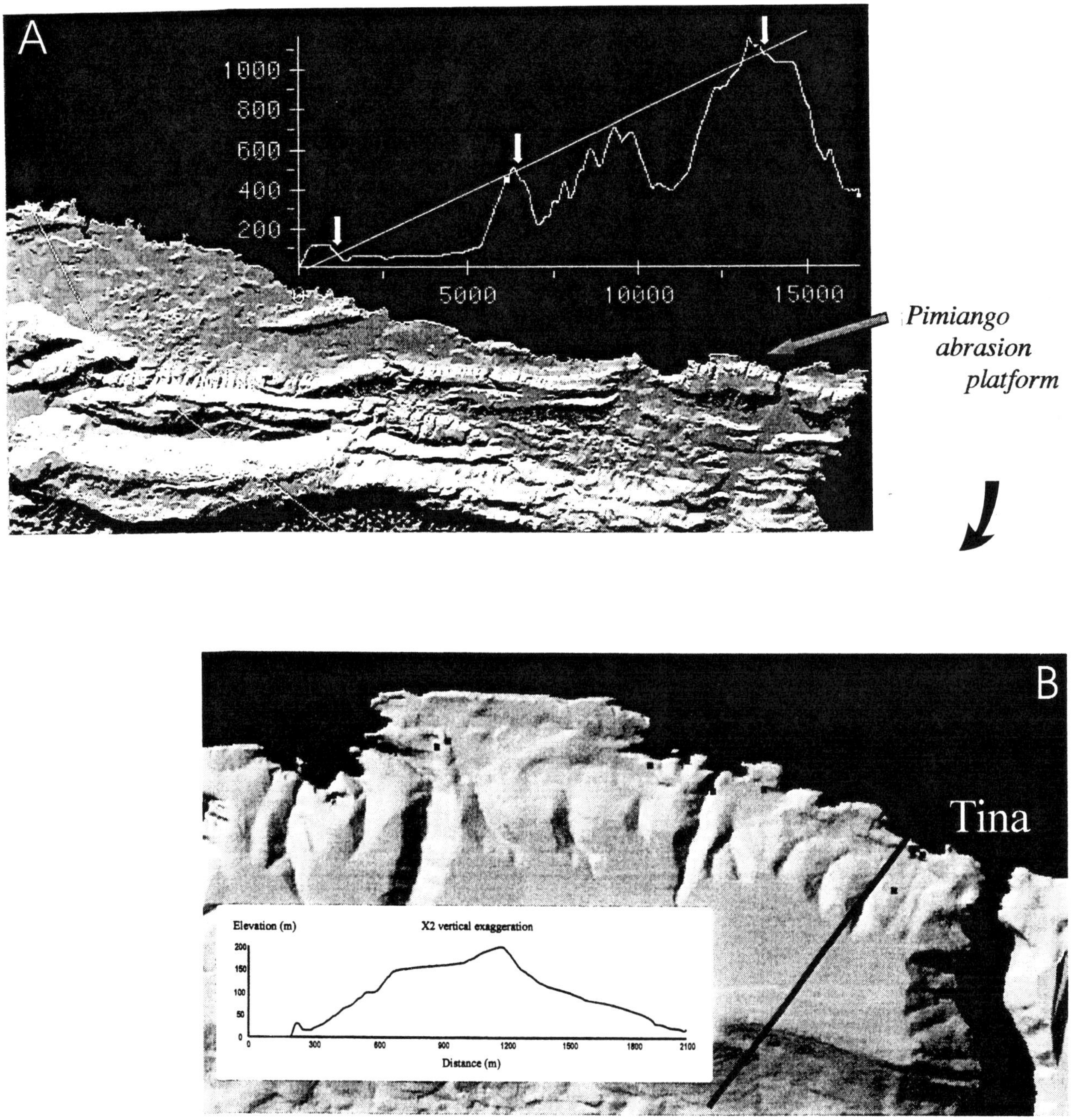

Figure 14.3. The insulation factor plays a key role in analyzing the conditions of habitability for settlements located in regions of abrupt topography such as eastern Asturias. In this area, site exposures in relation to geographic North can be largely irrelevant, given topographic screening (A). This is the case with several Mesolithic sites in the Tina Menor and Mayor river valleys (B); they open to the south, but the proximity of the *rasa* (coastal platform) of Pimiango radically modifies the insulation of the zone due to the effects of topographic screening.

Aside from favorable winter exposure, the information from other variables (environmental conditions prevailing during the period of site formation, settlement location with respect to the local topography, the location of evidence for human occupation in the caves, type of occupations, and estimates of the energy expended on the transport of aquatic resources) suggested that caves and rockshelters were loci of habitation, and consequently that the shell middens themselves were habitation areas (and not merely garbage heaps) (Fano 2002).

Isolated human remains recovered in the Asturian shell middens of Colomba and Poza l'Egua in 2000 and 2001 are also new elements in the debate about the significance of these deposits (Arias & Fano n.d.a, Arias et al. n.d.a, b). These isolated remains, like those previously recovered at Balmori (Clark 1974), are difficult to interpret, mainly because we cannot discount the possibility of previously disturbed contexts or erosion that might have destroyed funerary structures. To date, only one other single grave has been discovered in an Asturian context and this burial at the Molino de Gasparín rockshelter (Colombres), was unfortunately excavated at the beginning of the last century using crude techniques (Carballo 1926). Other human remains of allegedly Mesolithic age, like the calvarium from Cuartamentero (currently in process of AMS dating), either were recovered by local farmers (and thus had no contextual information), or came from disturbed deposits with limited stratigraphic information (e.g., the Mazaculos and Balmori mandible fragments (Arias & Fano n.d.a).

The Paleolithic-Mesolithic Transition and the Arrival of the Neolithic

Many of the new directions in Mesolithic research over the past 25 years were directly linked to the processual approach of the 'New Archaeology', and emphasized the evaluation of hypotheses formulated during the1970s. Since the early 1990s, however, there has been a resurgence of other theoretical viewpoints to explain the significance of the Cantabrian Mesolithic in relation to other Epipaleolithic societies. In addition, the arrival of the Neolithic way of life is becoming increasingly prominent as a research focus, albeit from divergent theoretical points of view. For the earlier transition, the excavation of Azilian and Mesolithic sequences in Cantabria, to the east of Asturias, is also forcing us to reformulate the aspects of continuity and divergences with respect to the Azilian not only for the classic Asturian sites but for other regional Mesolithic groups. Excavations between 1986 and 1991 at El Perro rockshelter, in eastern Cantabria, uncovered a sequence of levels from Late Magdalenian to Mesolithic (González Morales & Díaz Casado 1992). Shell collecting is present during the Magdalenian with limpets (*Patella* spp.) being the dominant species exploited. There is one radiocarbon date from these deposits: 12,140±180 BP (13,233-11,895 cal BC).[4] The overlying Azilian shows a significant increase in periwinkles (*Littorina littorea*); the contact between the two Azilian levels is dated to 10,160±110 BP (10,317-9,490 cal BC). Shellfish exploitation in the Early Mesolithic shows a sharpl increase in the incidence of mussels and oysters, and the usual shift to the topshell (*Monodonta lineata*).

The terrestrial faunas at El Perro are roughly similar in terms of general bone counts during the Magdalenian and Azilian occupations, but decrease dramatically during the transition to the Mesolithic. This follows the same trend well documented at other Asturian sites. The same thing happens with the lithic industries: both Magdalenian and Azilian levels contain the usual flint tools, including lots of microliths (backed bladelets, Azilian points), along with the common suite of late Upper Paleolithic retouched pieces (endscrapers, burins, etc.). These artifacts all but disappear in the Asturian. Bone tools recovered from the site also exhibit this pattern. Antler *sagaies* (spear points) were recovered in the Magdalenian deposits (several are decorated) and Azilian flat harpoons are present in both Azilian levels. But, again, the Mesolithic shell-rich layers are almost completely devoid of lithic and antler tools. Taken together, the lithic and faunal patterns at El Perro suggest that, away from the sites with typical Asturian middens and picks, there was transitional continuity from the Azilian to the coastal Mesolithic way of life. The 1991-6 excavations at the nearby site of La Fragua Cave confirmed the idea of a connected transition through a similar sequence of Magdalenian and late Azilian deposits, very rich in land snails (9,600±140 BP; 9,224-8,740 cal BC), and a Mesolithic shell midden, which accumulated over the course of about a millennium (7,530±70; 6,442-6,264 cal BC to 6,650±120 BP; 5,663-5,479 cal BC) (González Morales 2000).

The Mesolithic-Neolithic Transition, and the Appearance of Megalithic Tombs

Until recently the notion that there was a Cantabrian Neolithic prior to the appearance of megalithic tombs was based on the existence of an appreciable chronological gap. This gap was created by the comparison of more recent Mesolithic dates to older ones associated with megalithic monuments, as well as the sequence at the site of Arenaza, in Vizcaya, where level IC2 yielded fragments of impressed ceramics that have been attributed to the first half of the 5th millennium cal BC (Arias 1991, 1994, 1995, 1996, 1997).

The proposal for a shorter chronology for the beginning of the Neolithic transition has been also supported (González Morales 1992, 1996; see also Díez Castillo *et al.* 1995, Díez Castillo 1996; Serna 1997). First, by the continuity documented at a series of Mesolithic sites dating to the 5th millennium cal BC (e.g., Herriko Barra [Mariezkurrena & Altuna 1995], Pico Ramos [Zapata 1995], Tarrerón [Apellániz 1971 and La Trecha [González Morales 2002]) and, more recently, Cubío Redondo (Ruiz

& Smith 2001) and Los Canes caves (Arias & Altuna 1999). Second, by the absence of archaeological sites with domestic faunas dated to the first half of the 5th millennium cal BC. In fact, for a long time, the oldest evidence for domestic faunas came from level 1 at Marizulo cave at the very end of the 5th millennium cal BC. However, some recently published data suggests an older age for the process of animal and plant domestication in the region (Arias et al. 1999).

The chronometric resolution for this crucial period has become more precisely defined recently because of a series of radiocarbon dates from El Mirón cave, in the eastern Cantabria highlands. El Mirón has so far yielded ten radiocarbon dates from its short sequence of Neolithic levels. Two of them constitute evidence for the earliest Neolithic occupation in the area (5,790±90; 4772-4504 cal BC and 5,690±50 BP; 4576-4459 cal BC). Both the dated samples were taken from an unambiguous stratigraphic contexts that included rich economic faunas dominated by domestic sheep, goat and cattle; typical flint microliths, and plain undecorated pottery (González Morales & Straus 2000, Straus & González Morales 2003).

In the early 1990s, Arias (1991) proposed a continuity scenario wherein local groups of hunter gatherers would have been the main actors in the process of Neolithization. He has also argued that the Neolithization of the Cantabrian region was a gradual process that involved the coexistence, over a long period of time, of mixed predation-domestication economies, the components of which differed from site to site depending upon local ecology (Arias 1991). This view has been contested by González Morales (1996), who links the beginning of the Neolithic to the arrival of groups from outside the region, bringing with them new technologies (pottery), new ways of subsistence, and new a emphasis on the colonization of the Cantabrian *montaña.* The apparent absence of a well defined shell midden phase with pottery immediately after the 6th millennium cal BC (Fano 2000, Arias et al. n.d.a), and the persistence of Mesolithic predation economies beyond that date are two of the cornerstones of his argument (González Morales 1996). In addition, new evidence from the excavations at El Mirón show a very high incidence of domestic animals and a corresponding (and dramatic) reduction in game species which fall below 10% in layers dated to midway through the 5th millennium cal BC. The high incidence of domestic animals recovered from El Mirón and Arenaza (Arias & Altuna 1999) invite a reconsideration of the gradualist Neolithic transition model initially proposed by Arias (Arias & Fano n.d.a).

Paradigmatic Changes

The nature of the Mesolithic/Neolithic transition is currently being explored from an explicitly historical, materialist perspective that departs in some respects from the processualism of the 'New Archaeology' (González Morales 1991, 1996; Arias 1991) A central issue for future investigation is the regional integration of hunter-gatherer and tribal societies during the late Mesolithic-early Neolithic transition. The apparently rapid territorial expansion of Neolithic settlers, as attested by the early appearance of megalithic monuments, needs to be explained. This is especially true of coastal areas, where Mesolithic populations appear to have been densely concentrated (at least in the 'nuclear' areas of central and eastern Asturias, Cantabria and the Basque Country). The problems come when we try to understand the interaction among different populations in a restricted social and geographic space. We are of the opinion that social issues like these cannot be addressed by processualism, with its emphasis on resource exploitation, cost-effective strategies, and so on. Late Mesolithic foraging societies were quickly challenged by a new type of social order that included completely different ways of organizing both subsistence and territorial resources. These changes can be identified only a few centuries after their arrival through what might be called 'the monumentalization of the landscape' (see also Jarman et al. 1983). Local Mesolithic populations must have played a significant role in these changes, but we are only now beginning to appreciate the necessity for novel approaches to understand these questions (Arias & Fano n.d.a).

Acknowledgments

M. A. Fano presently holds a postdoctoral fellowship of the Department of Education, Universities and Research, in the government of País Vasco.

Endnotes

1. By 'Cantabrian region' or 'Cantabrian coast' we mean the Spanish provinces of Asturias, Cantabria, Vizcaya and Guipúzcoa, and not the modern administrative province, here called 'Cantabria'.
2. The Azilian is present across northern Spain from the 11th to the 9th millennium cal BC, with a very typical lithic industry dominated by small thumbnail endscrapers and microliths, including large quantities of pointed ('Azilian points') and backed bladelets. Bone and antler tools are mainly represented by flat harpoons, usually perforated, and a smaller number of spear points or *sagaies,* with a significant reduction or total disappearance of decoration (Fernández-Tresguerres 1995).
3. Liencres has never been dated radiometrically.
4. In this paper calibrated dates are expressed as the mean ± 1s.

References

Apellániz, J. M.
1971 El Mesolítico de la cueva de Tarrerón y su datación por el C-14. *Munibe (Antropologia-Arkeologia)* 23: 91-104.

Aranzadi, T., J. M. Barandiarán & E. Eguren
1931 *Exploraciones de la Caverna de Santimamiñe (Basondo, Cortézubi). 2ª Memoria. Los Niveles con Cerámica y el Conchero.* Bilbao: Imprenta de la Excma. Diputación de Vizcaya.

Arias Cabal, A.
1991 *De Cazadores a Campesinos. La Transición al Neolítico en la Región Cantábrica.* Santander: Servicio de Publicaciones de la Universidad de Cantabria.

Arias Cabal, P.
1994 El Neolítico de la región cantábrica. Nuevas perspectivas. In *1º Congresso de Arqueologia Peninsular* (Porto 1993), Vol. III, pp. 91-118. Porto: Sociedade Portuguesa de Antropologia e Etnologia.
1995 La cronología absoluta del Neolítico y el Calcolítico de la región cantábrica. Estado de la cuestión. *Eusko Ikaskuntza – Cuadernos de Sección Prehistoria-Arqueología* 6: 15-39.
1996 Los concheros con cerámica de la costa cantábrica y la neolitización del norte de la Península Ibérica. In *El 'Hombre Fósil' 80 Años Después: Volumen Conmemorativo del 50 Aniversario de la Muerte de Hugo Obermaier* (A. Moure, ed.), pp. 391-415. Santander: Universidad de Cantabria, Fundación Marcelino Botín e Instituto de Investigaciones Préhistoricas.
1997 ¿Nacimiento o consolidación? El papel del fenómeno megalítico en los procesos de neolitización de la región cantábrica. In *O Neolítico Atlántico e as Orixes do Megalitismo: Actas do Coloquio Internacional (Santiago de Compostela, 1996)* (A. Rodríguez Casal, ed.), pp. 371-389. Santiago de Compostela: Servicio de Publicacións e Intercambio Científico da Universidade de Santiago de Compostela.

Arias Cabal, P. & J. Altuna
1999 Nuevas dataciones absolutas para el Neolítico de la cueva de Arenaza (Bizkaia). *Munibe (Antropologia-Arkeologia)* 51: 161-171.

Arias Cabal, P., J. Altuna, A. Armendáriz, J. E. González, J. J. Ibáñez, R. Ontañón & L. Zapata
1999 Nuevas aportaciones al conocimiento de las primeras sociedades productoras en la región cantábrica. In *Actes del II Congrés del Neolític a la Península Ibèrica (València, 1999)* (J. Bernabeu & T. Orozco, eds.), pp. 549-557. València: Universitat de València *(Saguntum, Special No. 2)*.

Arias Cabal, P. & M. A. Fano Martínez
n.d.a Shell middens and megaliths: Mesolithic funerary contexts in Cantabrian Spain and their relation to the Neolithic. In *Stones and Bones. Formal Disposal of the Dead in Atlantic Europe during the Mesolithic-Neolithic Interface 6000-3000 BC.* (G. Burenhult, ed.). Archaeological Conference in Honour of the Late Professor Michael J. O'Kelly. Oxford: British Archaeological Reports.
n.d.b Le role des ressources marines dans le Mésolithique de la région Cantabrique (Espagne) : l'apport des isotopes stables. *Bulletin de la Société Préhistorique Française* 101(3).

Arias Cabal, P., M. A. Fano Martínez, A. Armendáriz, E. Álvarez, M. Cueto, Mª.D. Garralda, C. Mensua & L. C. Teira
n.d.a Programa de sondeos en concheros holocenos del oriente de Asturias. *Excavaciones Arqueológicas en Asturias, 1999-2002* 5.

Arias Cabal, P., J. A. Fernández-Tresguerres, E. Álvarez, A. Armendáriz, M. Cueto, M. A. Fano Martínez, Mª.D. Garralda, C. Mensua & L. C. Teira
n.d.b Excavación arqueológica de urgencia en la cueva de la Poza l'Egua (Lledías, Llanes). *Excavaciones Arqueológicas en Asturias, 1999-2002* 5.

Binford, L. R.
1965 Archaeological systematics and the study of culture process. *American Antiquity* 31: 203-210.

Carballo, J.
1926 *El Esqueleto Humano Más Antiguo de España.* Santander: Published privately by the author.
1960 *Investigaciones Prehistóricas II.* Santander: Publicaciones del Museo Provincial de Prehistoria y Arqueología.

Clark, G. A.
1971 The Asturian of Cantabria: subsistence base and the evidence for post-Pleistocene climatic shifts. *American Anthropologist* 73: 1244-1257.
1972 El Asturiense de Cantabria: bases sustentadoras y evidencias de los cambios climáticos post-Pleistocenos. *Trabajos de Prehistoria* 29: 17-30.
1974 Excavations in the Late Pleistocene cave site of Balmori. *Quaternaria* 18: 383-426.
1975a *Liencres: una Estación del Aire Libre de Estilo Asturiense cerca de Santander.* Bilbao: Universidad de Deusto, Cuadernos de Arqueología No. 3.

1975b El hombre y su ambiente a comienzos del Holoceno en la Región Cantábrica. *Boletín del Instituto de Estudios Asturianos* 85: 363-387.

1976 *El Asturiense Cantábrico*. Madrid: Consejo Superior de Investigaciones Científicas, Bibliotheca Praehistórica Hispana 13.

1979a Liencres, an open station of Asturian affinity near Santander, Spain. *Quaternaria* 21: 249-286, 300-304.

1979b Spatial association at Liencres, an early Holocene open site on the Santander coast, north-central Spain. In *Computer Graphics in Archaeology* (S. Upham, ed.), pp. 121-144. Tempe: Arizona State University Anthropological Research Papers No. 15.

1983a *The Asturian of Cantabria – Early Holocene Hunter-Gatherers in Northern Spain*. Tucson: Anthropological Papers of the University of Arizona No. 41.

1983b Una perspectiva funcionalista de la prehistoria de la región cantábrica. In *Homenaje al Prof. Martín Almagro Basch, Vol. I* (A. Balil et al., eds.), pp. 155-170. Madrid: Ministerio de Cultura.

1983c Boreal phase settlement-subsistence models in Cantabrian Spain. In *Hunter-Gatherer Economy in Prehistory: a European Perspective* (G. N. Bailey, ed.), pp. 96-110. Cambridge: Cambridge University Press.

1986 El nicho alimentício humano en el norte de España desde el Paleolítico hasta la Romanización. *Trabajos de Prehistoria* 43: 159-184.

1989 Site functional complementary in the Mesolithic of northern Spain. In *The Mesolithic in Europe* (C. Bonsall, ed.), pp. 589-603. Edinburgh: John Donald Publishers.

1990 Algunas reflexiones sobre la espistemología: una respuesta a Utrilla. *Trabajos de Prehistoria* 47: 369-374.

1995 Complementariedad funcional en el Mesolítico del norte de España. In *Los Últimos Cazadores. Transformaciones Culturales y Económicas durante el Tardiglaciar y el Inicio del Holoceno en el Ámbito Mediterráneo* (V. Villaverde, ed.), 45-62. Alicante: Instituto de Cultura Juan Gil-Albert y Diputación de Alicante.

2000 Thirty years of Mesolithic research in Atlantic coastal Iberia (1970-2000). *Journal of Anthropological Research* 56: 17-37.

Clark, G. A. & L. G. Straus

1983 Late Pleistocene hunter-gatherer adaptations in northern Spain. In *Hunter-Gatherer Economy in Prehistory: a European Perspective* (G. N. Bailey, ed.), pp. 131-148. Cambridge: Cambridge University Press.

Deith, M. & N. Shackleton

1986 Seasonal exploitation of marine molluscs: oxygen isotope analysis of shell from La Riera cave. In *La Riera Cave. Stone Age Hunter-Gatherer Adaptations in Northern Spain* (L. G. Straus & G. A. Clark, eds.), pp. 299-313. Tempe: Arizona State University Anthropological Research Papers No. 36.

Diez Castillo, A.

1996 *Utilización de los Recursos en la Marina y Montaña Cantábricas: una Prehistoria Ecológica de los Valles del Deva y Nansa* (Illunzar 3, 96/97). Gernika: Asociación Cultural de Arqueología AGIRI.

Díez Castillo, A., Y. Díaz Casado & G. Robles

1995 La neolitización de las comarcas de Liébana y Polaciones (Cantabria): implicaciones socio-económicas. In *1° Congresso de Arqueologia Peninsular (Porto 1993), vol. VI*, pp. 55-71. Porto: Sociedade Portuguesa de Antropologia e Etnologia.

Fano Martínez, M. A.

1996 El Mesolítico en Asturias: delimitación cronológica y espacial. *Complutum* 7: 51-62.

1998a Algunas reflexiones acerca de la historia de la investigación sobre el Mesolítico en el extremo occidental de la región cantábrica: a propósito de Jordá. In *Las Culturas del Pleistoceno Superior en Andalucía* (J. L. Sanchidrián & Mª. D. Simón, eds.), pp. 381-395. Nerja: Patronato de la Cueva de Nerja.

1998b *El Hábitat Mesolítico en el Cantábrico Occidental. Transformaciones Ambientales y del Medio Físico durante el Holoceno Antiguo*. Oxford: BAR International Series No. 732.

2000 Después del Asturiense: ocho décadas de incertidumbre acerca del inicio del Neolítico en el Cantábrico occidental. *Veleia* 17: 9-30.

2001 Habitability of the prehistoric settlements: proposal for the study of one of the elements involved and first results for the Cantabrian Mesolithic, north of Spain. *Journal of Iberian Archaeology* 3: 25-34.

2002 L'étude des conditions d'habitabilité des établissements préhistoriques: à propos des amas coquilliers mésolithiques de la région cantabrique occidentale (Nord de l'Espagne). *L'Anthropologie* 106: 411-422.

n.d. Hugo Obermaier and the research on the Mesolithic of Northern Spain. *Quartär* 52/53.

Fernández-Tresguerres, J. A.
1976 Azilian burial from Los Azules I, Asturias, Spain. *Current Anthropology* 17: 769-770.
1980 *El Aziliense en las Provincias de Asturias y Santander.* Madrid: Centro de Investigación y Museo de Altamira, Monografías No. 2.
1995 El Aziliense en la región cantábrica. In *El Final del Paleolítico Cantábrico* (A. Moure & C. González, eds.), pp. 199-224. Santander: Universidad de Cantabria.

Garralda, Ma.D.
1982 New human remains of the Spanish Mesolithic. *Historical Anthropology* 9: 375-378.

González Morales, M. R.
1978 Excavaciones en el conchero de la Cueva de Mazaculos II (La Franca, Ribadedeva, Asturias). *Boletín del Instituto de Estudios Asturianos* 93-94: 369-383.
1982 *El Asturiense y Otras Culturas Locales. La Explotación de las Áreas Litorales de la Región Cantábrica en los Tiempos Epipaleolíticos.* Santander: Ministerio de Cultura.
1991 From hunter-gatherers to food producers in northern Spain: Smooth adaptive shifts or revolutionary change in the Mesolithic? In *Perspectives on the Past – Theoretical Biases in Mediterranean Hunter-Gatherer Research* (G. A. Clark, ed.), pp. 204-216. Philadelphia: University of Pennsylvania Press.
1992 Mesolíticos y megalíticos: la evidencia arqueológica de los cambios en las formas productivas en el paso al megalitismo en la costa cantábrica. In *Elefantes, Ciervos y Ovicaprinos: Economía y Aprovechamiento del Medio en la Prehistoria de España y Portugal* (A. Moure, ed.), pp. 185-202. Santander: Servicio de Publicaciones de la Universidad de Cantabria.
1995 Memoria de los trabajos de limpieza y toma de muestras en los yacimientos de las Cuevas de Mazaculos y El Espinoso (La Franca, Ribadedeva) y La Llana (Andrín, Llanes) en 1993. *Excavaciones Arqueológicas en Asturias, 1991-94* 3: 65-78.
1996 La transición al Neolítico en la costa cantábrica. In *I Congrés del Neolític a la Península Ibèrica. Formació i Implantació de les Comunitats Agrícoles* (Gavà-Bellaterra, 1995), Vol. 2, pp. 879-885. Gavà: Museu de Gavà.
1999 Costa e interior: algunas observaciones sobre el uso de las cuevas en el Mesolítico. In *De Oriente a Occidente – Homenaje al Dr. Emilio Olávarri,* pp. 237-246. Salamanca: Universidad Pontificia de Salamanca.
2000 La Prehistoria de las Marismas: Excavaciones en la cueva de la Fragua (Santoña). Campañas de 1990, 1991, 1993, 1994 y 1996. In *Actuaciones Arqueológicas en Cantabria 1984-1999* (R. Ontañon, ed.), pp. 177-179. Santander: Consejería de Cultura y Deporte.

González Morales, M.R., Y. Díaz Casado & C. Yudego Arce
2002 Excavaciones en la Cueva de la Trecha de la Fuente la Corredora (Islares, Castro Urdiales). In *Actuaciones Arqueológicas en Cantabria 1987-1999* (R. Ontañon, ed.), pp. 49-53. Santander: Consejería de Cultura y Deporte.

González Morales, M. R. & Y. Díaz Casado
1991-92 Excavaciones en los abrigos de la Peña del Perro (Santoña, Cantabria). Estratigrafía, cronología y comentario preliminar de sus industrias. *Veleia* 8-9: 43-64.

González Morales, M. R. & M.C. Márquez Uría
1978 The Asturian shell midden of Cueva de Mazaculos II (La Franca, Asturias, Spain). *Current Anthropology* 19: 614-615.

González Morales, M. R., Mª. C. Márquez Uría, T. E. Díaz, J. A. Ortea & K. Volman
1980 Informe préliminar de las excavaciones en el conchero Asturiense de la cueva de Mazaculos II (La Franca, Asturias): campañas de 1976-78. *Noticiario Arqueológico Hispánico (Prehistoria)* 9: 35-62.

González Morales, M. R. & L. G. Straus
2000 El depósito arqueológico de la prehistoria reciente de la Cueva del Mirón (Ramales, Cantabria): bases estratigráficas para una secuencia regional. In *Actas del III Congreso de Arqueologia Peninsular* (V. O. Jorge, ed.), pp. 49-63. Porto: ADECAP, Vol. 4.

Jarman, M., G. Bailey & H. Jarman
1983 *Early European Agriculture: Its Foundation and Development.* Cambridge: Cambridge University Press.

Jordá Cerdá, F.
1959 Revisión de la cronología del Asturiense. In *V° Congreso Nacional de Arqueología,* pp. 63-66. Zaragoza: Secretaría General de los Congresos Arqueológicos Nacionales.

Llopis Lladó, N.
1953 Estudios hidrogeológicos y prehistóricos en Posada (Llanes). *Speleón* 4: 266.
1970 *Fundamentos de Hidrogeología Cárstica.* Madrid: Blume.

Mariezkurrena, K. & J. Altuna
1995 Fauna de mamíferos del yacimiento costero de Herriko-Barra (Zarautz, País Vasco). *Munibe (Antropologia-Arkeologia)* 47: 23-32.

Márquez Uría, Mª.C.
1988 El Conde de la Vega del Sella (1870-1941) y la comisión de investigaciones paleontológicas y prehistóricas. In *1907-1987, la Junta para Ampliación de Estudios e Investigaciones Científicas 80 Años Después.* (J. M. Sánchez, coord.), pp. 485-500. Madrid: Simposio Internacional (Madrid 1987), Vol. II, Consejo Superior de Investigaciones Científicas.

Obermaier, H.
1916 *El Hombre Fósil.* Madrid: Comisión de Investigaciones Paleontológicas y Prehistóricas No. 9.

Papalas, C., G. Clark & K. Kintigh
2003 Liencres revisited – the significance of spatial patterning revealed by unconstrained clustering. In *Mesolithic on the Move* (L. Larsson et al., eds.), pp. 253-261. Oxford: Oxbow Books.

Ruíz Cobo, J. & P. Smith
2001 El yacimiento del Cubío Redondo (Matienzo, Ruesga): una estación mesolítica de montaña en Cantabria. *Munibe (Antropologia-Arkeologia)* 53: 31-55.

Serna González, Mª.R.
1997 Ocupación megalítica y proceso de neolitización en la cornisa cantábrica. In *O Neolítico Atlántico e as orixes do Megalitismo: Actas do Coloquio Internacional (Santiago de Compostela, 1996* (A. Rodríguez Casal, ed.), pp. 353-369. Santiago de Compostela: Servicio de Publicacións e Intercambio Científico da Universidade de Santiago de Compostela.

Straus, L. G.
1979 Mesolithic adaptations along the northern coast of Spain. *Quaternaria* 21: 305-327.

1981 On marine hunter-gatherers: a view from Cantabrian Spain. *Munibe (Antropologia-Arkeologia)* 33: 171-173.

1985 Chronostratigraphy of the Pleistocene-Holocene transition: the Azilian problem in the Franco-Cantabrian Region. *Palaeohistoria* 27: 89-122.

1986 A comparison of La Riera assemblages with those from contemporary sites in Cantabrian Spain. In *La Riera Cave – Stone Age Hunter-Gatherer Adaptations in Northern Spain.* (L. G. Straus & G. A. Clark, eds.), pp. 219-236. Tempe: Arizona State University Anthropological Research Papers No. 36.

1992 *Iberia Before the Iberians. The Stone Age Prehistory of Cantabrian Spain.* Albuquerque: University of New Mexico Press.

1995 A través de la frontera Pleistoceno-Holoceno en Aquitania y en la Península Ibérica: cambios ambientales y respuestas humanas. In *El Final del Paleolítico Cantábrico* (A. Moure & C. González Sainz, eds.), pp. 341-363. Santander: Servicio de Publicaciones de la Universidad deCantabria.

Straus, L. G., G. A. Clark, J. Altuna, M. R. González Morales, H. Laville . Arl. Leroi-Gourhan, M. Menéndez & J. Ortea
1983 Excavaciones en la Cueva de La Riera (1976-1979): un estudio inicial. *Trabajos de Prehistoria* 40: 9-57.

Straus, L. G. & G. A. Clark
1986 Synthesis and conclusions-Part II: The La Riera excavation, chronostratigraphy, paleoenvironments and cultural sequence in perspective. In *La Riera Cave – Stone Age Hunter-Gatherer Adaptations in Northern Spain.* (L. G. Straus & G. A. Clark, eds.), pp. 367-383. Tempe: Arizona State University Anthropological Research Papers No. 36.

Straus, L.G. & M. R. González Morales
2002 The Mesolithic in the Cantabrian Interior: Facts or Fantasy? In *Mesolithic on the Move* (L. Larsson et al., eds.), pp. 347-356. Oxford: Oxbow Books.

2003 El Mirón Cave and the ^{14}C chronology of Cantabrian Spain. *Radiocarbon* 45: 41-58.

Vega del Sella, Ricardo Duque de Estrada, Conde de la
1914 *La Cueva del Penicial (Asturias).* Madrid: Comisión de Investigaciones Paleontológicas y Prehistóricas No. 4.

1916 *Paleolítico de Cueto de la Mina (Asturias).* Madrid: Comisión de Investigaciones Paleontológicas yPrehistóricas No. 13.

1923 *El Asturiense. Nueva Industria Préneolítica.* Madrid: Comisión de Investigaciones Paleontológicas y Prehistóricas No. 32 (Serie Prehistórica No. 27). Reprinted as *La Cueva del Penicial y el Asturiense,* in Biblioteca de Autores Asturianos, Vol. 15 (1991). Gijón: Auseva.

1930 *Las Cuevas de la Riera y Balmori (Asturias).* Madrid: Comisión de Investigaciones Paleontológicas y Prehistóricas No. 38 (Serie Prehistórica No. 29).

Watson, P. J., C. Redman & S. LeBlanc
1971 *Explanation in Archaeology.* New York: Columbia University Press.

Zapata, L.
1995 La excavación del depósito sepulcral calcolítico de la Cueva Pico Ramos (Muskiz, Bizkaia). La industria ósea y los elementos de adorno. *Munibe (Antropologia-Arkeologia)* 47: 35-90.

15

Coastal Paleoeconomies and Paleoenvironmental Trends: Asturian and Australian Middens Compared

Geoff Bailey
University of Newcastle

Alan Craighead
Robert's Creek, British Columbia

Abstract

We examine some of the controversies that have surrounded the interpretation of the Asturian shell middens in northern Spain, and place this case study in a wider context by comparing it with similar controversies in the interpretation of shell mounds in northern Australia. In both cases the crux of the problem is to find methods that can disentangle paleoenvironmental from cultural influences on the representation of mollusc shells in the archaeological record. We demonstrate that many of the characteristics that have been interpreted as evidence of changing human pressure on molluscan resources can be explained equally well in terms of environmental change operating independently of human activity. This poses a serious challenge to interpretation. We illustrate the kinds of evidence that can be used to discriminate between cultural and natural influences, with particular emphasis on the analysis of the limpet shells from the La Riera cave of northern Spain, and demonstrate the dominating effect of environmental processes on long-term changes in shell-gathering behaviour visible in the late Pleistocene/early Holocene archaeological record.

Introduction

The Asturian shell middens *(concheros)* of northern Spain (Fig. 15.1) have given rise to a variety of interpretations over the past 25 years (Bailey 1973, 1978, 1983; Clark 1971, 1983; Clark & Yi 1983, Clark & Straus 1983, Fano 1998, González Morales 1982, Straus 1979, Straus & Clark 1983, 1986). From a paleoeconomic point of view there are two principal issues which have given rise to debate:

1) Are the shell deposits with their preponderance of mollusc shells representative of the subsistence activities practised by their creators, or are they partial and biased indicators given undue prominence because of their archaeological visibility compared with the residues of other subsistence activities?

(2) Is the rate of shell accumulation in the Asturian shell middens significantly greater than in previous periods, and if so, does this indicate economic intensification in the form of a general shift in subsistence strategies to greater emphasis on the marine littoral (e.g., Clark & Straus 1986), or the increased archaeological visibility of midden deposits with changes in sea level (e.g., Bailey 1983), or changes in the type of littoral resources available for exploitation as a result of paleoenvironmental change, or some combination of all three possibilities?

The first issue focuses on the potential over-representation of mollusc remains, and, by extension, on the potential over-representation of shell-midden deposits especially when accumulated in the protective environment of a cave or rockshelter, and can be resolved into three further questions:

- Is the alleged over-representation of mollusc remains due to differential preservation within the *conchero* deposits (i.e., due to the fact that shells preserve better than animal bones or plant remains)?
- Is it due to *intrasite* spatial variation – that is spatial differentiation of discard within the immediate context of the settlements, the shells and other large items of debris being heaped up as garbage dumps in the back of caves and rockshelters, while other items were discarded on a living area in front of the cave entrance which has left little or no archaeological trace, as Straus (1979) has suggested?
- Is it due to *intersite* spatial variation, that is to the over-representation of midden deposits with respect to other locations in the landscape? In other words, are the shell middens but one type of site in a mobile or sedentary-cum-mobile economy that also incorporated exploitation of resources and discard of residues in the hinterland?

The second issue has been widely debated in the literature. Clark and Straus (1986) have advocated the hy-

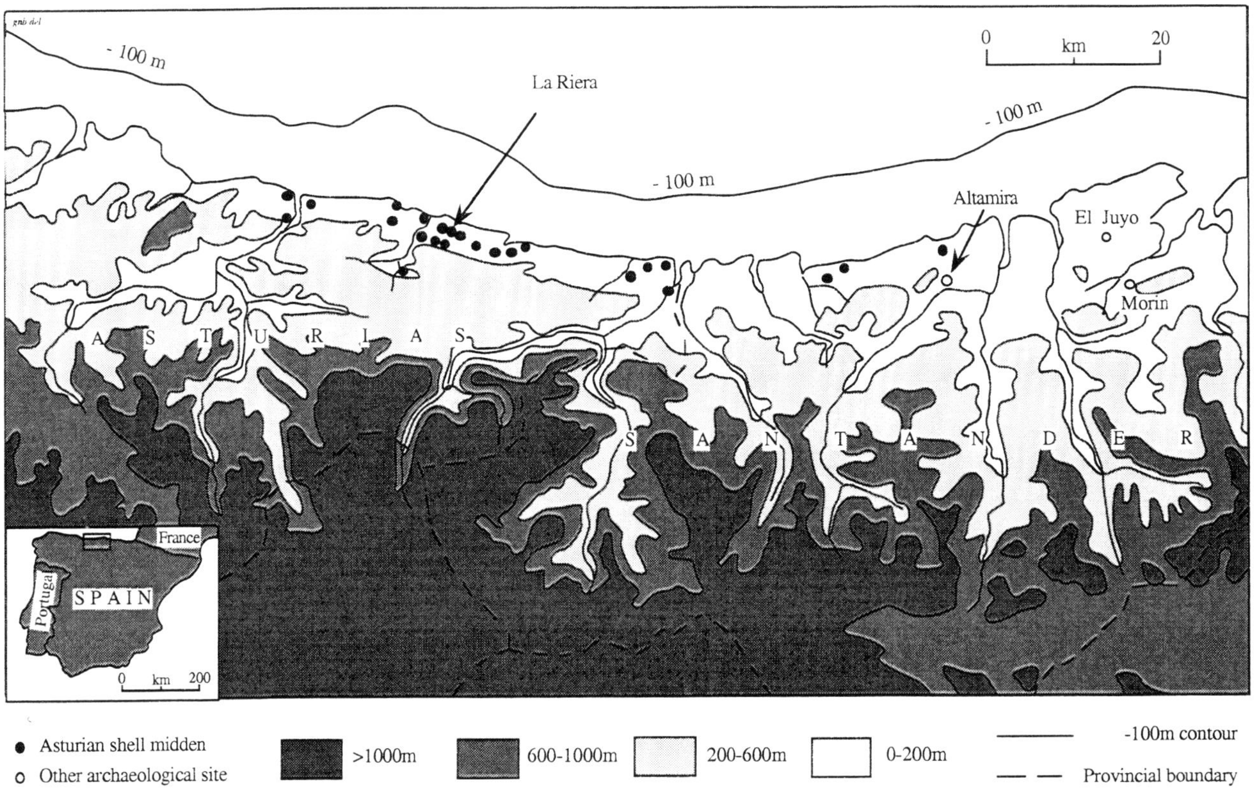

Figure 15.1 Distribution of Asturian shell middens and other sites in northern Spain (from Bailey 1975, González Morales 1982, Straus & Clark 1986).

pothesis of intensification under population pressure, citing in support not only the increased quantity of shells, but also the decrease in shell size of *Patella vulgata*, and the exploitation of the smaller limpet species, *Patella depressa,* growing in less easily accessible littoral environments, all of which are taken to imply greater pressure of human demand on the resource base. Bailey (1978, 1983) has argued in favour of changes in archaeological visibility, citing the potential impact of changes in sea-level and the reduced distance from sea shore to archaeological site as a key factor in the increased quantity of mollusc shells and midden deposits visible after 9000 bp and arguing that environmental change needs to be eliminated as a potential cause of changes in growth rates and species representation before such changes can be confidently ascribed to increased predation pressure. The data currently available do not allow unequivocal discrimination between these alternatives. The lack of independent evidence of environmental changes that could affect molluscan populations is a critical gap.

In this paper we propose to bring fresh light to bear on these issues from two different perspectives. First, we shall briefly examine the arguments and evidence surrounding a case study from Australia where one of us (Bailey) has engaged in renewed fieldwork. Here the relationship between the occurrence of mid-Holocene shell mounds and geomorphological changes on rapidly prograding mangrove-lined shorelines has been a major focus of inquiry. Such an exotic comparison may seem out of place in a discussion of European archaeology, but the Australian case study raises in a very different context many of the same issues that are the focus of attention in Cantabrian Spain. The area has also been the subject of an earlier comparison with northern Spain (Bailey 1983), and fresh field data and interpretations cast the comparison in a new light. As Beaton has recently noted:

> The impact of the last marine transgression arcs over the Pleistocene-Holocene Transition, and archaeologists working on coastlines everywhere . . . are confronted by similar and challenging issues. Now-familiar themes unite studies of archaeology in such distant places as the Americas, Africa, Asia, Europe and Australia (1995, pg. 804).

Second, we shall present the results of recent analytical work by one of us (Craighead 1995) on the limpet remains from La Riera cave, which have a direct bearing on the evidence for climatic changes and their effects on molluscan habitats, on the reasons for size changes in limpet shells, on the impact of human predation pressure, and on seasonality patterns.

Shell Mounds of Tropical Australia

First we turn to the tropical coastlines of northern Australia, where some of the largest shell mounds still surviving anywhere in the world have been recorded. The best known group of such sites is on the west coast of the Cape York Peninsula in northern Queensland, where there are some 500 sites recorded on the four main rivers flowing into Albatross Bay near Weipa (Fig. 15.2). The largest of these sites are over 10 m high and contain up to 10,000 metric tons of shell, although most of the sites are smaller mounds less than 1 m thick and some are barely more than surface scatters. The sites are associated with a mangrove-lined estuarine environment, and the dominant shell species is a bivalve, a type of cockle *(Anadara granosa),* that thrives in estuarine intertidal mudflats (Bailey 1977, 1983, 1993, 1994, 1999; Bailey et al. 1994; Wright 1971).

The total quantities in the area have been estimated at about 200,000 metric tons of shell, or 10,000 million mollusc shells. Notwithstanding these vast quantities and the size of many of the mounds, the hypothesis that they are natural deposits, either thrown up by wet-season storms (Stanner 1961), or scraped up into mounds by the nesting activities of the orange-footed scrub turkey (Stone 1989, 1993), has been comprehensively refuted. Storm-created shell deposits, or cheniers, forming linear features up to 3 m thick, and mounded scrub-fowl nests, sometimes located on cheniers and containing large quantities of shells, are both present on tropical coastlines. But, detailed comparative analysis in terms of location, size, shell content, sedimentological characteristics, stratigraphic features, and presence or absence of artefacts, fish bones and other evidence of human activity, has demonstrated quite conclusively that the mounds were created by human activity, not by storm waves or scrub turkeys (Bailey et al. 1994).

In terms of food, both archaeological (Bailey 1977, 1994) and ethnographic (Meehan 1982) data demonstrate that the molluscs would have contributed less than 20% of caloric requirements. Nevertheless, they played an important role as an easily available, regular and predictable food supply, and helped to tide over daily and seasonal fluctuations in other resources.

Other important groups of *Anadara*-dominated shell mounds with similar characteristics to the Weipa sites have been recorded in Princess Charlotte Bay on the east coast of the Cape York Peninsula (Beaton 1985), on the Love and Ward Rivers near Aurukun south of Weipa (Cribb 1986), in the Northern Territory at Castlereagh Bay in the area of Meehan's (1982) ethnographic study, and at Darwin harbour and Hope Inlet (Bourke 2000), on the island of Milingimbi (Roberts 1991), and in Western Australia (O'Connor 1999, Veitch 1999).

Large numbers of radiocarbon dates are now available, and one of the most striking features is that the great majority of these shell mounds were formed within the last two millennia. Since sea levels had stabilised at about the present level by about 6000 years ago, there appears to be a very considerable time lag between the period when shell mounds should theoretically have begun to accumulate and to have remained visible to the present day, and the period when people first started to consume molluscs in quantity.

Both environmental and cultural explanations have been proposed for this time lag. On the environmental side it is possible that the ecological window of opportunity for large intertidal beds of *Anadara* molluscs is quite short-lived on a time-scale of millennia. In geological terms, the shorelines of the tropical north are relatively unstable, even in the absence of sea-level change. The region is characterised by a tropical monsoon climate with high levels of seasonal rainfall, shorelines with shallow gradients fringed by mangroves and backed by bare seasonally inundated mudflats, and rivers which carry large quantities of sediment into a low-energy coastal environment. Sediment is thus concentrated onshore, with fairly rapid coastline progradation during the Holocene. As Chappell (1982) has observed, there can be time lags of up to several thousand years as sedimentary processes seek to re-establish equilibrium with a new shore profile created by a change of sea level. In the Weipa area, it has been argued that the relatively late appearance of the shell mounds reflects time lags in the accumulation of an intertidal mudflat habitat, following the drowning of the river valleys by the late Pleistocene-early Holocene sea-level rise (Bailey 1983).

Alternatively the time lag may reflect a delay in population growth or in the development of more complex forms of social organisation that resulted in intensification of resource exploitation (cf. Lourandos 1999). The demographic variant of this hypothesis has been argued most forcefully by Beaton (1985) in relation the Princess Charlotte Bay shell mounds. The coastal plain here, as in many other coastal areas of the tropical north, is characterised by well-developed systems of chenier ridges. The general trend of sediment infilling and coastal progradation is accompanied by episodic storm surges during particularly severe monsoonal storms which drive coarse shell debris ashore to form ridges separated by intervening mudflats, each ridge being successively younger as one moves towards the coast. Mounded shell middens of the Weipa type are located on some of these cheniers. The cheniers can be independently dated by radiocarbon, and contain large quantities of fragmented shells. Thus they provide a check on the availability of molluscs independent of the midden evidence. Since the earliest shell middens post-date by nearly two millennia the earliest cheniers, there appears to be a significant time lag between the appearance of abundant supplies of molluscs and their human exploitation, a delay which Beaton (1985) attributes to a time lag in human population growth in the region. On the other hand, the cessation of midden accumulation after about 400 years ago probably reflects the disappearance of the shell beds because of progressive infilling of the bay with sediment, rather than a reduction in human population densities or the reduced intensity of shell gathering activity (Chappell & Grindrod 1984).

In the Weipa area the large river estuaries which flow into Albatross Bay create a more protected marine environment less susceptible to the full force of storm surges. Cheniers do occur along the inner margins of the estuaries,

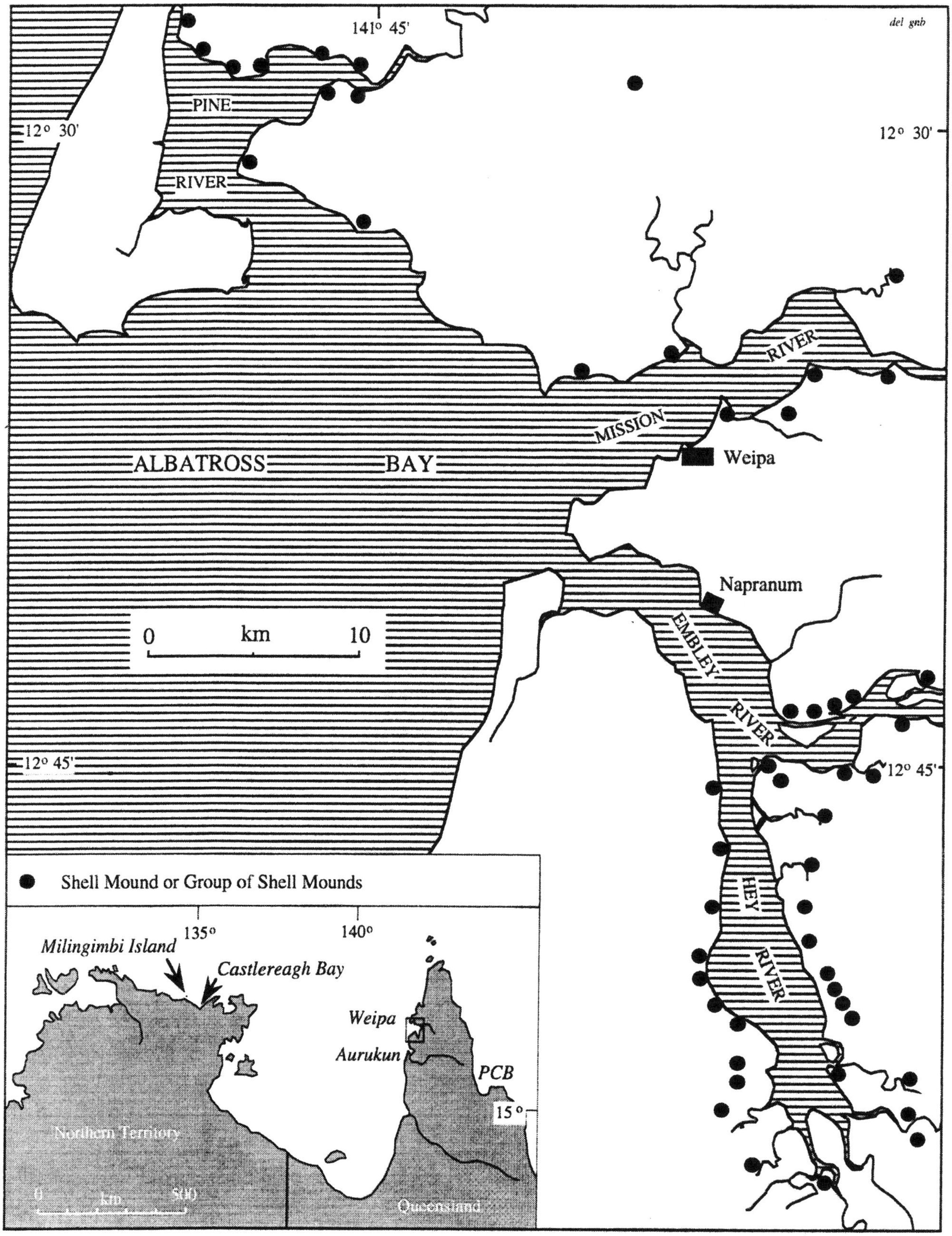

Figure 15.2 Distribution of shell middens at Weipa on the Cape York peninsula. The inset shows the location of other areas in northern Australia with significant concentrations of shell mounds of similar type to the Weipa deposits (PCB = Princess Charlotte Bay).

and the mounded shell middens are often located on them. In this case, however, the cheniers usually form a single line, are quite low, and are composed mainly of sand with very few shells. Since they also predate the mounds, albeit by an unknown interval, this would seem to suggest that relatively few edible bivalve molluscs were available in the millennia preceding the onset of midden formation. In the Weipa area, then, an environmental explanation for the delayed onset of shell gathering seems to work better than at Princess Charlotte Bay.

More recent studies of *Anadara* mounds elsewhere have produced a similar range of hypotheses to account for their formation. In Western Australia, O'Connor (1999) and Veitch (1999) have argued, respectively, for environmental and social factors as principal determinants of the onset of intensive shell accumulation, while Bourke (2000) has recently provided detailed evidence in support of an environmental hypothesis for the Hope Inlet mounds in the Northern Territory.

Wider Relevance of the Australian Evidence

These examples underline a number of points that have more general relevance. In the first place they emphasise the sensitivity of mollusc populations to small environmental changes, and the increasing likelihood of such changes occurring on an expanded time scale. Natural *Anadara* beds are known to shift location from year to year (Pathansali & Soong 1960) and are also vulnerable to periodic destruction by monsoonal storms, as the evidence of chenier ridges on many northern Australian coastlines amply demonstrates (Beaton 1985, Meehan 1982). Moreover, this periodicity is itself likely to vary from region to region, for example from the more sheltered environment of the Weipa estuaries to more exposed open embayments such as Princess Charlotte Bay. Such variability on interannual and longer time spans requires flexible scheduling strategies that must have important implications for consideration of longer-term paleoeconomic trends.

Obviously the ecological tolerance of the molluscs will vary considerably according to species and environment, and one can hardly extrapolate the ecological limiting factors operating on bivalve molluscs on tropical coastlines to the gastropods and rocky shorelines of northern Spain. But the general principle applies as much in the latter case. Molluscs generally have many of the features of *r*-selected resources: a high intrinsic rate of population increase, opportunistic life strategies, and vulnerability to periodic mortality crashes (e.g., Mackay & White 1987). The number of free-floating spat that survive predation and come to rest in a suitable habitat is highly variable from year to year. In many parts of the world shell beds are vulnerable to mass mortalities from unusual seasonal extremes of temperature. On a smaller scale, variations in the degree of exposure of the rocky shoreline can affect the availability or accessibility of molluscs such as limpets, while small changes of temperature, exposure or substrate can affect the particular species available and their patterns of growth. Limpets, for example, cannot survive in waters where a heavy sediment load resulting from high run-off blankets the rocky substrate. They also have a limited tolerance of suspended sediments because of the primitive gill arrangement (Yonge & Thompson 1976: 58) and because the sediments restrict the penetration of light needed for the plant growth on which they feed (Boyle 1981: 21).

Secondly, the Australian examples underline the point that local geomorphic or environmental (and hence paleoeconomic) responses can vary from one area to another even within a region which is otherwise apparently uniform in its general environmental and geomorphological history. In other words we should expect to find not only greater paleoenvironmental and paleoeconomic variability in time, but greater variability in space as well.

Finally, the examples discussed above demonstrate the critical importance of independent paleoenvironmental data as a check on archaeological interpretations. This is perhaps the most important lesson to be drawn from the Australian experience. The tropical coastlines are prograding coastlines of sediment accumulation, which by definition offer opportunities for the accumulation and preservation of geomorphological features that can provide independent data on environmental change. On rocky shorelines like those of northern Spain, where the general trend is towards erosion or submergence of shoreline features, such opportunities are likely to be much rarer. In this case, it will be necessary to rely on physical and chemical signals embodied in the shell carbonate of the exploited molluscs. As we show below, these can offer valuable independent data on paleoenvironmental and paleoeconomic trends.

Northern Spain

Shell Quantities

We first consider some general trends in the quantities of molluscs represented in Upper Paleolithic and Mesolithic deposits of the region. Mollusc shells have been recorded not only in the Asturian *concheros*, but also in earlier deposits, mostly Upper Paleolithic in age. The latter are few in number. Some are stratified beneath Asturian middens in caves and rockshelters, but others are located over a broader territory including the adjacent province of Santander. It is rather difficult to compare quantities in a systematic way between sites or between periods, because adequate quantitative data are rarely presented. Many of the Asturian deposits have also been partly removed by erosion, or by excavation earlier in the 20th century.

The largest reported Asturian midden was some 50 m long by 12 m high according to Obermaier (1925: 386). Taking into account the likely original dimensions of the deposit and the density of shells per unit volume, the total quantity of shells in this one deposit has been estimated to be of the order of one million shells (Bailey 1975). In contrast, none of the shell deposits reported from earlier contexts, for example from the Magdalenian levels at El Juyo in the neighbouring province of Santander (Barandiarán et al. 1985), appears to represent more than a few hundreds or at most thousands of shells. It is obviously rather hard to generalise on the basis of such inadequate quantitative records, but our opinion, based on the number and size of deposits that have been recorded, is that the rate of accumulation of mollusc shells in the middens of the Asturian period as a whole is orders of magnitude greater than in earlier periods.

For a more detailed study we rely on the evidence from the cave site of La Riera (Straus & Clark 1986), which has a cultural sequence extending from the beginning of the Solutrean period at about 20,700 bp (level 2) and possibly earlier, through to the end of the Asturian period at about 6500 bp. This site offers one of the most detailed systematic studies of a long sequence available anywhere in the region, and is the only site to have produced a systematic quantification of molluscan remains through a sequence spanning the late Pleistocene and early Holocene. Level 29 (together with the overlying stalagmitic crust of level 30) represents a classic Asturian shell midden deposit, which originally filled the whole of the cave entrance and extended out in front of it. As with so many Asturian shell middens, much of the original deposit has been removed by erosion and by excavation earlier in the 20th century, and only fragments cemented to the cave wall and ceiling still survive.

We use Ortea's figures (1986, table 15.2), expressed as percentages of shell numbers by stratigraphic level. Since these figures do not allow for variations in the volume of excavated sediment by layer, we also show Straus et al.'s (1986, table 9.3) weights of shell per cubic metre. We can also combine Ortea's figures with Straus et al.'s volumes of excavated deposit to give numbers of shells per cubic metre (Table 15.1, Fig. 15.3). We believe that these figures are a useful proxy measure of differential rates of shell deposition, but with some qualifications. Ortea studied 85–90% of the total shells collected in the 1976–79 excavations. It is not certain that these are all the shells recovered in excavation (rather than a non-systematic sub-sample), although it seems reasonable to assume so in the absence of information to the contrary. In addition, Ortea augmented the figures for the Asturian levels with a 16% sample from a small test pit excavated in 1969. On the other hand, it is likely that both the number and weight of shells in the Asturian level at La Riera may be considerably under-represented because of the fragmented nature of the surviving deposit and the heavily cemented nature of parts of it. Finally, we note that numbers and weights of shell per unit volume are a measure of geometric density (cf. Bailey et al. 1983) and do not necessarily accurately reflect rates of accumulation per unit time (or time density).

A sharp and sustained increase in shell deposition is apparent towards the top of the sequence, beginning at the top of the Magdalenian deposits (level 26, dated to about 13,000 bp and peaking in the Azilian (level 28) dated to about 10,000 bp (Clark & Straus 1986, table 2.2). The drop in the Asturian (levels 29 and 30) is surprising, but we suspect that this may reflect the problems of sampling cemented deposits noted above, and in any case needs to be seen in the context of the much greater total volume of shell deposit compared with earlier periods. There is another quite sizeable peak of shell deposition in levels 4–5 dated to about 19–20,000 bp, and lesser peaks in level 13 (c. 16,000–17,000 bp), and perhaps also in level 17 (c. 16,000 bp).

Sea Level Change

In order to test the relationship between shell quantities and changing distance to the sea shore, we plot the quantitative representation of mollusc shells at La Riera by stratigraphic level alongside Chappell and Shackleton's (1986) data on eustatic sea-level variation (Fig. 15.3). The key point here is that the earlier part of the Riera sequence, between 20,000 and 16,000 bp, coincided with a period when sea level was at its lowest at the Last Glacial Maximum (LGM). From about 15,000 bp onwards, there was a sustained sea level rise, which slowed somewhat at about 8,000 bp and approximated modern sea level at about 6,000 bp. The final sea level rise from about –25 m to the present level coincided with the period of accumulation of the Asturian shell middens.

In plotting the local position of the shoreline, we start with the figure of –125 m for global eustatic lowering of sea level at the LGM, but note that plotting the position of local shorelines is not simply a case of identifying the appropriate bathymetric contour (cf. Van Andel 1989). Accumulations of sediment on the seabed in subsequent millennia, hydro-isostatic and glacio-isostatic adjustments of the earth's crust, and other tectonic movements can introduce significant variations. Even allowing for these variables, there are additional inaccuracies and uncertainties in the data available to plot the depth of the seabed. Lambeck (1997) has modelled isostatic effects in NW Europe and shown that along the Atlantic coast of France the position of the LGM shoreline relative to modern sea level ranged from c. –90 m in the Calais region to –65/75 m in the area of Biarritz, the degree of isostatic displacement decreasing along the north to south gradient. This is substantially less than estimates based solely upon eustatic lowering of global sea levels. As the northern coast of Spain lies at the same distance from the centres of continental glaciation as Biarritz, we observe that the –75m isobath is probably a more accurate guide to locating the LGM shoreline while noting that the correct figure may differ by a margin of up to ± 30 m (Fig. 15.4). We also show bathymetric contours at –50 m and –18 m, approximating sea level positions at about 12,000 bp and 9,000 bp respectively. Again we note that these are not precise shoreline positions or dates, but emphasise that the key point is the relative change in distance from seashore to the cave site.

In judging how far people are likely to carry shells for consumption, it is generally assumed that bulk processing takes place as close as possible to the shore because of the high transport costs involved. Detailed studies of site locations (Bailey & Craighead 2003, see also Fano 1998) and experimental and ethnographic data (Bettinger & Malhi 1997, Bird & Bliege-Bird 1997, Meehan 1982) suggest that a daily limit for mollusc gathering may be 5 km or less, compared with the typical limit of 10 km widely assumed for hunting and gathering in general. Other factors, however, may over-ride distance constraints and encourage people to carry molluscs further inland, for example the desire for a sheltered spot such as that provided by south-facing rockshelters in which so many of the Asturian deposits are found, availability of freshwater or factors of social aggregation. Trade or barter may also result in the transportation of molluscs over much longer distances, but this usually applies to smaller quantities, to molluscs that have been preserved in some way after removal of the shells, or to use of draft animals or mechanised transport. In general the fall-off in quantities with increasing distance is likely to be sharp and 10 km is assumed to be an absolute maximum for daily subsistence trips, with the probability of much lower figures for most molluscs because of the high ratio of shell to usable meat.

With these factors in mind it is difficult to resist the conclusion that the increase in shell deposition visible at the top of the Riera sequence is closely correlated with the final stages of sea level rise. The shoreline would have come within the critical 5 km distance from about 12,000 bp onwards, which is about when the sharp increase in shellfood representation becomes apparent (Fig. 15.3). By the beginning of the Asturian period at 9,000 bp, sea level would have been close to the −18 m contour, and would have approximated the present position by the end of the Asturian period.

However, this is clearly not the whole story, because there is also an earlier peak of shell deposition in levels 4 and 5 at a time when the seashore must have been at its most distant. This accords well with evidence from other Solutrean and Magdalenian sites, notably El Juyo, and suggests that the number of shells, though small relative to the Asturian period, appears significantly greater than in earlier periods. If this can be substantiated, it is a point of some interest, because the increased representation of shells occurs at a time when the nearest shorelines would have been as much as 10 km away. Reduction in the distance from seashore to cave site can hardly be invoked to explain the increased representation of limpet shells in Solutrean and Magdalenian levels with respect to earlier periods. We note the presence at La Riera in level 4 of *Patella rustica* and the land snails *Oestophorella bouvinieri* and *Cochlostoma berilloni*, which imply a warmer climate closer to that of the present day than to maximum glacial conditions. But we doubt that this can be equated with an equivalent sea level rise, unless it was so short lived that it is obscured in the deep-sea record of changing ice volumes.

Unfortunately, the La Riera sequence does not extend far enough back in time to the period of higher sea levels at about 28,000 years and earlier (Fig. 15.3). If varying distance to the contemporaneous shoreline were the only factor affecting the numbers of shells deposited at the La Riera location, we should expect to find significantly increased numbers of shells in deposits of these earlier periods. For the moment we conclude that reduced distance with rising sea level is a major factor in the increased shell deposition apparent in the upper levels at La Riera, but not sufficiently overwhelming to eliminate other possibilities.

Species Representation

Detailed data are available on relative species representation, thanks to Ortea's (1986) study of the material at La Riera (Fig. 15.5). Four species dominate the La Riera collections: *Patella vulgata* Linnaeus 1758; *Patella intermedia* Murray 1758 (*Patella depressa* Pennant 1777); *Littorina littorea* Linnaeus 1758; and *Monodonta lineata* da Costa 1778. Throughout the greater part of the sequence *P. vulgata* dominates, with the addition of small percentages of *L. littorea*, but the most striking time trend is the demise of these species in favour of *P. depressa* and *M. lineata* in the late Azilian and Asturian levels (levels 27–30). The replacement of *L. littorea* by *M. lineata* has long been recognised as the result of postglacial climatic warming. *M. lineata* is a southern species with a present-day distribution extending from southwest Ireland to Morocco and Madeira, whereas *L. littorea* has a more northerly range from the White Sea on the north Russian coast to northern Spain and Portugal (Hayward & Ryland 1990). Indeed the complete absence of *L. littorea* in Asturian deposits has been interpreted as evidence of temperatures warmer than the present, since today *L. littorea* is present on the northern Spanish shorelines along with *M. lineata* (Clark 1971). *Patella rustica* is also a thermophile limpet species with a present-day range from Biarritz to the Atlantic coasts of North Africa (Campbell 1976, Fischer-Piette & Gaillard 1959) and is present in very small numbers in the La Riera sample. Ortea (1986, table 15.3) records it in level 4, corresponding to the end of the Laugerie Interstadial. In a re-examination of a sample of 883 limpet shells from levels 5 to 30 sent to Cambridge for oxygen isotope analysis, we have identified *P. rustica* as being certainly present in levels 23/21 and 29. These data would seem to confirm a climatic warming trend towards the top of the La Riera sequence.

The rise to dominance of *P. depressa* similarly can be explained in climatic terms. It has a more southerly geographical distribution than *P. vulgata*, ranging from the coast of North Wales to North Africa, whereas *P. vulgata* ranges from northern Norway to Portugal (Hayward & Ryland 1990). *P. depressa* tolerates more exposed shore conditions than *P. vulgata*, but unlike *P. vulgata* cannot tolerate low salinities or locations high in the intertidal zone (Bowman & Lewis 1977, Fischer-Piette 1948, Fretter & Graham 1976). Accordingly, where the two species are in competi-

Table 15.1 Representation of mollusc shells by Level at La Riera.

Level	Volume m^3	Shells N	%	Shells N/m^3	Shell %	kg/m^3	%
30*	0.13	492	2.51	3785	8.3	1.43	0.50
29	0.21	653	3.34	3110	6.8	5.3	1.85
28	0.255	3279	16.8	12859	28.0	39.87	13.90
27	2.00	8383	42.8	4192	9.1	17.15	6.00
26	0.125	545	2.78	4360	9.5	33.58	11.70
25						14.98	5.24
24	0.60	71	0.36	118.3	0.3	1.97	0.69
2123	2.25	94	0.48	41.78	0.1	0.51	0.18
1920	0.52	208	1.06	400	0.9	3.68	1.29
20	0.40	206	1.05	515	1.1	3.76	1.31
19	0.56	56	0.29	100	0.2	3.68	1.29
18	0.56	66	0.34	117.9	0.3	6.56	2.29
17	0.1875	211	1.08	1125	2.5	12.46	4.36
16	1.20	147	0.75	122.5	0.3	3.46	1.21
15	0.32	188	0.96	587.5	1.3	4.9	1.71
14	0.92	571	2.92	620.7	1.4	11.26	3.94
13	0.092	209	1.07	2272	5.0	19.22	6.72
12	0.0825	66	0.34	800	1.7	5.01	1.75
11	0.165	18	0.09	109.1	0.2	3.48	1.22
10	0.35	106	0.54	302.9	0.7	2.69	0.94
9	0.40	117	0.60	292.5	0.6	10.78	3.77
8	0.405	589	3.01	1454	3.2	9.63	3.37
7	0.85	587	3.00	690.6	1.5	11.85	4.14
6	0.16	127	0.65	793.8	1.7	5.29	1.85
5	0.385	1380	7.05	3584	7.8	24.1	8.43
4	0.33	1050	5.36	3182	6.9	27.67	9.68
3	0.858	99	0.51	115.4	0.3	0.82	0.2
2	0.128	23	0.12	179.7	0.4	0.66	0.23
1	1.500	34	0.17	22.67	0	0.19	0.07
Totals	15.943	19575	100	45852	100	285.9	100

*Level 30 represents a cemented crust sealing the prehistoric deposits and shell quantities for this and the underlying *conchero* of level 29 should be treated as minimum estimates.

tion, *P. vulgata* tends to be restricted to estuaries or high shore positions. In Portugal, close to the southern limit of the *P. vulgata* range, *P. depressa* replaces *P. vulgata* as the dominant species on both sheltered and exposed shores (Guerra & Guadencio 1986). In addition it is possible that rivers would have been subject to heavier sediment loads resulting from climatic change and increased run-off at the beginning of the Holocene; as noted earlier, *P. vulgata* in estuarine habitats are liable to mass mortalities caused by high sediment loads and silting up of rocky substrates. These factors, together with the independent evidence noted above for temperatures warmer than the present, provide a *prima facie* case for attributing the change in limpet species to climatic warming.

Ortea (1986), however, suggests an alternative explanation. *P. depressa* is generally smaller than *P. vulgata* and also prefers more exposed shorelines. In cost-benefit terms it is a less desirable resource for human consumption than *P. vulgata* and may be ignored until rising demand increases human pressure on the food supply. Thus its appearance in quantity may be due to a widening of the range of shell gathering activity to include more exposed (and presumably less accessible) stretches of shoreline and smaller limpets. The implication of this argument is that exposed shorelines with abundant populations of *P. depressa* were equally available in earlier periods but neglected by human shell gatherers because of a lack of population pressure or other motivations to exploit less abundant and less accessible resources.

To these climatic and demographic hypotheses can be added a third, and that is that the character of the coastline changed significantly about 9000 years ago to one much more dominated by exposed shores favouring *P. depressa* at the expense of *P. vulgata*. We consider this to be the weakest of the three hypotheses, since the modern coastline is exposed with steep rocky coasts and cliffs but nev-

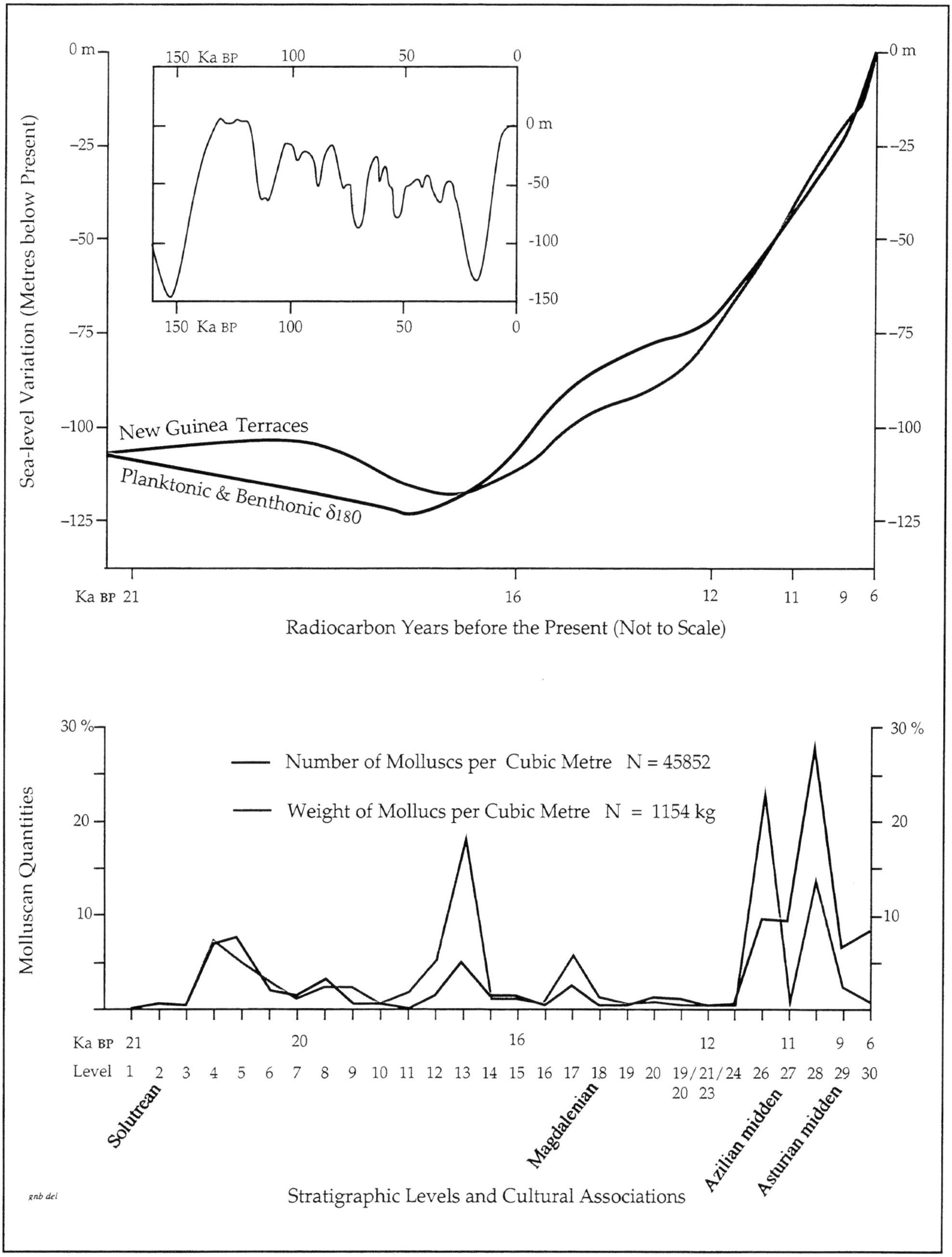

Figure 15.3 Marine mollusc shell frequencies at La Riera, plotted against sea-level variation. Data from Ortea (1986), Straus et al. (1986) and Chappell & Shackleton (1986). See text for discussion of shell quantities and Table 15.1 for raw data.

Table 15.2 Metrical data for *Patella vulgata* samples from the La Riera sequence. Data from Craighead (1995) and Ortea (1986).

Level	Mean Length		Mean Length			Mean Width			Mean Height			Low Shore		Sheltered Shore	
	mm (Ortea)	n	mm	s.e.	n	mm	s.e.	n	mm	s.e.	n	%	n	%	n
30.1	31.0	68	35.8	1.67	15	33.1	0.87	7	12.9	0.66	14	79	15	100	6
29	26.8	34	42.4	3.32	5	36.7	3.77	3	16.6	1.65	5	100	5	100	3
28	28.1	1466	31.8	0.46	107	25.7	0.43	107	11.4	0.24	107	63	107	46	107
27	35.2	3154	37.7	0.38	183	31.4	0.37	179	14.2	0.46	183	76	183	91	176
27 Upper	24.8	502	-	-	-	-	-	-	-	-	-	-	-	-	-
27 Lower	27.9	3630	-	-	-	-	-	-	-	-	-	-	-	-	-
26	34.2	474	33.2	0.50	126	26.8	0.44	122	12.4	0.23	126	60	126	50	122
25	-	-	33.6	0.96	27	27.7	0.84	24	12.9	0.43	27	46	27	54	23
24	33.8	58	31.4	0.71	49	25.8	0.88	26	11.8	0.38	49	68	49	46	24
21/3	43.0	81	39.3	1.24	45	33.3	1.31	42	13.5	0.63	44	82	45	79	39
20	42.5	194	43.4	0.78	43	38.2	0.80	43	15.9	0.39	37	95	43	100	36
19/20	44.5	176	42.7	0.66	64	37.0	0.73	62	15.8	0.37	56	91	64	100	60
19	45.0	47	41.8	0.79	92	37.3	0.69	80	14.8	0.41	64	89	92	93	75
18	45.0	56	43.7	0.50	114	38.7	0.57	107	16.6	0.24	72	87	114	98	105
17	46.2	210	44.7	0.40	119	39.2	0.42	118	16.3	0.33	97	97	119	100	115
16	41.3	144	41.5	0.35	142	36.1	0.37	135	16.1	0.26	74	91	142	99	134
15	38.6	188	38.7	0.84	43	33.1	0.89	42	15.1	0.58	28	67	43	95	42
14	40.1	567	42.0	0.69	112	36.5	0.69	110	16.4	0.39	61	87	112	87	109
13	44.5	208	47.8	1.38	15	42.5	1.45	15	18.3	0.66	3	100	15	100	15
12	45.4	66	44.7	1.76	10	38.5	1.86	10	-	-	-	100	10	100	10
11	40.8	18	47.0	1.91	10	40.9	2.10	10	-	-	-	100	10	100	10
10	35.0	106	38.5	1.55	22	33.0	1.55	22	15.8	0.72	5	68	22	100	22
9	40.3	117	44.6	0.61	76	38.9	0.62	73	16.7	0.44	31	92	76	92	73
8	38.4	576	43.6	0.62	78	37.8	0.63	77	17.6	0.76	27	87	78	93	75
7	44.3	582	43.2	0.52	167	37.3	0.55	157	17.3	0.32	108	80	167	95	152
6	44.4	127	44.0	1.73	16	38.2	2.21	15	20.1	2.14	4	88	16	93	15
5	42.4	1370	43.6	0.60	135	38.2	0.60	129	17.2	0.83	32	91	135	94	126
4	46.1	1045	43.4	0.41	247	37.4	0.41	246	17.8	0.28	109	85	247	84	246
3	42.8	97	35.4	1.04	61	29.4	1.00	61	10.4	0.91	15	79	61	61	61
2	42.3	23	-	-	-	-	-	-	-	-	-	-	-	-	-
1	41.5	34	35.2	1.08	5	27.8	0.87	5	-	-	-	100	5	60	5
Totals		11286			2128			2027			1378		2123		1986

ertheless supports abundant populations of both the principal limpet species. But we concede that this hypothesis is very difficult to test by means of independent evidence and cannot be categorically excluded.

We also consider the demographic hypothesis to be weak because it seems improbable that a human population under pressure to collect more food would ignore *P. vulgata* shells if they were available, unless we suppose that populations of *P. vulgata* had been virtually wiped out by heavy overexploitation. Nevertheless this hypothesis is testable, because serious overexploitation should lead to significant reductions in size and age structure, both of which are measurable in principle. Similarly, additional paleotemperature evidence can be obtained from the shells themselves to further test the climatic hypothesis.

Size Changes

By far the most comprehensive series of measurements on limpets has been recorded by Ortea (1986, table 15.7). As far as we can tell, Ortea measured all of 15,487 identifiable *P. vulgata* shells, and included not only complete specimens but also estimated the maximum length of broken specimens from the thickness of the apex. We cannot evaluate the reliability of these measurements because Ortea does not give details of the formula for estimating length from apical thickness, or the proportion of speci-

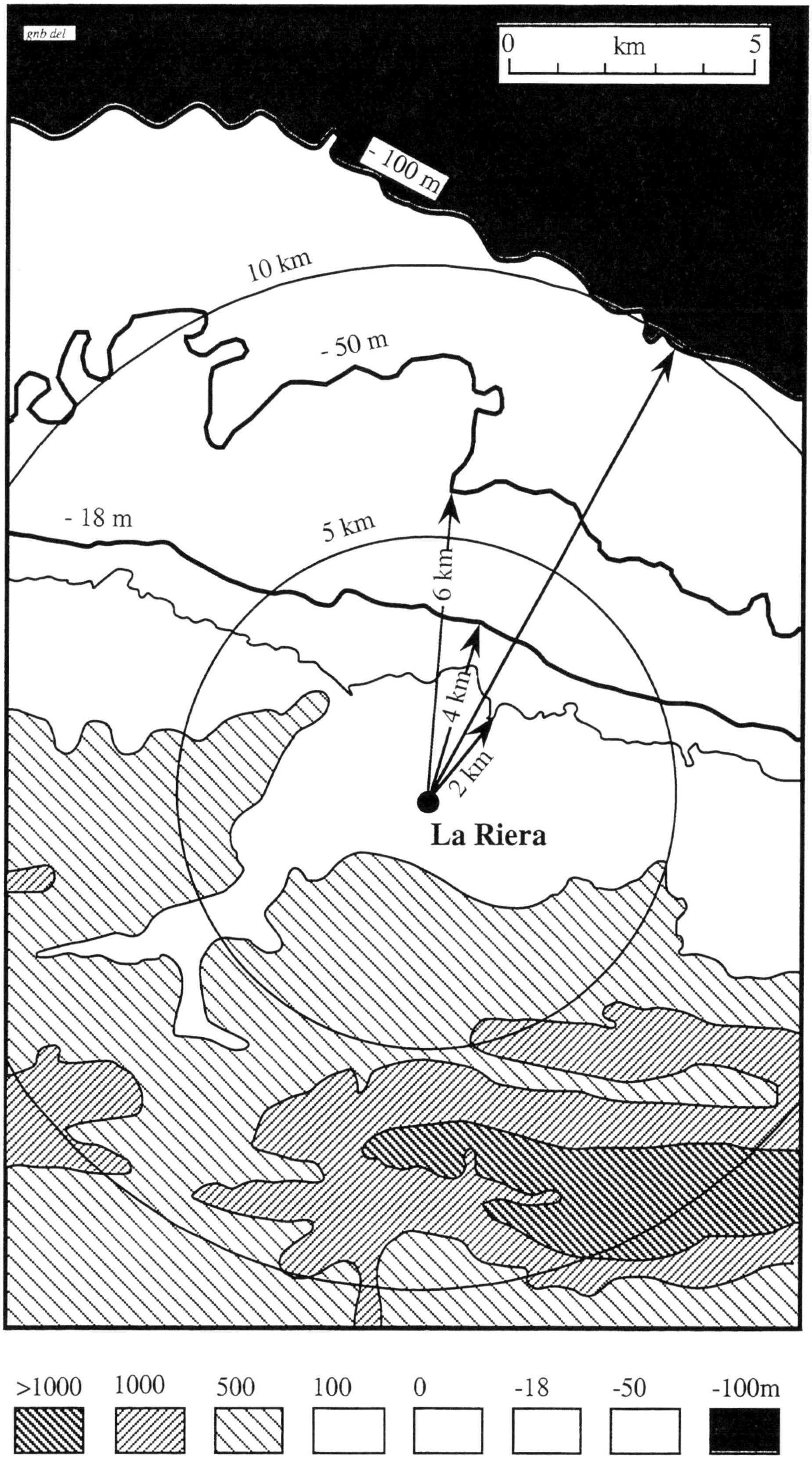

Figure 15.4 La Riera in its local setting, showing simplified topography and bathymetric contours. The –100 m contour is shown as a guide to the LGM coastline but the true level may have been somewhat less than this (see text for further discussion). From Bailey (1975), Straus & Clark (1986).

mens measured in this way. We have, therefore, set out to replicate his data by taking independent measurements on a sample of the La Riera material comprising 1610 shells from the original collections stored in the Oviedo Museum plus 518 shells from the Cambridge sample (Craighead 1995). Only complete shells were measured and all measurements were taken with Vernier calipers to the nearest 0.5 mm. The results are set out in Table 15.2 alongside Ortea's original data, and in Figure 15.6. Comparison of the means of both samples shows no significant difference (Student's t = 0.85, less than the significant value of t = 2.056 using a 2-tailed test at the 0.05 significance level with 26 degrees of freedom). In broad terms both data show a general tendency for shells to be smaller in the Azilian and Asturian levels, although both data sets, interestingly, show that the trend to small shell size begins somewhat earlier, in level 24. There are, however, some notable differences for individual levels, the most glaring being level 29, where Ortea's data give a mean of 26.8 mm and Craighead's data a mean of 42.4 mm. It is regrettable that only 5 complete specimens of *P. vulgata* were available for measurement from this level (Ortea records measurements from 34 shells in his sample), especially since it is a level so critical to the interpretation of the sequence. Further evaluation of this difference for level 29 is seriously compromised by the small sample sizes.

Interpreting size changes in samples of limpet shells from archaeological sequences is a complex issue because a reduction in the mean maximum dimension (length) may be the result of any one of four independent variables: (1) climate change resulting in a reduction in mean annual sea surface temperatures; (2) increased exposure, involving collection of shells from more exposed shores; (3) a change of zonation, that is collection of shells from higher in the intertidal zone; or (4) a shift to a predominance of younger age classes because of increased predation pressure (whether human or otherwise). The impact of these variables on limpet size is well established, the reduction in growth rates of limpets on exposed shores and high in the intertidal zone being due to the reduced time available for feeding (Ebling et al. 1962, Fretter & Graham 1976, Jones 1984, Lewis & Bowman 1975). Additional habitat variables such as salinity variations in estuaries or variations in food supply may have additional effects, but the above four variables comprise the principal sources of variation. To these should be added uncertainties of measurement because of sample bias and measurement inaccuracies. We comment further on size variation after consideration of other lines of evidence.

Paleotemperature Variation

The best known technique for identifying paleotemperature variation is the use of oxygen isotope ratios. However, as Deith & Shackleton (1986) have shown, the technique produces erratic and unreliable results when applied to *P. vulgata* shells from the La Riera deposits be-

Table 15.3 Aragonite/calcite ratios, estimated sea surface temperatures, and mean limpet ages in selected levels of the La Riera sequence. Data from Craighead (1995).

Level	A/C Ratio	N	SST °C	*Patella vulgata* Mean Age Years	N	*Patella depressa* Mean Age Years	N	Mean Length mm
30	0.77	8	8.8	-	-	-	-	-
29	1.06	12	15.8	-	-	4.0	34	25.1
28	0.76	30	8.6	4.4	36	4.4	5	28.2
27	0.81	36	9.6	5.2	40	4.6	23	30.9
26	0.74	14	8.2	4.5	29	-	-	-
21/23	0.79	10	9.2	-	-	-	-	-
19	0.84	32	10.3	5.2	32	-	-	-
18	0.92	18	12.2	-	-	-	-	-
17	0.95	10	12.9	-	-	-	-	-
16	1.04	36	15.2	4.5	34	-	-	-
15	0.87	16	11.0	3.8	15	-	-	-
14	0.95	16	12.9	5.0	10	-	-	-
13	0.99	17	13.9	5.1	14	-	-	-
12	0.94	11	12.6	5.0	10	-	-	-
11	0.95	13	12.9	4.3	11	-	-	-
10	0.89	9	11.4	4.2	9	-	-	-
9	0.90	33	11.7	4.9	42	-	-	-
8	0.86	20	10.7	5.6	30	-	-	-
7	0.93	33	12.4	5.0	51	-	-	-
6	0.90	10	11.7	4.7	11	-	-	-
5	1.25	9	21.3	4.8	12	-	-	-

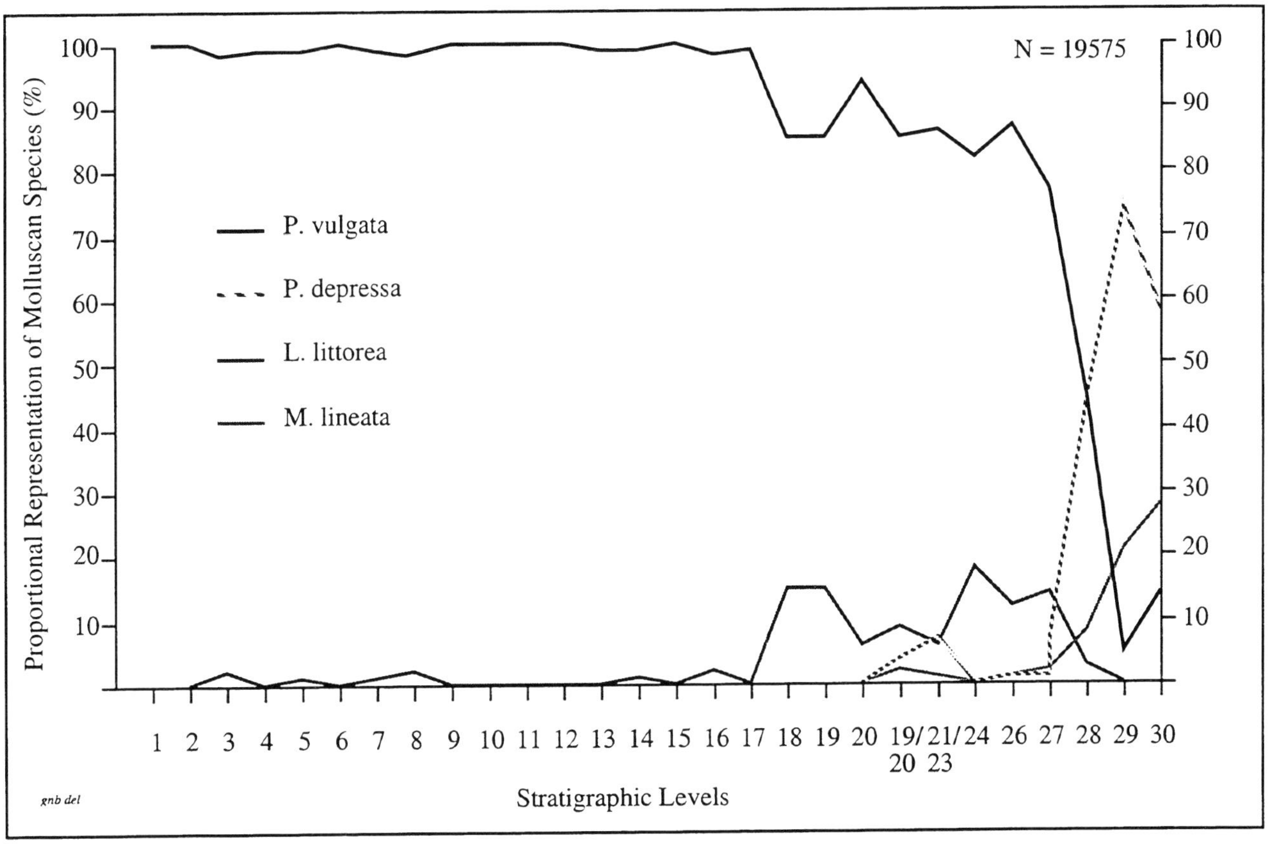

Figure 15.5 Graphical representation of changes in the proportions of the principal molluscan species in the La Riera sequence, after Ortea (1986).

cause of irregular growth patterns and shell recrystallisation and post-depositional diagenesis. An alternative method has been established by Cohen & Branch (1992) on the South African limpet, *Patella granularis.* They have shown that the interior surface of the limpet shell is composed of separate, sometimes overlapping, bands of calcite and aragonite. These bands are defined with respect to the *m* band, the myostracum or muscle attachment scar. The two inner bands (*m-1* and *m-2*) are composed of aragonite and calcite respectively, while the first outer band (*m+1*) is composed of aragonite, and the second and third outer bands (*m+2* and *m+3*) are composed of calcite. They have demonstrated that the overall percentages of aragonite and calcite in the shell structure as measured by X-ray diffractometry are directly proportional to variations in sea-surface temperature, and that there is also a strong correlation (r = 0.91) between sea-surface temperature and the width of the *m+1* aragonite band relative to the *m+2* and *m+3* calcitic bands.

Craighead (1995) has tested this relationship on 224 modern *P. vulgata* shells collected from eight locations around European coastlines with local data on mean annual sea surface temperatures. Shells were cleaned and a stripe of reagent solution, prepared in accordance with Feigle and Anger (1972: 542, 543), applied with a watercolour brush to the interior of each shell. This solution is absorbed more rapidly by the aragonite portions of the shell, which show up as a dark silver colour. The widths of the bands were then measured to the nearest 0.01 mm using a dendrochronology increment measurer incorporating a 10 x microscope and designed by V.R. Switsur for counting and measuring tree rings. Because of difficulties in identifying the muscle attachment scar, a modified formula has been used to measure the aragonite/calcite (A/C) ratio, namely the ratio of the width of the aragonite layers (*m* and *m+1*) to the width of the calcite layers (*m+2* and *m+3*). Because of a slight fuzziness in the boundary between bands, and variation in measurements taken to the outer edge of the shell because of indentations in the shell edge, the A/C ratio for each shell is based on the mean of 4 sets of measurements. The results produce a very strong correlation between the A/C ratio and mean annual sea surface temperatures (r = 0.96), and the relationship between the two is expressed by the following formula:

(1) SST °C = Exponent [1.83 (Ln A/C ratio) + 2.65]

This technique was applied to 393 shells from the La Riera sequence, drawn from the Cambridge sample. The average A/C ratio was calculated from up to four measurements on each shell and the mean determined for each of 21 stratigraphic levels (Table 15.3). As a further check on the method, 19 specimens from five levels (levels 14–17 and level 18) were selected for oxygen isotope analysis. Recrystallised calcitic shell and aragonite material were ground away from the edges and surfaces of each specimen and up to eight samples were taken along the axis of growth at about 2.0 mm intervals to cover at least one year's growth. The mean d ^{18}O values were compared with the A/C ratios from the corresponding levels and show a moderately strong correlation ($r = -0.84$). We conclude that A/C ratios provide a reliable guide to paleotemperature variation, but we caution against placing too much reliance on the absolute values of the estimated sea-surface temperatures. Many of the archaeological specimens have slightly abraded edges, and removal of a few mm of edge material may result in underestimates of temperature by as much as 1–2°C. Cohen and Branch (1992) also draw attention to other minor measurement errors. They also note that the formation of calcite and aragonite crystals may be affected by other factors such as variable microhabitat conditions (resulting in inter-individual variation of A/C ratios by as much as 0.18), differential solubility of aragonite and calcite at different temperatures, and different phylogenetic histories. For all these reasons quite large margins of error should be attached to the temperature estimates, of the order of ± 1–2°C. Some of the absolute values in Table 15.2 are questionable, notably the result from level 5, which produces an anomalously high value well outside the range of any other climatic indicators for this period. The value for level 16 is also strikingly high. This level is associated with the middle of the Lascaux Interstadial, so that a relatively high value is not unreasonable, especially given the presence of the warmth-loving snail *Oestophorella bouvinieri*. However, these figure seems too high in relation to level 29 (or the level 29 value too low in relation to levels 5 and 16), since other evidence from level 29, noted above, suggests temperatures warmer than the present day. We conclude that the A/C ratios provide a useful guide to paleotemperature variation but that the absolute values should be treated with caution.

Figure 15.6 compares the A/C ratios with the mean maximum size of *P. vulgata* from the Craighead sample, and shows that the two variables track each other very closely, the direction of change being the same for all levels except levels 7 and 17. If the level 5 outlier is excluded from the calculation, the two variables are quite strongly correlated ($r = 0.76$), from which we conclude that the mean size of *P. vulgata* was quite strongly dependent on sea surface temperatures. However, these results do not exclude the influence of other factors, especially if Ortea's figure for the mean size of *P. vulgata* shells in level 29 is preferred.

Exposure and Zonation

Exposure refers to the effect of wave action, in turn a function of wind velocity and fetch, and the steepness and geometry of the coastline, and can be measured systematically according to criteria defined by Thomas (1986). Zonation refers to the subdivision of the intertidal zone. It is well known that shells of *P. vulgata* on exposed shores or in high-shore positions, in addition to being smaller than those in sheltered or low-shore positions, tend also to have a more conical shape (Fretter & Graham 1976; Jones 1984; Orton 1928a; 1928b).

In an attempt to establish whether there are consistent metrical relationships between length/height and length/width ratios of limpet shells and habitat variation, Craighead (1995) studied 1325 modern *P. vulgata* shells gathered from three stretches of coastline around the British Isles. From each area shells were collected from high-shore and low-shore positions on exposed and sheltered littorals, and measurements were taken of length, height and width. It was found that low-shore shells tend to be longer, flatter and wider than high-shore animals. Discriminant analysis gives an unstandardised discriminant function, which assigns shells to high shore and low shore positions with a success rate of 86%, using the following formula:

(2) Zone = Length (–0.123) + Height (0.52) + Width (0.191) + 2.289

Values more positive than – 0.25 are assigned to high-shore locations, while more negative values are assigned to low-shore locations.

Similarly, exposed-shore limpets tend to be slightly narrower and flatter than those from sheltered shores, and discriminant analysis produces an unstandardised discriminant function as follows:

(3) Exposure = Length (0.142) – Height (0.06) + Width (0.0489) – 5.328

Values more positive than –0.15 are associated with sheltered shores and values more negative with exposed shores. Use of these values resulted in shells being assigned to their correct position with a 68% success rate.

It was also found that length/width and length/height ratios vary with latitude and climate, but that the nature of the variation is similar to that caused by variations in exposure. Thus, it is possible to discriminate between high-shore, low-shore, exposed-shore and sheltered-shore limpets in a given sample, provided that climate is constant. If we assume that climate remained constant within each stratigraphic level of the La Riera sequence, then the discriminant functions derived from the modern study can be applied to the raw data from La Riera to produce reasonably reliable measures of the source of origin of the *P. vulgata* shells from the various levels (Table 15.3).

Figure 15.6 plots the variation in percentage of low shore and sheltered shore specimens against variations in the A/C ratio. It is apparent that these variables track each other quite closely. Regression analysis gives moderately strong coefficients of correlation: for the relationship between the A/C ratio and low shore specimens $r = 0.76$, and for the relationship between the A/C ratio and sheltered shore specimens $r = 0.79$ (excluding in both comparisons the anomalous A/C value for level 5). The implication of

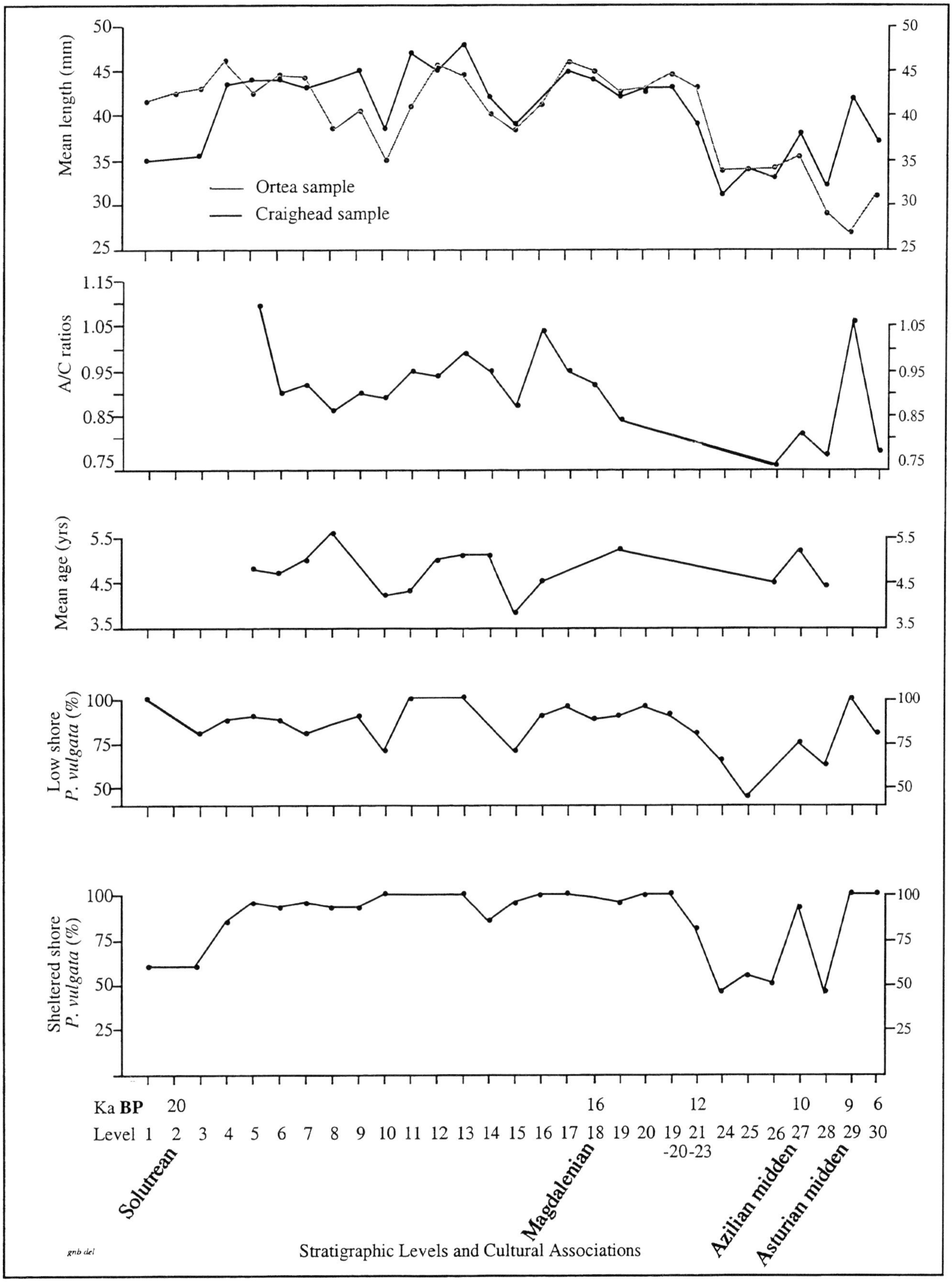

Figure 15.6 Data on *P. vulgata* from the La Riera sequence: mean length, A/C ratio, low shore and sheltered shore specimens, and mean age (see Tables 15.2 and 15.3 for raw data).

these results is that during colder periods limpet gathering tended to focus more on high-shore locations and exposed shores.

In cost-benefit terms we would expect low-shore limpets on sheltered shores to be the preferred target of shell gathering since this is where the larger limpets tend to grow. Increased proportions of high shore limpets might indicate greater pressure on food resources, as predicted by the demographic hypothesis, or more restricted access to the low shore zone because of cold temperatures and bad weather. The latter is certainly a plausible hypothesis, and we consider this the more likely explanation on two grounds: firstly, because there are periodic decreases in the proportion of low shore limpets throughout the La Riera sequence, and not solely towards the end of the sequence as the demographic hypothesis would require, and secondly because of the moderately strong correlation with variations in temperature.

The reasons for the increase in exposed shore specimens are less certain. As is clearly shown in Figure 15.6, the preference for sheltered shore specimens is very strongly marked throughout the La Riera sequence, with sharp reductions being apparent only towards the end of the sequence, notably in levels 26 and 28. It is possible that we are seeing here the additional effect of a widening of shell gathering activity to include less favourable stretches of shoreline. However, there are two other possibilities. The first is that reduced temperatures are associated with a general increase in stormy conditions and poor weather, which have the effect of increasing the exposure of the shoreline as a whole. A second possibility is that the figures are an artefact of the measuring technique. As noted above, the discriminant function for distinguishing exposed shore and sheltered shore limpets is very similar to that which discriminates between shells growing in cold climates from those growing in warmer climates. It is not possible to disentangle the two effects unless one is known to be constant. The interpretation of the La Riera L/W and L/H ratios is based on the assumption that climate remained constant within each stratigraphic level. If climate were changing within a stratigraphic level, which is quite possible given the span of time encompassed in relation to the rapidity of climatic changes, which are known to have swung between climatic extremes over as little as a few decades (Johnsen et al. 1992, Taylor et al. 1993), it is possible that the effects of climatic change and differential exposure are being confounded in our data. We conclude that changes in the proportions of sheltered shore limpets are primarily the result of climatic changes, either acting directly on the L/W and L/H ratios, or indirectly on the availability or accessibility of sheltered shore locations. We cannot fully rule out independent changes in shell gathering strategies towards more intensive exploitation of the littoral zone. But if these changes were taking place, their effect on the metrical characteristics of the limpet shells is too weak to be detected with confidence in our data.

Age Data

Incremental growth structures in *P. vulgata* show a similar pattern to other molluscan species. Microscopic banding is related to periodic tidal immersion, and macroscopic bands visible on the shell surface represent a break in normal growth due to cold sea-surface temperatures in winter (Ekaratne & Crisp 1982, 1984). The winter growth check may show up as a dark ring and provides a basis for estimating the age at death of individual specimens. Rings representing minor growth checks may also occur as a result of spawning, or as a result of high summer temperatures near the southern limits of the geographical range, but these checks are usually less pronounced and more irregularly spaced and can be distinguished from the winter growth checks. The winter growth checks form a regularly spaced series, the interval between rings showing a slight but successive decrease as the growth rate of the individual slows down with increasing age. A total of 448 limpets from 17 La Riera levels was available for study, and visible winter growth checks were counted on each shell and the average minimum age for each level established (Table 15.3, Fig. 15.6). In 60 cases the apices of the shell were missing, and the ages of these shells were estimated using a formula to derive the missing growth increments from the ratio of growth increments between the final three years' of growth, or the final two years' growth in the case of younger shells (Hamai 1937, Monks & Johnston 1993). This formula was applied to *P. vulgata* specimens of known age and growth characteristics from a variety of habitats (Craighead 1995) and found to give a close correlation between observed and estimated age ($r = 0.93$ where the estimated age was based on the final three years' of growth and $r = 0.77$ where based on the final two years' of growth).

P. vulgata is a protandrous hermaphrodite (Fretter & Graham 1976). The animal becomes a mature male at an average age of nine months with a shell length of 10–15 mm. Sex change to female generally occurs during the second and third years, and thus the vast majority of older individuals are female with size ranges for 2-year olds and 3-year olds respectively of 11–25 mm and 13–38 mm. If human gatherers removed large quantities of the bigger 2 and 3 year old females, the remaining population would be predominantly male and recruitment drastically reduced.

The La Riera data shows that mean age in the levels studied never falls below four years and therefore that limpet populations during these periods were not subjected to dangerously high levels of predation. Nor is there any significant correlation between shell ages and mean lengths in the sequence as a whole (Fig. 15.6), as would be expected if reduction in length were solely a function of reduction in the mean age of the collected shells. On the other hand, it is worth noting that the dip in mean age in levels 15, 26 and 28 is correlated with a dip in mean length, suggesting, at least for these levels, the possibility of a link between the two variables. But the age reduction is quite slight in all three cases and certainly not sufficient to endanger the recruitment of subsequent generations of limpets. In any case, since the drop in mean age also correlates with a drop in estimated sea surface temperatures, we consider climatic change to be the more likely cause of the size changes. We should also emphasise the regrettable lack of ageable *P. vulgata* specimens in the samples available from levels 29 and 30. In the absence of age data we have to fall back on indirect inferences from the length data. In this regard it is worth noting that one of the smallest mean lengths in the whole sequence in both Ortea's and Craighead's samples is in level 28 (28.1mm and 31.8 mm respectively, Table 15.2). Yet the mean age of shells in this level is 4.4 years (compared with a mean age from the whole sequence of 4.7 years). Clearly the reduction in mean length in this level is not primarily related to increased predation pressure. This is also the level in which the proportion of *P. depressa* shells draws nearly level with the proportion of *P. vulgata* shells (Figure 15.5), and when we might therefore expect the effects of overexploitation on *P. vulgata* populations to become apparent, if that were the reason for the increase in *P. depressa* populations. As for levels 29 and 30, although we have no age data for *P. vulgata*, the mean length of *P. vulgata* is actually higher in level 29 than in level 28 in both Ortea's and Craighead's sample (Table 15.2). It seems most unlikely that limpet populations that had been subjected to greatly intensified levels of predation with respect to those in level 28 would show an *increase* in mean length; the expected trend under heavy predation pressure would be a decrease. We conclude that there is no evidence to support the hypothesis that the greatly reduced proportions of *P. vulgata* in the uppermost levels of the La Riera sequence are due to overexploitation by human shell gatherers.

Equally we should observe that, although both the mean length and age of *P. vulgata* are highly variable through time, there is a slight decrease in mean size and mean age in the upper levels, which may be plausibly interpreted as the consequence of a slight but progressive increase in exploitation pressure.

Finally we present age and length data for samples of *P. depressa* from levels 28–30 (Table 15.3). In this species the sexes remain separate throughout life and both sexes mature in their second year (Guerra & Guadencio 1986, Orton & Southward 1961). Individuals of both sexes are found in all size ranges with the bulk of the breeding population in the 24–30 mm length range. As the sexes remain separate and the smaller limpets can reproduce, it is likely that recovery would be rapid even after the removal of large numbers of the breeding population. As it is, our data on both age and mean length suggest that the level of human predation fell far short of endangering recruitment in any of the levels for which we have measurements.

Seasonality

Deith and Shackleton (1986) have provided a comprehensive review of seasonality issues at La Riera, based on the application of the oxygen isotope technique to shells of *M. lineata* from the upper levels. Their results, however, are limited by their inability to obtain reliable measurements from *Patella* shells because of slow and erratic growth patterns, recrystallisation of shell carbonate, and irregularities in the growing edge. In a more recent study Craighead (1995) has estimated seasonality by measuring the amount of growth relative to the last winter growth check on a sample of 413 limpet shells from 17 La Riera levels (Table 15.4). Several difficulties should be recognised with this technique. Very small variations in measurement can substantially alter the estimate of season, particularly in limpets, where the shell edge is sometimes deeply serrated and measurement from the last winter growth line to the shell edge can vary considerably depending on the point from which the measurement is taken. In this study a conscious effort was made to use the mid-point of an indentation on the shell edge. Another potential source of error is the loss of material from the shell edge by abrasion. Accordingly the seasonality estimates should be treated with some caution, although they appear to provide consistent general trends, even allowing for these measurement errors.

The results in Table 15.4 show that *P. vulgata* shells were collected throughout the year but with the majority falling in the summer and autumn periods, whereas the majority of the *P. depressa* shells were collected in autumn and winter. If these figures are adjusted to allow for a loss of up to 0.5 mm of material from the shell edge, this advances the estimates by as much as a season. For the most part winter and spring estimates remain low for *P. vulgata*, while the percentage of winter estimates for *P. depressa* increases. Even a loss of 1.0 mm of edge material does not substantially alter the general shape of the seasonality distributions for the two species.

These differences between the two species can be explained by reference to seasonal variations in meat weight, which are closely related to the timing of spawning. *P. vulgata* meat weight values are lowest in winter, building up to a maximum just before the spawning period in autumn (Jones 1984). Conversely *P. depressa* spawns in winter/spring and again in summer/autumn (Bowman & Lewis 1986), so that optimum meat values would occur in winter and again in autumn. If human shell gatherers were optimising their subsistence activities in cost-benefit terms, we would expect them to concentrate their collection of *P. vulgata* in the summer and early autumn, and of *P. depressa* in the winter and autumn. On the whole this is what the archaeological results show, although *P. depressa* seems to have been ignored during the potentially favourable autumn period. These results provide an interesting comparison with *M. lineata*, which, according to Deith and Shackleton's (1986) isotope analyses, were collected mainly in winter. *M. lineata* spawns in spring (Fretter & Graham 1977), so that winter would be predicted as the optimum season of collection.

Notwithstanding the uncertainties of the method, it produces consistent results within quite wide margins of error, from which two general conclusions can be drawn. Firstly, it is clear that La Riera was visited at all periods of the year, and was not confined to a narrow season of use or a narrowly specialist function. This agrees with the aggregate faunal analyses of Ortea, Altuna and others (Straus & Clark 1986). This does not of course exclude the use of sites and resources in the hinterland, but it does strengthen the case for associating sites such as La Riera with residential bases. Secondly, the seasonality data suggest that the exploitation of different molluscan species was scheduled so as to concentrate their collection in those seasons which would have generated the optimum return of food.

Discussion

The main outcome of the above analyses is to show that the changes in molluscan representation and metrical characteristics observable in the La Riera sequence are most likely the result of climatic and habitat changes, rather than the result of increased pressure of human populations on the food supply. There are hints of slight changes that might be attributable to human predation pressure, but these are weak and certainly insufficient to account for the major changes observed throughout the sequence.

To the extent that there are remaining ambiguities in the Spanish data, this is as much due to the inadequacy of some of the samples available for study as to the inadequacy of the analytical techniques. For example, it would certainly be desirable to examine larger samples of *P. vulgata* shells from the deposits of the Asturian period. Nevertheless we do not rule out the possibility that human populations were increasing in the late Pleistocene–early Holocene period – there is of course a range of other food resources and paleoeconomic and paleoclimatic indicators represented at La Riera and discussed by Clark & Straus (1986), which we have not attempted to incorporate into this discussion. What our results do indicate is that, if there were a genuine time trend towards increased rates of shellfood consumption during Azilian and Asturian times, whether stimulated by human population increase or other cultural factors, this had little discernible impact on the size, growth rates or age structure of the exploited limpet populations.

Conclusions

Shell middens and the evidence of shell gathering activity provide some of our most ubiquitous and durable evidence of prehistoric coastal activity and settlement, not only on the coastlines of Mesolithic Europe, but worldwide. It is not surprising that this sort of evidence should figure prominently in discussions of change at the Pleistocene/Holocene boundary. Moreover, it is a widely accepted proposition that the gathering, consumption and use of molluscs, and the deposition and accumulation of the discarded shells, provide sensitive indicators of social and cultural behaviour across a very wide range of issues, times and places (Claassen 1998). However, our analysis and comparison of the evidence from Spain and Australia indicates that there is an equally wide and subtle range of environmental, climatic, ecological and taphonomic factors that can influence variations in the characteristics of middens and molluscs, and indeed their visibility in the archaeological record. It is, perhaps, no surprise that in both the Australian and Spanish case studies considered here, the same pattern of variability in the occurrence of shell middens through time has been interpreted by one set of commentators as evidence of population-resource imbalances leading to more intensive exploitation of shellfish, and by another set of commentators as evidence of environmental or ecological variation operating independently of human influence. Our own conclusions may seem to emphasise the latter, but this does not mean that we exclude the former. On the contrary, we believe that both factors are likely to be implicated in the interpretation of long-term trends. Rather our point has been to focus on the methodological underpinnings of such interpretations, and the need to take into account a wide range of possible environmental factors and techniques for their analysis. In this respect, we believe that serious analytical work has barely begun, and that there is now a wide range of problems opening up for investigation. We single out two areas in need of further investigation by way of conclusion.

The first is the need for an improved understanding of the ways in which sea-level change has affected not only the character and ecology of the coastal environment and the resources that it can support, but also the visibility of archaeological evidence. The Australian evidence discussed above draws particular attention to the former set of charactcristics, and in particular the ways in which a rapidly prograding coastline combined with changes in relative sea level can alter the nature of the available intertidal habitats, and hence the availability and abundance of molluscs. The Spanish case study highlights the latter issue.

It is now increasingly accepted that late Pleistocene and early Holocene coastal shell middens could have existed on shorelines that are now submerged, and may well still be available for study if we can find them. On many coastlines, the earliest appearance of large and abundant open-air shell mounds coincides with the period at which sea-level rise approximated the modern level at about 5000 to 6000 bp. Significantly earlier coastal sites or shell deposits are now coming to light on coastlines that have undergone tectonic or isostatic uplift, notably in Norway (Rowley-Conwy 2001) and the Red Sea basin (Walters et al. 2000). Deep coastal cave sequences in South Africa have produced evidence of substantial shell gathering associated with much earlier periods of high sea level (Deacon & Shuurman 1992, Henshilwood et al. 2001). Underwater exploration is also beginning to identify coastal sites at least from the early Holocene, sometimes with exceptional conditions of preservation (Fischer 1995). It is now increasingly accepted that the utilisation of intertidal, marine and aquatic resources was not confined to the Holocene, and that such resources may have played a far more significant

Table 15.4 Seasonality estimates of limpet shells from La Riera.

Unadjusted					Adjusted for 0.5 mm edge loss				Adjusted for 1.0 mm edge loss				
Level	W %	Sp %	Su %	Au %	W %	Sp %	Su %	Au %	W %	Sp %	Su %	Au %	Total N
29*	5.9	8.8	35.3	**50.0**	41.8	-	14.7	**44.1**	41.2	-	8.8	**50.0**	34
28*	20.0	20.0	20.0	**40.0**	**60.0**	-	20.0	20.0	**60.0**	-	20.0	20.0	5
28	2.8	25.0	**38.9**	33.3	22.2	-	33.3	**44.4**	25.0	-	27.8	**47.2**	36
27	12.5	7.5	**55.0**	25.0	25.0	-	20.0	**55.0**	37.5	2.5	17.5	**42.5**	40
26*	**35.7**	14.2	21.4	28.6	**64.3**	7.1	21.4	7.1	**63.4**	7.1	24.1	7.1	14
26	-	26.3	**47.4**	26.3	36.3	5.3	**42.1**	26.3	**37.6**	5.3	**37.6**	**37.6**	19
19	13.3	20.0	**50.0**	16.7	26.7	10.0	26.7	**36.6**	26.7	-	26.7	**46.6**	30
16	9.5	11.9	**61.9**	23.8	23.8	2.4	35.7	**38.1**	33.3	2.4	26.2	**38.1**	42
15	-	24.3	**54.5**	18.2	-	9.1	36.4	**54.5**	18.2	9.0	**36.4**	**36.4**	11
14	6.7	13.3	33.3	**46.7**	26.7	-	30.0	**53.3**	**46.7**	-	13.3	40.0	15
13	6.2	12.5	25.0	**58.3**	25.0	6.2	31.2	**37.5**	**37.4**	-	31.3	31.3	16
12	11.1	22.2	22.2	44.4	**44.4**	-	**44.4**	11.1	**33.3**	11.1	**33.3**	22.2	9
11	-	30.0	**60.0**	10.0	-	-	**60.0**	40.0	10.0	-	40.0	**50.0**	10
10	9.1	36.4	**54.5**	-	9.1	-	**63.6**	27.3	18.2	-	35.4	**45.4**	11
9	29.3	7.3	**56.1**	7.3	34.1	4.9	24.4	**36.6**	39.0	2.4	17.1	**41.5**	41
8	19.2	15.4	19.2	**46.2**	53.2	-	19.2	26.9	**57.7**	-	15.4	26.9	26
7	17.6	14.7	**52.9**	14.7	29.4	-	**35.3**	**35.3**	**35.3**	-	29.4	**35.3**	34
6	-	11.1	**55.6**	33.3	11.1	-	**44.4**	**44.4**	22.2	-	33.3	**44.4**	9
5	9.1	18.2	**45.3**	27.3	27.2	-	**36.4**	**36.4**	36.4	-	**45.4**	18.2	11

* = results for *Patella depressa*; all other figures are for *Patella vulgata*. Figures in bold highlight the seasons with the highest percentages. Winter = January to March. Spring = April to June. Summer = July to September. Autumn = October to December.

role throughout prehistory than has commonly been recognised (Bailey & Milner 2003, Erlandson 2001).

Even if sites themselves have not been preserved underwater or are difficult to locate, the benefits of underwater survey in terms of improved understanding of paleogeographical and paleoenvironmental change are incalculable. Moreover, a wide range of techniques is increasingly being deployed in such investigations (Fischer 1995, Masters & Flemming 1983, Van Andel & Lianos 1984). However, the challenges of such work should not be underestimated. Reconstructions of the coastal environment with changes in sea level are difficult enough. Identifying their impact on human subsistence and settlement activity is likely to be even more challenging, as is clear from the controversy surrounding Shackleton and Van Andel's (1986) pioneering attempt to compare changes in species abundance at Franchthi cave with reconstructions of the adjacent paleoshoreline (see Thomas 1987).

The second area of investigation is in the development of analytical methods for extracting paleoenvironmental and paleoeconomic evidence from the mollusc shells themselves. The Spanish results discussed above demonstrate that analysis and interpretation of long-term shell gathering trends associated with a period of changing and in all likelihood quite rapidly oscillating climate is a complex procedure involving simultaneous consideration of a number of variables. Nevertheless we have shown that analytical techniques based on the metrical, chemical and physical characteristics of the exploited mollusc shells, backed up with studies of modern control populations, can be devised to disentangle the relative influence of paleoclimatic, paleoecological and paleoeconomic factors. Features of archaeological sequences such as reduction in the size of molluscs or changes in taxonomic representation are commonly observed and often taken to indicate evidence of overexploitation resulting from economic intensification and population growth. However, our results demonstrate that climatic and environmental changes may play an equally powerful role through their impact on habitat change and on growth rates and age structure, alternative hypotheses that are all too rarely explored let alone tested against independent evidence (Claassen 1998).

In all these examples the issue that emerges above all others is the critical importance of independent, paleoenvironmental data as a check on archaeological interpretations. We have shown above some of the techniques that can be used to obtain such information and some of the pitfalls in their interpretation. The wider deployment of these techniques to investigate problems in human prehistory has barely begun.

Acknowledgements

The analysis of the La Riera material in the Oviedo Museum would not have been possible without the help and advice of various institutions and individuals, notably the Consejería de Cultura y Educación, Principado de Asturias, Dña. Matilde Escortell Ponsoda, former Director of the Museo Arqueológico Provincial (Asturias), and the co-directors of the La Riera excavations, Geoff Clark (Arizona State University), Lawrence Straus (University of New Mexico), and Manuel R. González Morales (University of Cantabria). We are grateful to all parties for permission to examine the collections. Thanks are also due to Manolo González Morales for advice and assistance in Spain; to Nick Shackleton, Roy Switsur and Mike Hall for making available the facilities of the Godwin Laboratory, Cambridge, and for assistance with the analytical work; to Steve Hawkins for assistance with the identification of archaeological molluscs and advice on nomenclature; to Rosemary Bowman, Mike Kendall, J.D. Fish, Mick Wood, Louis Cabioch, Jean-Louis Birrien, Ole Lönne, Jon-Arne Sneli, Mary Seddon and Glen Jamieson for advice and assistance in the collection of modern limpets and information on water temperatures and limpet ecology, and to Nick Winder and Jack Nance for advice on statistical procedures. We are also grateful to the British Academy for funding the isotope analyses. Lawrence Straus, Cheryl Claasen, Jon Erlandson and Patricia Bourke provided useful comments on an earlier version of this paper.

References

Bailey, G.N.
1973 Los concheros del norte de España: una hipótesis préliminar. *Congreso Nacional de Arqueología – 1971,* Jaén, pp. 12, 73-84. Zaragoza: Secretaría General de los Congresos Arqueológicos Nacionales.
1975 *The Role of Shell Middens in Prehistoric Economies*. PhD Dissertation, Department of Archaeology, University of Cambridge.
1977 Shell mounds, shell middens and raised beaches in the Cape York Peninsula. *Mankind* 11: 132–43.
1978 Shell middens as indicators of postglacial economies: a territorial perspective. In *The Early Postglacial Settlement of Northern Europe* (P. A. Mellars, ed.). pp. 37–63. London: Duckworth.
1983 Problems of site formation and the interpretation of spatial and temporal discontinuities in the distribution of coastal middens. In *Quaternary Coastlines and Marine Archaeology* (P. M. Masters & N. C. Flemming, eds.), pp. 559–82. London: Academic Press.
1993 Shell mounds in 1972 and 1992: reflections on recent controversies at Ballina and Weipa. *Australian Archaeology* 37: 1–18.
1994 The Weipa shell mounds, natural or cultural? In *Archaeology in the North, Proceedings of the 1993 Australian Archaeological Association Conference* (M. Sullivan, S. Brockwell & A. Webb, eds.), pp. 107–29. Darwin: North Australian Research Unit (ANU).
1999 Shell mounds and coastal archaeology in northern Queensland. In *Australian Coastal Archaeology* (J. Hall & I. McNiven, eds.), pp. 105–112. Canberra: Occasional Papers in Prehistory Series No. 30, Division of Archaeology and Natural History, Research School of Pacific and Asian Studies, the Australian National University.

Bailey, G. N., J. C. Chappell & R. L. Cribb
1994 The origin of *Anadara* shell mounds at Weipa, North Queensland, Australia. *Archaeology in Oceania* 29: 69–80.

Bailey, G.N. & A. Craighead
2003 Late Pleistocene and early Holocene coastal paleoeconomies: a reconsideration of the molluscan evidence from Northern Spain. *Geoarchaeology* 18: 175–204.

Bailey, G.N. & N. Milner
2003 Coastal hunters and gatherers and social evolution: marginal or central? *Before Farming: the Archaeology of Old World Hunter-Gatherers* 3–4: 1–15. (available online at *http://waspjournals.com*).

Barandiarán, I., L. G. Freeman, J. González Echegaray & R. G. Klein
1985 *Excavaciones en la Cueva del Juyo.* Madrid: Monografías del Centro de Investigación y Museo de Altamira No. 14.

Beaton, J. M.
1985 Evidence for a coastal occupation time-lag at Princess Charlotte Bay (North Queensland) and implications for coastal colonization and population growth theories for Aboriginal Australia. *Archaeology in Oceania* 20: 1–20.
1995 The transition on the coastal fringe of Greater Australia. *Antiquity* 69: 798–806.

Bettinger, R.L. & R. Malhi
1997 Central place models of acorn and mussel processing. *Journal of Archaeological Science* 24: 887–899.

Bird, D.W. & R.L. Bliege-Bird
1997 Contemporary shellfish gathering strategies among the Meriam of the Torres Straits Islands, Australia: testing predictions of a central place foraging model. *Journal of Archaeological Science* 24: 39–63.

Bourke, P.M.
2000 *Late Holocene Indigenous Economies of the Tropical Australian Coast, an Archaeological Study of the Darwin Region*. Ph.D Dissertation, Department of Archaelogy. Darwin: Northern Territory University.

Bowman, R. S. & J. R. Lewis
1977 Annual fluctuations in the recruitment of *Patella vulgata. Journal of the Marine Biological Association of the United Kingdom* 57: 793–815.

Bowman, R. S. & J. R. Lewis
1986 Geographical variation in the breeding cycles and recruitment of *Patella* spp. *Hydrobiologia* 142: 41–56.

Boyle, P. R.
1981 *Molluscs and Man.* London: Arnold.

Campbell, A. C.
1976 *The Hamlyn Guide to the Seashore and the Shallow Seas of Britain and Europe*. London: Hamlyn Publishing.

Chappell, J. C.
1982 Sea levels and sediments: some features of the context of coastal archaeological sites in the Tropics. *Archaeology in Oceania* 17: 69–77.

Chappell, J. C. & J. Grindrod
1984 Chenier plain formation in northern Australia. In *Coastal Geomorphology in Australia* (B. G. Thom, ed.), pp. 197–231. Sydney: Academic Press.

Chappell. J. C. & N. J. Shackleton
1986 Oxygen isotopes and sea level. *Nature* 324: 137–40.

Claassen, C.
1998 *Shells*. Cambridge: Cambridge University Press.

Clark, G. A.
1971 The Asturian of Cantabria: subsistence base and the evidence for post-Pleistocene climatic shifts. *American Anthropologist* 73: 1244–1257.

1983 Boreal phase settlement-subsistence models for Cantabrian Spain. In *Hunter-Gatherer Economy in Prehistory* (G. N. Bailey, ed.), pp. 96–110. Cambridge: Cambridge University Press.

Clark, G. A. & L. G. Straus
1983 Late Pleistocene hunter-gatherer adaptations in Cantabrian Spain. In *Hunter-Gatherer Economy in Prehistory* (G. N. Bailey, ed.), pp. 131–48. Cambridge: Cambridge University Press.

1986 Synthesis and conclusions – part I: Upper Paleolithic and Mesolithic hunter-gatherer subsistence in northern Spain. In *La Riera Cave – Stone Age Hunter-Gatherer Adaptations in Northern Spain* (L. G. Straus & G. A. Clark, eds.), pp. 351–365. Tempe: Arizona State University Anthropological Research Papers No. 36.

Clark, G.A. & S. Yi
1983 Niche-width variation in Cantabrian archaeofaunas: a diachronic study. In *Animals and Archaeology, Vol. 1, Hunters and their Prey* (J. Clutton-Brock & C. Grigson, eds.), pp. 183–208. Oxford: British Archaeological Reports International Series No. 163.

Cohen, A. L. & G. M. Branch
1992 Environmentally controlled variation in the structure and mineralogy of *Patella granularis* shells from the coast of southern Africa: implications for paleotemperature assessments. *Paleogeography, Paleoclimatology, Paleoecology* 91: 49–57.

Craighead, A. S.
1995 *Marine Mollusca as Paleoenvironmental and Paleoeconomic Indicators in Cantabrian Spain*. Ph.D Dissertation, Department of Archaeology, University of Cambridge.

Cribb, R. L.
1986 Archaeology in the Aurukun Region: a report on work carried out in 1985. *Queensland Archaeological Research* 3: 133–158.

Deacon, H.J., & R. Shuurman
1992 The origins of modern people: the evidence from Klasies River. In *Continuity or replacement: Controversies in* Homo sapiens *evolution* (G. Brauer & F.H. Smith, eds.), pp. 121–129. Rotterdam: Balkema.

Deith, M. R. & N. J. Shackleton
1986 Seasonal exploitation of marine molluscs: oxygen isotope analysis of shell from La Riera Cave. In *La Riera Cave – Stone Age Hunter-Gatherer Adaptations in Northern Spain* (L. G. Straus & G. A. Clark, eds.), pp. 299–313. Tempe: Arizona State University Anthropological Research Papers No. 36.

Ebling, F. J., J. F. Sloane, J. A. Kitching & H. M. Davies
1962 The ecology of Lough Ine XII: the distribution and characteristics of *Patella* species. *Journal of Animal Ecology* 31: 457–470.

Ekaratne, S. U. K. Crisp & D. J. Crisp
1982 Tidal micro-growth bands in intertidal gastropod shells: an evaluation of band-dating techniques. *Proceedings of the Royal Society, London, Series B,* 214: 305–323.

Ekaratne, S. U. K. Crisp & D. J. Crisp
1984 Seasonal growth studies of intertidal gastropods from shell micro-growth band measurements, including a comparison with alternative methods. *Journal of the Marine Biological Association of the United Kingdom* 64: 183–210.

Erlandson, J.M.
2001 The archaeology of aquatic adaptations: paradigms for a new millennium. *Journal of Archaeological Research* 9: 287–350.

Fano Martínez, M.A.
1998 *El Habitat Mesolítico en el Cantábrico Occidental*. Oxford: British Archaeological Reports International Series No. 732.

Feigle, F. & V. Anger
1972 *Spot Tests in Inorganic Analysis (6th ed.)*. Amsterdam: Elsevier.

Fischer, A. (Editor)
1995. *Man and Sea in the Mesolithic: Coastal Settlement Above and Below Present Sea Level*. Oxford: Oxbow Books.

Fischer-Piette, E.
1948 Sur les éléments de prospérité des patelles et sur leur spécificité. *Journal de Conchyliologie* 88: 45–96.

Fischer-Piette, E, & J. M. Gaillard
1959 Les patelles au long des côtes Atlantiques Ibériques et Nord-Morocaines. *Journal de Conchyliologie* 99: 135–200.

Fretter V. & A. Graham
1976 The Prosobranch molluscs of Britain and Denmark. *Journal of Molluscan Studies, Supplement 1.*

González Morales, M. R.
1982 *El Asturiense y Otras Culturales Locales.* Santander: Monografías del Centro de Investigación y Museo de Altamira No. 7.

Guerra, M. T. & M. J. Guadencio
1986 Aspects of the ecology of *Patella* spp. on the Portuguese coast. *Hydrobiologia* 142: 57–69.

Hamai, I.
1937 Some notes on relative growth, with special reference to the growth of limpets. *The Science Reports of the Tohoku Imperial University 4th Series (Biology)* 12: 71–95.

Hayward, P. J. & J. S. Ryland (Editors)
1990 *The Marine Fauna of the British Isles and North-West Europe.* Oxford: Clarendon Press.

Henshilwood, C.S., J. Sealy, R. Yates, K. Cruz-Uribe, P. Goldberg, F. Grine, R. Klein, C. Poggenpoel, K. van Niekerk & I. Watts
2001 Blombos Cave, Southern Cape, South Africa: preliminary report on the 1992-1999 excavations of the Middle Stone Age levels. *Journal of Archaeological Science* 28: 421–448.

Johnsen, S. J., H. B. Clausen, W. Dansgaard, K. Fuhrer, N. Gundestrup, C. U. Hammer, P. Iversen, J. Jouzel, B. Stauffer & J. P. Steffensen,
1992 Irregular glacial interstadials recorded in a new Greenland ice core. *Nature* 359: 311–313.

Jones, D. A.
1984 *An Ecological Interpretation of the Mesolithic Shellfish Remains on the Island of Oronsay, Inner Hebrides.* Ph.D. Thesis, Department of Archaeology, University of Sheffield.

Lambeck, K.
1997 Sea-level change along the French Atlantic and Channel coasts since the time of the last glacial maximum. *Paleogeography, Paleoclimatology, Paleoecology* 129: 1–22.

Lewis, J. R. & R. S. Bowman
1975 Local habitat-induced variations in the population dynamics of *Patella vulgata* L. *Journal of Experimental Marine Biology and Ecology* 17: 165–203.

Lourandos, H.
1998 *Continent of Hunter Gatherers: New Perspectives in Australian Prehistory.* Cambridge: Cambridge University Press.

Mackay, R. & J. P. White,
1987 Musseling in on the NSW coast. *Archaeology in Oceania* 22: 107–11.

Masters P.M., & N.C. Flemming (Editors)
1983 *Quaternary Coastlines and Marine Archaeology.* New York: Academic Press.

Meehan, B.
1982 *Shell Bed to Shell Midden.* Canberra: Australian Institute of Aboriginal Studies.

Monks, G. & R. Johnston
1993 Estimating season of death from growth increment data: a critical review. *Archaeozoologia* 5: 17–40.

Obermaier, H.
1925 *El Hombre Fosíl (2nd ed.).* Madrid: Comisión de Investigaciones Paleontológicas y Prehistóricas, Memoria No. 9.

O'Connor, S.
1999 A diversity of coastal economies: shell mounds in the Kimberley region in the Holocene. In *Australian Coastal Archaeology* (J. Hall & I. McNiven, eds.), pp. 37–50. Canberra: Occasional Papers in Prehistory No. 30, Division of Archaeology and Natural History, Research School of Pacific and Asian Studies, Australian National University.

Ortea, J.
1986 The malacology of La Riera Cave. In *La Riera Cave, Stone Age Hunter-Gatherer Adaptations in Northern Spain* (L.G. Straus & G.A. Clark, eds.), pp. 289–98. Tempe: Arizona State University Anthropological Research Papers No. 36.

Orton, J. H.
1928a Observations on *Patella vulgata.* Part 1. Sex-phenomena, breeding and shell growth. *Journal of the Marine Biological Association of the United Kingdom* 15: 851–862.

1928b Observations on *Patella vulgata.* Part 2. Rate of growth of shell. *Journal of the Marine Biological Association of the United Kingdom* 15: 863–874.

Orton, J. H. & A. J. Southward
1961 Studies on the biology of limpets – IV. The breeding of *Patella depressa* Pennant on the north Cornish coast. *Journal of the Marine Biological Association of the United Kingdom* 41: 653–662.

Pathansali, D. & M. K. Soong
1960 Some aspects of cockle (*Anadara granosa* L.) culture in Malaya. *FAO Indo Pacific Fisheries Council Proceedings,* 8th session, section III, pp. 26–31.

Roberts, A.
1991 *An Analysis of Mound Formation at Milingimbi Island, N.T.* Armidale: M.Litt. Thesis, Department of Archaeology, University of New England.

Rowley-Conwy, P.
2001 How original is the 'Original Affluent Society'? Time, change and the archaeology of hunter-gatherers. In *Hunter-gatherers: an Interdisciplinary Perspective* (C. Painter-Brick, R.H. Layton & P. Rowley-Conwy, eds.), pp. 39–72. Cambridge: Cambridge University Press.

Shackleton, J.C. & T.H. Van Andel
1986 Prehistoric shore environments, shellfish availability, and shellfish gathering at Franchthi, Greece. *Geoarchaeology* 1: 127–143.

Stanner, W. E. H.
1961 The Weipa shell mounds. *The Etruscan* 11: 8–12.

Stone, T.
1989 Origins and environmental significance of shell and earth mounds in Northern Australia. *Archaeology in Oceania* 24: 59–64.
1993 Birds, boffins and blunders. *Australian Geographic* 29: 25–26.

Straus, L. G.
1979 Mesolithic adaptations along the northern coast of Spain. *Quaternaria* 21: 305–327.

Straus, L. G. & G. A. Clark
1983 Further reflections on adaptive change in Cantabrian prehistory. In *Hunter-Gatherer Economy in Prehistory* (G. N. Bailey, ed.), pp. 166–167. Cambridge: Cambridge University Press.

Straus, L. G. & G. A. Clark (Editors)
1986 *La Riera Cave, Stone Age Hunter-Gatherer Adaptations in Northern Spain*. Tempe: Arizona State University Anthropological Research Papers No. 36.

Straus, L. G., G. A. Clark, J. Ordaz, L. Suárez & R. Esbert
1986 Patterns of lithic raw material variation at La Riera. In *La Riera Cave, Stone Age Hunter-Gatherer Adaptations in Northern Spain* (L. G. Straus & G. A. Clark, eds.), pp. 189–208. Arizona State University Anthropological Research Papers No. 36.

Taylor, K. C., G. W. Lamorey, G. A. Doyle, R. B. Alley, P. M. Grootes, P. A. Mayewski, J. W. C. White & L. K. Barlow
1993 The 'flickering switch' of late Pleistocene climate change. *Nature* 361: 432–436.

Thomas, M. L. H.
1986 A physically derived exposure index for marine shorelines. *Ophelia* 25: 1–13.

Thomas, K.
1987 Prehistoric coastal ecologies: a view from outside Franchthi Cave, Greece. *Geoarchaeology* 2: 231–240.

Van Andel, Tj. H.
1989 Late Quaternary sea-level changes and archaeology. *Antiquity* 63: 733–745.

Van Andel, Tj. H. & N. Lianos
1984 High-resolution seismic reflection profiles for the reconstruction of postglacial transgressive shorelines, an example from Greece. *Quaternary Research* 22: 31–45.

Veitch, B.
1999 Shell middens on the Mitchell Plateau, a reflection of a wider phenomenon? In *Australian Coastal Archaeology* (J. Hall & I. McNiven, eds.), pp. 51–64. Canberra: Occasional Papers in Prehistory Series No. 30, Division of Archaeology and Natural History, Research School of Pacific and Asian Studies, Australian National University.

Walters, R.C., R.T. Buffler, J.J. Bruggemann, M.M.M. Guillaume, S.M. Berhe, B. Negassi, Y. Libsekal, H. Cheng, R.L. Edwards, R. von Gosel, D. Neraudeau & M. Gagnon
2000 Early human occupation of the Red Sea coast of Eritrea during the Last Interglacial. *Nature* 405: 65–69.

Wright, R. V. S.
1971 Prehistory in the Cape York Peninsula. In *Aboriginal Man and Environment in Australia* (D. J. Mulvaney & J. Golson, eds.), pp. 133–140. Canberra: Australian National University Press.

Yonge, C. M. & T. E. Thompson
1976 *Living Marine Molluscs*. London: Collins.

16

The Iberian Mesolithic in the European Context

Geoffrey A. Clark
Arizona State University

Abstract

This chapter has three principal objectives. First, I summarize recent trends in European Mesolithic research drawn from an analysis of the contents of the six international symposium volumes published to date. Next, I attempt to situate Iberian Atlantic coastal Mesolithic research in the European context, and outline some of the principal features that characterize it. Finally, I propose a scenario based on evolutionary ecology in order to explain what I take to be the salient features of the Iberian Mesolithic.

Introduction

As some readers may know, it has been my contention for some years now that there are definable differences and similarities in the ways in which researchers from different intellectual traditions go about their work (see, e.g., papers in Clark 1991, Clark & Willermet 1997, Clark 2002a). These differences and similarities are grounded in overlapping but partly discrete 'packages' of implicit preconceptions and biases about the nature of the human past (in this case, Mesolithic adaptations), and how we should go about studying it. There is also an historical component to archaeological research traditions in that they change over time, both within and across regions (Clark 1993). If this is so, it should be possible to identify the main characteristics which set Iberian Atlantic coastal Mesolithic research apart from that of other European coastal regions.

The Proceedings of the International Symposia: a Trend Analysis

Those of you who read the contribution by Neeley and Clark in the 1990 Leuven volume will recognize some of what follows. Building on the earlier data base, I classified the contents of the 1990 and 1999 volumes according to the criteria we used to describe the contents of the 1973, 1981 and 1989 volumes (Table 16.1). The 2003 proceedings arrived while I was undertaking these revisions. It has its own organizational framework and, although it was scrutinized for content, I did not attempt a systematic reclassification. There is, of course, considerable overlap in our analytical categories, with many essays quite justifiably assignable to several of them. The categories are intentionally broad. The intent was to pick out the major theme of each essay and assign it to one of 11 groups. The earlier survey focused on Mesolithic social complexity – how it might be defined and recognized archaeologically. The present effort seeks to identify the major variables considered to be important in Mesolithic research over the past 25 years (Table 16.2). Some interesting trends are apparent.

One is that European Mesolithic research is healthy and vigorous, and is getting stronger all the time. In fact, the growth of the international symposia was the primary motivation for smaller, regional meetings like this one – face-to-face interaction was becoming difficult. The 1989 volume contained almost three times as many contributions as the 1973 effort (62 vs 23), and a much more diverse range of papers. The 1981 volume is almost exactly 'in the middle' on both criteria (44 papers). The 1990 volume consisted of a modest 40 essays, but the stringent submission deadline limited the number of essays actually published since the editors sought to have the proceedings available at the time of the meetings. Ninety scholars registered for the Leuven symposium, *vis à vis* 88 at Edinburgh, so they were about the same size.

The Leuven volume was the first to divide the contributions into specific, intentionally broad categories: (1) ecology, land use and site structure (6 papers, 15%), (2) physical anthropology, burials and social issues (5, 12.5%), (3) culture change and the Mesolithic-Neolithic transition (5, 12.5%), and (4) regional studies (24, 60%). While its successor, the Grenoble volume, had an overall theme (Epipaleolithic and Mesolithic paleoenvironments and (re)colonization), the 68 essays that comprised it were organized geographically into five regions, with an emphasis on the Alps, Italy, the Massif Central and Corsica (25 papers, 37%). The most recent book, the proceedings of the Nynashamn conference, returns to a thematic organizational format. It is the largest of the six books in the series (92 articles by 136 authors from 23 countries) and the corresponding meetings also had the largest attendance (c.

Table 16.1 The six Mesolithic Symposium volumes.

Title	Editor(s)	Venue	Year Mtg.	Year Pub.	Publisher
The Mesolithic in Europe	Stefan Kozlowski	Warsaw	1973	1973	Warsaw University Press, Warsaw
Mesolithikum in Europa	Bernhard Gramsch	Potsdam	1978	1981	VEB Deutscher Verlag der Wisssenschaften, Berlin
The Mesolithic in Europe	Clive Bonsall	Edinburgh	1985	1989	John Donald Publishers, Ltd., Edinburgh
Contributions to the Mesolithic in Europe	Pierre Vermeersch & Philip van Peer	Leuven	1990	1990	Leuven University Press, Lueven
L'Europe des Derniers Chasseurs	André Thévenin & Pierre Bintz	Grenoble	1995	1999	Comité des Travaux Historiques et Scientifiques, Paris
Mesolithic on the Move	Lars Larsson, Hans Kindgren, Kjel	Nynäshamn	2000	2003	Oxbow Books, Oxford

200 registrants, of which slightly less than half gave oral or poster presentations). The meetings at Nynashamn, in east-central Sweden, were also the first to have parallel sessions, necessary because of the large number of registrants and the limited time and space available. Introduced by a general essay that assessed the current status of Mesolithic research (Kozlowski 2003) and by three summaries of recent Swedish work, papers were allocated to one of eight topics or sessions: (1) colonization processes (9 papers, 9.8%), (2) enculturating the landscape (12, 13%), (3) social relations and group formation (5, 5.4%), (4) intrasite spatial organization (11, 12%), (5) territoriality and regionalization (12, 13%), (6) exchange and communication (8, 8.7%), (7) ritual and symbolic behavior (13, 14.1%), (8) hunters in transition (9, 9.8%), and a ninth miscellaneous category (9, 9.8%). Each session was chaired by an archaeologist chosen by the conference organizers. These chairpersons also wrote short introductions to the articles grouped in their respective sessions.

A second trend is an apparent decline in the importance of typological systematics. In the 1973 and 1981 volumes, nearly half the papers (48%, 50%) were directed

Table 16.2 Major categories of Mesolithic Symposium contributions from the Warsaw (1973), Potsdam (1978), Edinburgh (1985), Leuven (1990) and Grenoble (1995) Conferences.

Category	Year of Publication									
	1973		1981		1989		1990		1999	
	n	%	n	%	n	%	n	%	n	%
culture history, chronology, typology,	11	47.8	22	50.0	17	27.4	11	27.5	16	24.0
raw material characteristics, distribution	2	8.7	2	4.5	6	9.7	1	2.5	6	8.3
technology	3	13.0	1	2.3	6	9.7	5	12.5	7	10.4
settlement, subsist. economy, population	5	21.7	8	18.2	13	21.0	8	20.0	11	15.6
colonization, settlement dist., seasonality	-	-	-	-	2	3.2	2	5.0	9	12.5
sedentism, structures	-	-	2	4.5	8	12.9	1	2.5	2	3.1
sedentism, storage facilities, pits, features	-	-	1	2.3	-	-	-	-	1	2.0
cemeteries, burials	1	4.3	-	-	1	1.6	2	5.0	4	6.2
art, ornaments	1	4.3	2	4.5	-	-	-	-	2	3.1
style, symbolism	-	-	-	-	1	1.6	2	5.0	-	-
the Mesolithic-Neolithic transition	-	-	6	13.6	-	-	6	15.0	5	7.3
miscellaneous categories	-	-	-	-	8	12.9	2	5.0	5	7.3
Total	23	99.8	44	99.9	62	99.9	40	99.9	6	99.8

primarily toward culture historical ends. Chronology building, typology and the identification of culture groups were primary objectives. In the 1989, 1990, and 1999 volumes, such essays account for only 27% (1989, 1990) and 24% (1999) respectively. An estimate for the 2003 volume is about 13%. There is a probable geographical component to this pattern, in that eastern European papers make up a larger proportion of the Warsaw and Potsdam volumes than they do in the Edinburgh and Leuven books. This could be taken to mean that typological systematics remain more strongly entrenched in the East than they do in the West (although, since eastern European contributions are underrepresented in the 1989 and 1990 volumes, there is no way to evaluate this bias directly). It is reasonable to expect that the regional research traditions of Europe would tend to emphasize different aspects of Mesolithic research differentially. A good example of this is a content analysis of the Grenoble and Nynashamn volumes. The Grenoble volume (1999) contains many essays on the paleoenvironment, stratigraphy, chronology, subsistence and regional culture history, reflecting the importance of the roles played by natural history and regional expertise in the French intellectual tradition. On the other hand, the Nynashamn compendium (2003), dominated by researchers who publish extensively in English (Scandinavians, British, Belgians) has a stronger emphasis on inter- and intrasite social organization, territoriality, regionalization, exchange, ritual, and the ideational realm.

A third trend is a partial inverse of the second: as a concern with culture history and typological systematics is de-emphasized, workers are apparently more willing to look to technology and raw material characteristics for explanations of observed patterns of variability. This probably has a lot to do with the increasing emphasis on forager mobility strategies, most easily monitored by technology and by tracing the distributions of raw materials from known sources (e.g., Geneste 1985, Boëda et al. 1990; Kuhn 1992, 1995, Neeley 1997, al-Nahar 2000). The shift in emphasis is not demonstrable quantitatively, however, since 21.7% of the papers in the 1973 volume address aspects of technology and raw material variability, versus 18.7% in the 1999 book. The impression that this is a real trend is partly attributable to similar shifts in emphasis in European paleolithic archaeology, and to the tendency of many researchers to work in both areas. However, the very considerable implications of uncoupling typology and technology have not been widely explored (Clark 1999a, 2002b).

Essays on settlement, subsistence, population and economy account for 21.7% of the 1973 volume and 28.1% of the 1999 volume, a modest increase that might be related to the widespread adoption of Anglo-American adaptationist biases and the realization that hunter-gatherer adaptations are regional phenomena that require the collection of regional settlement survey data. It should be noted that regional survey has not been a traditional focus of European Mesolithic research, although survey-based systematics seem to be somewhat stronger in the North and East than in the West. An exception is the United Kingdom, where the regional perspective of influential Cambridge prehistorians (e.g., Grahame Clark, Eric Higgs) has had an impact on all subsequent workers. The numerous essays in all six volumes concerned with aspects of the subsistence economy mirror a traditional emphasis of Mesolithic research, and the conviction that the subsistence economy is in some sense primary for hunter-gatherer adaptations in general. As might be expected, there is evidence of a sharp increase in methodological sophistication in coming to grips with pattern in the paleoeconomy when the 1973 contributions are compared with those in the 2003 volume. This probably reflects the rise to prominence of archaeozoology, taphonomic research and, to a lesser extent, computer-aided statistical analyses and simulations in the West over the past two decades. Since Mesolithic sites often preserve rich and varied archaeofaunas, and since the perturbing effects of natural site formation processes are reduced when compared with the paleolithic, the Mesolithic is a natural laboratory for testing these promising new 'high-tech' methodologies. Quantitative research protocols in general are widely adopted now in the Low Countries, Scandinavia, and the United Kingdom.

A fifth trend has to do with evidence for sedentism. It has become commonplace in recent years to assign to sedentism a major role in promoting social complexity - usually because of the implication of reduced birth spacing, increased survivorship and consequent increases in the rate of population growth. Although the linkages are often imprecise, recognition of this role is indicated by the 10 papers (16.1%) devoted to sedentism, structures, storage facilities and cemeteries in the 1989 volume. Essays on these topics are scarcely represented at all in the earlier conference proceedings (4.3%, 6.8% respectively). In the 1990 Leuven book, they account for a respectable 12.5%, but they decline again in the 1999 volume (5.1%), then increase sharply in the 2003 volume (17.4%).

Although they are rare, spectacular cemeteries like those at Skateholm, Zvejniecki, Oleoneostrovski and Cabeço de Amoreira show up for the first time in the later phases of the European Mesolithic, after about 6500 years ago (Clark & Neeley 1987). A comparatively direct monitor of social complexity, it could be argued that the analytical potential of cemeteries for discerning aspects of social differentiation has been largely unrealized. Only four papers in the first four volumes are primarily devoted to the analysis of mortuary data; three other essays provide incidental accounts of burial patterns and grave goods. There are five papers (7.3%) on these topics in the Grenoble proceedings (1999), and 10 in the Nynashamn volume (2003), perhaps signaling increased interest over the past decade and/or the fact that some of the best 'modern-quality' mortuary data come from Scandinavian sites.

There is a sprinkling of essays concerned with aspects of art and style in the first five volumes, but no trends are apparent. The 2003 proceedings contain 12-15 essays (13.6-17%) devoted wholly or in part to ritual and symbolic behavior manifest in art and style, most of them grouped in a section that formally recognizes these themes. Since there is an abundance of 'art' in Mesolithic contexts, the lack of pattern could be indicative of the ambivalence that many workers feel regarding our capacity to recover symbolic behavior – an ambivalence reflected in numerous and very different conceptions of what 'style' is or does, and how stylistic variables might be defined operationally in different kinds of archaeological contexts (cf., e.g., papers in Conkey & Hastorf 1990, Carr & Neitzel 1995). To the best of my knowledge, the only systematic compendium of the time-space distributions of the only Mesolithic artifact class arguably numerous enough to provide quantitative information on art and style – ornaments made on organic blanks – is the somewhat controversial *An Inquiry into the Ethnic Resolution of Mesolithic Regional Groups* (Newell et al. 1990).

Finally, there are a small number of 'transition' papers, directed at defining either the beginning or the end of the Mesolithic. In the first two volumes, the former are more common than the latter. They do not appear to exhibit any clear-cut pattern although, prior to the mid-1980s, the appearance of microlith-dominated assemblages was often (and erroneously) taken to mark the LUP-Mesolithic transition. Conceptual papers concerned with the 'end' of the Mesolithic (i.e., the nature of the Mesolithic-Neolithic transition) exhibit a bimodal distribution, and are essentially confined to the 1981 volume (13.6%, where that topic was identified as a focus of the meeting) and the 1990 volume (15%, where it was not). The corresponding frequencies for the 1999 and 2003 volumes are 7.3% and 9.8% respectively. Since the early 1980s, a fairly broad consensus has emerged in regard to the temporal boundaries of the Mesolithic. Its beginning is usually defined according to geochronological criteria (10 kyr BP, or the Pleistocene-Holocene boundary as recognized by Quaternary science), and its ending corresponds to the local appearance of domestication economies (occasionally, as in eastern Europe, the appearance of pottery). This definition was formalized and expanded somewhat in the 1989 (Edinburgh) volume, although perhaps the best current treatment of the topic along the coasts of Atlantic Europe is Arias (1999).

Cross-cutting these thematic issues, and to a large extent inseparable from them, are what might be called 'regional studies'. Geographical regions (and – sometimes – the boundaries of modern nation states!) are used as organizing principles in all the proceedings except the first (1973) and last (2003) ones. They constitute the sole organizational framework in the 1989 (Edinburgh) and 1999 (Grenoble) volumes and, in the 1990 (Leuven) volume, make up 60% (24 papers) of the 40 papers published. While we have no reason to believe that Mesolithic adaptations ever corresponded to modern political boundaries, the regionalism reflects the regional character of hunter-gatherer adaptations in general and that of the different research traditions involved in Mesolithic research. Despite it, there is an impression that the parochialism that has sometimes afflicted Mesolithic research in the past is beginning to break down, and, especially since the mid-1980s, a greater awareness of some of the theoretical perspectives that affect all hunter-gatherer research. Although strict empiricist biases are still very much a part of Mesolithic studies, and there is, as yet, little explicit concern with epistemology, there seems to be an emergent consensus about what problems are significant to investigate (e.g., social complexity issues, the transition to domestication economies) and a certain confidence that we can eventually resolve some of them. A pattern of regional variability linked to differences in adaptation, ultimately explicable by reference to environmental and resource diversity and to population-resource imbalances, probably underlies the mosaic configuration documented by the regional studies in the four symposium volumes.

Iberian Mesolithic Research

How does Iberian Mesolithic coastal research fit into the European context? Well, to judge only from the contents of the earlier symposium volumes, it hardly fits at all. There were no Iberian contributions in the first two proceedings, seven papers (11.3%) in the Edinburgh volume, and two (5%) in the Leuven volume. The decrease between the 1985 and 1990 meetings is interesting in light of what one might expect to be a bias in favor of contributions from Latin Europe on the part of the Leuven conference organizers. However, Bonsall went to great lengths to include as many papers as possible in the Edinburgh proceedings (and at the cost of a long delay in publication), whereas Vermeersch and van Peer put such stringent time constraints on manuscript submission that less than half of the 90 symposium participants were able to provide manuscripts quickly enough to meet the deadline. There are 12 Iberian papers (17.6%) in the Grenoble volume, reflecting the close geographical proximity of the meeting venue, and the tendency for Iberian researchers to be fluent in French, and to collaborate extensively with French scholars. The incidence of Iberian contributions plummets to four (4.5%) in the Nynashamn volume, which is – understandably – heavily dominated by Scandinavian research.

Fortunately, there have been a number of recent syntheses of Iberian Mesolithic research, most notably those of Arnaud (1985, 1989), Clark (1983a, 1989, 2000), González-Morales and Arnaud (1990), Straus (1991a), Vierra (1992, 1995), Bicho (1992) and Arias (1999), as well as various overview papers in Spanish and Portuguese journals and collected works (e.g., Arnaud 1990, 1993), and general essays referring to western Europe (e.g., Price 1987) or parts thereof (e.g., Zvelebil & Dolukhanov 1991, Larsson 1990, Street et al. 2001). I do not intend to recapitulate the contents of these papers here, since they are probably familiar to most readers. I do, however, want to address the second objective noted at the beginning of this essay. Posed as a question, what are the main features that characterize Iberian Mesolithic research in the 1990s, and how do they differ from those of other European research traditions? In keeping with the theme of this conference, I will confine my remarks to Mesolithic research along the Atlantic coasts of Spain and Portugal.

'Modern' era research on the Vasco-Cantabrian Mesolithic can be said to date from the early 1970s, that in Portugal to about a decade later. Both regions see much accelerated work during the 1980s and 1990s, share a central 'package' of research biases that can broadly be defined as 'ecological' and 'adaptationist' and are characterized by research protocols that represent a departure from the lithic-centered typological approaches of earlier generations. The intellectual mandate for this kind of research appears to have originated in the heavily ecological economic prehistory school at Cambridge, dominated by Grahame Clark (e.g., 1972) and Eric Higgs (e.g., 1972, 1975), and most evident in the Spanish context with the work of Geoff Bailey (e.g., 1983), and in the interdisciplinary natural science approaches of Scandinavian workers (e.g., Welinder 1978, Broadbent 1978), which have always been somewhat buffered against the typological excesses of Latin Europe. These approaches first began to have an impact in Spain during the late 1960s, with the collaborative efforts of Joaquín González Echegaray and Leslie Freeman (e.g., 1971, 1973) at Cueva Morín (Santander) and with work during the 1970s by their students (Clark, Moure, Straus etc.). The end of the dictatorship coincided with a resurgence of foreign collaboration which now, unhappily, has diminished somewhat, partly as a consequence of the political realignment resultant from the formation of the regional governments in 1984 (which tends to put political obstacles in the way of collaboration), and partly from the emergence of an substantial cadre of Spanish prehistorians now in the most active phases of their careers. This process was delayed by about a decade in Portugal, where work by the Canadians Lubell, Jackes and Meiklejohn (e.g., 1989) and, later, by Marks and Straus, acted as a stimulus for wider application of ecofunctional approaches by a new generation of Portuguese prehistorians, including Arnaud, Zilhão and Bicho.

Here as elsewhere, the justification for studying the Mesolithic as a problem domain is rarely made explicit, but seems to depend upon the idea that the deglaciation was a period of accelerated climatic change that forced changes in human adaptation across a wide range of distinct environments. Although these changes were less marked in Iberia than in more northerly latitudes, there were nevertheless shifts in the distributions of the plant and animal populations upon which Mesolithic foragers depended for food, and a gradual rise in sea levels that reduced, to some extent, land areas available for settlement. Research efforts in both Spain and Portugal have emphasized (1) refining the chronological framework (most of the radiocarbon determinations from Mesolithic sites post-date 1980), (2) reconstructing the sequence of local and regional environmental changes, often accompanied by efforts to place Iberian data into dated, north European pollen sequences (although the rationale for doing so is rarely questioned), (3) the study of the subsistence economy, grounded in efforts to understand relationships of site and resource distributions over time, (4) determinations of seasonality, where adequate faunal data (marine gastropods, ungulate dentitions) are available, (5) activity-area oriented intrasite spatial analyses (esp. of open sites where extensive horizontal exposure is a possibility), (6) identifying the regional characteristics of functionally categorized site distributions (e.g., residential base camps, game lookouts, specialized hunting stations and other limited activity sites etc.), (7) evidence for technological changes that might have influenced the subsistence economy (e.g., mass hunting techniques, fishing, fowling), and (8) evidence for population growth, resulting in population-resource imbalances. Although most research has been directed toward reconstructing regional ecosystems over time, and toward identifying the human component in them, little effort has been expended so far in addressing general questions of process and causality, nor has the nature of the basic analytical units been subjected to much critical scrutiny.

Spain

In terms of normative typological systematics, most of the first half of the Cantabrian Preboreal is characterized by backed-bladelet dominated Azilian assemblages, which in default of the diagnostic harpoons, are indistinguishable from the late Magdalenian of the region. Dated Azilian sites extend down in time to c. 9400 BP, when they give way to the coastally-distributed Asturian, associated with limpet-dominated *concheros* and heavy-duty quartzite tools and flake assemblages dated mainly to the Boreal and early Atlantic periods. Ceramics show up in shell midden contexts around the end of the Atlantic period, at c. 5700 BP, but the 'neolithization' of Cantabria was very partial and very late (Clark 1987). Although consensus has remained elusive (cf. González Morales 1982), I continue to believe that there is substantial evidence for overlap between the Azilian and the Asturian for the 9500-8500 BP interval if standard deviations for individual radiocarbon dates are used to establish contemporaneity (cf. Straus 1979, Straus & González Morales 2003 with Clark 1983b, 1989, 2000). A block of 10 determinations comprising five Azilian samples, three Asturian ones, and two that cannot be assigned to either analytical unit fall into this interval and suggest the possibility of functional complementarity between these two very different assemblage types. Of some interest is the observation that all 'late' Azilian sites are located inland, whereas all Asturian sites are coastal, implying a bimodal pattern of site location during the period of overlap. As González Morales (1989) has pointed out, however, grand means and standard deviations for the two series of dates do not overlap (see also Straus & González Morales 2003).

In my opinion, the Azilian and the Asturian comprise two major kinds of Mesolithic assemblage represented along the Cantabrian coastal strip. Other kinds of assemblages thought to date to the Dryas III-Boreal interval (10.8-7.5 kyr BP) are extremely rare in coastal contexts, although they show up inland in some frequency in the Basque country (e.g., Barandiarán & Cava 1989). Thirty-two fairly credible Asturian sites lie at low elevations ($\overline{x}$ = 36.5 m) at a mean distance to the present shoreline of 1.4 km, with very little evidence of midden accumulation greater than 3 km inland. The corresponding mean for 26 credible Azilian sites is 10.3 km, and might have been as great as 15 km during the Dryas III interval, to which most Azilian sites are dated. Mean elevation is 122 m, indicating the existence of substantial numbers of sites located inland in relatively high, mountainous terrain. In regard to surface area, Asturian sites are unimodally distributed and are, on average, 'small' (<36 m^2) and 'medium'-sized (36-60 m^2) on average. Azilian site areas are bimodally distributed, with the larger sites (>60 m^2) found in coastal contexts and with small sites (<36 m^2) occurring most often in high alpine gorges (e.g., Los Azules). The latter tend to have ibex-dominated faunas, lending credence to the idea that these sites owe their existence to monitoring, procuring and processing these wary alpine caprids. Azilian site locational data reported in Clark (1989) indicate possible residential camp-hunting station dyads for eight pairs of Azilian sites (i.e., Santimamiñe/Atxeta (Ría Güernica), Lumentxa/Ermittía (Deva), Ekain/Urtiaga (Urola), Morín/Piélago and Rascaño (Solia-Miera), Río/Azules (Sella), Valle/Mirón (Asón) and Candamo/Sofoxó (Nalón-Nora)). While these elements in an Azilian settlement-subsistence system are not as distinct as might be desirable, it is significant that they form part of a general Upper Paleolithic pattern in which such site functional relationships appear to be redundant and to persist over time (Clark & Straus 1986, Straus & Clark 1986). The fact that the Asturian configuration is quite distinct implies a fundamental re-articulation of the elements in the system at some point after c. 8500 BP, since there are few indications of inland occupation after that date (although cf. Arias & Perez 1995, Díez 1999).

The Mesolithic data presented here only begin to make sense when viewed from the larger perspective of patterns visible in the Cantabrian Upper Paleolithic. Recent comparative evaluations of post-Aurignacian Upper Paleolithic and Mesolithic assemblages by Straus (1983, 1985) and me (1983, 1989, Clark & Straus 1983, 1986) have determined the existence of two major Vasco-Cantabrian assemblage types that seem to persist and recur throughout the latter part of the last glacial and into the early post-glacial. These assemblage types are dominated on the one hand by quartzite notches and denticulates, side-scrapers and crude heavy-duty tools, and on the other by retouched (esp. backed) bladelets with substantial numbers of small endscrapers and burins made of flint. The Asturian is in general an example of the former; most Azilian collections (and probably Liencres) are examples of the latter. We believe these assemblage types can most readily be explained in functional terms. They seem to represent two rather generalized and flexible technologies which persist for about 15 millennia (c. 21-7 kyr BP) and which probably reflect in some imperfect way two distinct, yet broadly defined, sets of activities or behaviors (Clark 1989, 2000).

There are suggestions of periodicity in the occurrence of these assemblage types which hint at recurrent shifts in adaptation, perhaps coupled with changes in the distribution and availability of key resources. These changes have traditionally been explained by invoking changes in climate, with concomitant changes in sea level and in the extent of exposure of the continental shelf (e.g., Bailey 1983a, 1983b). While the impact of climatic change on human behavior and resource distributions cannot be dismissed in the case of the north Spanish Upper Paleolithic, it seems to have been of relatively minor importance from the more restricted perspective of the Pleistocene-Holocene boundary assemblages of interest here.

To the extent that trends in the human food niche can be monitored over the long term (c. 45-2 kyr BP), patterns of increasing resource intensification and diversification seem to be documented, at least from c. 20 kyr BP on, with the most diverse economic faunas occurring during the Asturian, and the most convincing examples of intensification corresponding to the established domestication economies of the Bronze and Iron Ages. In the absence of evidence to the contrary, Straus and I tend to explain these trends as due to imbalances created by a growing regional population and its traditional resource base (Straus et al. 1980, Clark & Straus 1983, Clark 1987).

Portugal

Research on the Portuguese Mesolithic can be said to date to 1863, with the discovery by F.A. Pereira da Costa and Carlos Ribeiro of the famous open-air *concheiro* sites of the Tagus Valley (Muge estuary - Moita do Sebastião, Cabeço da Amoreira, Cabeço da Arruda), intensively investigated during the 1950s and 1960s by the Abbé Jean Roche (e.g., 1972), and best known for their spectacular cemeteries and structures (e.g., Ferembach 1974). In the 1930s, somewhat less impressive open-air shell middens were reported by Hipólito Cabaço from the Magos estuary, another southern tributary of the Tagus, and by Lereno Antunes from the Río Sado, some 100 km to the south. These last have been the focus of long-term excavations first, during the 1950s and 1960s, by the staff of the National Archaeological and Ethnological Museum (Lisboa) under the direction of Manuel Heleno (these were never published) and, after 1981, by a multidisciplinary team headed by J.E.Morais Arnaud (1989). Of the Sado valley sites, Cabeço do Pez and Poças de São Bento are probably the most important. At about the same time, a Canadian team headed by David Lubell and Mary Jackes (1985, 1988) began a program of survey and excavation at Medo Tojeiro and Samouqueira in the Mira valley on the Alentejo coast, in the extreme southwestern part of the country. Other, smaller projects were also undertaken in the Alentejo during the 1980s (e.g., Straus 1991b) and in the Algarve during the 1990s (e.g., Bicho et al. 2000, Stiner et al. 2003). So – 'modern era' research is very recent indeed, and approximately 10-15 years behind comparable Spanish work. It could accurately be described as still in the descriptive stages, with much effort expended on chronology-building, paleoenvironmental reconstruction, lithic-based time-space systematics (although, despite the important influence of French typological schemes manifest in the long-term research efforts of Georges Zybszewski and the Abbé Breuil himself, more oriented toward technology), studies of settlement patterns and subsistence strategies, and dietary reconstructions. Some individual sites are beginning to be fairly extensively reported (e.g., Vidigal - Straus [1991b], Areeiro - Bicho [1991]). The first synthetic, explanatory efforts post-date 1990 (e.g., Vierra 1992, 1995; Bicho 1992, 1994). In light of strong evidence for continuity in adaptation between the late Upper Paleolithic and the early postglacial, some Portuguese workers (esp. Bicho) have proposed the term 'Epipaleolithic' to denote sites of Holocene age with LUP characteristics (10.5-8 kyr BP), whereas 'Mesolithic' denotes sites dating to the Atlantic period (8.5-6 kyr BP), when technologies, subsistence economies and settlement patterns all changed dramatically. Other workers (e.g., Arnaud) divide the 10.5-6 kyr BP interval into an 'early' and a 'late' Mesolithic but on the same criteria (i.e., major shifts in technology, subsistence, settlement post-dating 8 kyr BP).

Portuguese sites considered to be 'early' Mesolithic date from c. 10-8.4 kyr BP, and include *concheiros*, open-air lithic scatters and cave deposits. Many are situated in fossil dunes along the coasts of Estremadura and Alentejo. They can be dominated by either microlithic chert artifacts or macrolithic cobble tools (rarely both), thus reflecting the fundamental dichotomy observed for the north Spanish epipaleolithic. As was true of the Spanish Asturian, there has been some debate over whether some or all of the macrolithic cobble assemblages are of paleolithic or Mesolithic age. These include 'Asturian' assemblages found in fossil beaches and river gravels along the north Portuguese coast (and extending into Galicia) (da Serpa Pinto 1928, Clark 1976), the 'Languedocian' of the central and south Portuguese littoral, and the 'Mirian', a supposed Languedocian facies or variant, found in the Rio Mira estuary and on the Alentejo coast near Vila Nova de Mil Fontes (Breuil & Zybszewski 1942, 1946). Mirian 'diagnostics' include unifacial cobble choppers, picks, the eponymous 'axes', disc cores, net sinkers (stone donuts) and weights (waisted pebbles) typically made of a local greywacke – essentially the same kinds of artifacts that characterize the Cantabrian Asturian. While distinctions among these assemblages are problematic, the question of whether the 'Mirian', at least, actually dated to the paleolithic or the epipaleolithic was resolved by Raposo and Penalva's excavations at Palheirões do Alegra, on the Alentejo coast near the mouth of the Mira, which produced both typical Mirian cobble tools and microliths, dated to 8400 ± 70 BP (Raposo 1988, Raposo et al. 1989). This Mirian assemblage, the only one dated so far, is thus contempraneous with the north Spanish Asturian *concheros*, which also consist mainly of cobble picks and unifaces (but which lack microliths). Unlike the Spanish sites, Mirian and Portuguese Asturian sites are not associated with shell middens. None of the Portuguese Asturian sites have been systematically excavated or dated, but Maury (1977) suggests that it, like its Cantabrian counterpart, is also of post-glacial age.

Late Mesolithic sites tend to be almost exclusively microlithic and date from approximately 7.5-5.1 kyr BP. Ceramics appear in the upper levels of some of these middens at around 6.0 kyr BP. In contrast to the early sites, late Mesolithic *concheiros* are typically situated along river systems in estuarine settings – true not only of the Muge and Sado site clusters, but also of smaller site groups along minor drainages in the Alentejo (south) and Estremadura (central) regions of Portugal. These occupations coincide with the onset of the Flandrian transgression at c. 6.5 kyr BP, which reached its maximum between c. 5.7 and 4.1 kyr BP. Given the relatively low relief of the central and southern Portuguese coastal plain, the Flandrian transgression resulted in the creation of new estuarine habitats which extended as much as 30 km upstream from present-day shorelines. The few pollen sequences known from central and southern Portugal indicate the presence of pine forests which, during the 'late' Mesolithic, were being replaced by oak park-forest woodlands (Mateus 1989). The Mesolithic-Neolithic transition, defined on the basis of pottery (and on the introduction of ovicaprines at Gruta do Caldeirão and Casa da Moura) coincides on the central and south coast with the maximum extent of the Flandrian marine transgression at c. 6-5 kyr BP. There is little evidence of a late Mesolithic occupation in northern Portugal (Vierra 1992), but a strong possibility of forager-farmer contemporaneity in the south and central parts of the country. Two hypotheses for the neolithization have been proposed: (1) maritime pioneer colonization from the central coast, marked by an abrupt intrusion of a fully-developed farming and herding adaptation, followed by replacement (Zilhão 2000, 2001), and (2) a complex and longstanding mosaic of forager and farmer adaptations, with gradual acculturation of the former by the latter, and/or exchange between indigenous Mesolithic foragers and agropastoralists (Silva & Soares 1987, Carvalho 2002).

Relative to Spain, dietary information is still very scarce for the Portuguese late paleolithic and Mesolithic. Sparse data on the LUP and epipaleolithic have been summarized by Bicho (1993); there are some extensive reports and dietary syntheses for the Mesolithic shell midden sites by Arnaud (1990, 1993) and by members of the Canadian team (Lubell & Jackes 1985, Lubell et al. 1989). In the LUP, red deer, aurochsen, wild boar and horse were the dominant economic species, with significant rabbit and hare remains reported from Caldeirão cave (Zilhão 1987, 1992). Nuts, berries and fruit were available, as indicated by the tree species present in pollen samples from Cabeço de Porto Marinho (Figueiral 1993) and implied by the presence of grinding stones from various sites dated to c. 11 kyr BP (Bicho 1993).

Between 11-8 kyr BP (Preboreal, Boreal periods) there are apparently few significant changes in the major elements of the diet, except for the occasional appearance of ibex and chamois. Molluscs are present in abundance in both inland (estuarine) and coastal sites. The main species represented are the common cockle, oyster, various clams, mussels, limpets, periwinkles and topshells, indicating exploitation of several estuarine habitats and of the rocky, open coastline. The same range of shellfish species are found in Atlantic age sites, except for the cold-water adapted periwinkle *(Littorina littorea).* Roche (1982) reports unidentified fish remains from Lapa do Suão, while an estuarine-coastal sparid, gilthead, is present at Caldeirão and at other sites dated to the Atlantic period (Toledo, Samouqueira, Arapouco), along with meagre, ray and shark (Arnaud 1986). Bird bones show up in the Muge *concheiros*, and duck, pigeon and partridge are common enough to suggest that they were probably of dietary significance (Arnaud 1993).

Overall, the Portuguese data suggest a pattern similar to that of Cantabrian Spain with a trend, which probably began in the LUP, towards subsistence intensification through both specialization (e.g., Bicho [1993] argues that in central Portugal, aurochsen and red deer were probably hunted by driving and surrounds) and diversification, including increasing emphasis on a wide range of aquatic resources such as estuarine and coastal shellfish, birds and marine fish (there are no salmonids in Portuguese sites). Dietary diversity apparently reached a maximum in the Atlantic age shell middens, and included the long-term dietary staples – large and medium-sized ungulates like red deer, aurochs, roe deer, boar and horse; small species that were probably hunted collectively with nets, like rabbit and hare; birds, and both estuarine and open coastal shellfish, crustaceans, echinoderms, estuarine and marine fish, and occasional marine mammals (e.g., otter). Plant remains are known from some sites (e.g., Medo Tojeiro, Samouqueira), and include oak, pine and pistachio. The stable isotope analysis of several human skeletons from Muge has indicated a diet balanced between terrestrial and aquatic resources (Lubell et al. 1989, Jackes & Lubell 1997, 1999; Bamforth et al. 2003).

What Does It All Mean?

Despite differences of scale and timing, differences in the specifics of the regional ecologies, and differences in the tempo and history of investigation, it appears to me that broadly similar processes are taking place over the 20-7 kyr BP interval along the Atlantic coasts of Spain and Portugal. Moreover, I think these processes can be generalized for large parts of western coastal Europe, including France and the Low Countries, the British Isles and Scandinavia. The big question is: *What do these apparent similarities in pattern tell us about hunter-gatherer adaptations in general and Mesolithic adaptations in particular, and where do the latter fit in the long-term evolutionary ecology of western coastal Europe?* An arbitrary 'slice' of a temporal continuum, the Mesolithic only makes sense when viewed from the perspective of the pleniglacial maximum. The salient features that must be explained are (1) vectored changes in the paleoeconomy in the direction of resource intensification manifest in increasing dietary diversity, (2) the common aspects of lithic technologies – notably the widely-observed dichotomy between large, crude, cobble and flake dominated assemblages made on quartzite and diminutive microlithic bladelet industries (more generally, the appearance of microlithic technologies), (3) the comings and goings of the 'art' (why does it show up when and where it does in the European context, and why does parietal art disappear at the LUP-Mesolithic boundary?), and (4) the transition to domestication economies, different in timing and detail but similar in pattern over large areas of western Europe.

As I pointed out earlier, post-1970 Mesolithic research seems to be characterized by broadly similar research protocols emphasizing ecological-functionalist explanations for pattern – explanations that can be subsumed under evolutionary ecology and that are grounded in 'adaptationist' biases (Mithen 1993). While this shared perspective is a considerable strength of the Mesolithic research traditions, it is also a potential weakness. If it is acknowledged that we have adopted an adaptationist program, we should ask ourselves what that really means. Are there any viable alternatives to the ecofunctionalist biases that underlie Iberian Mesolithic research? Do these biases preclude the possibility of getting at certain kinds of social information? Do they allow us to access the ideological domain? More particularly, what does the apparent disappearance of parietal art and aggregation sites in the Mesolithic mean, given their presence in the LUP, and given the continuation of portable art? What does the widespread and penecontemporaneous appearance of microlithic technologies after c. 20 kyr BP imply and what does their elaboration during the Mesolithic mean in terms of human adaptation? Most generally, is it possible to come up with a comprehensive scenario for Mesolithic adaptations that explains their salient features and that is at the same time more compelling than the piecemeal post hoc accommodative arguments that have prevailed in the literature so far? I believe the answer is a qualified 'yes', acknowledging the many empirical insufficiencies that preclude putting such a scenario to a rigorous test.

Regional Demography and Subsistence Intensification

The cornerstore of this scenario is the idea that patterns observed in Iberia over the 20-7 kyr BP interval were ultimately determined by changes in regional demography and by the consequent population-resource imbalances that they entailed. The proximate cause of these patterns is population displacements in northern and central Europe brought about by climatic severity during the pleniglacial maximum (21-13 kyr BP). As many workers have noted, this would have led to increases in population densities in southwestern (and possibly eastern) European refugia, the best documented of which is the Franco-Cantabrian one, comprising a 340,000 km^2 area of southwestern France and the topographically-bounded north Spanish coastal strip (19,800 km^2). These geographically constrained areas already sustained relatively high population densities prior to the LGM and the influx of immigrants would have created selective pressures for ways to broaden the human food niche, setting in motion a process which culminated (in terms of dietary diversity) in the 'broad spectrum' economies of the Mesolithic. Seen from this perspective, the neolithization is nothing more than a 'worst case scenario' that comes about when foragers have run out of options for further diversification, so that domestication (perhaps the ultimate form of intensification) is the only alternative (see, e.g., Richards et al 2001).

The 'Neolithization' as a Last Resort

It is no accident that the neolithization of much of Atlantic Europe is very partial and late *vis à vis* that of Mediterranean Europe. The reason has relatively little to do with the westward march of neolithic colonizers (although there was undoubtedly significant population movement), but rather with the fact that foragers are very flexible in terms of dietary choice and, under conditions of increasing stress, select resources in a predictable way so that energy expended in doing so is kept to a minimum (the least cost selection assumption of many ecologists). This means that foragers will respond to stress by diversifying their resource base by the sequential addition of higher cost, lower yield resources (like shellfish) and by overexploitation (often to the point of diminishing returns) of traditional dietary staples (like red deer) (Redding 1988). This process invariably occurs at the expense of efficiency in resource extraction, but generally does not approach the massive labour inputs of full-blown agricultural societies (Clark 1999b).

The Iberian data document this process of diversification and intensification as well as might be expected, given the vagaries of preservation, sampling error, tempo of research etc. The timing of the onset of these processes after 20 kyr BP is scarcely coincidental. They were set in motion by the pleniglacial depopulation of large parts of northern and central Europe, and the resultant influx of immigrants into the Franco-Cantabrian refugium. Evidence for precocious social complexity (e.g., sedentism, structures, cemeteries) seen in the Tagus and Sado estuaries – unique in the Iberian context – can be explained by social circumscription and the closing of social networks due to 'packing' of an already densely populated, resource-rich estuarine environment. The cemeteries probably represent the appearance of corporate descent groups 'staking a claim' to the right of eminent domain over an increasingly-circumscribed economic territory, as Saxe argued in the early 1970s (e.g., Saxe 1971).

Microlithic Technologies, Time-Stress and Risk Management

The widespread appearance of microliths after c. 20 kyr BP, and the increasingly microlithic character of Mesolithic technologies, can also be explained from an adaptationist perspective by invoking a model that relates them to scheduling constraints and time-stress resultant from dietary intensification due to population-resource imbalances. This is an idea that can be traced in the present context to Robin Torrence (1983, 1989; see papers in Elston & Kuhn 2002), and ultimately to David Clarke (1976), who proposed that microlithic technologies were highly flexible, "plug -in, pull-out" technologies that were more efficient than macrolithic ones in certain contexts and that were not confined ethnographically to hunting pursuits.

Drawing on principles of evolutionary ecology, Torrence (1983) suggested that selection should favor a technology that maximized the efficient use of time by reducing resource acquisition and processing costs. She assumed that fitness is a function of foraging efficiency as measured by some currency (e.g., input-output ratios, or the net rate of energy return). She modified this position in 1989 by adding a risk-management component: although time constraints are often the proximate causes for changes in technological systems, risk management is the ultimate cause, risk being minimized by managing the efficient use of time (1989: 60). Time-stress is a function of the duration of time that a resource is available for exploitation, determined by the kind of resource being exploited and where that resource 'fits' among the competing demands of other resource procurement systems. Efficiency should, therefore, vary as a function of resource availability and accessibility.

The time-stress model defines more specialized technologies as those with greater tool diversity and complexity, and predicts that they would be found in highly seasonal environments characterized by limited periods of resource availability and less accessible resources (i.e., mobile animals). More generalized technologies are those with reduced tool diversity and complexity, expected to occur most commonly in less or non-seasonal (tropical) environments with longer periods of resource availability and more accessible resources (i.e., immobile plants). Ignoring the criterion of latitude as an index of resource availability but retaining Torrence's definitions, microlith-dominated industries would be the quintessential example of a specialized technology, comprised of interchangeable parts, maximally flexible and very probably produced by a broadly generalizable reduction sequence (Neeley & Barton 1994). In contrast, simple cobble and flake-dominated assemblages like the Asturian would correspond reasonably well to Torrence's definition of a generalized technology, perhaps oriented toward the procurement of plants.[1]

Myers (1989) provides a case study using the perspective advocated by Torrence (1983, 1989). He was interested in explaining technological change from the early to late Mesolithic in Britain by identifying changes in selective pressures for minimizing resource procurement risks. The observed pattern was a shift from the use of a few large non-geometric microliths per tool in the early Mesolithic (c. 10.3-8.7 kyr BP) to the use of small geometrically-shaped microliths mounted in series on composite tools in the late Mesolithic (8.7-5.4 kyr BP). He argued that late Mesolithic technologies comprised a "maintainable weapon system" *(sensu* Bleed 1986) in which tools were continually used and maintained 'in the field' due to unpredictability in the temporal and spatial distribution of food resources. The geometric microlithic technologies of the late Mesolithic might have been designed to maximize flexibility, and were used in a wide range of procurement and processing activities. He suggested that the composite nature of microlithic tools made them more easily repaired and/or modified according to circumstances, and thus more efficient (because of their flexibility) under conditions of time-stress and resource unpredictability.

Vierra (1992, 1995) has taken this approach considerably further, applying it to temporally vectored changes in subsistence and technology in the Portuguese Mesolithic. He suggested that, although time-stress could be related to variations in environment and resource distributions, it could also arise from increasing subsistence diversification and the scheduling conflicts that a diversified subsistence strategy necessarily entails. Straus (1977) and I (Clark 1983a, 1987) have argued that increasing diet breadth is characteristic of the Iberian epipaleolithic and Mesolithic, and that it probably results mainly from regional population growth and the 'packing' phenomenon. Dietary diversity would be especially important in seasonal environments like that of mid-latitude Europe, where populations would have needed to store food, or to have had access to 'insurance resources' like shellfish to get them over the winter lean season. Given a density-dependent situation, foods appear to be added to the diet in decreasing order of net rate of energy return (i.e., as more easily acquired 'high yield' resources become less abundant in the environment, more 'costly' foods in terms of greater acquisition and processing costs are added to the diet). This means greater reliance on smaller food package sizes (e.g., shellfish) and on small game and aquatic resources.

It seems likely, therefore, that increasing diet breadth would also lead to scheduling conflicts, resolved by special task groups with specialized technologies at their disposal geared toward the efficient procurement of a particular resource. Population growth, subsistence diversification, time-stress and technological complexity should all be highly correlated with one another and, in fact, Vierra (1992, 1995) achieved reasonably good agreement between predicted and observed patterns in the Portuguese Mesolithic data that constituted the basis for his doctoral dissertation.

Art from an Information-Exchange Perspective

Finally, there is the art - why does it come and go when and where it does? An inability to deal with non-technological aspects of adaptation like 'art' is frequently cited as a defect in the ecofunctionalist program (e.g., González Morales 1991). What does the apparent disappearance of parietal art in the Mesolithic imply, given a continuation of the portable art? Barton, Cohen and I have taken up this question in two recently published papers (Barton et al. 1994, Clark et al. 1996). Basically, we argue that the florescence of parietal art in the Franco-Cantabrian LUP, and its disappearance in the Mesolithic, are related aspects of a single evolutionary process that involved the initial pleniglacial intensification, and subsequent tardiglacial relaxation of selective pressures favoring the expression of certain kinds of symbolic behavior in response to changing population densities in the southwestern European refugium over the 20-10 kyr BP interval.

Given our adaptationist biases, our approach to the art is essentially a functionalist one that views it as the remains of communication systems involving the exchange of information. This is an idea that, in its present context, originated with Polly Wiessner (1983), who views the creation of art as an act of social communication defined, at various levels and scales, by style. Style in turn has its behavioral basis in a fundamental human cognitive process: the personal and social identification of images through visual comparison. Anglophone researchers in recent years have studied style in order to try to identify the different functions that it can fulfill at different scales and in different social contexts. In sharp contrast to the pattern searches that have prevailed on the Continent, style is usually defined here by its determining processes, rather than by its material conditions.

The 25-10 kyr BP interval in Europe brackets the pleniglacial maximum, the tardiglacial and the initial phases of the deglaciation, marked initially by virtual abandonment of northern and central Europe (25-21 kyr BP), and by population concentration in the Franco-Cantabrian refugium (21-13 kyr BP), followed by recolonization of areas formerly abandoned or depopulated during what Gamble (1986, 1991) has called the 'upturn' (13-10 kyr BP). Both parietal and portable art are heavily concentrated in the southwestern refugium, and generally date to 18-13 kyr BP. Jochim (1983, 1987) argued that, as populations moved into southwestern Europe from elsewhere, increases in population density would have led to socioeconomic problems of various kinds, as increasing numbers of people were packed into topographically constrained regions like the Dordogne and the north Spanish coastal strip. The art would have played a role in conflict resolution as the social geography of the region came to be characterized more and more by relatively closed social networks. This, he argued, would have selected for alliance formation and negotiation, essentially because these populations would have had nowhere else to go (i.e., they couldn't have resolved conflicts in the traditional forager way – by emigration). Conkey's studies of the diversity of design attributes on Cantabrian portable art suggest that sites with high diversity indices were places where people from many different identity-conscious social units came together periodically for a multiplicity of reasons (1980, see also 1984, 1985). Called 'aggregation sites', their presence lends support to an information-exchange type explanation of the art as linked to rituals of demarcation and boundary negotiaton (possibly of economic territories), conflict resolution in general, the sharing of subsistence information and a host of other factors. It is noteworthy that during eight or nine previous Middle and Upper Pleistocene glacial-interglacial fluctuations, Europe, and presumably other middle-latitude temperate regions of the Old World, had been largely or completed abandoned by human populations. Thus the appearance of art and aggregation sites is a significant change in the way human foragers responded to severe environmental stress. It signals the appearance of an adaptive complex not seen before among Upper Pleistocene foragers.

Portable art is only one part of the picture, however. That parietal art is concentrated when and where it is also demands an explanation. Although one could argue a priori that the time-space distribution of parietal art is probably connected in some way to the demographic factors which I take to be causal, we have to explain why it is almost exclusively a refuge period, Franco-Cantabrian phenomenon. Here it is useful to invoke Polly Wiessner's distinction between *assertive* and *emblemic* style and, recalling Conkey's argument for aggregation sites, suggest that channels of information flow revealed in the distributions of parietal and portable art are overlapping but distinct. In other words, the two broad art categories are monitoring two different aspects of Franco-Cantabrian social geography, manifest at two different scales. We think that the portable art, mainly ornaments, probably indicates the presence of Wiessner's assertive style – formal variation in material culture which is "personally-based and which carries information supporting individual identity" (1983: 258). It functions at an idiosyncratic level to differentiate individuals from similar others. Assertive style is found in all societies and, indeed, throughout Europe and across the time interval of interest here. It cross-cuts identity-conscious social units. Since it functions on both intra- and intergroup levels, it is a universal phenomenon in modern human contexts, and is completely unconnected to the relative degree of closure of social networks.

We think the parietal art – not necessarily found at sites with lots of portable art – probably indicates Wiessner's emblemic style: "formal variation in material culture that has a distinct referent and that transmits a clear message to a defined target population" (1983: 257). By definition, emblemic style always carries a symbolic loading, whereas assertive style does not necessarily do so. It could thus be argued that emblemic style manifest in parietal art facilitates intergroup communication – messaging across social unit boundaries. If indeed it does this, we would expect examples of emblemic style to be much more common under conditions of high population density and social network closure than they would be under low population density and more open social networks. Parietal art is a Franco-Cantabrian phenomenon, heavily concentrated in the 18-13 kyr BP interval. It disappears in the refugium after c. 10 kyr BP, suggesting that the conditions that selected for its appearance were no longer present after the LGM. The ornaments, however, persist into the postglacial, indicating a return to the open social networks characteristic of the early last glacial (i.e., 30-21 kyr BP) (Newell et al. 1990). In other words, the model postulates open systems with relatively low population densities and little parietal art for the 30-21 kyr BP interval, followed by a closing of social networks because of increases in population density, and by a high incidence of both portable and parietal art during the refuge phase (20-13 kyr BP). The deglaciation (13-10 kyr BP) sees the reappearance of more open social networks, the disappearance of the parietal art, and a dispersal of the concentrated 'refuge phase' population during the Mesolithic.

The information exchange approach to the art taken here is wholly consistent with the adaptationist biases that underlie the evolutionary ecology paradigm. It depends upon differentiating the art (i.e., making a distinction between assertive and emblemic style), arguing that emblemic style is, among other things, a monitor of demographic stress, and that the appearance of parietal art is an indication of social network closure under conditions of increasing population density. We are not suggesting that the appearance of parietal art is the only response of foraging peoples to demographic stress. However, its occurrence might mark episodes of accelerated demographic change, especially if those changes are in the direction of higher population densities relative to carrying capacity. The population-resource imbalances so created might select for increased emphasis on emblemic style traditions to the extent that it becomes visible in the archaeological record.

Summary

At the beginning of this chapter, I identified four salient features of the Iberian Mesolithic that call for explanation:

(1) vectored changes in the paleoeconomy in the direction of resource intensification manifest in increasing dietary diversity,

(2) the common aspects of the lithic technologies – notably the widely observed dichotomy between large, crude, cobble and flake tool-dominated assemblages made in quartzite, and microlithic bladelet industries made in flint,

(3) the comings and goings of the art – why it shows up when and where it does, and why parietal art apparently disappears at the LUP-Mesolithic boundary, and

(4) the transition to domestication economies – different in timing and detail but similar in pattern throughout much of Atlantic coastal Iberia.

Each of us has his or her own answers to these questions and, in this essay, I attempt to develop an integrated approach that explains them all. Whether or not it is realistic to attempt to do this is arguable, of course, and readers must judge for themselves the adequacy of the empirical support marshaled in support of these contentions. Since I see archaeology as a 'science-like' endeavor, I regard general explanations as a 'regulatory ideal' – something to strive for (Clark 1993). Others, who view archaeology as 'history-like', might not agree that it is possible or even worthwhile to do this.

The answers I would propose to the four questions posed above are as follows:

Vectored Paleoeconomic Change

I believe that resource intensification manifest as increasing dietary diversity in Spain and Portugal is forced upon Mesolithic foragers by population-resource imbalances caused by a combination of environmental change (esp. Holocene climatic amelioration, marine transgression) and population growth. Over time, these changes forced foragers along parts of the Atlantic façade to go further down the trophic pyramid, procuring greater quantities of 'expensive' (high cost, low yield) resources, and searching for ways to wring more nutrients out of 'traditional' resources in order to feed a growing regional population. In Cantabria, this process was set in motion by an influx of 'refuge period' immigrants from France, who packed the narrow coastal strip with dense human populations that essentially had nowhere to go, and that remained in the region during the deglaciation and on into the Holocene. Relatively high population densities first become evident in Cantabria and Asturias during the Solutrean, around 20,000 years ago, and they increase sharply in the Preboreal, Boreal and Atlantic phases. In Portugal and Galicia, perhaps less affected by LGM migration than Cantabria, a principal cause of population-resource imbalances was marine transgression that flooded the broad, shallow and exceptionally productive estuarine habitats of the Tejo, Minho and Douro rivers, and decreased the yield from these rich environments.

Common Aspects of the Lithic Technologies

The dichotomy between large, crude, quartzite 'heavy duty' tools, and small, microlith-dominated flint industries is a pattern that goes back in time to the LGM (c. 20 kyr BP), if not before.[2] Among other things, it reflects different activity suites, and the constraints on artifact form imposed by differences in mobility mitigated by a fundamental raw material dichotomy. Microliths always occur in combination with macroliths (i.e., no wholly microlithic assemblages are known), and the macrolithic component of an assemblage can remain fairly stable over very long periods of time while the incidence of microliths can changes radically. The replaceable elements of complex, multicomponent tools made from organic materials like wood, bone, antler, the appearance of assemblages in which microliths are important or dominant can most satisfactorily be explained by time-stress, which would have selected for flexible, 'plug-in, pull-out' technologies in order to mitigate 'risk' of artifact failure at the moment of application, and/or 'cost' of manufacture and maintenance in the field (Torrence 1983, Bamforth & Bleed 1997). Microliths occur in small numbers in some of the earliest Upper Paleolithic industries both in Europe and the Levant.[3] However, in Iberia, the demographic and social conditions that resulted in their appearance apparently did not exist much before c. 20 kyr BP, underscoring the historically contingent nature of their comings and goings. There is probably no single explanation for the appearance of microlithic technologies (other than the general one proposed above), nor for their proliferation at the end of the Pleistocene.

Pattern in the Evidence for Art

I believe this is most satisfactorily explained by adopting an information-exchange perspective that views art as a monitor of the volume of information flow circulating through channels cut by alliance networks of various kinds. The density of alliance networks is a function of population density. It is essential to differentiate parietal from mobiliary art in terms of the social contexts in which the two broad art categories functioned. Pattern in the parietal art – used in *intergroup* messaging – is caused by (i) the closing of social networks during the LGM in the Franco-Cantabrian refugium due to an influx of people, with a concomitant increase in population density that exacerbated resource competition and resulted in increasing territoriality, followed by (ii) the subsequent relaxation of selective pressures favoring the expression of Wiessner's 'emblemic style' (the parietal art is an example of emblemic style) during the post-glacial. So-called 'sanctuary sites' with concentrations of parietal art show some correlation with economic territories bounded by drainage systems (Straus 1987). Ornaments (beads, mostly) are the most common kind of mobiliary art. Used in *intragroup* messaging, they persist in quantity throughout the LUP and into the Mesolithic and beyond, and probably represent an example of Wiessner's 'assertive style', which functions to differentiate individuals from similar others.

The Transition to Domestication Economies

On any index of efficiency, fully developed domestication economies are inefficient and labour-intensive whereas humans are fundamentally lazy – they do not invest any more effort in the food quest than they have to. The adoption of and eventual reliance upon domestication economies comprises a temporal-spatial mosaic in the Iberian context, although generally much delayed along the Atlantic façade when compared with that of the Mediterranean coast. Whatever the case in a particular region, I view domestication as a 'last resort' forced upon those foragers who have run out of options for further dietary diversification and intensification. Domestication is perhaps the ultimate form of dietary intensification and – eventually – reliance upon domesticates becomes the only remaining option both globally and in Iberia, given a knowledge of domesticated plants and animals introduced from elsewhere. The domestication process itself and its proximate causes are, in my view, most adequately modeled by Zvelebil's availability, substitution and consolidation phases (e.g., 1986). The ultimate cause of the transition is population-resource imbalances that, in Iberia, can be traced to the spread of agropastoral economies first along the Mediterranean littoral, and subsequently inland, following the valley of the Ebro River.

Endnotes

1. *Pace* Madariaga (1976), it is virtually certain that the large, crude Asturian picks, made on quartzite cobbles, were not used to detach limpets – an activity most easily accomplished with a wood or bone splinter, a flake or a blade.
2. Microliths occur in low frequencies as early as 40,000 years ago in some EUP assemblages, both in Europe (e.g., Straus 2002) and the Levant (e.g., Belfer-Cohen & Goring-Morris 2002). Thus their high incidence in some Mesolithic and Epipaleolithic assemblages does not represent technological innovation, but rather a frequency shift over time related to a frequency shift in the (multifunctional) armatures in which they were mounted.
3. Supposedly an Aurignacian diagnostic, Dufour bladelets are microliths. Their occurrence in small numbers in EUP industries throughout western Eurasia might simply reflect inadequate recovery techniques in use prior to the 1960s (i.e., failure to water screen). Microliths are clearly less common in the EUP than they are in the LUP or Mesolithic, perhaps signaling a shift over time toward more reliance upon composite tools, weapons and armatures.

References

Al-Nahar, M.
2000 *The Upper and Epipaleolithic Transition in the Southern Levant: Microlith Typology versus Technology.* Ph.D dissertation, Department of Anthropology, Arizona State University. Ann Arbor: University Microfilms.

Arias, P.
1999 The origins of the Neolithic along the Atlantic coast of continental Europe: a survey. *Journal of World Prehistory* 13: 403-464.

Arias, P. & C. Pérez
1995 Excavaciones arqueológicas en Arangas, Cabrales: las cuevas de Los Canes, El Tiu Llines y Arangas. In *Excavaciones Arqueológicas en Asturias 1991-1994,* pp. 79-92. Oviedo: Principado de Asturias.

Arnaud, J.
1985 Mesolithic in Portugal: a report on recent research. *Mesolithic Miscellany* 6: 11-15.
1986 Post-glacial adaptations in southern Portugal. World Archaeological Congress. Southampton (preprint).
1989 The Mesolithic communities of the Sado Valley, Portugal, in their ecological setting. In *The Mesolithic in Europe* (C. Bonsall, ed.), pp. 614-631. Edinburgh: John Donald Publishers.
1990 Le substrat mésolithique et le processus de néolithisation dans le sud du Portugal. In *Rubané et Cardial* (D. Cahen & M. Otte, eds.), pp. 437-446. Liège: ERAUL No. 39.
1993 O Mesolítico e a neolitização: balanço e perspectives. In *O Quaternário em Portugal: Balanços e Perspectivas* (G. Carvalho, A. Ferreira & J. Senna-Martínez, eds.), pp. 173-184. Lisboa: Edições Colibri.

Bailey, G.
1983a Long term economic change: demographic and environmental factors. In *Hunter-Gatherer Economy in Prehistory: a European Perspective* (G. Bailey, ed.), pp. 127-130. Cambridge: Cambridge University Press.
1983b Economic change in late Pleistocene Cantabria. In *Hunter-Gatherer Economy in Prehistory: a European Perspective* (G. Bailey, ed.), pp. 149-165. Cambridge: Cambridge University Press.

Bamforth, D. & P. Bleed
1997 Technology, flaked stone technology, and risk. In *Rediscovering Darwin: Evolutionary Theory and Archeological Explanation* (C. M. Barton & G. A. Clark, eds.), pp. 109-139. Arlington: AP3A No. 7.

Bamforth, F., M. Jackes & D. Lubell
2003 Mesolithic-Neolithic population relationships in Portugal: the evidence from ancient mitochondrial DNA. In *Mesolithic on the Move* (L. Larsson et al., eds.), pp. 581-587. Oxford: Oxbow Books.

Barandiarán, I. & A. Cava
1989 The evolution of the Mesolithic in the north-east of the Iberian Peninsula. In *The Mesolithic in Europe* (C. Bonsall, ed.), pp. 572-581. Edinburgh: John Donald Publishers.

Barton, C. M. & M. P. Neeley
1996 Phantom cultures of the Levantine Epipaleolithic. *Antiquity* 70: 139-147.

Barton, C. M., G. A. Clark & A. Cohen
1994 Art as information: explaining Upper Paleolithic art in western Europe. *World Archaeology* 26: 185-207.

Belfer-Cohen, A. & A. N. Goring-Morris
2002 Why microliths? Microlithization in the Levant. In *Thinking Small: Global Perspectives on Microlithization* (R. Elston & S. Kuhn, eds.), pp. 57-68. Arlington: AP3A No. 12.

Bicho, N.
1991 Areëiro III, an open air site dated to 8,850 BP (Río Maior, Portugal. *Mesolithic Miscellany* 12: 1-10.
1992 *Technological Change in the Final Upper Paleolithic of Río Maior, Portuguese Estremadura.* Ph.D Dissertation, Department of Anthropology, Southern Methodist University. Ann Arbor: University Microfilms.
1993 Late glacial prehistory of central and southern Portugal. *Antiquity* 67: 761-775.
1994 The end of the paleolithic and the Mesolithic in Portugal. *Current Anthropology* 35: 664-674.

Bleed, P.
1986 The optimal design of hunting weapons: maintainability or reliability. *American Antiquity* 51: 737-747.

Boëda, E., J.-M. Geneste & L. Meignen
1990 Identification de chaînes opératoires lithiques du paléolithique ancien et moyen. *Paleo* 2: 43-80.

Bonsall, C. (editor)
1989 *The Mesolithic in Europe.* Edinburgh: John Donald Publishers.

Breuil, H. & G. Zybszewski
1942 Contribution à l'étude des industries paléolithiques du Portugal et leurs rapports avec

la géologie du Quaternaire. *Comunicações dos Serviços Geológicos de Portugal* 23: 1-394.

1946 Contribution à l'étude des industries paléolithiques des plages quaternaires de l'Alentejo littoral. *Comunicações dos Serviços Geológicos de Portugal* 27: 269-334.

Broadbent, N.

1978 Prehistoric settlement in northern Sweden: a brief survey and a case study. In *The Early Postglacial Settlement of Northern Europe* (P. Mellars, ed.), pp. 177-204. London: Gerald Duckworth.

Carr, C. & J. Neitzel (editors)

1995 *Style, Society & Person.* New York: Plenum.

Carvalho, A.

2002 Current perspectives on the transition from the Mesolithic to the Neolithic in Portugal. In *Neolithic Landscapes of the Mediterranean* (E. Badal et al., eds.), pp. 235-250. València: Saguntum, Papeles del Laboratoro de Arqueología de València, Universitat de València.

Clark, G. A.

1976 *El Asturiense Cantábrico.* Madrid: Consejo Superior de Investigaciones Científicas, Ministerio de Cultura, Bibliotheca Praehistorica Hispana No. 13.

1983a *The Asturian of Cantabria: Early Holocene Hunter-Gatherers in Northern Spain.* Tucson: Anthropological Papers of the University of Arizona No. 41.

1983b Boreal phase settlement-subsistence models in Cantabrian Spain. In *Hunter-Gatherer Economy in Prehistory: a European Perspective* (G. Bailey, ed.), pp. 96-110. Cambridge: Cambridge University Press.

1987 From the Mousterian to the Metal Ages: long-term change in the human diet of Cantabrian Spain. In *The Pleistocene Old World: Regional Perspectives* (O. Soffer, ed.), pp. 293-316. New York: Plenum.

1989 Site functional complementarity in the Mesolithic of northern Spain. In *The Mesolithic in Europe* (C. Bonsall, ed.), pp. 589-603. Edinburgh: John Donald Publishers.

Clark, G. A. (editor)

1991 *Perspectives on the Past.* Philadelphia: University of Pennsylvania Press.

Clark, G. A.

1993 Paradigms in science and archaeology. *Journal of Archaeological Research* 1: 203-234.

1999a Modern human origins: highly visible, curiously intangible. *Science* 283: 2029-2032; 284: 917.

1999b Paleolithic population growth. *Science* 284: 1467.

2000 Thirty years of Mesolithic research in Atlantic coastal Iberia. *Journal of Anthropological Research* 56: 17-38.

2002a Observations on paradigmatic bias in French and American paleolithic archaeology. In *The Role of American Archeologists in the Study of the European Upper Paleolithic* (L. Straus, ed.), pp. 19-26. Oxford: BAR International Series No. 1048.

2002b Neandertal archaeology – implications for our origins. *American Anthropologist* 104: 50-67.

Clark, G. A., C. M. Barton & A. Cohen

1996 Explaining art in the Franco-Cantabrian refugium: an information exchange model. In *Debating Complexity: Proceedings of the 26th Annual Chacmool Conference* [D. Meyer, P. Dawson & D. Hanna, eds.], pp. 241-253. Calgary: Archaeological Association of the University of Calgary.

Clark, G. A. & M. P. Neeley

1987 Social differentiation in European Mesolithic burial data. In *Mesolithic Northwest Europe: Recent Trends* (P. Rowley-Conwy, M. Zvelebil & H. P. Blankholm, eds.), pp. 121-127. Sheffield: John Collis Ltd.

Clark, G. A. & L. G. Straus

1983 Late Pleistocene hunter-gatherer adaptations in Cantabrian Spain. In *Hunter-Gatherer Economy in Prehistory: a European Perspective* (G. Bailey, ed.), pp. 131-148. Cambridge: Cambridge University Press.

Clark, G. A. & L. G. Straus

1986 Synthesis and conclusions - Part I: Upper Paleolithic and Mesolithic hunter-gatherer subsistence in northern Spain. In *La Riera Cave: Stone Age Hunter-Gatherer Adaptations in Northern Spain* (L. Straus & G. Clark, eds.), pp. 351-366. Tempe: Arizona State University Anthropological Research Papers No. 36.

Clark, G. A. & C. M. Willermet (editors)

1997 *Conceptual Issues in Modern Human Origins Research.* New York: Aldine de Gruyter.

Clark, J. G. D.

1972 *Star Carr: a Case Study in Bioarchaeology.* Reading: Addison-Wesley Anthropology Module No. 10.

Clarke, D.

1976 Mesolithic Europe: the economic basis. In *Problems in Economic and Social Archaeology* (G. Sieveking, I. Longworth & K. Wilson, eds.), pp. 449-481. London: Gerald Duckworth.

Conkey, M.

1980 The identification of prehistoric hunter-gatherer aggregation sites: the case of Altamira. *Current Anthropology* 21: 609-620.

1984 To find ourselves: art and social geography of prehistoric hunter-gatherers. In *Past and Present in Hunter-Gatherer Studies* (C. Schrire, ed.), pp. 253-276. New York: Academic Press.

1985 Ritual communication, social elaboration, and the variable trajectories of paleolithic material culture. In *Prehistoric Hunter-Gatherers: the Emergence of Cultural Complexity* (T. D. Price & J. Brown, eds.), pp. 299-323. New York: Academic Press.

Conkey, M. & C. Hastorf (editors)

1990 *The Uses of Style in Archaeology.* Cambridge: Cambridge University Press.

Díez Castillo, A.

1999 *Utilización de los Recursos en la Marina y Montaña Cantábricas.* Gernika.

Elston, R. & S. Kuhn (editors)

2002 *Thinking Small: Global Perspectives on Microlithization.* Arlington: AP3A No. 12.

Ferembach, D.

1974 *Le Gisement Mésolithique de Moita do Sebastião, Muge, Portugal. II Anthropologie.* Lisboa: Publicações do Instituto de Alta Cultura.

Figueiral, I.

1993 Cabeço de Porto Marinho: une approche paléoécologique – prémiers résultats. *Estudios sobre el Cuaternario* 8: 167-172.

Gamble, C.

1986 *The Palaeolithic Settlement of Europe.* Cambridge: Cambridge University Press.

1991 The social context for European palaeolithic art. *Proceedings of the Prehistoric Society* 57: 3-16.

Geneste, J.-M.

1985 *Analyse Lithique des Industries Moustériennes du Périgord: une Approche Technologique du Comportement des Groupes Humains au Paléolithique Moyen.* Ph.D dissertation, University of Bordeaux I.

González Echegaray, J. & L. G. Freeman

1971 *Cueva Morín: Excavaciones 1966-1968.* Santander: Diputación Provincial de Santander.

1973 *Cueva Morín: Excavaciones 1969.* Santander: Diputación Provincial de Santander.

González Morales, M.

1982 *El Asturiense y Otras Culturas Locales.* Santander: Centro de Investigación y Museo de Altamira, Monografías No. 10.

1989 Asturian resource exploitation: recent perspectives. In *The Mesolithic in Europe* (C. Bonsall, ed.), pp. 604-606. Edinburgh: John Donald Publishers.

1991 From hunter-gatherers to food producers in northern Spain: smooth adaptive shifts or revolutionary change in the Mesolithic. In *Perspectives on the Past* (G. A. Clark, ed.), pp. 204-216. Philadelphia: University of Pennsylvania Press.

González Morales, M. & J. Arnaud

1990 Recent research on the Mesolithic of the Iberian Peninsula: problems and perspectives. In *Contributions to the Mesolithic in Europe* (P. Vermeersch & P. van Peer, eds.), pp. 451-461. Leuven: Leuven University Press.

Gramsch, B. (editor)

1981 *Mesolithikum in Europa 2. Internationales Symposium Potsdam.* Berlin: Deutscher Verlag der Wissenschaften.

Higgs, E. (editor)

1972 *Papers in Economic Prehistory.* Cambridge: Cambridge University Press.

1975 *Paleoeconomy.* Cambridge: Cambridge University Press.

Indrelid, S.

1978 Mesolithic economy and settlement patterns in Norway. In *The Early Postglacial Settlement of Northern Europe* (P. Mellars, ed.), pp. 147-176. London: Gerald Duckworth.

Jackes, M. & D. Lubell

1999 Human skeletal biology and the Mesolithic-Neolithic transition in Portugal. In *L'Europe des Derniers Chasseurs* (A. Thévenin, ed.), pp. 59-64. Paris: Éditions du CTHS.

Jackes, M., D. Lubell & C. Meiklejohn

1997 On physical anthropological aspects of the Mesolithic-Neolithic transition in the Iberian Peninsula. *Current Anthropology* 38: 839-846.

Jochim, M.

1983 Palaeolithic cave art in ecological perspective. In *Hunter-Gatherer Economy in Prehistory: a European Perspective* (G. Bailey, ed.), pp. 212-219. Cambridge: Cambridge University Press.

1987 Late Pleistocene refugia in Europe. In *The Pleistocene Old World: Regional Perspectives* (O. Soffer, ed.), pp. 317-331. New York: Plenum.

Kozlowski, S.
2003 The Mesolithic: what do we know and what do we believe? In *Mesolithic on the Move* (L. Larsson et al., eds.), pp. xvii-xxi. Oxford: Oxbow Books.

Kozlowski, S. (editor)
1973 *The Mesolithic in Europe.* Warsaw: Warsaw University Press.

Kuhn, S.
1992 On planning and curated technologies in the Middle Paleolithic. *Journal of Anthropological Research* 48: 185-214.
1995 *Mousterian Lithic Technology.* Princeton: Princeton University Press.

Kuhn, S. & R. Elston
2002 Introduction: thinking small globally. In *Thinking Small: Global Perspectives on Microlithization* (R. Elston & S. Kuhn, eds.), pp. 1-8. Arlington: AP3A No. 12.

Larsson, L.
1990 The Mesolithic of southern Scandinavia. *Journal of World Prehistory* 4: 257-310.

Larsson, L., H. Kindgren, K. Knutsson, D. Loeffler & A. Akerlund (editors)
2003 *Mesolithic on the Move.* Oxford: Oxbow Books.

Lubell, D. & M. Jackes
1985 Mesolithic-Neolithic continuity: evidence from chronology and human biology. In *Actas da I[a] Reunião do Quaternario Iberico,* pp. 113-133. Lisboa: Instituto Nacional de Investigação Cientific.
1988 Portuguese Mesolithic-Neolithic subsistence and settlement. *Rivista de Antropologia* 66: 231-248.

Lubell, D., M. Jackes & C. Meiklejohn
1989 Archaeology and human biology of the Mesolithic-Neolithic transition in southern Portugal: a preliminary report. In *The Mesolithic in Europe* (C. Bonsall, ed.), pp. 632-640. Edinburgh: John Donald Publishers.

Madariaga, B.
1976 Consideraciones acerca de la utilización del 'pico marisquero' del Asturiense. In *XL[o] Aniversario del Centro de Estudios Montañeses,* pp. 435-451. Santander: Diputación Provincial.

Mateus, J.
1989 Lagoa Travessa: a Holocene pollen diagram from the southwest coast of Portugal. *Revista de Biología* 14: 17-94.

Maury, J.
1977 *The Asturian in Portugal: Typology and Chronology.* Oxford: BAR International Series No. 21.

Mithen, S.
1993 Individuals, groups and the palaeolithic record: a reply to Clark. *Proceedings of the Prehistoric Society* 59: 393-398.

Neeley, M. P.
1997 *Assigning Meaning to Lithic Variability in the Epipaleolithic of the Southern Levant.* Ph.D dissertation, Department of Anthropology, Arizona State University. Ann Arbor: University Microfilms.

Neeley, M. P. & C. M. Barton
1994 A new approach to interpreting late Pleistocene microlith industries in southwest Asia. *Antiquity* 68: 275-288.

Neeley, M. P. & G. A. Clark
1990 Measuring social complexity in the European Mesolithic. In *Contributions to the Mesolithic in Europe* (P. Vermeersch & P. van Peer, eds.), pp. 127-138. Leuven: Leuven University Press.

Newell, R., D. Kielman, T. Constandse-Westermann, W. van der Sanden & A. van Gijn
1990 *An Inquiry into the Ethnic Resolution of Mesolithic Regional Groups: the Study of Their Decorative Ornaments in Time and Space.* Leiden: E. J. Brill.

Price, T. D.
1987 The Mesolithic of western Europe. *Journal of World Prehistory* 1: 225-306.

Raposo, L.
1988 A pré-história antiga na Costa Sudoeste. *Anais da Real Sociedade Arqueológica Lusitana* 2: 17-36.

Raposo, L., C. Penalva & J. Pereira
1989 Noticia da descoberta da estação Mirense de Palheirões do Alegra, Cabo Sardão (Odemira, Portugal). In *Actas de la 2[a] Reunión del Cuaternario Ibérico,* pp. 25-29. Madrid: Sociedad Española del Quaternario.

Redding, R.
1988 A general explanation for subsistence change: from hunting and gathering to food production. *Journal of Anthropological Archaeology* 7: 56-97.

Richards, M., P. Petitt, M. Stiner & E. Trinkaus
2001 Stable isotope evidence for increasing dietary breadth in the European mid-Upper Paleolithic. *Proceedings of the National Academy of Science* 98: 6528-6532.

Roche, J.
1972 Les amas coquilliers (concheiros) mésolithiques de Muge (Portugal). *Fundamenta* 3: 72-107.
1982 A gruta chamada Lapa do Suão (Bombarral). *Arqueologia* 5: 5-18.

Saxe, A.

1971 Social dimensions of mortuary practices in a Mesolithic population from Wadi Halfa, Sudan. In *Approaches to the Social Dimensions of Mortuary Practices* (J. Brown, ed.), pp. 39-57. Washington: Memoirs of the Society for American Archaeology No. 25.

Serpa Pinto, R.

1928 O Asturiense em Portugal. *Trabalhos da Sociedade Portuguesa de Antropología e Etnología* 4: 1-47.

Silva, C. & J. Soares

1987 Les Communautés du Néolithique Ancien dans le Sud du Portugal. In *Les Premières Communautés Paysannes en Méditerranée Occidental,* pp. 663-671. Paris: Éditions du CNRS.

Stiner, M. N. Bicho, J. Lindly & R. Ferring

2003 Mesolithic to Neolithic transitions: new results from shell-middens in the western Algarve, Portugal. *Antiquity* 77: 75-86.

Straus, L. G.

1977 Of deerslayers and mountain men: paleolithic faunal exploitation in Cantabrian Spain. In *For Theory Building in Archaeology* (L. Binford, ed.), pp. 41-76. New York: Academic Press.

1979 Mesolithic adaptations along the northern coast of Spain. *Quaternaria* 21: 305-327.

1983 Terminal Pleistocene faunal exploitation in Cantabria and Gascony. In *Animals and Archaeology I: Hunters and their Prey* (J. Clutton-Brock & C. Grigson, eds.), pp. 209-225. Oxford: BAR International Series No. 163.

1985 Chronostratigraphy of the Pleistocene-Holocene boundary: the Azilian problem in the Franco-Cantabrian region. *Palaeohistoria* 27: 89-122.

1986 The paleolithic cave art of Vasco-Cantabrian Spain. *Oxford Journal of Archaeology* 6: 149-163.

1991a Epipaleolithic and Mesolithic adaptations in Cantabrian Spain and Pyrenean France. *Journal of World Prehistory* 5: 83-104.

1991b The 'Mesolithic-Neolithic transition' in Portugal: a view from Vidigal. *Antiquity* 65: 899-903.

2002 Selecting small: microlith musings for the Upper Paleolithic and Mesolithic of western Europe. In *Thinking Small: Global Perspectives on Microlithization* (R. Elston & S. Kuhn, eds.), pp. 69-82. Arlington: AP3A No. 12.

Straus, L. G. & G. A. Clark

1986 Synthesis and conclusions - Part II: the La Riera excavation, chronostratigraphy, paleoenvironments and cultural sequence in perspective. In *La Riera Cave: Stone Age Hunter-Gatherer Adaptations in Northern Spain* (L. Straus & G. Clark, eds.), pp. 367-384. Tempe: Arizona State University Anthropological Research Papers No. 36.

Straus, L. G. & M. González Morales

2003 The Mesolithic in the Cantabrian interior: fact or fantasy? In *Mesolithic on the Move* (L. Larsson et al., eds.), pp. 359-368.

Straus, L., G. Clark, J. Altuna & J. Ortea

1980 Ice age subsistence in northern Spain. *Scientific American* 242: 142-152.

Street, M. et al.

2001 Final Paleolithic and Mesolithic in reunified Germany. *Journal of World Prehistory* 15: 365-453.

Torrence, R.

1983 Time budgeting and hunter-gatherer technology. In *Hunter-Gatherer Economy in Prehistory* (G. Bailey, ed.), pp. 11-22. Cambridge: Cambridge University Press.

1989 Re-tooling towards a behavioral theory of stone tools. In *Time, Energy and Stone Tools* (R. Torrence, ed.), pp. 57-66. Cambridge: Cambridge University Press.

Vermeersch, P. & P. Van Peer (editors)

1990 *Contributions to the Mesolithic in Europe.* Leuven: Leuven University Press.

Vierra, B.

1992 *Subsistence Diversification & the Evolution of Microlithic Technologies: a Study of the Portuguese Mesolithic.* Ph.D Dissertation, Department of Anthropology, University of New Mexico. Ann Arbor: University Microfilms.

1995 *Subsistence and Stone Tool Technology: an Old World Perspective.* Tempe: Arizona State University Anthropological Research Papers No. 47.

Welinder, S.

1978 The concept of 'ecology' in Mesolithic research. In *The Early Postglacial Settlement of Northern Europe* (P. Mellars, ed.), pp. 11-26. London: Gerald Duckworth.

Wiessner, P.

1983 Style and social information in Kalahari San projectile points. *American Antiquity* 48: 253-276.

Zilhão, J.

1992 Estratégias de povoamento de subsistência no paleolítico e no mesolítico de Portugal. In *Elefantes, Ciervos y Ovicaprinos* (J. A. Moure Romanillo, ed.), pp. 149-162. Santander: Prensa Universitaria de la Universidad de Cantabria.

2000 From the Mesolithic to the Neolithic in the Iberian Peninsula. In *Europe's First Farmers* (T. D.

Price, ed.), pp. 144-182. Cambridge: Cambridge University Press.

2001 Radiocarbon evidence for maritime pioneer colonization at the origins of farming in west Mediterranean Europe. *Proceedings of the National Academy of Science* 98: 14180-14185.

Zilhão, J., E. Carvalho & A. Araújo
1987 A estação epipaleolítica da Ponta da Vigia. *Arqueologia* 16: 8-18.

Zvelebil, M.
1986 Mesolithic prelude and Neolithic revolution. In *Hunters in Transition* (M. Zvelebil, ed.), pp. 5-16. Cambridge: Cambridge University Press.

Zvelebil, M. & P. Dolukhanov
1991 The transition to farming in eastern and northern Europe. *Journal of World Prehistory* 5: 233-278.

Zvelebil, M. & P. Rowley-Conwy
1986 Foragers and farmers in Atlantic Europe. In *Hunters in Transition* (M. Zvelebil, ed.), pp. 67-93.

17

Iberian Foragers and Funerary Ritual – a Review of Paleolithic and Mesolithic Evidence on the Peninsula

Pablo Arias & Estéban Alvárez-Fernández
University of Cantabria

Abstract

A critical assessment of the evidence for paleolithic and mesolithic funerary ritual is attempted for the Iberian Peninsula. Although recording is often inadequate and/or uneven, we show that Iberian late Mesolithic mortuary data is actually relatively common compared with those from the rest of Europe. A sharp increase in the incidence of intentional burial occurs during the 6th millennium cal BC, when the first evidence for agropastoralism shows up along the Mediterranean coast of Spain. The first true cemeteries also appear during this time, suggesting reduced mobility and population increases in some areas, accompanied by some evidence for increased social complexity.

Introduction

Although extremely rare, evidence for funerary ritual is one of the most important sources of knowledge about prehistoric cognitive evolution and social organization. Since the inception of the discipline in the last half of the 19th century, archaeologists have made the study of burials, burial attributes, grave contents and contexts a central part of their research programs (Bartel 1982). However, it has not been until recently that its potential for shedding light on the more elusive aspects of the past has been fully realized. The tomb is much more than a place to dispose of the dead; it is also intended "for the living" – as Fleming's (1973) excellent title so aptly states. Funerary behavior is one of the arenas in which social statuses and roles are negotiated. The grave is a context for the public display of symbols of wealth, power, and spirituality (Renfrew 1994). Since the 1970s, the 'archaeology of death' (Chapman et al. 1981) has provided the discipline with new methods and conceptual frameworks for the study of the social, ideological, and religious organizations of prehistoric communities. In fact, from both 'processual' and 'postprocessual' points of view, funerary ritual formalized in mortuary programs has become one of the main sources for social archaeology (Riel-Salvatore & Clark 2001).

Paleolithic and Mesolithic hunter-gatherers in Iberia have, so far, occupied a marginal position in this kind of research.[1] Yet, as we show here, evidence from this part of Europe is both more abundant and relevant than generally acknowledged in the literature on this topic. Our intention in this chapter is to present an overview of the current evidence concerning funerary behavior of prehistoric Iberian hunter-gatherers. A comprehensive, critical review is attempted that will encompass all evidence for mortuary ritual found in Upper Paleolithic and Mesolithic contexts on the Peninsula. We also discuss the implications of Iberian research for improved understanding of prehistoric hunter-gatherer symbolic expression and social organization.

Death in Iberia during the Ice Age

Funerary remains from the Iberian Upper Paleolithic are very scarce (Fig. 17.1). Thus far, burials of this age have been reported only in four sites despite the hundreds of sites that have been excavated since the mid-19th century in a part of Europe with a long tradition of archaeological research.[2]

Cueva Morín

The oldest relatively credible evidence comes from an early Aurignacian context, layer 8a at Cueva Morín (Cantabria, Spain).[3] There, J. González Echegaray and L.G. Freeman (1973) describe four trenches originally covered with elongated earthen barrows, in which they observed pseudomorphs (positive casts) of human and animal bodies, apparently preserved by a process of saponification. In Morín I, the excavators uncovered the cast of a cadaver of a tall human male, perhaps decapitated and with the feet amputated prior to burial. The body was lying on its left side, with the arms flexed over the breast, and the legs slightly bent. Several offerings were deposited in the grave: the pseudomorph of a small ungulate (an adult roe deer or a juvenile red deer, ibex or chamois) had been placed over

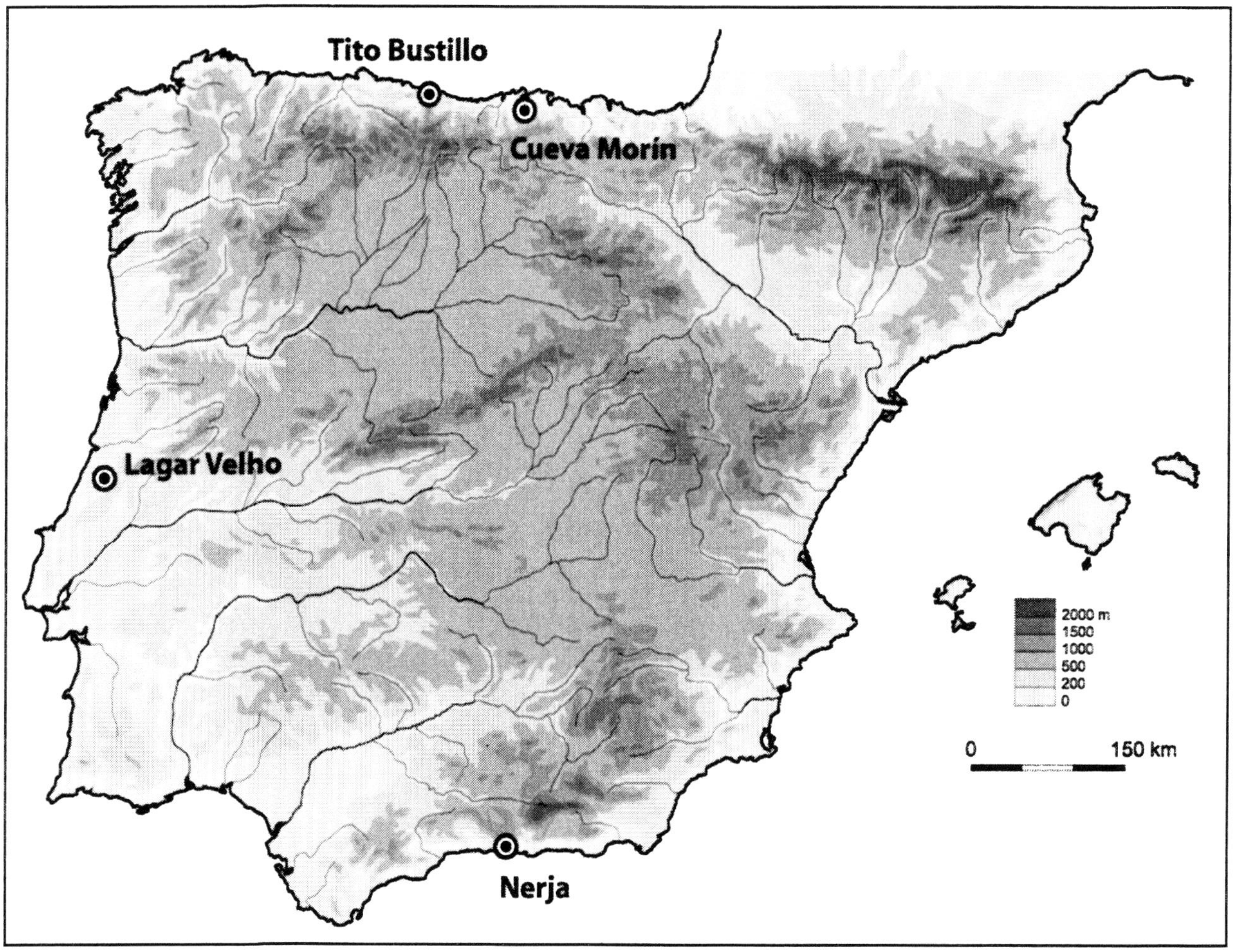

Figure 17.1 Iberian sites where Upper Paleolithic burials have been described.

the head, and some ribs of a large mammal and several artifacts (a quartzite blade, a flint sidescraper) had been placed alongside the body. The bottom of the trench connected to a little pit, supposedly dug for the purpose of containing additional offerings. A hearth dated to c. 29500 cal BC was probably associated with the grave.[4]

Morín III was interpreted as a fragment of an older burial, partly destroyed by the excavation of the pit containing Morín I. Pseudomorphs of both legs of an adult individual – again with the feet amputated and partially charred by fire – were uncovered by the excavators. A radiocarbon determination dates this feature to 28515 ± 1324 BP (SI-956).[4] The other two features (Morín II and IV) were intentionally excavated (digging stick marks were visible in the cave clay), oval shaped depressions which resembled the shallow pits containing the Morín I and III pseudomorphs, but without any bones or positive body casts (Freeman & Echegaray 1973: 221-254; Madariaga 1973: 256-265).

In recent years, considerable skepticism has arisen about the significance of the Morín pseudomorphs. In fact, they are not even mentioned in some general works on mortuary practices during the paleolithic (e.g., May 1986). The validity of the interpretation of the excavators has also been called into question (González Morales 1997, Djindjian et al. 1999: 170). There is some basis for those reservations. Although the 1973 monograph proposes an explanation of how the process of saponification might have occurred, the chemical analysis published by Madariaga (1973) is not conclusive, and no clear evidence has been forthcoming on the existence during the Aurignacian of the unusual, extremely wet, conditions that are required for the saponification process. It could be argued that the putrefaction of corpses in ditches dug into dense clays during the wet period that corresponds to layer 8a might have produced very localized processes leading to the transformation of a corpse into adipocere, but actually demonstrating this would probably require an experimental approach. Of course, that does not mean that the excavators' interpretation was necessarily wrong. One has to wonder what highly localized process might have destroyed bones and teeth in their entirety, given well-preserved bones and teeth less than a meter away. An experimental approach might be possible thanks to the foresight of the excavators and the generous funding of the Smithsonian Institution. Morín I is currently preserved in a block of transparent plastic in the National Museum of Altamira (Cantabria), so a reanalysis is a possibility, and probably advisable to dispel the doubts that have arisen over the years.

Lagar Velho

The best evidence for Upper Paleolithic mortuary practices on the Peninsula comes from the recently excavated Gravettian burial of Lagar Velho (Estremadura, Portugal), dated to around 25000/24500 BP (c. 27000/26400 cal BC) (Fig. 17.2) (Zilhão & Trinkaus 2002).[5] The skeleton of a child about five years old was found near the wall of the rockshelter, at the bottom of a pit some 30 cm deep. The corpse lay in the extended supine position, with the head slightly inclined to the left, towards the rockshelter wall. The legs were slightly flexed, with the feet together; the right hand rested alongside the hip. Several aspects of the grave context suggest ritualized behavior. It seems that a single pine branch *(Pinus silvestris)* was burned at the bottom of the grave prior to interment, as suggested by charcoal remains under the child's legs. Both the sediments and the bones were strongly stained with ochre, interpreted by the excavators as the remains of a painted shroud in which the corpse would have been wrapped. The semi-articulated skeleton of an immature rabbit *(Oryctolagus cuniculus),* found beside the right tibia, is perhaps the most obvious funerary offering. Two pelves of male red deer *(Cervus elaphus)* were recovered near the head and feet respectively, perhaps the remnants of meat offerings, although the excavators also suggest that they might be have been related to the construction of the grave. Personal adornment comprised four red deer canine pendants associated with the skull fragments, and two perforated marine gastropod shells *(Littorina obtusata)*, one of them found near the cervical vertebrae, and the other above the burial.

Figure 17.2 Gravettian burial at Lagar Velho (from Zilhão & Trinkaus 2002).

Nerja Cave

Solutrean mortuary data consist of four relatively complete skeletons (three adults – a male, a female and an individual of indeterminate sex – and a child) found in 1963 in Nerja cave, close to the coast of Málaga (Andalucía, Spain).[6] Unfortunately, the archaeological record from these excavations is very poor. Re-examination of the photographs and section drawings led González-Tablas (1990) to conclude that the cadavers were originally placed on the cave floor, without any pit or trench. He attributed the three adults, all laid out in the extended supine position, to the Solutrean; the child, in a fetal position, was assigned to the Upper Magdalenian. However, the possibility that the skeletons were buried in pits dug from upper, later layers cannot be excluded, as a later reappraisal acknowledges (Aura et al. 1994: 241). The cave was also used as a mortuary during the Mesolithic (see below), the Neolithic, and the Chalcolithic, raising the possibility of a postglacial age for the burials. Lending support to this suggestion is a direct radiocarbon date from human bone, UBAR-134: 7360 ± 830 BP (7308-5473 cal BC) (Turbón et al. 1994), too imprecise to be really useful, but clearly post-Paleolithic. New excavations directed by Francisco Jordá in 1984 recovered more human remains (metatarsals and phalanges from a human foot, probably corresponding to one of the adults) in what has been considered an undisturbed Solutrean layer (Aura et al. 1998). Three new radiocarbon dates for this context (NV8: 15990 ± 260 BP, 17940 ± 200 BP and 18420 ± 530 BP) and one for the layer above (NV7: 12130 ± 130 BP) unfortunately do not dispel all the doubts, since the relationship of the samples dated to the human remains is not clear (the dates all come from charcoal flecks gathered over a wide area). Although Aura and colleagues (1980) emphasize the existence of an erosional episode between NV8 and NV7, and suggest that the graves were opened between the deposition of the two layers, a later (post-Paleolithic) age cannot be excluded. The radiocarbon dates are inconclusive, since it is perfectly possible that post-Paleolithic burials dug into Solutrean and Magdalenian deposits resulted in the fortuitious association of Pleistocene charcoal with early Holocene skeletons. In this case, an AMS date on the human bones themselves would be highly desirable.

Tito Bustillo

The cave of Tito Bustillo (Asturias, Spain), one of the key sites for the Iberian Magdalenian, has recently provided the first funerary evidence for this period. Among the large accumulation of stones that blocks the original entrance, excavations directed by Rodrigo de Balbín have uncovered some small earthen mounds that, in at least one case, have yielded a Magdalenian burial (Balbín & Alcolea n.d.).

Possible Secondary Burial Sites

There are also some sites where human remains have been found associated with structures that are not, strictly speaking, graves, but that might be related to some kind of secondary interment. That is the case with *Cova Beneito* (Alicante, Spain), where two fragmentary crania were found in the early 1990s associated with a hammerstone, two cores, a large ochre nodule, a perforated shell, a blade and an endscraper. The crania and the associated artifacts were enclosed by limestone blocks in a Solutreo-Gravettian layer dated to 16560 ± 480 BP (Ly-3593; 17508 ± 529 cal BC). This feature has been interpreted as a secondary burial (Iturbe et al. 1993: 73).

More doubtful is the case of layer 9 of the cave of *Lapa do Suão* (Estremadura, Portugal), where Jean Roche found a pit containing two human teeth, ochre fragments, perforated shells (40 *Littorina obtusata*, some *Trivia monacha*, and one *Phalium granulatum*)[7] and three perforated teeth of *Lynx pardina*. This pit and its contents were interpreted as an intentional deposition (Roche 1982). The layer in which it appeared was attributed by Roche to the Magdalenian, but a recent revision by Zilhão (1997: 596-602) suggests that it should be re-classified as Solutrean.

New Times, New Practices: the Mesolithic Evidence

The Epipaleolithic Azilian

Epipaleolithic and Mesolithic funerary evidence is much more common than that of the Upper Paleolithic (Fig. 17.3). The earliest evidence comes from an Azilian grave in the *Los Azules* cave, in the Picos de Europa near Cangas de Onís (Asturias, Spain) (Fernández-Tresguerres 1976, 1980). Dated to the first half of the 9th millennium cal BC,[8] it consisted of a complete skeleton of an adult male buried in the extended supine position in an oval pit measuring c. 2 x 1 x 0.4 m, and associated with several Azilian painted cobbles, ochre, the skull of badger (*Meles meles*), unperforated marine shells (*Modiolus barbatus),* some artifacts – harpoons, endscrapers, burins – and manufacturing debris resulting from their fabrication. (Fig.17.4)[9] Fernández-Tresguerres suggested (plausibly enough) that the deceased had been provided with a kind of generalized 'daily tool kit', raw material for making other artifacts, and tools for making tools (1976, 1980).

Less precise information was reported long ago by Hernández-Pacheco (1923) from another key site for the Cantabrian Azilian, the cave of *La Paloma,* also in Asturias. The skeleton of a child with the deciduous molars still in place (thus likely < 10 years old) was recovered from what might have been a burial context excavated from the Azilian layer in this site, and extending into the Magdalenian layers below (Hernández-Pacheco 1923: 37). Unfortunately, the preservation of the human remains was very poor, and they can no longer be located.[10]

The Mediterranean Middle and Late Mesolithic

The cave of *Nerja* (see above) has provided the only early-middle Mesolithic burial in Mediterranean Spain, from a layer dated to 8260 ± 360 BP (7599-6698 cal BC, published with no laboratory reference).[11] The skeleton of a young adult female lying on her right side, with her arms and legs flexed, was protected by limestone blocks and, apparently, was associated with some hearths. The only possible offering described is a piece of ochre, found near the frontal, and some marine mollusc shells (Pellicer & Acosta 1986: 446-450; García 1982, González-Tablas 1990).

The open air site of *El Collado* (Valencia) dates to a later phase of the Mediterranean Middle Mesolithic. Although this most interesting site was excavated more than 15 years ago, no detailed report is available yet. Nevertheless, some preliminary papers (Aparicio 1990, 1992, Chimenos et al. 1992) describe what might be the first Mesolithic cemetery in the Spanish *Levante.* Fourteen individual graves in pits were found. The skeletons were all flexed or semi-flexed, and were accompanied by marine and terresrial molluscs (including the ornamental species *Columbella rustica*, *Nassarius reticulatus* or *Cerithium vulgatum*), some mammal bones and lithics (chipping debris, endscrapers, burins, backed bladelets, geometric microliths) (Fig. 17.5). Small freshwater gastropods (*Melanopsis dufouri*) were found beside the feet and around some of the skulls. Two ^{14}C dates from human bone (No. XIII – UBAR-280: 7570 ± 160 BP and UBAR-281: 7640 ± 120 BP; pooled mean: 7615 ± 96 BP; 6589-6273 cal BC) indicate an age in the Mediterranean Late Mesolithic.

The last reference for the Mediterranean Mesolithic appears to correspond to the period of the Mesolithic-Neolithic transition. In *Cingle del Mas Nou,* an open air site located in the mountains of Castellón near some of the most important concentrations of Levantine rock art, a recent excavation documents a so-far unique mortuary practice. A narrow, shallow, oval pit at the bottom of a late Mesolithic-early Neolithic (cardial stamped pottery) stratigraphic sequence contained the complete skeleton of a young adult male, in the extended supine position, with disarticulated bones of five individuals (skulls, some long bones) placed over his legs. The excavators interpret this as a secondary burial interred over a primary inhumation, although whether the two are associated is not clear from preliminary publications. The double burial feature, dated to c. 7000-6750 BP (5875-5650 cal BC), pertains to the Late Mesolithic occupation at the site (Olària & Gusi 2003).

The Cantabrian Mesolithic

The coastal Cantabrian Mesolithic, the Asturian (c. 8000-5000 cal BC) has also yielded intentional human interments. The most convincing of them was excavated in the 1920s at the Molino de Gasparín rockshelter, near the hamlet of *Colombres* (Carballo 1926: 131-138). It con-

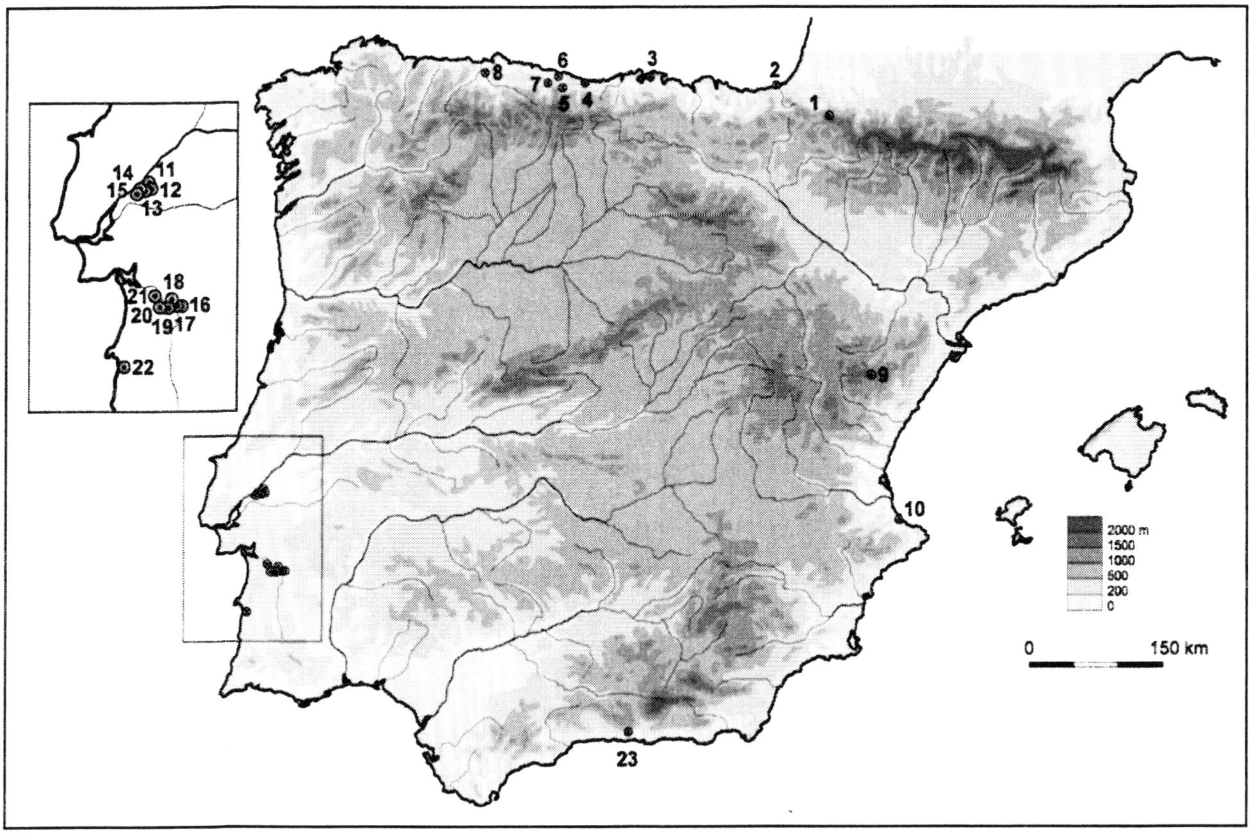

Figure 17.3 Iberian sites with Mesolithic burials. Key: 1– Aizpea, 2– Jaizkibel (J3), 3– El Truchiro, 4 – Molino de Gasparín (Colombres), 5 – Los Canes, 6 – Colomba, 7 – Los Azules, 8 – La Paloma, 9 – Cingle del Mas Nou, 10 – El Collado, 11 – Cabeço da Arruda, 12 – Cabeço da Amoreira, 13 – Moita do Sebastião, 14 – Fonte do Padre Pedro, 15 – Cova da Onça, 16 – Cabeço do Pez, 17 – Vale de Romeiras, 18 – Várzea da Mó, 19 – Amoreiras, 20 – Poças de São Bento, 21 – Arapouco, 22 – Samouqueira, 23 – Nerja.

sisted of the badly preserved but complete skeleton of an adult female c. 50 years of age, apparently buried in a shallow oval pit dug into an occupation site, surrounded by limestone blocks, and covered by a small earthen mound. The corpse was placed in the extended supine position, with the head turned toward the right, and was associated with a red deer tibia and three Asturian picks, the latter contained in a small circular rock enclosure above the head (Clark 1976: 90-94). Arias notes that the picks are almost certainly grave goods because one was still fresh, whereas most Asturian picks recovered from the *concheros* are extremely abraded at the tip and along the edges due to heavy use (1991: 218-220).[12]

During a program of test excavation in Asturian *concheros* carried out in 2001 and 2002, some bones from the right leg of an adult male were found in a shell midden in the *Colomba* rockshelter (Asturias): a tibia, a fibula and a talus (Arias & Fano 2003). A radiocarbon date of 7090 ± 60 BP (TO-10223; 6010-5890 cal BC) was obtained from a collagen sample from the tibia. Although the bones were not articulated, they were nevertheless close to their correct anatomical positions, so it seems likely that they come from the same leg, and might represent part of a disturbed burial in a shell midden. Isolated human bones found in Mesolithic contexts along the coast of eastern Asturias (e.g., Cuartamentero, Balmori, Mazaculos, Poza l'Egua) might be the result of disturbed graves dug into the loose or weakly consolidated shell middens that are the hallmark of the Asturian (Arias et al. n.d.).

Inland in the same region, three late Mesolithic burials were excavated in the entrance of the cave of *Los Canes* (Asturias) (Arias 1991, 1999; Arias & Garralda 1996). Only two of them were related stratigraphically (Feature II, which cuts into Feature III) (Fig. 17.6). Feature III (Fig. 17.7) has been dated to 6930 ± 95 BP (AA-6071; 5959-5720 cal BC). It consists of an oval pit (120 x 83 x 54 cm) filled with earth and stones arranged in an orderly fashion. An adult male had been placed at the bottom lying on his back. His legs were strongly flexed so that the knees were some 30 cm higher than the pelvis. His head was slightly inclined to the right, the arms extended along the sides of the body, with the left elbow flexed to permit his hand to rest on the abdomen, and the right one lying alongside the right leg. In the grave fill, immediately above the knees of the adult male, the skeleton of a small child was recovered in association with loose bones of chamois *(Rupicapra rupicapra)*, red deer *(Cervus elaphus)* and wild boar *(Sus scropha)*. It appears that the grave was reopened to bury the child, al-

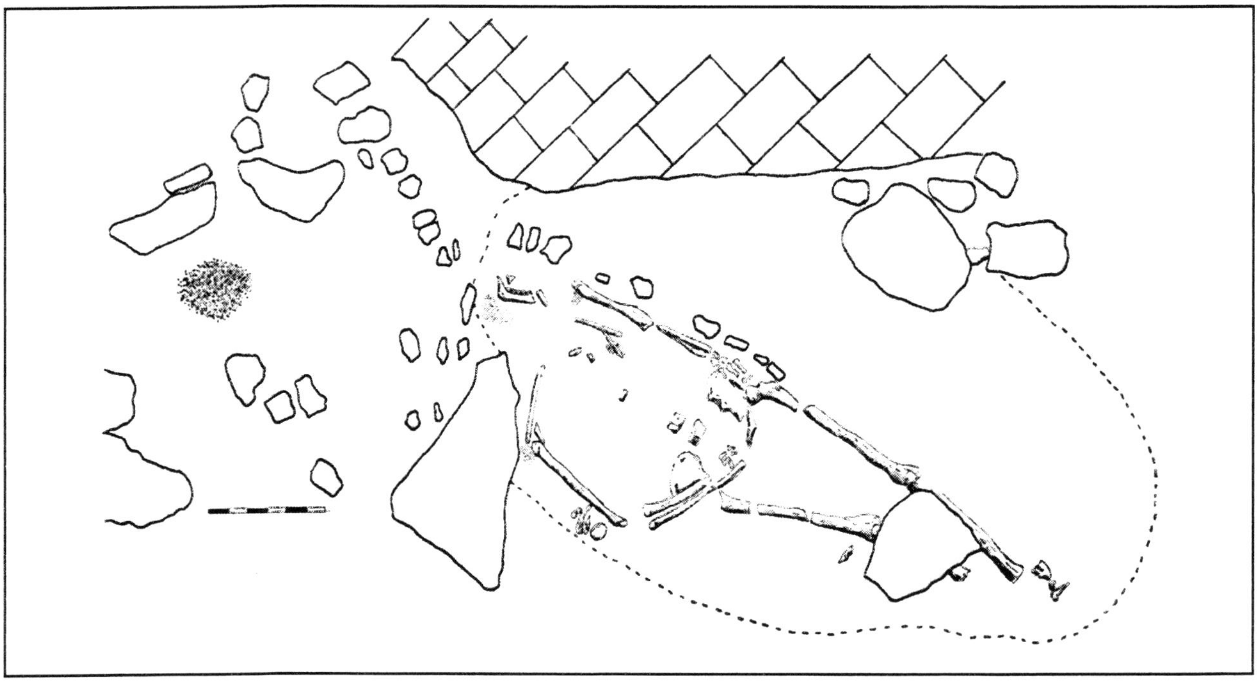

Figure 17.4 Azilian grave at Los Azules rockshelter (from Fernández-Tresguerres 1980).

beit after a relatively short interval. This in turn suggests that the earlier interment was either marked or still visible on the ground surface.

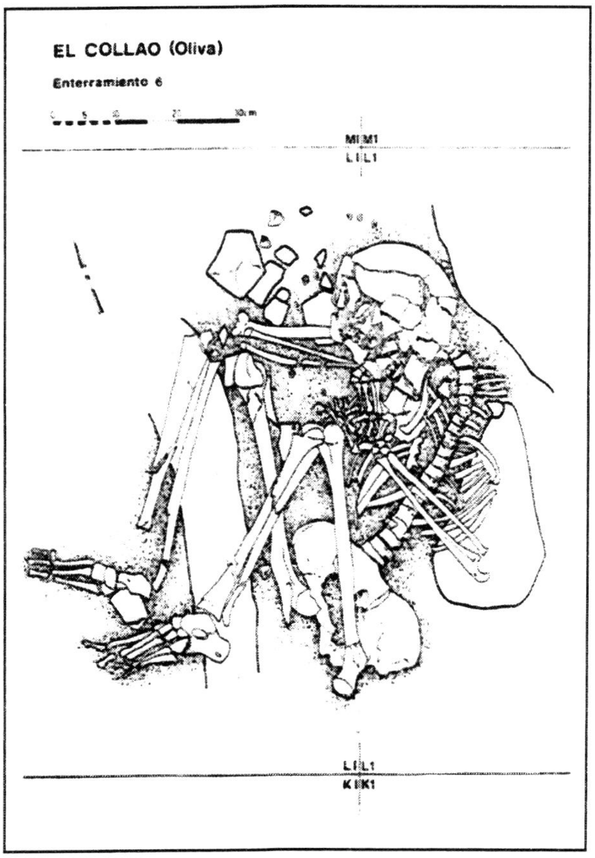

Figure 17.5 El Collado—Burial Number 6 (from Aparicio 1990).

Feature II (Fig. 17.8) was also an oval grave (measuring 160 x 75 cm) containing the remains of two individuals: the nearly-complete skeleton of a young male, and the feet from a second individual, placed in the grave in the opposite direction, near the head of the first. This feature has been dated by three ^{14}C determinations: two from the skeleton of the young man (AA-5296: 6770 ± 65 BP; 5718-5628 cal BC; AA-11744: 7025 ± 80 BP; 5988-5807 cal BC) and the third from the isolated feet (AA-5295: 6860 ± 65 BP; 5799-5665 cal BC). It seems plausible that those isolated feet pertain to an earlier burial, removed almost in its entirety by the opening of the grave for the young man. The body of the latter was placed on its left side, with the arms extended along the trunk and the legs flexed. The skull, dorsal vertebrae and pelvis were displaced from their anatomical position, possibly because of decomposition in a hollow space.[13] This individual was accompanied by several items that can be interpreted as grave goods: (1) a very long pointed bone underneath the skull, (2) a red deer antler with an oval perforation, (3) a flaked cobble (possibly a rough representation of a human head) and a perforated shell (*Callista chione*) placed above the femora, (4) two adult female ibex frontals *(Capra ibex)* near the right forearm, and (5) a oblong cobble with traces of ochre. Perforated shells were very abundant. Sixty-one *Trivia* sp., three *Littorina fabalis* and one Naticidae appeared in the burial, most of them around and in back of the head and trunk, suggesting that they might have been a necklace and/or sewn onto what might have been a cloak. A perforated red deer canine and some perforated *Trivia* beads found under the feet of the complete individual could be an anklet, or possibly items related to the head of the earlier burial. The edible molluscs – land snails (*Cepaea nemoralis*) and lim-

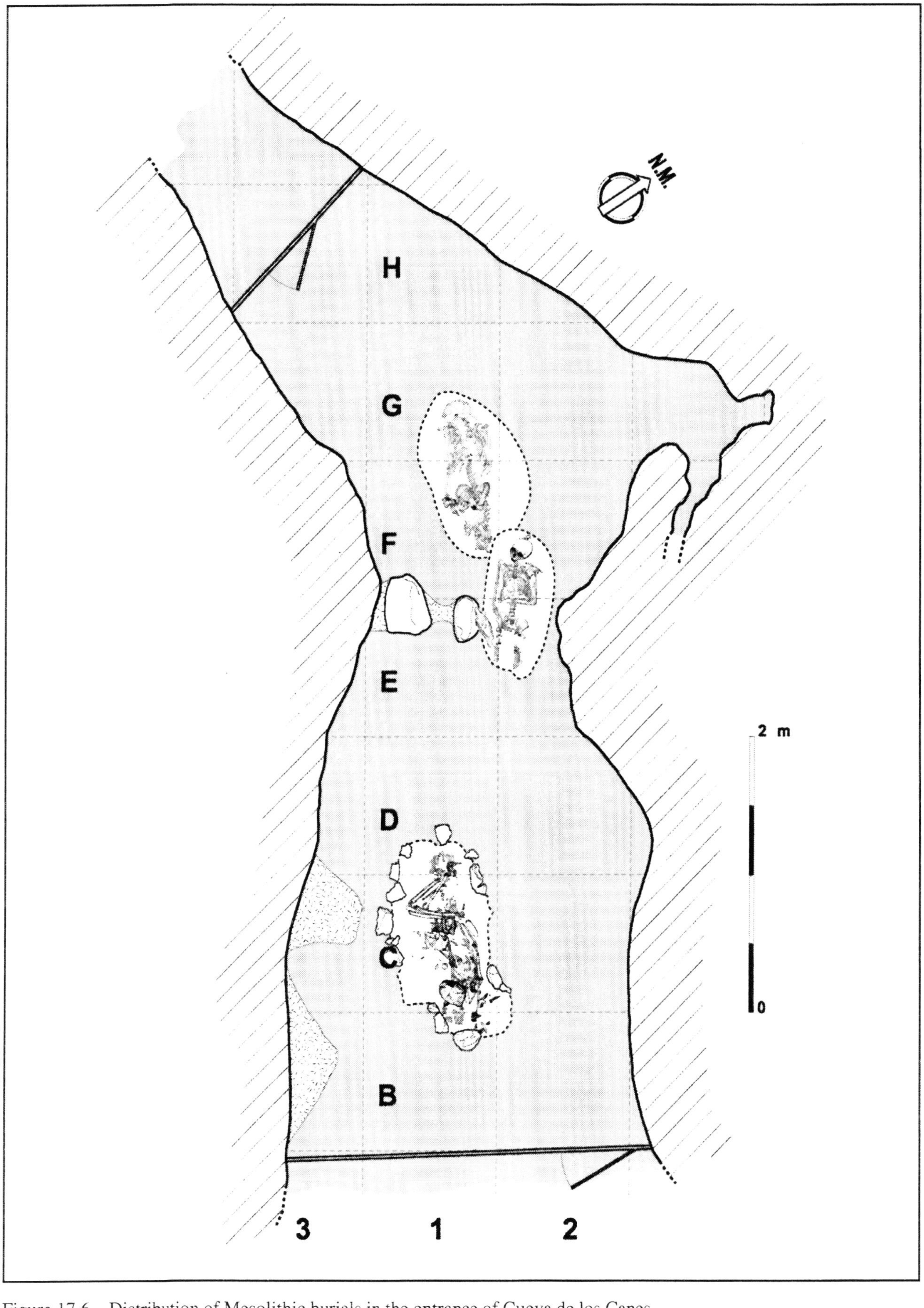

Figure 17.6 Distribution of Mesolithic burials in the entrance of Cueva de los Canes.

Figure 17.7 Cueva de Los Canes—Burial III.

pets (*Patella* sp.) – found in the grave fill were probably economic species.

Feature I is another oval pit (108 x 50 cm) delimited by the north wall of the cave and a large stone block. It contained the bones of an elderly woman interred on her back with her legs drawn up over her pelvis in the same position as Feature III.[14] The upper part of the feature had been damaged by a later prehistoric pit, so that the legs had disappeared and only the feet (close to the pelvis) were preserved (Fig. 17.9). The head was slightly inclined toward the right. The arms, placed alongside the body, had the right elbow flexed, so the hand was resting on the pelvis, while the left forearm crossed the abdomen, as if the hand had been supported by the right forearm, over the hip. Several objects that can be considered grave offerings appeared beside the skeleton: (1) a red deer scapula, placed almost vertically near the left foot; (2) an ungulate rib, found over the right shoulder blade, and (3) three perforated red deer canines, alongside the outer edge of the left foot. Numerous land snail (*Cepaea nemoralis*) shells were recovered from the fill of the burial pit; it seems likely that this abnormal accumulation was intentionally placed in the grave.

Another late Mesolithic burial has been found recently in one of the sites of the Archaeological Zone of La Garma (Cantabria), the cave of *El Truchiro* (Arias et al. 2003). This grave, not completely excavated yet, is currently being studied by A. Armendariz. The skeleton, probably that of a young woman, was buried at the entrance of a narrow gallery near the mouth of the cave. It was associated with several perforated shells of *Cerastoderma edule*,

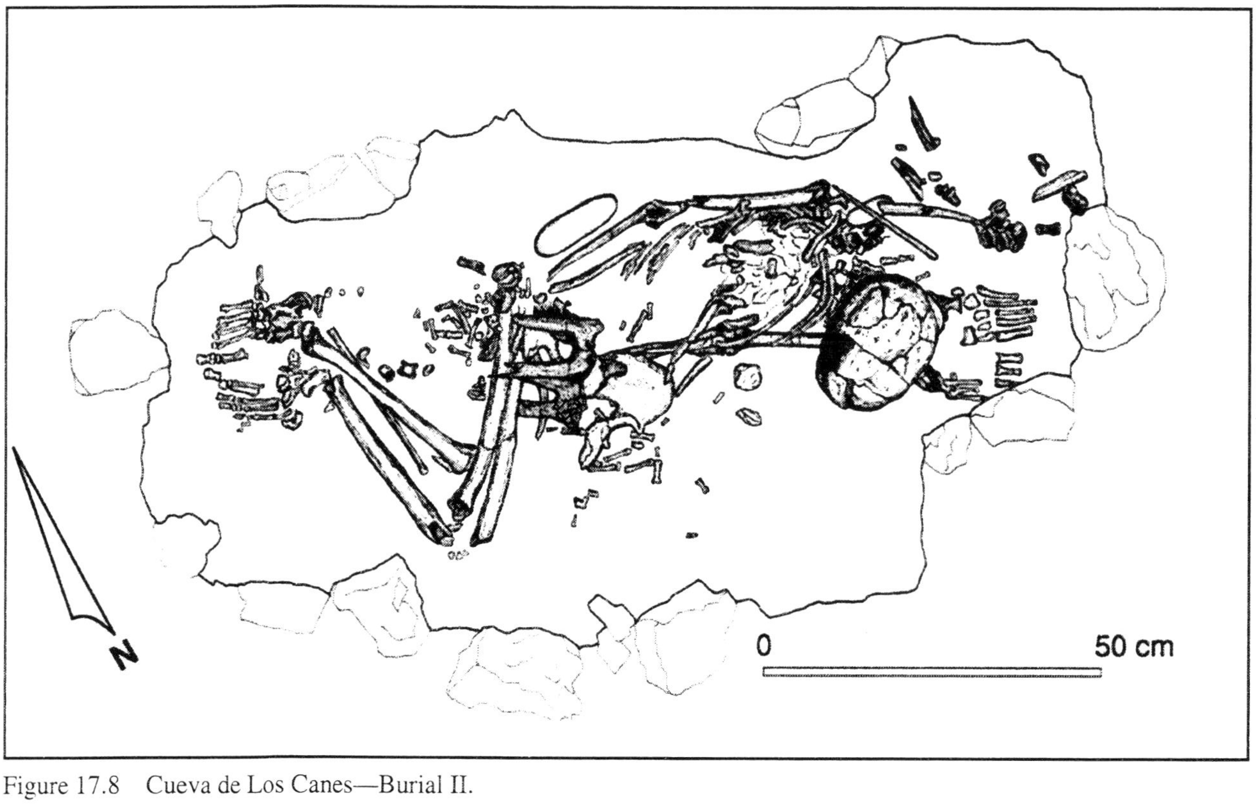

Figure 17.8 Cueva de Los Canes—Burial II.

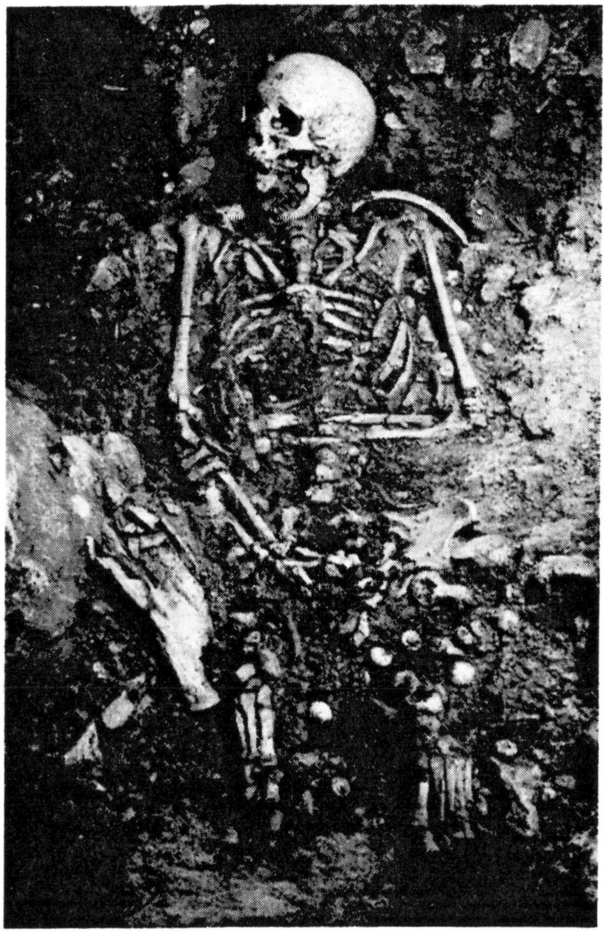

Figure 17.9 Cueva de Los Canes—Burial I.

probably the remains of a necklace. A direct date of 6470 ± 70 BP (TO-10912; 5481-5363 cal BC) has been obtained from a sample taken from the left humerus. The presence of human remains in other Mesolithic contexts on the same hill, such as La Garma B (dated to 7165 ± 65 BP; OxA-7300; 6156-5925 cal BC), the La Garma A shell midden (seven dates between 6500 and 5700 cal BC) and possibly La Garma C suggest extensive use of caves as burial places during the Mesolithic, as also documented at the roughly coeval Asturian sites along the coast (see above).

The Basque Country

The burial of an adult male, lying on its side in a flexed position, was discovered recently in a shell midden outside the coastal town of *Jaizkibel,* near San Sebastián (Guipúzcoa, Spain) (Iriarte et al. 2003). Although the site, named *J3,* is still being studied, the current evidence suggests that it is also a Mesolithic grave, probably dated to the eight millennium cal BC.

In the Pyrenees, another late Mesolithic burial has been reported recently at the *Aizpea* rockshelter (Navarra, Spain) (Fig. 17.10). In this case, the interment consisted of the skeleton of a woman of about 30 years of age placed, in a flexed position, directly on the living surface, and covered with a mound of rocks No obvious grave goods or offerings are noted by the excavators (Barandiarán & Cava 2001). A direct radiocarbon determination dates the skeleton to 6600 ± 50 BP (GrA-779; 5608-5484 cal BC), and is consistent with the geometric microliths found in the level with which the human remains are associated.

The Portuguese Mesolithic Cemeteries – Ribatejo

Without doubt, the most important funerary assemblages on the Iberian Peninsula (in fact, in Europe) are the late Mesolithic cemeteries in central and southern Portugal. The best known sites are shell middens *(concheiros)* in the Muge valley (Ribatejo), in the vast estuary of the Tagus River. Nearly three hundred human skeletons have been excavated since the 1860s at *Moita do Sebastião, Cabeço da Arruda* and *Cabeço da Amoreira,* and probably from lesser known shell middens in the Muge-Magos area [15]. A long series of absolute dates for those sites has been published, most of them corresponding to the sixth millennium cal BC.[16]

The Muge burials are normally individual inhumations in shell middens.[17] Burial pits have only occasionally been recognized in the excavations, although it seems likely that the corpses were originally placed in them, since otherwise it is hard to account for anatomical articulation. However, Roche's observations at Moita do

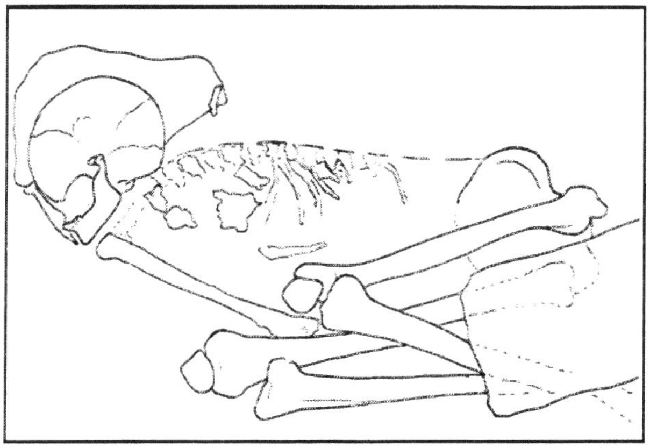

Figure 17.10 Mesolithic burial at Aizpea (from Barandiarán & Cava 2001).

Figure 17.11 Burials exposed in the 1880 excavation at Cabeço da Arruda (from Cartailhac 1886).

Sebastião led him to claim that, in most cases, the deceased were simply deposited in shallow natural depressions in the sand, at the base of the shell midden (Roche 1960). On the other hand, infants (1-7 years old) were fairly consistently buried in circular or oval pits there, on average 0.85 m in diameter and 0.23 m depth. At Moita, most of the adults were buried lying on their backs, with the head intentionally raised, the arms alongside the trunk, and the legs flexed to one side (Roche 1972), but it seems that there was some variation, since the 19th century reports also describe flexed and semi-flexed skeletons (the most common positions, according to Cartailhac [1886]; see Fig. 17.11), a practice also found at Cabeço da Amoreira (Roche & Ferreira 1967).[18] In respect of the mortuary programs followed at the Muge estuary sites, it is usually assumed, following Roche, that the standard treatment of the deceased was primary, usually individual, inhumation. Yet, in the 19th century literature, several references allude to heaps of bones (see, e.g., the burial on the left in Fig. 17.11).[19] Whether secondary burial, or indeed any post-mortem manipulation of the dead was practiced, or whether these bone accumulations resulted from accidental disturbance of earlier burials cannot be ascertained at present.

The burials in the Muge estuary sites were not distributed randomly throughout the shell middens, but were instead concentrated in relatively well-defined cemetery areas, among the earliest known. Arthur Saxe (1970) proposed long ago that the appearance of cemeteries (communal burial precincts) might be related to the appearance of corporate descent groups, membership in which is validated or legitimatized in terms of access to resources in a defined economic territory. He argued that the physical presence of the groups' ancestors in the form of their skeletal remains constituted a symbolic claim to the right of eminent domain over that economic territory (Clark & Neeley 1987: 122). Communal burial precincts were first identified in the 1880 excavations at Cabeço da Arruda, where it was observed that the skeletons were concentrated in the southwestern part of the site (Oliveira 1892), a pattern that was more definitively established by Roche (1960, 1989) at Moita do Sebastião. In this site, the burials recovered in the 1950s were arranged in two clusters suggestive of age-grading. The adults were concentrated outside the main living area, and infant burials were found inside the dwelling zone among the huts, arranged in a semi-circle (Fig. 17.12). On the other hand, there is little consistency between the late 19th and mid-20th century phases of excavation in respect of orientation. While Ribeiro (1880) noted that the Cabeço da Arruda skeletons appeared to be aligned toward the cardinal points, Roche's excavations at Moita do Sebastião showed no consistency in orientation (Roche 1960).

It is not easy to discern grave lots, since the limits of the burials were seldom precisely determined. At Moita do Sebastião personal ornaments (mainly perforated shells of the freshwater gastropod *Teodoxus fluviatilis*) were associated with the human remains, while at Cabeço da Arruda utilitarian objects were the most common kind of offering (endscrapers above the left forearm in Burial II; a pointed bone under the left hand of Burial VII) (Roche 1972). In Moita do Sebastião Burial XVI, a cylindrical quartzite cobble was found close to the skull, and a trapeze was located over the chest of the skeleton of Burial XXXII. Ochre (Moita do Sebastião VI, XI and XXX) and traces of fire (Moita do Sebastião I, III and XXXIV) are also described (Roche 1960). There were possible food offerings at Moita do Sebastião: a concentration of closed bivalves *(Lajonkairia lajonkairia)* under the skeleton of Burial III, and the abnormal abundance of *Helix pisana* in Burial XII, and of *Scrobicularia plana* around Burial XXXIII.

The Portuguese Mesolithic Cemeteries – Baixo Alentejo

The shell middens of the Sado valley (Baixo Alentejo) appear to be later than those of the Muge. The radiocarbon dates (unfortunately mostly from shell – there are none from human bone) range from the beginning of the 6th millen-

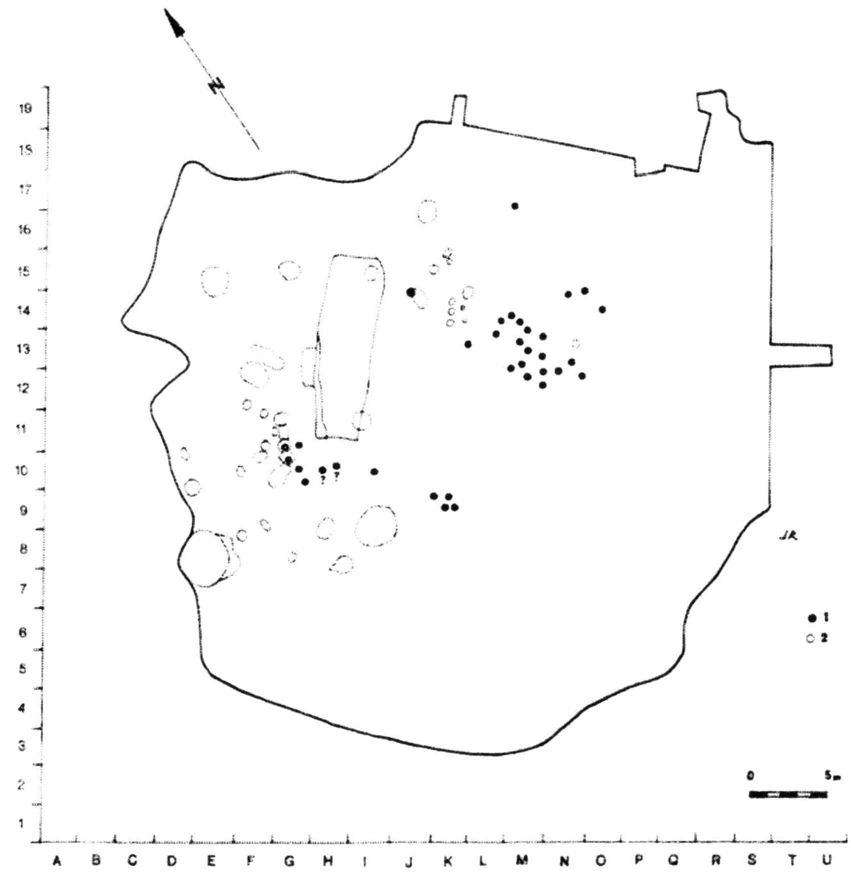

Figure 17.12 Distribution of burials and dwelling structures at Moita do Sebastião (from Roche 1972). Key: 1—adult, 2—infant.

nium cal BC to the first quarter of the 5th (Araújo 2003). Although they were excavated in the 1950s and 1960s, the contextual information is very poor. The excavations remained unpublished for decades, and most of the field documentation has been lost. Nevertheless, hundreds of thousands of artifacts and faunal remains are stored in the Museu Nacional de Arqueologia e Etnologia in Lisbon, together with more than one hundred human skeletons preserved in wooden cases where they were cast *in situ* in blocks of paraffin wax.[20] A program of study of the museum collections has allowed a first approximation of pattern in this important site cluster (Arnaud 1989).

Perhaps the most obvious feature of the Sado Valley cemetery sites is the high density of burials. Except in the largest shell midden *(Cabeço do Pez),* the estimated density is always greater than 0.1 burial per square meter, and in *Vale de Romeiras* that value reaches 0.41. There is considerable variation in regard to the mortuary programs at individual cemeteries. Whereas at *Amoreiras* the skeletons were found scattered at random across the midden, they were consistently oriented E/W (IV, VI, VII, VIII) or W/E (III, V), at *São Bento* there was no consistent pattern of disposition or orientation. The burials at Romeiras were placed radially in a semi-circle facing the river. Arnaud (1989) notes that, even in sites with high burial densities like Romeiras (26 burials in only 54 m^2), superpositions are very rare, suggesting a rough contemporaneity for the use of the cemetery, and/or some kind of surface marker to indicate the position of a grave.

The Sado burials almost always occur dug into the sandy, nearly sterile layer at the base of the shell middens. As at Muge, no clear evidence of burial pits was found. The cadavers were consistently placed in the grave lying on one side in a flexed or semi-flexed position (Fig. 17.13). As no clear limits for the graves were recognized, it is difficult to distinguish any intentional grave offerings. However, most of the perforated shells (*Teodoxus fluviatilis, Trivia* sp., *Nassarius reticulatus*) were found close to the skeletons. Occasionally, microliths (which are present throughout the middens) have been claimed to be offerings, but – given their ubiquity in these sites – that association might well be fortuitious (Arnaud 1989).

Finally, parts of the skeletons of two adult males were found in the site of *Samouqueira,* on the Baixo Alentejo coast (Lubell & Jackes 1985). It is most likely that they come from individual graves, disturbed by ploughing and

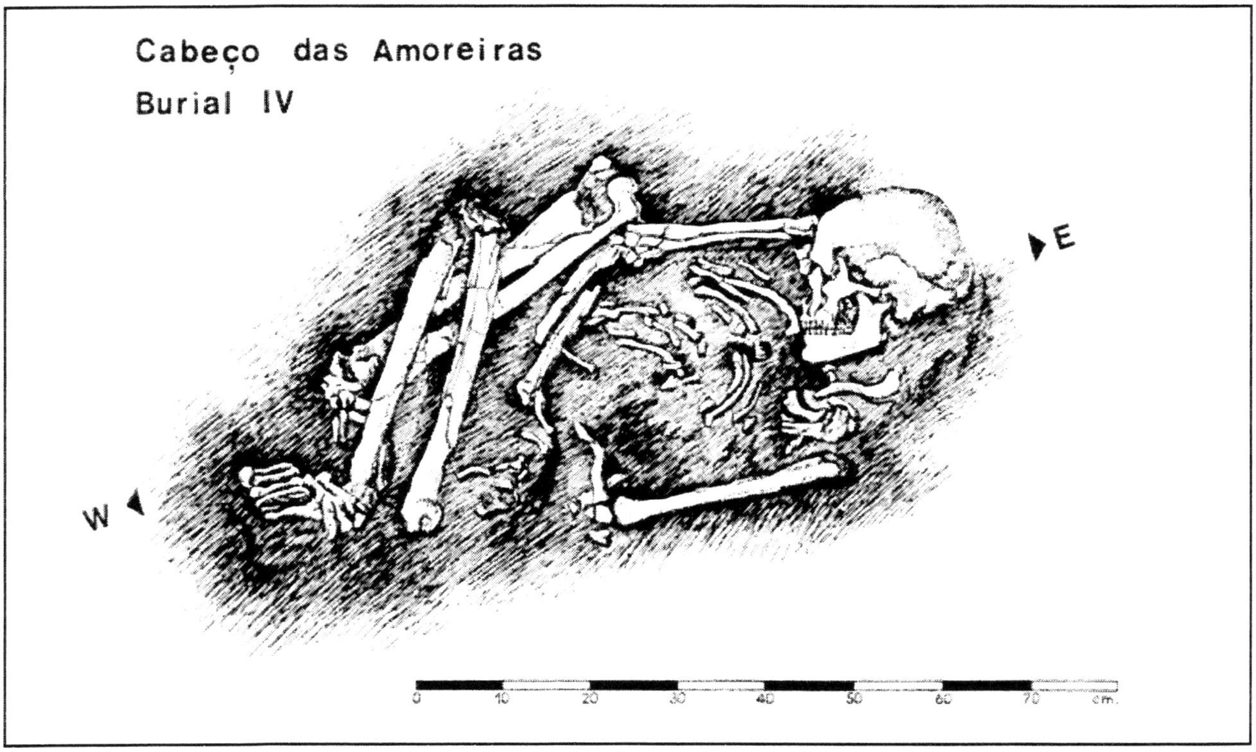

Figure 17.13 Amoreiras—Burial IV (from Arnaud 1989).

natural erosion. One was buried in the extended supine position, while the other was laid in the grave in a flexed position, after rigor mortis had relaxed. A radiocarbon date in the second half of the 6th millennium cal BC was obtained from one of the skeletons (TO-130: 6370 ± 70 BP; 5382-5238 cal BC).

General Trends and Patterns

It is difficult to draw general conclusions from the Iberian paleolithic and Mesolithic funerary data we have presented here. Indeed, it would be unsound intellectually to try to reach firm conclusions from such temporally and spatially uneven data, now lost – in many cases – and poorly documented in many others. As Wüller (1999) has remarked in regard to the Magdalenian, consistent mortuary programs for 'millennial' time scales could never have existed. Mortuary data from Iberia span, at a minimum, 25,000 years, and the fact that they all correspond to foraging societies does not justify treating them as if they were all homogeneous. In addition, mortuary analysis is a much more complex endeavor than many archaeological approaches (e.g., and esp., processualism) have acknowledged. As the ethnographic evidence shows, human social responses to death vary enormously and are expressed materially in a multitude of ways (Ucko 1969, Testart 2001). Funerary behavior does not merely 'reflect' the social organization of a group in a fairly direct way; it is instead a context for the display of symbols, and also an occasion for the expression of personal or group feelings towards the deceased. Without denying that death rituals everywhere are highly structured and systematic, driven by ideology, there is also much room for contingency, and for personal or collective choice. For example, the objects recovered from a grave might have been deposited as a viaticum, as goods required by the deceased, either literally or metaphorically, in another life; yet they can also be personal mementos, offerings from relatives or friends, or elements that represent the decedent's activities in life. They can also represent a political or social tribute to an individual and/or his lineage. Burying objects in a grave can also be a way of getting rid of them (e.g., if the possessions of the deceased were considered impure or dangerous, as they are in many ethnographic contexts). More mundanely, objects recovered archaeologically could simply have been attached to a shroud, or garments worn by the dead. If this were the case, the tangible remains of a grave lot could indicate ritualized behavior prior to death (in societies that reserved a particular item of clothing for burial), the custom of dressing the deceased 'correctly' or just the fortuitous garments and adornment the individual happened to be wearing at the time of death.

Distribution by Context

Acknowledging the potential complexity of pattern in mortuary contexts, some very general trends can be highlighted. Among them, we note that most hunter-gatherer funerary evidence is associated with living sites, although we acknowledge that, in many cases, it is not possible to demonstrate contemporaneity of interment with site occupation. Also, the probability of discovering a non-monumental burial outside an excavated site is very low, as indicated by the frequency with which ancient human fossils are recovered (this despite ongoing, massive construction projects throughout the whole of Europe).

Distribution by Mode of Disposal

A second regularity is the predominance of individual inhumations in pits. Although it is often suggested that cadavers were simply abandoned on the floors of caves and rockshelters, it is practically impossible to account for the preservation of articulated skeletons without postulating some form of protection (a pit, a mound, a wooden structure, etc.). In fact, most cases of supposedly unprotected graves come either from poorly documented excavations (e.g., Nerja), or from sites where sedimentary peculiarities make it extremely difficult to recognize features (e.g., the Portuguese shell middens).

Nearly all the skeletons were found in *dorsal decubitus* (extended supine) or lying on their sides in flexed or semi-flexed positions. There is no clear pattern concerning the positions of the corpses. *Dorsal decubitus* appears to be the dominant position in the paleolithic, but the sample is too small to be conclusive. In the Mesolithic, the lateral flexed position becomes more common, but in most sites and burials the corpses continued to be interred lying on their backs.[21]

The skeletons are usually articulated, but there are some interesting cases of changes of position that cannot be attributed to post-depositional disturbance. Possibly at Cova Beneito, and very likely at Los Canes, Cabeço da Arruda and Cingle del Mas Nou, some individuals underwent some kind of post-mortem manipulation, especially at Cingle, where a secondary burial occurs in association with a primary one. Post-mortem manipulation of human remains appears to be more frequent in Mesolithic Europe than commonly acknowledged (Cauwe 1998, 2001; Orshiedt 1999.

Distribution of Grave Goods

Probable grave goods are quite variable, and rather difficult to study. The fact that most Iberian graves have been dug in archaeological layers, together with inadequate reporting, makes it very difficult to distinguish those items left intentionally and those which are part of the grave fill. Nevertheless, there is a tendency over time to include more ordinary utilitarian artifacts (occasionally unused) and what are taken to be food offerings (e.g., the joints of ungulates, quantities of shellfish). The high frequency of (marine) shell beads should be interpreted either as personal adornment (necklaces, bracelets, anklets etc.) or attached to the clothing. As time goes on, red ochre (hematite), fairly common in paleolithic and early Mesolithic burials, becomes less common. Although ochre is one of the things that archaeologists typically look for, meaningful concentrations are reported only at Lagar Velho, Los Azules, and in some of the Muge burials. At Lagar Velho, it appears to have been used as a colorant on the shroud in which the Gravettian child was wrapped (Duarte 2002). There is also occasional evidence of the use of fire in the funeral ritual. Hearths, or at least pieces of wood charcoal, have been found associated with the grave contexts at Cueva Morín, Lagar Velho, and at some of the burials in the Portuguese shell middens.

Distribution by Age and Sex

Because of preservation, lapses in recording, or simply lack of relevant expertise, it has not often been possible to establish with certainty the sex of the human remains. This has precluded the isolation of patterned variation due to differential treatment of male and female adults. It also true that most of the Iberian sites consist only of a single, or at most a few, individuals. At Los Canes, the most complex burial (Feature II) is that of a male, but the sample is too small to generalize. Unfortunately, the data from the Portuguese shell middens lack the necessary detail to address this question. It appears that the sex ratio in the Muge cemeteries of Muge approximates the natural proportion at birth (Vallois 1930, Ferembach 1974, Cunha & Cardoso 2002-2003) but this information cannot be taken further because, as noted, it is very difficult to establish the limits of the burials and the offerings corresponding to each individual.

Although better control of taphonomic factors would obviously be very desirable, there are hints that some selection related to age at death was a common feature. During most of the period analyzed here, there are very few children, far fewer than would be expected from the actual mortality of that age group. This is especially true of neonates and children <2 years old, suggesting some kind of an age-grade distinction – for most Iberian hunter-gatherers, full membership in the community was attained some years after birth (perhaps at weaning which, in ethnographic contexts, can be as late as 4-5 years). However, when infants are found, differences in the mortuary program are minimal (unlike other parts of Europe [Vanhaeren & D'Errico 2001]). Although an isolated find, the Lagar Velho child resembles those of other mid-Upper Paleolithic European adult burials. At Nerja the only difference is in the lateral position of the child, but it is uncertain whether the child and the adult burials at that site are contemporaneous.

The mortuary program for small children appears to change in the late Mesolithic. At Los Canes, Moita do Sebastião and Cabeço da Arruda (Ferembach 1974, Jackes & Meiklejohn n.d.), children under five years of age (including neonates) are documented. The percentage distribution of age-at-death recently provided by Jackes and Meiklejohn (n.d.) for the Portuguese cemeteries indicates that the burial sample there probably reflects the demographic structure of those populations. Although information for most cases is inadequate, some differences in the funerary treatment of children also arise. The Los Canes child was buried in the fill of an adult's grave, whereas at Moita do Sebastião separate burial precincts were maintained for adults and children and, according to Roche (1960), the shapes of the graves were different.[22]

Distribution by Space

There is also a difference between Iberia, on the one hand, and the rest of Europe, on the other, concerning the relative density of burials and burial sites. While Iberia has one of the highest concentrations of Mesolithic burials on the Continent, [23] its density for the Upper Paleolithic is comparatively low, especially when compared with those of France, Italy or Moravia. However, compared to the European paleolithic average (c. 5.2 per 1 million square kilometers), the Iberian average is c. 10 burials per 1 million square kilometers (compared with c. 25.7 for France, c. 56.5 for Italy). Yet, many other regions have none or just a few graves. As a matter of fact, the Iberian density is higher than the average value for the Continent, even if we exclude the northern part that was under the ice sheets during much of the Pleistocene.[24] Despite the existence of large numbers of Iberian paleolithic and Mesolithic sites, only few funerary sites have been recorded. What is the meaning of fewer than ten cases among hundreds of sites, representing 30,000 years of human activity? Certainly, many funerary features might have been destroyed by subsequent human activity, but this argument alone cannot account for the poverty of funerary evidence in Iberia. One alternative explanation is that Iberian foragers might have buried their dead outside of settlement contexts (or at least outside caves and rockshelters, which are the most common kind of paleomesolithic site). Given the rarity of open-air sites, we have no inkling of possible methods of disposal of the dead if they occurred away from caves and rockshelters.

Distribution over Time

There is an apparent increase in the frequency of burial towards the end of the Mesolithic, during the 6th millennium cal BC, an interval which contains 66.7% of the sites and 93.9% of the individual interments (Fig. 17.14, Table 17.1). It is evident that this fact cannot be related to any statistical artifact or bias in the research.[25] Although demographic factors might account for the general increase in frequency over time, no population can experience so dramatic a change as that of the 6th millennium cal BC unless there was massive immigration, for which there is no evidence at all. It seems clear, then, that there was a significant change in the mortuary behavior of Iberian Mesolithic foragers, and that the change was a general one, since 6th millennium sites are found in most regions with a significant sample of Mesolithic settlement. This appears to be a pan-European phenomenon (Clark & Neeley 1987, Grünberg 2000). And if we include data from other continents (e.g., the north American Archaic), it seems clear that the practice of burial is becoming more prevalent among early Holocene forgers generally.

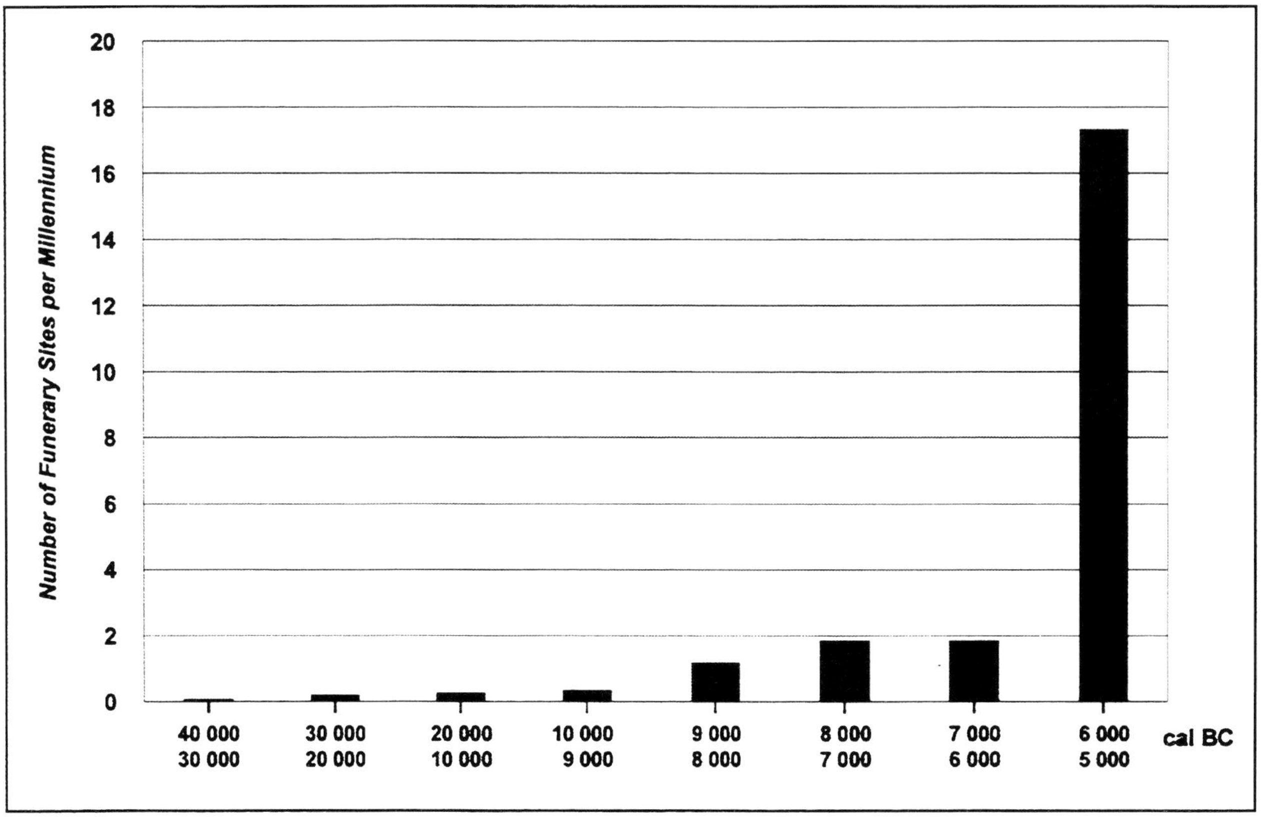

Figure 17.14 Incidence of funerary sites by time in the Iberian Upper Paleolithic and Mesolithic.

Table 17.1 Buried individuals per Millennium on the Iberian Peninsula during the Upper Paleolithic and Mesolithic.

Years cal BC	40000-30000	30000-20000	20000-10000	10000-9000	9000-8000	8000-7000	7000-6000	6000-5000
Index of Burial Frequency	0	0.30	0.55	0.33	1.17	1.83	14.83	408.33

Cemeteries

Sometimes, these burials are concentrated in cemeteries (i.e., specialized burial precincts for the disposal of the dead – Goldstein 1981). In pre-Neolithic Europe, cemeteries are clearly related to the more complex foraging societies, and – arguably – those living at the highest population densities. Most of them are late Mesolithic groups, living on ecotones with very abundant – and usually seasonal – resources, and showing a tendency towards sedentism, as well as some evidence for vertical social differentiation (Clark & Neeley 1987). Some of the more compelling examples are those found in typical 'affluent forager' contexts (e.g., the Iron Gates of the Danube, in Romania [Srejovic 1969, 1972; Srejovic & Letica 1978] or at the island site of Oleneostrovski, in Karelia [Gurina 1956, O'Shea & Zvelebil 1984]).

Most relevant for Iberia are the cemeteries of (1) Téviec and Hoëdic, in Brittany (Péquart et al. 1937, Péquart & Péquart 1954), dated between c. 6000 and 4500 cal BC (Schulting 1998), and (2) the western Baltic sites of Henriksholm-Bøgebakken, at Vedbæk (Albrethsen & Brinch Petersen 1976), and Nederst (Brinch Petersen 1989, 1990) in Denmark, and Skateholm (Larsson 1988) and Tågerup (Karsten & Knarrström 2001) in Scania (Sweden), dated mainly to the early 5th millennium cal BC. Like the Muge and Sado cemeteries, they are open air sites with dozens of graves, mainly individual burials in oval pits. Yet, a considerable variability in funerary practices exists, marked by the appearance of collective graves, secondary burial, and the earliest cremations. Red deer antler offerings, wooden or stone structures above the burials, cenotaphs and dogs' burials also appear (Larsson 1990).

Social Complexity?

For Iberia, as elsewhere, the frequently cited relationship between late Mesolithic funerary behavior, cemeteries and social complexity remains unclear. The high incidence of burials shown in Figure 17.14 and Table 17.1 comes from sites with a single, or at most only a few interments (e.g., Aizpea, Samouqueira, Cingle del Mas Nou, Molino de Gasparín, El Truchiro). Nor is there any clear linkage of true cemeteries (e.g., some of the Tagus and Sado sites, El Collado) with other, independent evidence for social complexity. Complex hunter-gatherers are usually characterized ethnographically by sedentism, high population densities, a social division of labor, socio-economic inequality, developed systems of exchange, warfare, public ritual and complex political organization (Testart 1982, Price & Brown 1985, Ingold et al. 1988, Arnold 1996). Thus far, however, there is no evidence for these traits for most Iberian Mesolithic groups. It could be argued that the high density of Asturian shell middens (c. 130 sites on just 50 km of coastline) might indicate an increase in population density, although differential preservation compared to sites occupied during the pleniglacial maximum might also be a factor (Fano 1998). The obvious differences in complexity and offerings between Los Canes II, the burial of a young man, and the other burials of the site might be evidence of social inequality. However, it is a real stretch to call the late Mesolithic societies of eastern Asturias 'complex' (Arias 1999). The same can be said of the Mediterranean, Pyrenean and Andalusian late Mesolithic.

The case of the Portuguese *concheiros* is more complicated. Recent analyses show that they developed broad spectrum economies, in some cases with diets evenly balanced between terrestrial and aquatic foods (Lubell & Jackes 1988, Lubell et al. 1994, Cunha et al. 2003). The large size of the Tagus shell middens and the huts at Moita do Sebastião suggest that they might have lived in permanent or semi-permanent settlements, although that cannot be taken as fully confirmed. However, there is little to indicate social complexity so far as the rest of the traits are concerned. The main evidence for high population densities are the cemeteries themselves – clearly a circular argument.[26] Moreover, evidence for social inequality, long distance exchange, presence of prestige goods, warfare, and ceremonial activity is completely absent. Therefore, the evidence from the Tagus-Sado Mesolithic is rather ambiguous. Whereas some few aspects of the archaeological record might suggest social complexity, most suggest otherwise. Thus it would be inappropriate to label them as 'complex', if by that we mean the same thing as we do when we refer to their Breton or Baltic counterparts.

Territoriality

As Saxe pointed out long ago (1970), the appearance of cemeteries might signal the appearance of corporate descent groups that are expressing rights (or claims) of eminent domain over the use and control of crucial resources and the economic territories that contain them. In an scenario of growing populations and increasing resource competition, cemeteries might represent a tangible symbolic 'statement' of a claim to land ownership, legitimated or reinforced by the practice of burying the dead in a single spot on that economic landscape. There is some evidence to suggest that this factor might account for the main concentrations of mortuary evidence in late Mesolithic Iberia: Cantabrian Spain and the Tagus-Sado estuaries. In the north Spainish case, recent research has provided new data that seem to confirm patterns suggested by the study of lithic technology and the nearly exclusive use of poor quality, local raw materials (Arias 1991, 1999). Stable isotope analyses (^{13}C and ^{15}N) of skeletons at Los Canes (a site located in the Cantabrian Mountains c. 11 km from the modern coast) show a dramatic difference from those from coetaneous coastal ('Asturian') sites in the same region (e.g., Poza l'Egua, Colomba). Whereas Los Canes samples present ^{13}C values ranging between -20 ‰ and -19 ‰ (typical of a terrestrial protein diet), those from the coastal sites show higher values (-17 to -15 ‰), along with ^{15}N readings of 12-13 ‰, indicating a mixed terrestrial-marine protein diet, probably rich in fish. We interpret this to suggest the possible existence of social boundaries that might have denied mountain foragers access to marine coastal resources (Arias & Fano n.d.).

The case of the lower Tagus valley *concheiros* might be similar. As in Cantabria, paleoecological and subsistence data suggest the exploitation of a limited territory from permanent or semi-permanent base camps (Araújo 2003), whereas stable isotope analyses shows a sharp contrast between the diet of hunter-gatherers and that of Neolithic groups living at roughly the same time in neighboring areas (Lubell & Jackes 1988, Zilhão 1993). Thus, it appears likely that territoriality was becoming an important factor in the social evolution of late Mesolithic groups in these two exceptionally productive parts of Iberia.

Mesolithic-Neolithic Overlap

Reinforcing an apparent shift towards more territorial behavior among some Iberian hunter-gatherers is their proximity to contemporary Neolithic agropastoralists. It is perhaps significant that these 6th millennium Mesolithic groups appear to represent an 'availability phase' (i.e., they were living near Neolithic groups practicing domestication economies – Zvelebil & Rowley-Conwy 1986), and it is likely that they had some kind of contact with those first farmers. This seems to be the case with (1) the Muge shell middens, where radiocarbon dates overlap with those of the neighboring Cardial Neolithic of the Estremadura limestone massif (Zilhão 1993); (2) with the latest Mesolithic burials of Cantabrian Spain (Los Canes, El Truchiro) and the Pyrenees (Aizpea), both of which are roughly contemporaneous with the expansion of the Cardial Neolithic up the Ebro valley, and especially with the late Mesolithic of Cingle del Mas Nou, dated to a time when the early Neolithic is widespread in Mediterranean Spain.

These cases of Mesolithic-Neolithic temporal overlap raise questions about the nature of the relationships between late Mesolithic mortuary practices and the processes involved in the Neolithization (Arias 1999). In Atlantic Europe, there appears to be a connection between areas where the Neolithization is retarded or delayed and the appearance of cemeteries. Furthermore, some regions where this relationship exists (e.g., southern Portugal) are among the earliest to exhibit another important focus of mortuary behavior – the earliest appearance of megalithic tombs (Scarre et al. 2003). This association has long been noted in the archaeological literature of Brittany, where similarities with the local megalithic sites are apparent in inferred funerary ritual and the architecture of Mesolithic structures at Téviec. In the last few years, this issue has been revisited, and some scholars propose a connection between the development of the architectural monumentality and the complex processes of social competition that appear to have played a role in the Neolithization of the hunter-gatherers of Atlantic Europe (Sherratt 1990, 1995; Scarre 2002).

Final Remarks

As we show from the descriptive part of this essay, the funerary evidence from the Iberian paleolithic and Mesolithic is more relevant than is usually acknowledged. Four sites and eight graves are known for the Upper Paleolithic, whereas the Mesolithic sample (23 sites, c. 427 individuals) is one of the most important in Europe for that time period. Some Iberian sites, especially those excavated recently (e.g., Lagar Velho [Zilhão & Trinkaus 2002]), have provided highly detailed information that sets a standard for future work.

Among the more important aspects of pattern shown in this analysis are the dramatic changes that occurred in the late Mesolithic. The last hunter-gatherers on the Peninsula buried their dead around the edges of, or in some cases inside, settlements, and cemeteries are well documented in the Portuguese *concheiro* sites, and at El Collado, on the coast of Valencia. Small children were more frequently buried, and there is more evidence for post-mortem manipulation of human remains. Although the reasons for these 6th millennium BC transformations are not clear, territoriality and contact with early Neolithic groups almost certainly played important roles. Mortuary ritual is one of the principal sources of information about the social and ideological realms of the prehistoric societies but, in the Iberian case, much research is still required. Although the quality of the data and of the conceptual frameworks used to analyze them have improved in the last few years, important sites like El Collado remain almost unpublished. As the recent re-analyses of the Portuguese *concheiros* show, even sites excavated early on can still provide valuable data when analyzed using novel methods and modern conceptual frames.

Acknowledgements

We acknowledge the help of our colleagues Ana Cristina Araújo, Álvaro Arrizabalaga, Rodrigo de Balbín, María José Iriarte, Mary Jackes, David Lubell, José Rolão and João Zilhão, who shared with us unpublished information and/or provided graphic material for this essay. We also thank Luis C. Teira, who prepared all the figures.

Endnotes

1. We discuss only funerary evidence corresponding to anatomically modern humans, although we acknowledge that it might antedate the apperance of *H. sapiens* (e.g., and especially, Atapuerca, where the accumulation of human remains in the Sima de los Huesos has been interpreted as the result of intentional funerary behavior [Arsuaga et al. 1997, Carbonell et al. 2003]).

2. On very imprecise and unverifiable grounds (many perforated shells associated with a humerus, now lost), a disturbed burial was claimed to have been found in the "Aurignaco-Perigordian" layer of Abric Romaní (Barcelona, Spain) (Ripoll 1959, cf. Carbonell et al. 1996: 423-424 for a modern, critical assessment). Old excavations at another EUP site in Catalunya, Reclau Viver (Gerona), yielded human bones (left and right femoral diaphyses, a fragment of a right femur, a talus, a metatarsal, a first phalanx) and some teeth that were claimed to represent an interment (Corominas 1949).

3. The Pleistocene dates cited in this paper have been calibrated with the curve CalPal2004_jan curve, using the program CalPal (Weninger *et al.* 2003). For the Holocene dates we used the INTCAL98 curve (Stuiver et al. 1998), with the aid of the program Calib 4.4.1 (Stuiver & Reimer 1993). Except for rough approximations like this, the range of the calibrated age, plus and minus a single standard deviation ($\pm$1s), is reported.

4. Evaluating the radiocarbon dates from Cueva Morín (one of the first Spanish paleolithic sites where this method was used) is rather complicated. Two *Radiocarbon* references have been published for this hearth: SI-952: 28435 ± 556 BP and SI-952A: 28155 ± 757 BP (Stuckenrath 1978). Both of them actually come from the same charcoal sample, the latter corresponding to the NaOH soluble fraction. Furthermore, they were originally published in a 'corrected' version (i.e., 3% was added to the determinations in order to convert the Libby half-life to the 'real' half-life of 5730 calendar years). Unfortunately, it has not been possible to determine the Libby half-life for those dates (i.e., no round figure results from dividing them by 1.03), so we can only cite them as they appear in the original paper. Their conversion to the Libby half-life would correspond to 27607 ± 540 (29427 ± 847 cal BC); if we include the NaOH soluble fraction, a pooled mean age of 27233 ± 435 BP can be obtained. The date for Morín III (too inaccurate to be useful) would correspond, with the same method, to 29791 ± 1436 cal BC. However, there are several problems concerning the chronometric age of this layer. The dates are certainly too recent for a context described as 'early Aurignacian.' There are contradictions with the stratigraphic sequence, since the upper layer 7 yielded older radiocarbon dates. For a critical reappraisal of the chronology of this site, see Laville & Hoyos (1994).

5. Three reliable radiocarbon determinations are directly associated with the burial: OxA-8422: 23920 ± 220 BP (rabbit vertebra), OxA-8421: 24660 ± 260 BP (red deer pelvis near the shoulder) and GrA-13310: 24860 ± 200 BP (*Pinus silvestris* charcoal under the legs). Another date comes for a red deer phalanx found near the edge of the burial, but outside it (OxA-8423: 24520 ± 240 BP). Unfortunately, four human bone collagen samples that were processed failed to produce usable determinations.

6. In the Lower Solutrean layers of Parpalló (Valencia), Luís Pericot found the skull of a young individual associated with a humeral epiphysis (Pericot 1942: 273). Although he found no evidence of a grave, the association of the bones might suggest the former presence of a burial.

7. These species were originally cited as *Cyprea europaea* and *Cassis undulata,* updated here after the CLEMAM Taxonomic Database on European Marine Mollusca of the Musée National d'Histoire Naturel (Paris) (available at *http:// www.somali.asso.fr/clemam).*

8. Excavation of the pit is bracketed by two ^{14}C dates: CSIC-260: 9540 ± 120 (9140-8740 cal BC) and CSIC-216: 9430 ± 120 BP (9112-8482 cal BC).

9. Offerings were particularly evident in cluster I, found beside the right elbow, consisting of a piece of antler and a unused harpoon, with a burin, a burin spall, two endscrapers, a *pièce esquillée*, a flint core, a flint cobble, three flakes (2 in quartzite, 1 in flint) and a quartzite hammerstone.

10. This Azilian grave has been largely ignored in the literature, probably because there were also Neolithic burials in the site, and the old excavations at La Paloma have suffered a (probably unjustified) bad reputation. Hernández-Pacheco's field notes make it clear that he had no doubt about the stratigraphic context of the skeleton (Martínez Navarrete & Chapa 1980: 118-119). An unusually large number of perforated red deer canines in the Azilian layer might be an indicator of a grave, as noted by Martínez Navarrete and Chapa (1980: 139).

11. In the literature of the 1980s there are several references to another early Mesolithic burial in Roc del Migdia rockshelter (Barcelona). A direct radiocarbon date on the skeleton (UA-2550: 1560 ± 85 BP; 423-597 cal AD) indicates that it was actually an early medieval grave dug into Mesolithic deposits.

12. Carballo (1926: 24) reports a large trepanation on the left side of the skull involving the temporal and the inferior border of the parietal, but the poor preservation of the skeleton and the doubtful technical quality of the excavation advise caution.

13. A large limestone block was found in an oblique position atop the ribs of this skeleton. It is likely that it was originally placed horizontally above the head and trunk of the corpse.

14. There are two radiocarbon dates for this skeleton: AA-5294: 6265 ± 75 BP and OxA-7148: 6160 ± 55. As they are statistically consistent, it is possible to use their pooled mean: 6197 ± 44 BP (5230-5063 cal BC).

15. Human remains have also been attributed to some of the other shell middens of the region, but the available information on them is very imprecise. The clearest evidence of Mesolithic burials are Ribeiro's (1880) allusion to four skeletons at Fonte do Padre Pedro (Muge) and the presence of an important osteological collection from the site of Cova da Onça, in the valley of the Magos River, in the Museu Geológico of Lisbon (Cunha & Cardoso 2002-2003). It is very difficult to arrive at an accurate figure for the number of burials actually excavated, mainly because of the lack of detail and inadequate publication of the extensive 19th century research. A review of the (often obscure) literature by Grünberg (2000) estimates a minimum of 220 graves and 270 individuals for the three main sites. A recent reappraisal of the anthropological material preserved at various museums permitted E. Cunha and F. Cardoso (2002-2003) to calculate a minimum figure of 197 individuals (21

from Cabeço da Amoreira, 63 from Cabeço da Arruda, 51 from Moita do Sebastião, 32 from Cova da Onça and 30 with unknown provenance), excluding the Moita do Sebastião remains preserved in Lisbon (either 96 individuals according to D. Ferembach (1974), or 79 after M. Jackes (Lubell & Jackes 1985). Thus, Cunha and Cardoso (2002-2003) consider 300 individuals a reasonable estimate.

16. See Araújo (2003) for a current review of the radiocarbon dates for the Portuguese Mesolithic. The direct dates for human remains range from 6100 to 5700 cal BC for Moita do Sebastião, around 6000 cal BC for Cova da Onça, 5900-5350 cal BC for Cabeço da Arruda (there is also a problematic isolated date of c. 6400), and 5725-5500 for Cabeço da Amoreira, when calibrated for atmospheric variation. Yet, as the detailed reappraisal by Jackes and Meiklejohn (n.d.) shows, if the reservoir effect is corrected with the stable isotope values for each individual, there are relevant changes: all the dates from Moita do Sebastião and Cova da Onça are located in the first half of the 6th millennium cal BC; those from Cabeço da Arruda in the second third of the 6th millennium (and some at the end of the 6th, and the beginning of the 5th), whereas the Cabeço da Amoreira burials are dated to the last third of the 6th millennium cal BC.

17. Roche (1960) reports the existence at Moita do Sebastião of several double (graves I, VII, XVIII) or multiple (grave XIII) burials, like those in the Breton cemeteries of Hoëdic and Téviec.

18. See photographs and field drawings in Cardoso & Rolão (2002-2003: Figs. 50, 51, 53).

19. Thus, Émile Cartailhac (1886: 57) states that *"on voit des ossements çà et là, mis en tas confus comme s'ils avaient été rapprochés une fois la chair disparue"* (see more references in Jackes & Meiklejohn n.d.).

20. A recent restudy of the human skeletal remains has established a minimal number of 116 individuals (36 from Cabeço do Pez, 26 from Vale de Romeiras, 15 from Poças de São Bento, 32 from Arapouco, 6 from Amoreiras, and 1 from Várzea da Mó) (Cunha & Umbelino 1995-1997).

21. The lack of good information on most of the Portuguese Mesolithic burials hinders accurate study of this issue.

22 There is pretty good evidence for age grading. Infants and pre-adolescent children were typically buried inside, and adults outside, the dwelling area.

23. The Peninsula has produced 14.4% of the 160 funerary sites of the European Mesolithic, and 23.1% of the 1850 individuals buried (Grünberg 2000), although Iberia accounts for only 5.9% of the European land mass. The indices used here are based on general and/or regional catalogs of paleolithic burials (May 1986, Binant 1991a,b; Wüller 1999, Palma di Cesnola 2003, Mussi 2004), and are only rough approximations since there are many debatable aspects (esp. chronology) to these claims. Also, many paleolithic burials were discovered early in the history of archaeological research, when a concern for taphonomy was not very common (see, e.g., recent reassessments of Paviland [Aldhouse-Green & Pettitt 1998], Cro-Magnon [Henry-Gambier 2002], Arène Candide [Pettitt et al. 2003] and Barma Grande 6 (Formicola et al. 2004).

24. The density for the whole continent would be c. 5.2 burials/ 1,000,000 km^2 and roughly 8.3 if the photosynthetic desert (comprising northern Russia, the Baltic and Scandinavian countries and Ireland) is excluded.

25. The undated sites La Paloma and Molino de Gasparín (Colombres) have been distributed across the millennia that correspond to the chronology of the period to which they have been assigned (c. 11500-8500 cal BC for the Azilian; 8000-5000 cal BC for the Asturian). Even excluding the Portuguese shell middens, the 6th millennium shows the highest concentration of funerary evidence (42.2 % of the sites, 38.0 % of the individuals). In fact, excavation bias runs in the opposite direction, since archaeological research has focused more intensively on the Upper Paleolithic than on the Mesolithic.

26. Araújo (2003) suggests that the late Mesolithic data indicate a reorganization of the settlement systems, rather than evidence of substantial population growth.

References

Albrethsen, S.E. & E. Brinch Petersen

1976 Excavation of a Mesolithic cemetery at Vedbæk, Denmark. *Acta Archaeologica* 47: 1-28.

Aldhouse-Green, S. & P. Pettitt

1998 Paviland Cave: contextualizing the 'Red Lady'. *Antiquity* 72: 756-772.

Aparicio, J.

1990 Yacimientos arqueológicos y evolución de la costa Valenciana durante la prehistoria. *II Seminari sobre el Mediterrani-El Mare Nostrum. Les Costes Valencianes: Geografia Fisica i Humana. Gandia, Agost 1987,* pp. 26-40. València: Academia de Cultura Valenciana.

1992 Los orígenes de Oliva. In *Els Origens de les Nostres Ciutats,* pp. 75-143. València: Academia de Cultura Valenciana.

Araújo, A.C.

2003 Long term change in Portuguese early Holocene settlement and subsistence. In *Mesolithic on the Move: Papers presented at the Sixth International Conference on the Mesolithic in Europe. Stockholm 2000* (L. Larsson, H. Kindgren, K. Knutsson, D. Leoffler & A. Akerlund, eds.), pp. 569-580. Oxford: Oxbow.

Arias Cabal, P.

1991 *De Cazadores a Campesinos – La Transición al Neolítico en la Región Cantábrica.* Santander: Servicio de Publicaciones de la Universidad de Cantabria.

1999 The origins of the Neolithic along the Atlantic coast of continental Europe: a survey. *Journal of World Prehistory* 13: 403-464.

Arias Cabal, P. & M. A. Fano Martínez

2003 Shell middens and megaliths: Mesolithic funerary contexts in Cantabrian Spain and their relation to the Neolithic. In *Stones and Bones. Formal Disposal of the Dead in Atlantic Europe during the Mesolithic-Neolithic Interface 6000-3000 BC* (G. Burenhult, ed.), pp. 145-166. Oxford: British Archaeological Reports International Series No. 1201.

n.d. Le rôle des ressources marines dans le Mésolithique de la région Cantabrique (Espagne): l'apport des isotopes stables. *Bulletin de la Société Préhistorique Française* 101.

Arias Cabal, P., J. Fernández-Tresguerres, E. Álvarez Fernández, A. Armendariz, M. Cueto, M. A. Fano, M. D. Garralda, C. Mensua & L. C. Teira

n.d. Excavación arqueológica de urgencia en la cueva de La Poza l'Egua (Lledías, Llanes). *Excavaciones Arqueológicas en Asturias 1999-2002.* Oviedo: Servicio de Publicaciones del Principado de Asturias.

Arias Cabal, P. & M. D. Garralda

1996 Mesolithic burials in Los Canes cave (Asturias, Spain). *Human Evolution* 11: 129-138.

Arias Cabal, P., R. Ontañón, A. Armendariz & E. Pereda

2003 Zona arqueológica de La Garma (Ribamontán al Monte): La Garma A, cuevas sepulcrales y castro del Alto de la Garma. In *III Congreso del Neolítico en la Península Ibérica – Libro Guía de la Excursión. Preactas* (P. Arias, R. Ontañón, C. García-Moncó & L.C. Teira, eds.), pp. 42-57. Santander: Servicio de Publicaciones de la Universidad de Cantabria.

Arnaud, J.E.

1989 The Mesolithic communities of the Sado valley, Portugal, in their ecological setting. In *The Mesolithic in Europe* (C. Bonsall, ed.), pp. 614-631. Edinburgh: John Donald Publishers, Ltd.

1996 The archaeology of complex hunter-gatherers. *Journal of Archaeological Method and Theory* 3: 77-126.

Arsuaga, J.L., I. Martínez, A. Gracia, J.M. Carretero, C. Lorenzo, N. García & A.I. Ortega

1997 Sima de los Huesos (Sierra de Atapuerca, Spain) – the site. *Journal of Human Evolution* 33: 109-127.

Aura Tortosa, J.E., J. González-Tablas Sastre & S. Jiménez Brobeil

1998 Los enterramientos 'Solutrenses' de la Cueva de Nerja. In *Las Culturas del Pleistoceno Superior en Andalucía. I Simposio de Prehistoria* (J.L. Sanchidrián & M.D. Simón, eds.), pp. 237-249. Málaga: Patronato de la Cueva de Nerja.

Balbín Behrmann, R. de & J. J. Alcolea

n.d. Espace d'habitation, espace d'enterrement, espace graphique. Coïncidences et divergences dans l'art paléolithique de la corniche cantabrique. In *Colloque Comportements des hommes du Paléolithique moyen et supérieur en Europe: Territoires et milieux.* Paris: Muséum National d'Histoire Naturelle.

Barandiarán Maestu, I. & A. Cava Almuzara

2001 *Cazadores-recolectores en el Pirineo Navarro. El Sitio de Aizpea entre 8.000 y 6.000 Años antes de Ahora.* Vitoria: Servicio Editorial de la Universidad del País Vasco.

Bartel, B.

1982 A historical review of ethnological and archaeological analyses of mortuary practice. *Journal of Anthropological Archaeology* 1: 32-58.

Binant, P.
1991a *La Préhistoire de la Mort. Les Premières Sépultures en Europe*. Paris: France.
1991b *Les Sépultures Paléolithiques*. Paris: Errance.

Brinch Petersen, E.
1989 Late Palaeolithic and Mesolithic. *Arkæologiske Udgravningar i Danmark 1988*. København: København University Press.

Carballo, J.
1926 *El Esqueleto Humano Más Antiguo de España*. Santander: Published privately by the author.

Carbonell, E., A. Cebrià, E. Allué, I. Cáceres, Z. Castro, R. Díaz, M. Esteban, A. Ollé, L. Pastó, X.P. Rodríguez, J. Rosell, R. Sala, J. Vallverdú, M. Vaquero & J.M. Vergés
1996 Behavioural and organizational complexity in the Middle Palaeolithic from the Abric Romaní. In *The Last Neandertals, the First Anatomically Modern Humans* (E. Carbonell & M. Vaquero, eds.), pp. 385-434. Tarragona: Universitat Rovira i Virgili.

Carbonell, E.; M. Mosquera, A. Ollé, X.P. Rodríguez, R. Sala, J.M. Vergès, J.L. Arsuaga & J.M. Bermúdez de Castro
2003 Les premièrs comportements funéraires auraient-ils pris place à Atapuerca, il y a 350 000 ans? *l'Anthropologie* 107:1-14.

Cardoso, J.L. & J. Rolão
2002-2003 Prospecções e escavações nos concheiros mesolíticas de Muge e de Magos (Salvaterra de Magos): contribução para a história dos trabalhos arqueológicos efectuados. *Muge-Estudos Arqueológicos* 1: 7-169.

Cartailhac, É.
1886 *Les Ages Préhistoriques de l'Espagne et du Portugal*. Paris: Reinwald.

Cauwe, N.
1998 *La Grotte Margaux à Ansereme-Dinant*. Liège: Université de Liège, ERAUL No. 59.
2001 *L'Héritage des Chasseurs-cueilleurs dans le Nord-Ouest de l'Europe (10 000-3 000 avant Notre Ère)*. Paris: Errance.

Chapman, R., I. Kinnes & K. Randsborg (eds.)
1981 *The Archaeology of Death*. Cambridge: Cambridge University Press.

Chimenos, E., A. Malgosa & M.E. Subira
1992 Paleopatología oral y análisis de elementos traza en el estudio de la dieta de la población epipaleolítica de "El Collado" (Oliva, Valencia). In *Enfermedad y Muerte en el Pasado. Actas del I Congreso Nacional de Paleopatología. IV Reunión de la Asociación Española de Paleopatología*, pp. 177-188. San Sebastián: Sociedad de Ciencias Aranzadi *(Munibe*, Sup. No. 8).

Clark, G.A.
1976 *El Asturiense Cantábrico*. Madrid: Bibliotheca Praehistorica Hispana No. 13.

Clark, G.A. & M.P. Neeley
1987 Social differentiation in European mesolithic burial data. In *Mesolithic Northwest Europe: Recent Trends* (P. Rowley-Conwy, M. Zvelebil & H.P. Blankholm, eds.), pp. 121-127. Sheffield: John R. Collis, Ltd.

Corominas Planellas, J. M.
1949 El Paleolítico superior en la cueva de 'Reclau-Viver' de Seriñá (España). *Rivista di Scienze Prehistoriche* 4: 43-54.

Cunha, E., F. Cardoso & C. Umbelino
2003 Inferences about Mesolithic lifestyle on the basis of anthropological data. The case of the Portuguese shell middens. In *Mesolithic on the Move* (L. Larsson, H. Kindgren, K. Knutsson, D. Leoffler & A. Akerlund, eds.), pp. 184-188. Oxford: Oxbow Books.

Cunha, E. & F. Cardoso
2002-2003 New data on Muge shell middens: a contribution to more accurate numbers and dates. *Muge-Estudos Arqueológicos* 1: 171-183.

Cunha, E. & C. Umbelino
1995-1997 Abordagem antropológica das comunidades mesolíticas dos concheiros do Sado. *O Arqueólogo Português* Série IV, 13-15: 161-179.

Djindjian, F., J. Kozlowski & M. Otte
1999 *Le Paléolithique Supérieur en Europe*. Paris: Armand Colin.

Duarte, C.
2002 The burial taphonomy and ritual. In *Portrait of the Artist as a Child* (J. Zilhão & E. Trinkaus, eds.), pp. 187-201. Lisboa:Trabalhos de Arqueologia No. 22.

Fano Martínez, M.A.
1998 *El Hábitat Mesolítico en el Cantábrico Occidental: Transformaciones Ambientales y Medio Físico durante el Holoceno Antiguo*. Oxford: BAR International Series No. 732.

Ferembach, D.
1974 *Le Gisement Mésolithique de Moita do Sebastião, Muge, Portugal. II. Anthropologie*. Lisboa: Direcção Peral dos Asuntos Culturais.

Fernández-Tresguerres Velasco, J.A.
1976 Azilian burial from Los Azules I, Asturias, Spain. *Current Anthropology* 17: 769-770.
1980 *El Aziliense en las Provincias de Asturias y Santander*. Santander: Centro de Investigación y Museo de Altamira, Monografía No. 2.

Fleming, A.
1973 Tombs for the living. *Man* 8: 177-193.

Formicola, V., P. Pettitt & A. del Lucchese
2004 A direct AMS radiocarbon date on the Barma Grande 6 Upper Paleolithic skeleton. *Current Anthropology* 45: 114-118.

García Sánchez, M.
1982 El esqueleto epipaleolítico de la 'Cueva de Nerja' (Málaga). *Cuadernos de Prehistoria de la Universidad de Granada* 7: 37-71.

Goldstein, L.
1981 One-dimensional archaeology and multi-dimensional people: spatial organization and mortuary analysis. In *The Archaeology of Death* (R. Chapman, I. Kinnes & K. Randsborg, eds.), pp. 53-69. Cambridge: Cambridge University Press.

González Echegaray, J. & L.G. Freeman
1973 *Cueva Morín. Excavaciones 1969.* Santander: Publicaciones del Patronato de la Cuevas Prehistóricas de la Provincia de Santander.

González Morales, M.R.
1997 Changes in the use of caves in Cantabrian Spain during the Stone Age. In *The Human Use of Caves* (C. Bonsall & C. Tolan-Smith, eds.), pp. 63-69. Oxford: BAR International Series No. 667.

González-Tablas Sastre, F.J.
1990 La cueva de Nerja como santuario funerario. *Zephyrus* 43: 61-64.

Grünberg, J.M.
2000 *Mesolithische Bestattungen in Europa: Ein Beitrag zur vergleichenden Gräberkunde.* Rahden/Westfalen: Internationale Archäologie No. 40.

Gurina, N.N.
1956 *Oleneostrovskij Mogil'nik.* Moskvá-Leningrad: Materialy i Issledovanija po Archeologii SSSR No. 47.

Henry-Gambier, D.
2002 Les fósiles de Cro-Magnon (Les Eyzies-de-Tayac, Dordogne): nouvelles donées sur leur position chronologique et leur attribution culturelle. *Bulletin et Mémoires de la Société d'Anthropologie de Paris* 14: 89-112.

Hernández-Pacheco, E.
1923 *La Vida de Nuestros Antecesores Paleolíticos según los Resultados de las Excavaciones en la Caverna de La Paloma (Asturias).* Madrid: Museo Nacional de Ciencias Naturales (Comisión de Investigaciones Paleontológicas y Prehistóricas, Memoria No. 31).

Ingold, T., D. Riches & J. Woodburn (eds.)
1988 *Hunters and Gatherers.* Oxford: Berg.

Iriarte, M.J, A. Arrizabalaga, F. Etxeberria & L. Herraste
2003 La inhumación humana en conchero de J3 (Hondarribia, Guipuzkoa). In *III Congreso del Neolítico en la Península Ibérica – Libro Guía de la Excursión. Preactas* (P. Arias, R. Ontañón, C. García-Moncó & L. Teira, eds.), pg. 92. Santander: Servicio de Publicaciones de la Universidad de Cantabria.

Iturbe, G., M.P. Fumanal, J.S. Carrión, E. Cortell, R. Martínez, P.M. Guillem, M.D. Garralda & B. Vandermeersch
1993 Cova Beneito (Muro, Alicante): una perspectiva interdisciplinar. *Recerques del Museu d'Alcoi* 2: 23-88.

Jackes, M. & C. Meiklejohn
n.d. Building a method for the study of the Mesolithic-Neolithic transition in Portugal. In *Neolithic Studies 11 (Documenta Praehistorica* No. 31). Ljubljana: University of Ljubljana.

Karsten, P. & B. Knarrström
2001 Tågerup – Fifteen hundred years of Mesolithic occupation in western Scania, Sweden. *European Journal of Archaeology* 4: 165-174.

Larsson, L.
1988 *The Skateholm Project – I. Man and the Environment.* Stockholm: Almqvist & Wiksell *(Acta Regiae Societatis Humaniorum Litterarum Ludensis* No. 74).
1990 Dogs in fraction – symbols in action. In *Contributions to the Mesolithic in Europe* (P. Vermeersch & P. van Peer, eds.), pp. 153-160. Leuven: Leuven University Press.

Laville, H. & M. Hoyos
1994 Algunas precisiones sobre la estratigrafía y sedimentología de Cueva Morín (Santander). In *El Cuadro Geocronológico del Paleolítico Superior Inicial* (F. Bernaldo de Quirós, ed.), pp. 199-209. Madrid: Ministerio de Cultura.

Lubell, D. & M. Jackes
1985 Mesolithic-Neolithic continuity: evidence from chronology and human biology. In *Actas do I Reunião do Quaternário Ibérico,* pp. 113-133. Lisboa: Grupo de Trabalho Português para o Estudo do Quaternário.

Lubell, D. & M. Jackes
1988 Portuguese Mesolithic-Neolithic subsistence and settlement. *Rivista di Antropologia* 66: 231-248.

Lubell, D., M. Jackes, H. Schwarcz, M. Knyf & C. Meiklejohn
1994 The Mesolithic-Neolithic transition in Portugal: isotopic and dental evidence of diet. *Journal of Archaeological Science* 21: 201-216.

Madariaga, B.
1973 Nota préliminar acerca de los posibles fenómenos de conservación del hombre de Morín. In *Cueva Morín. Excavaciones 1969* (J. González Echegaray & L.G. Freeman, eds.), pp. 257-265. Santander: Publicaciones del Patronato de la Cuevas Prehistóricas de la Provincia de Santander.

Martínez Navarrete, M.I. & T. Chapa Brunet
1980 La industria prehistórica de la cueva de La Paloma (Soto de Las Regueras, Asturias). In *La Cueva de la Paloma, Soto de las Regueras (Asturias)* (M. Hoyos, M.I. Martínez Navarrete, T. Chapa Brunet, P. Castaños & F. B. Sanchiz, eds.), pp. 114-204. Madrid: Ministerio de Cultura (*Excavaciones Arqueológicas en España* No. 116).

May, F.
1986 *Les Sépultures Paléolithiques. Étude Critique.* Paris: Centre National de la Recherche Scientifique.

Mussi, M.
2004 East and South of the Alps: the MUP funerary and artistic record of Italy and Moravia compared. In *The Gravettian along the Danube* (J.A. Svoboda & L. Sedláková, eds.), pp. 252-269. Brno: Institute of Archaeology.

Olária, C. & Gusi, F.
2003 Un enterramiento meso-neolítico en el Cingle del Mas Nou (Ares del Maestre, Castellón). In *III Congreso del Neolítico en la Península Ibérica – Libro Guía de la Excursión. Preactas* (P. Arias, R. Ontañón, C. García-Moncó & L. Teira, eds.), pp. 92-93. Santander: Servicio de Publicaciones de la Universidad de Cantabria.

Oliveira, F. de Paulo
1892 Nouvelles fouilles faites dans les kjoekenmoeddings de la vallée du Tage (mémoire posthume). *Comunicações da Comissão dos Trabalhos Geológicos* 2(i): 57-81.

Orshiedt, J.
1999 *Manipulation of Human Skeletal Remains. Taphonomic Processes, Secondary Burials or Cannibalism?* Tübingen: Mo Vince (*Urgeschichtliche Materialhefte* No. 13).

O'Shea, J. & M. Zvelebil
1984 Oleneostrovski mogilnik: reconstructing the social and economic organization of prehistoric foragers in northern Russia. *Journal of Anthropological Archaeology* 3: 1-40.

Palma di Cesnola, A.
2003 Évolution des rites funéraires du Paléolithique Supérieur italien dans le temps et l'espace. In *Actes du Symposium International Préhistoire des Pratiques Mortuaires. Paléolithique-Mésolithique-Néolithique* (E. Derwich, ed.), pp. 131-139. Leuven: ERAUL No. 102.

Pellicer Catalán, M. & P. Acosta
1986 Neolítico y Calcolítico de la cueva de Nerja. In *La Prehistoria de la cueva de Nerja (Málaga)* (F. Jordá Pardo, ed.), pp. 339-450. Málaga: Patronato de la cueva de Nerja-Universidad de Málaga.

Péquart, M., S.J. Péquart, M. Boule & H. Vallois
1937 *Téviec. Station-nécropole Mésolithique du Morbihan.* Paris: Masson (*Archives de l'Institut de Paléontologie Humaine, Mémoire* No. 18).

Péquart, M. & S.J. Péquart
1954 *Hoëdic, Deuxième Station-nécropole du Mésolithique côtier Armoricain.* Antwerpen: De Sikkel.

Pericot García, L.
1942 *La Cueva del Parpalló (Gandía).* Madrid: Consejo Superior de Investigaciones Científicas.

Pettitt, P.B., M. Richards, R. Maggi & V. Formicola
2003 The Gravettian burial known as the Prince ('Il Principe'): new evidence for his age and diet. *Antiquity* 77: 15-19.

Price, T.D. & Brown, J.A. (eds.)
1985 *Prehistoric Hunter-Gatherers. The Emergence of Cultural Complexity.* San Diego: Academic Press.

Renfrew, C.
1994 The archaeology of religion. In *The Ancient mind: Elements of Cognitive Archaeology* (C. Renfrew & E.B.W. Zubrow, eds.), pp. 47-54. Cambridge: Cambridge University Press.

Ribeiro, C.
1880 Les kjoekkenmoedings de la vallée du Tage. *Comptes Rendues de la IXème Session du Congrès International d'Anthropologie et d'Archéologie Préhistoriques,* pp. 279-290. Lisboa: Typographie de l'Academie Royale des Sciences.

Riel-Salvatore, J. & G.A. Clark
2001 Grave markers: Middle and early Upper Paleolithic burials and the use of chronotypology in contemporary paleolithic research. *Current Anthopology* 32: 449-460.

Ripoll Perelló, E.
1959 Excavaciones en el Abrigo Romaní (Capellades, Barcelona). *Ampurias* 21: 247-248.

Roche, J.
1960 *Le Gisement Mésolithique de Moita do Sebastião. Muge. Portugal.* Lisboa: Instituto da Alta Cultura.

1972 Les amas coquilliers (*concheiros*) mésolithiques de Muge (Portugal). In *Die Anfänge des Neolithikums vom Orient bis Nordeuropa. Teil VII. Westliches Mittelmeergebiet und Britische Inseln*, pp. 72-107. Köln: Böhlau.

1982 A gruta chamada Lapa do Suão. *Arqueologia* 5: 5-18.

1989 Spatial organization in the Mesolithic sites of Muge, Portugal. In *The Mesolithic in Europe* (C. Bonsall, ed.), pp. 607-613. Edinburgh: John Donald.

Roche, J. & O. da Veiga Ferreira
1967 Les fouilles récents dans les amas coquilliers mésolithiques de Muge (1952-1965). *O Arqueólogo Português*, Série III, 1: 19-41.

Saxe, A.
1971 Social dimensions of mortuary practices in a Mesolithic population from Wadi Halfa, Sudan. In *Approaches to the Social Dimensions of Mortuary Practices* (J. Brown, ed.), pp. 39-57. Salt Lake City, UT: Memoirs of the Society for American Archaeology No. 25.

Scarre, C.
2002 Contexts of monumentalism: regional diversity at the Neolithic transition in north-west France. *Oxford Journal of Archaeology* 21: 23-61.

Scarre, C., P. Arias, G. Burenhult, M.A. Fano, L. Oösterbeek, R. Schulting, A. Sheridan & A. Whittle
2003 Megalithic chronologies. In *Stones and Bones. Formal Disposal of the Dead in Atlantic Europe during the Mesolithic-Neolithic Interface 6000-3000 BC* (G. Burenhult, ed.), pp. 65-111. Oxford: BAR International Series No. 1201.

Schulting, R.J.
1998 Slighting the sea: stable isotope evidence for the transition to farming in northwestern Europe. *Documenta Praehistorica (Porojilo o Raziskovanju Paleolitika, Neolitika in Eneolitika v Sloveniji)* 25: 203-218.

Sherratt, A.
1990 The genesis of megaliths: monumentality, ethnicity and social complexity in Neolithic western Europe. *World Archaeology* 22: 147-167.

1995 Instruments of conversion? The role of megaliths in the Mesolithic/Neolithic transition in North-West Europe. *Oxford Journal of Archaeology* 14: 245-260.

Srejovic, D.
1969 *Lepenski Vir. Nova praistorijska Kultura u Podunavlju*. Beograd: Srpska Knjilevna Zadruga.

1972 *Europe's First Monumental Sculpture. New Discoveries at Lepenski Vir*. London: Thames & Hudson.

Srejovic, D. & Z. Letica
1978 *Vlasac: Mezolitsko naselje u Derdapu*. Beograd: Srpska Akademija Nauka i Umetnosti.

Stuckenrath, R.
1978 Apéndice V. Dataciones de Carbono-14. In *Vida y Muerte en Cueva Morín* (J. González Echegaray & L.G. Freeman, eds.), pg. 215. Santander: Institución Cultural de Cantabria.

Stuiver, M. & P. Reimer
1993 Extended ^{14}C data base and revised CALIB 3.0 ^{14}C age calibration program. *Radiocarbon* 35: 215-230.

Stuiver, M., P. Reimer, E. Bard, J.W. Beck, G. Burr, K. Hughen, B. Kromer, F.G. McCormac, J. van der Plicht & M. Spurk
1998 INTCAL98 Radiocarbon age calibration, 24,000-0 cal BP. *Radiocarbon* 40: 1041-1083.

Testart, A.
1982 *Les Chasseurs-cueilleurs ou l'Origine des Inégalités*. Paris: Société d'Ethnographie.

2001 Deux politiques funéraires. *Trabalhos de Antropologia e Etnologia* 41: 45-66.

Turbón, D., A. Pérez Pérez & C. Lalueza
1994 Los restos humanos del nivel solutrense de la cueva de Nerja (Málaga). *Actas del VIII Congreso de la Sociedad Española de Antropología Biológica*, pp. 51-62. Madrid: Universidad Autónoma de Madrid.

Ucko, P.J.
1969 Ethnography and archaeological interpretation of funerary remains. *World Archaeology* 1: 262-280.

Vallois, H.
1930 Recherches sur les ossements mésolithiques de Mugem. *l'Anthropologie* 40: 337-389.

Vanhaeren, M. & F. D'Errico, F.
2001 La parure de l'enfant de La Madeleine (fouilles Peyrony). Un nouveau regard sur l'enfance au Paléolithique supérieur. *Paléo* 13: 201-237.

Vierra, B.
1995 *Subsistence and Stone Tool Technology: an Old World Perspective*. Tempe: Arizona State University Anthropological Research Papers No. 47.

Weninger, B., O. Jöris & U. Danzeglocke
2003 *CalPal-University of Cologne Radiocarbon Calibration Program Package*. Universität zu Köln: Institut für Ur- und Frühgeschichte. Text available at *http://www.calpal.de*

Wüller, B.
1999 *Die Ganzkörperbestattungen des Magdalénien*. Bonn: Habelt.

Zilhão, J.
1993 The spread of agro-pastoral economies across Mediterranean Europe: A view from the Far West. *Journal of Mediterranean Archaeology* 6: 5-63.

Zilhão, J.
1997 *O Paleolítico Superior da Estremadura Portuguesa*. Lisboa: Colibrí.

Zilhão, J. & E. Trinkaus (eds.)
2002 *Portrait of the Artist as a Child. The Gravettian Human Skeleton from the Abrigo do Lagar Velho and its Archeological Context*. Lisboa: Instituto Português de Arqueologia.

Zvelebil, M. & P. Rowley-Conwy
1986 Foragers and farmers in Atlantic Europe. In *Hunters in Transition – Mesolithic Societies of Temperate Eurasia and their Transition to Farming* (M. Zvelebil, ed.), pp. 67-93. Cambridge: Cambridge University Press.

Part IV

The Mesolithic of the Atlantic Façade

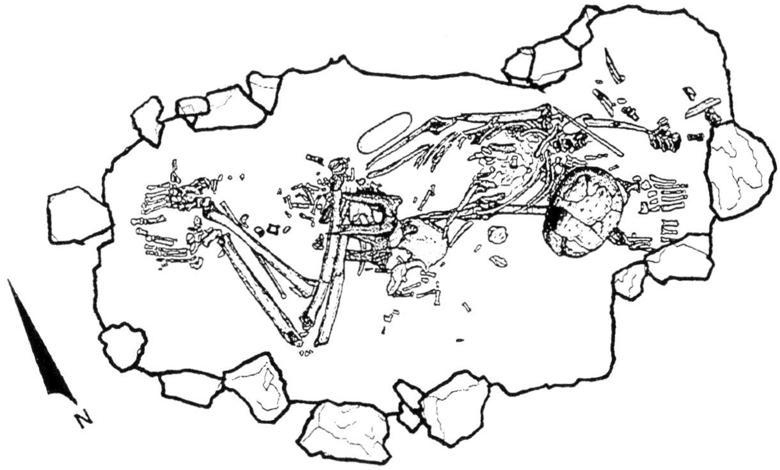

18

Transitions: Into and Out of Mesolithic Adaptations along the Atlantic Façade of Europe and Beyond

Lawrence Guy Straus
University of New Mexico

Introduction

One of the enduring, central problems of prehistoric archeology is how to explain strong patterns of similarity in material culture over broad expanses of space and within fairly closely defined periods of time. The widespread appearances of geometric microlithic industries on the one hand, and of simple macrolithic ones on the other in the early Holocene of Europe are problems of just this sort. The first common (late 19th century) explanations focused on the vitalistic notions of progress (as in the appearance of the Neolithic and food production) or (especially in the case of the European Mesolithic) degeneration within the overarching paradigm of self-assured Euro-American social evolutionism. These were generally supplanted early in the 20th century (following the crisis of confidence wrought by World War I) by appeals to movements (i.e., invasions, migrations of people and/or diffusion of objects and/or ideas) linked to the culture-historical paradigm which equated artifact types and other manifestations of cultural behavior with identifiable ethnic units. Of course V. Gordon Childe was the principal and enduring exponent of this approach. Novelties were usually seen as coming from somewhere else (coincidentally places where the archeology was generally less known), while local people were considered merely to be passive receptors.

By the mid-1960s, dissatisfaction with the sterility of essentialist culture-history and enthusiasm for developments in the anthropological study of pre-industrial economies and ecological theory all combined to promote functionalist-adaptationist explanations of causality with the archeological record. Application of uniformitarian principles led processualist or economic archeologists to propose that, faced with similar problems, prehistoric people would tend to invent similar solutions (technological, social, ideological), given demographic, technological and subsistence constraints, and given key assumptions about the operation of specific (as opposed to general) evolution and a high degree of reliance on the notion of convergent function through the determinism of systems theory.

Recent developments in archeological theory have, among other things, pointed out the danger of tautology in the adaptationist perspectives of many processualist studies and expressed dismay at the perceived (environmental, demographic, etc.) deterministic nature of explanatory scenarios in which the roles of the individual and of the sociocultural entity (with free will and the capacity for idiosyncratic behavior – including art and ideology) are ignored or minimized.

The symposium, 'Old People and the Sea: the Mesolithic of the Atlantic Facade,' organized by the other two members of the 'La Riera troika' (Geoff Clark and Manolo González Morales – the latter my partner in the ongoing excavation of El Mirón), was for me an occasion for reflection on my past work and theoretical perspectives on the Epipaleolithic/Mesolithic in Cantabrian Spain (e.g., Straus & Clark 1986, Straus 1992), Portuguese Estremadura-Alentejo (e.g., Carvalho, Straus et al. 1989, Straus 1989, 1991; Straus et al. 1988), and French Gascony (e.g., Straus 1995), and for comparison with my (then) current work in Belgian Wallonia (e.g., Otte & Straus 1997, Léotard, Otte & Straus n.d.). It helped me to focus on those aspects of our excavations (generally somewhat peripheral to my main Upper Paleolithic interests) that had led my colleagues and me to uncover, analyze and interpret materials dating to the first four millennia of the Holocene at La Riera Cave, at Escravalheira and Vidigal, at Abri Dufaure, at Bois Laiterie Cave and Abri du Pape. The symposium and resultant publication on the one hand helped to consolidate some of my long-held views (e.g., on the importance of first considering functional factors in artifact interassemblage variability; on the significance of demographic/regional population packing factors in subsistence and settlement change), but on the other hand forced me to recognized the over-simplicity or probably incorrectness of some of my once-cherished ideas (e.g., the relative unimportance of climatic/environmental explanatory factors *vis à vis* demographic ones [at least in the lower-middle latitudes]; the contemporaneity and functional complementarity of the Azilian and Asturian in north-coastal Spain; the non-limpet collecting function of Asturian picks; the unlikelihood of being able to define/identify prehistoric time-transgressive, geographically-bounded sociocultural entities). Many of these developments in my thinking actually were fermenting as a consequence of my

Solutrean and Magdalenian research into the contraction and reexpansion of the human range during and after the Last Glacial Maxium (e.g., Straus 1991b,c; Straus et al. n.d.), and because of my first-hand encounters (through excavation) with the Neolithic record in Portugal and Belgium (Straus et al. 1988, 1992, n.d.; Léotard, Otte & Straus n.d.).

Nonetheless, what follows is not the 'mea culpa of a fanatic', to paraphrase Emile Cartailhac's famous 1902 article recanting his rejection of the authenticity of the Altamira paintings. Instead, I propose to seek some degree of balance among the various possible explanations for Mesolithic similarities and differences in Western Europe and argue that different kinds of explanations may be suitable for various aspects of the record. Not everything stems from ethnicity and isocrestic style, yet not all phenomena are deterministically caused be some inherent trend toward subsistence efficiency in the case of stress and demographically driven necessity. And some developments are contingent on antecedent developments; technology is cumulative, taught, learned, and disseminated. Inventions – akin to genetic mutations – are sometimes successful and replace earlier forms when and if the conditions afford some 'selective advantage' to the people adopting them. But many technological elements – like neutral mutations – are functionally equivalent (interchangeable), or at least 'good enough' for the job. Some inventions spread at the expense of others due simply to the serendipity of human contacts or territorial boundaries and because of critical, but non-human conditioning factors (such as local lithology and ecology). Explanation in archaeology should be 'a big tent', to borrow a phrase from American politics and diplomacy of the early 1990s, yet the fact remains that the key determinant of forager behavior was and always will be the acquisition of food produced naturally in their regional environments sufficient to maintain group survival. This strongly implies central co-explanatory roles for the physical and demographic environments in all forager studies. These factors were undoubtedly crucial to the development of Mesolithic 'complexity' from the Baltic to the Tagus estuary, from Brittany to the Iron Gates gorge. Each local culture then went on to develop its own specific forms of more elaborate social organization (including social stratification) and of ideology suited to new circumstances of lessened mobility (or even sedentism) and denser full-time co-resident populations. Why 'complexity' developed in some areas and not in others can best be explained through a combination of resource and demographic factors. Unsurprisingly the areas where those factors converged were associated with lakes and rivers, and ocean coastlines and estuaries – the richest ecotonal habitats of mid-latitude Europe.

Northwest Coasts and Islands

Appropriately, this volume begins with a characteristically lucid discussion of Mesolithic complexity by Peter Rowley-Conwy (Chapter 1). He proposes a distinction between developmental and adaptational explanations, and, in essence, argues that the former is vitalistic and 'progress-oriented.' Population pressure is put in this category and is rejected as a non-explanation. In my view, however, Rowley-Conwy has set up a straw man and then has over-generalized and misrepresented the population pressure argument. It is specific, regional situations of imbalance between numbers of mouths to feed and resources that are argued to have caused intensification in the wild (and, later, domesticated) food quest through overall diversification and situational specialization and through changes in mobility, group size, organization and ideology. I am in sympathy with the notion that some degree of complexity occurred in some places and time intervals during the Upper Paleolithic, given specific conditions of relatively high regional population density (usually in geographically circumscribed areas like Cantabrian Spain). This was one of the main findings of the La Riera Paleoecology Project: Mesolithic-like broad-spectrum adaptations arose under specific conditions of the Solutrean and Magdalenian in this narrow coastal region of the Iberian refugium during the Last Glacial Maximum (Clark & Straus 1983). It is true that we lack evidence of late Upper Paleolithic settlement in some now inundated coastal strips, but there do seem to be quantitatively and qualitatively very complex adaptations in limited areas of the early Holocene world along some European coastlines – and they are the consequence of local adaptations by relatively human populations. I fail to see the need to separate the variable of local population density from the overall ecology of hunter-gatherers. Population density is one of the key variables in all ecological systems.

The chapter by Clive Bonsall (Chapter 2) evoked a sense of *déja vu* since nearly 20 years ago I made a similar argument attempting to put both microlithic (Azilian) and macrolithic (Asturian) industries in Cantabrian Spain together as part one system with different structural 'poses' (Straus 1979). While my own hypothesis seems to be on the ropes (for now anyway), I can applaud Bonsall's fairly convincing demonstration of contemporaneity and functional complementarity between the Obanian and the Scottish microlithic industries – always a viable alternative to be explored in opposition to unthinking cultural/ethnic 'explanations'. This is a clear example of what intensive new research and a concerted radiocarbon dating program can achieve by way of new understanding.

Marek Zvelebil (Chapter 3), as usual, does a splendid job of synthesizing the extent of variability embedded within the terms 'hunter-gatherer mode of production' or 'band society;' he shows us just how complex (and far removed from the 'little Bushman' model) things got along and near the Baltic Coasts in the Atlantic and Subboreal periods in terms forager social stratification. With delayed return (storage), these peoples were torn, he argues, between the egalitarian ideology of less complex foragers and incipient differences in wealth and power, the latter smoothing the way for the eventual adoption of the agricultural way of life with its very different ideological underpinnings. This is a fascinating, not so outlandish (although ultimately unproven) argument which goes a ways to bridge the gap between processualist and post-processualists approaches to the late, complex Mesolithic and to its eventual transformation into the Neolithic.

Peter Woodman (Chapter 4) who has done so much to renew and expand the study of the first colonisation and earliest prehistory of Ireland, that of the Mesolithic, once again provides us with significant questions and partial answers. In this case, the issue is 'where and what is the evidence for coastal resource exploitation in Ireland.' Ireland is a good case of data that are hard to come by, but that can eventually be teased out by dint of hard, well-planned work, and despite, drowned early coastlines. Eustatic rise, isostatic rebound, peat bog formation and intensive agriculture all make the task very difficult and variably successful from region to region. Yet, Woodman (at this point, at least) seems to be fairly certain that the coast was underutilized in the Mesolithic, especially relative to later periods. Not one to challenge his assessment, I would, however, sound a cautionary note, as underwater discoveries (like those of recent years off Denmark [see Fischer 1995]) could easily change the state of affairs, as did Woodman's excavations at Mount Sandel overlooking the Bann River estuary near Belfast (Woodman 1985). Recent and current survey and testing programs in several Irish counties could also change the picture.

In his contribution (Chapter 5), Christopher Tolan-Smith continues his documentation and discussion of the radiocarbon record for the recolonization of Britain. He observes that the Atlantic Façade was only recolonized when interglacial-type marine currents were once again re-established at the beginning of the early Holocene, bringing with them rich coastal resources exploitable by humans. These environmental factors, combined with demographic pressures, led to the development of Mesolithic coastal adaptations with all their attendant technological inventions and infrastructural investments, and hence social organizational changes leading to a relative degree of complexity. Obviously a sensible approach, multicausal in character and built up on a solid base of chronological and archeological facts.

Southwestern France, Northeastern and Eastern Spain

The papers in this section are a mixed bag, but with a relatively strong focus on typological and culture-historical matters. The coastal regions of Aquitaine are generally poor in Mesolithic evidence, probably because of the nature of the coast from the mouth of the Adour to the mouth of the Gironde (cover sands and dunes along a vast stretch where a huge, actively eroding area of the continental shelf was inundated during the Tardiglacial and early Holocene). It is not surprising that A. Turq and L. Detrain (Chapter 6) focus on litho-cultures and their possible phylogenetic relations, as there are few intact sites and few subsistence data. In truth, there is still a major 'hole' in our evidence for the early Holocene between the western Pyrenees and Brittany, making the late survival of a Mesolithic ('Sauveterrian') lifeway and late appearance of early Neolithic ('Cardial') during the Atlantic period at the site of Léde du Gurp on the Mèdoc Peninsula above Bordeaux all the more enigmatic and intriguing (e.g., Marambat & Roussot-Larroque 1989). Much more research, beginning with systematic survey (including core boring), clearly is needed. And this is exactly what is reported by Pascal Foucher and Cristina San Juan (Chapter 7). Their research project (conducted in association with González Morales and as mitigation preparatory to highway construction) has revealed a wealth of terminal Upper Paleolithic and Mesolithic sites in a region – Charente Maritime and Vendée – where before this intensive work there had been none at all. A record is being constructed. The next step will be to carry out larger excavations; to gather, analyze and interpret both archeological and paleoenvironmental/resource data. The rich marshlands and streams of the region between the Gironde and the Brittany Peninsula are likely to yield fascinating information about late forager and early farmer adaptations in the near future. The possibility (suggested by González Morales) that when the Neolithic finally 'arrived' along the Cantabrian Atlantic coast it was from the Aquitaine, and not directly from the Spanish Mediterranean via the Ebro Valley, is an interesting one (since the Neolithic had 'stalled' on the Ebro side of the Cantabrian Cordillera for a millennium), making the study of post-paleolithic southwestern France critical to understanding the neolithization of north-coastal Spain.

A brief excursion takes the reader away from the Atlantic Façade to the Mediterranean one, with articles by Emili Aura on Levante and Andalucia (Chapter 8) and by Pilar García-Argüelles (Chapter 10) on Catalunya. Parallels between the Mediterranean and Atlantic regions of Iberia are truly surprising, given the manifest environmental differences (both terrestrial and marine). Are they due to common causes and basic subsistence conditions leading to similar results (functional convergence) or can they be attributed to folk migrations, diffusion or some combination of all these mechanisms? Although attenuated *vis à vis* Vasco-Cantabria, the kinds of climatic and vegetational fluctuations in Mediterranean Spain (and Portugal) are quite similar, and cultural 'correlates' or 'results' are also broadly (and sometimes narrowly) analogous. Despite different terminology (microlaminar Epipaleolithic versus Azilian; geometric Epipaleolithic versus Mesolithic), there are striking similarities in the technological trajectory. There are even worked (and sometimes painted) cobbles in the Mediterranean world, as in the Atlantic one. And there are striking similarities in the development of marine resource exploitation (best shown by the long record from the Nerja Cave in Málaga). In the earlier periods (microlaminar/

Azilian) there is significant evidence of the slaughter of young red deer and ibex along both Iberian coastlines. Now there is even a cemetery associated with a shell midden (El Collado) in Valencia, with trace element evidence of a diet rich in marine and vegetal foods – so reminiscent of the *concheiros* of the Atlantic estuaries of south-central Portugal. Coincidence or result of similar circumstances of environment (sea level rise, reforestation, faunal change, dense local human populations) during the early Atlantic period on both sides of the Peninsula? Of course it is also true that the rather distinctive 'La Cocina' spurred isosceles triangles (first defined in Valencia) are likewise abundant in penecontemporaneous Portuguese sites (and are sometimes called 'Muge' triangles! [see Roche 1972]).

Ceramics appear in the shell middens of both coasts, even though the forager lifeway survived in some cases 'cheek-by-jowl' with full-fledged Neolithic communities, raising the possibility of all kinds of 'mutualistic' relationships between the two 'groups' that combined to form complex regional systems of trade or raid, intermarriage or intolerance, and eventual absorption or extinction. The overall picture is one of cultural and settlement continuity, making pivotal the question: Why did some Mesolithic foragers eventually opt for Neolithic food production when, in some cases, it was available to them for generations?

Portugal and Galicia

Chapter 11, by João Zilhão, is part of a much larger essay that has been published elsewhere and that is a *tour de force* (Zilhão 1993). It is clear that the coastal/estuarine zones of Portuguese Estremadura, Ribatejo and Alentejo saw the development of (semi-) sedentary foraging systems with extraordinarily broad spectrum subsistence and a degree of social complexity that rivaled those of the Danish Ertebølle, the Breton Teviecian and the Romano-Serb Iron Gates Gorge cultures (Lepenski Vir, Padina, Vlasac, Cuina Turcului – Radovanovic [1996]), As Zilhão shows, however, there are several types of Mesolithic settlements even at the zenith of this cultural phenomenon in the mid-Atlantic period, from the huge, permanent 'villages' of the Muge estuary to more modest, seasonally occupied sites such as Vidigal (first excavated by me with Bradley Vierra). That these people were heavily dependent upon aquatic food sources has recently been demonstrated through stable isotope analyses by David Lubell and Mary Jackes (e.g., Lubell et al. 1994).

The contrast with the early Neolithic folk of south-central Portugal could not be more stark: diets rich in C3 plants (as in the case of a 6000 year-old human from my excavations in Casa da Moura [Straus et al. 1988]). These were at least dietarily distinct populations, located close to one another, probably in contact (cf. the ceramics at Vidigal and in late levels at some of the Muge and Sado *concheiros),* but separate. The sea-borne Neolithic enclave model seems to be the most likely one (at least for now), although from where those immigrants came (and who they were – Andalusians, North Africans? [see Lewthwaithe 1986]) remains a mystery. Once again, one piece of critical new research has revolutionized the early Holocene archeology of a whole region (and in this case, the Upper Paleolithic prehistory as well), namely Zilhão's seminal excavation of Caldeirão Cave in the interior of Estremadura (Zilhão 1992). With an hiatus between the Magdalenian and the early (c. 6800-6200 BP) Neolithic, Caldeirão seems to argue for occupation by farmers of precisely those areas not favored by Mesolithic foragers. This makes it all the more crucial to obtain more credible dates and stratigraphic details for the supposed early Neolithic sites excavated by C. Tavares da Silva and J. Soares (Tavares da Silva et al. 1986, Tavares da Silva 1989) near the coast of the Sines Peninsula between the Muge, Sado and Mira *concheiro* clusters.

Portuguese Mesolithic collections made by Zilhão, J. Arnaud, myself and others were analyzed for a doctoral dissertation by Bradley Vierra under my direction (Vierra 1992, 1995). In Chapter 12, Vierra gives a brief résumé of the body of theory underlying his analysis, together with the principal results. He attacked one of the central questions of Mesolithic technology, namely 'why the shift to microliths?' Vierra sees increased subsistence diversification, especially the very broad-spectrum typical of the Atlantic period, as ultimately causing the culmination of the microlithization trend in the late Mesolithic (with abundant geometrics). He arrives at this conclusion through the application of theories of tool efficiency (under conditions of time stress and the need for reliability) as developed by David Clarke, Robin Torrence, Peter Bleed, Lawrence Keeley and others. This is one of the more theoretical of the chapters in the present volume. The strength of Vierra's ideas is first nicely tested through ethnographic analogy with reference to a survey of subsistence foods and subsistants *(sensu* Oswalt) from the Human Relations Area Files. Vierra's laudable goal is to go beyond the simple (but vitalistic) sort of statement that somehow microliths can be assumed to be 'more efficient' in the abstract. Under what conditions would they be selected for as the expendable elements of a 'pull-out, plug-in' technology for hunting (and other tasks such as reaping)? Yet the dangers of tautology still lurk in arguing that any particular invention is 'more efficient' under particular specified conditions. While the subsistence data from Vierra's sample sites are admittedly quite inadequate, his results are intriguing. But the fact remains that very similar kinds of geometrics (triangles, crescents, etc.) were adopted in all kinds of environments, with many different specific resource bases, local demographic densities, seasonal regimens, etc., from the Near East to northwest Europe (Clark 1996, Barton & Neely 1996). There is a great deal of equifinality in the rise of microliths – and they (in a general sense) had arisen before at different times under very different climatic conditions (e.g., in the late Solutrean, in the early and late Magdalenian – see papers in Elston & Kuhn 2003). Actually demonstrating that one type of tool is more 'reliable' than another is very difficult to do. What is clear is that we lack the really important elements in these technologies: the hafts (probably mostly made of wood). All may not be purely functional, as the spread of geometric microliths (which could be more or less isofunctional with other kinds of tools) could also be indicative of diffusion in the context of a general need for a technology (consisting of curated hafts and expendable bits of stone) that was cheap, effective, multi-purpose, easy to replace and maintain under a variety of conditions of broad-spectrum subsistence where diversity selected for flexibility.

A note in passing on the Portuguese Mesolithic is that microliths can be found in archeological association with macroliths (cobble picks, Mirian 'hoes', choppers, large flakes, and even large slab grinding stones and handstones). But they can also be found by themselves, as can the macroliths. When associated, as in the Muge *concheiros* or (less abundantly) at the Mirian coastal site of Palheirões do Alegra (Penalva & Raposo 1986), they suggest the co-existence of very different kinds of subsistence activities (e.g., hunting versus shell fishing versus plant collection and processing) that in other, less complex loci, were segregated. At Vidigal, we were able to show that all these activities were probably represented directly or indirectly (in the case of on-site refitting for off-site hunting) at the site, but in spatially separated activity/discard areas.

Moving up the Atlantic coast to 'land's end', the situation becomes murkier in some ways in Galicia. Although long known as the setting for coastal Mesolithic manifestations, one of the ironies of Galicia is that exploitation of marine/estuarine resources – often with 'rudimentary' cobble tools – has been a continuous fact of life ever since Mesolithic times (and before). This is a region where the possibility of technological convergence as a major explanation for formal archeological similarities reigns supreme, as has been shown by J.M. Vázquez Varela both here (Chapter 9) and in other publications. Unraveling the skeins of Galician lithic technology through the ages is a task reminiscent of (but far greater than) that faced by Geoffrey Clark when he set out to restudy the original 'Asturian culture' of Cantabrian Spain and to ultimately prove the Conde de la Vega del Sella correct in separating Lower Paleolithic materials from this truly Holocene Mesolithic phenomenon. For reasons of lithology and geography (a long, complex, resource-rich coastline abutting a relatively poor hinterland), Galicia has repeatedly seen its people turn to the sea, often with what seem to be simple, expedient technologies. Here, as in Portugal, the Neolithic (with its megalithic monuments and imported cereals) seems to have appeared from elsewhere – perhaps via the sea – and to have co-existed for a time with coastal Mesolithic foragers.

Cantabrian Spain

Quite appropriately, treatment of the Cantabrian geographic region1 begins with a consideration of the Azilian by Juan Fernández-Tresguerres, the foremost authority on this Epipaleolithic culture or phase (Chapter 13). Not only has he comprehensively studied the existing collections from Asturias and Cantabria, but Father Tresguerres has been, for a quarter-century, excavating and analyzing one of the most important Azilian sites in the region – Los Azules Cave in the mountainous interior of eastern Asturias. Long undifferentiated geographically and temporally, the Azilian is here shown to be more diverse than once portrayed, mainly because there has been so much new research, both in the classic regions of Asturias and Cantabria and in the Basque Country, and because there is now such an impressive list of high-quality radiocarbon dates for many sites, including the multilevel Los Azules. The temporal, artifactual, settlement, subsistence, and even artistic (geometric) overlaps with the terminal Magdalenian are now firmly established. The Azilian is technologically an Epimagdalenian, but it is also a phenomenon in its own right, spanning at least 2000 years straddling the Pleistocene-Holocene boundary. This was a time of considerable climatic instability and cultural change manifest in changes in art and ideology that are skillfully synthesized by Tresguerres. The correlation between the termination of full glacial climate and the disappearance of cave art and of realistic portable art in the Azilian is, for me, one of the strongest arguments in favor of at least a partially ecological-functional explanation for the great Upper Paleolithic art (Barton et al. 1994, Clark et al. 1996).

From my point of view, one of Tresguerres' most interesting findings is the suggestion of a shrinking in the size of territories and reduction in mobility in the Azilian in the Cantabrian region. This rings true with what has also been argued for France, although the time of the appearance of the Azilian (or of the disappearance of the classic Upper Magdalenian) is regionally variable – particularly early in Périgord (where reindeer went extinct by the beginning of Alleröd), particularly late along the edge of the western Pyrenees (where reindeer abundantly survived through Alleröd and even (albeit in much reduced numbers) into Dryas III and Preboreal), and slow, gradual and fairly late (end of Alleröd/Dryas III) in Vasco-Cantabria (where reindeer was never important to subsistence and where the main food sources [red deer, ibex, fish and limpets] remained unchanged across the Pleistocene-Holocene transition). Given this variability in timing and in ecological context, and despite Tresguerres' arguments for provinciality and 'stylistic' diversity in the Azilian, it is apparent that a degree of cultural unity does exist, tying the Azilian 'cultural sphere' together from Asturias to northeast France. The flat-section harpoons and the painted and engraved pebbles may have very uneven distributions and localisms of style, but do seem suggest at least a loose network of human relations (intermarriages, etc.) among Azilian bands. The question of the broader similarities to the Tjongerian and other Federmesser complex industries in northwest Europe, on the one hand, and to the microlaminar Epipaleolithic of Mediterranean Iberia on the other, is clearly another matter, in which convergence may have played more of a role than diffusion.

It is in Chapter 14 that I (almost) raise the white flag. Persistent research and increased numbers of radiocarbon dates seem to make it clear that González Morales convincingly demonstrates that the Asturian mainly or fully follows the Azilian and that it is not simply a functional/ depositional facies of the Cantabrian Azilian/Epipaleolithic. It would seem that my 1979 hypothesis is disproven on the strength of new discoveries (Straus 1979). In fact, preliminary excavation at El Mirón Cave, deep in the montane interior of eastern Cantabria, by González Morales and me in 1996, yielded evidence of an hiatus in occupation between the Azilian and the early Cantabrian Neolithic (i.e., between c. 10-9000 BP and 5200 BP) – at least in one small area of this huge cave (Straus & González Morales 1996). This is one more indication that the Cantabrian interior – abundantly occupied during the Magdalenian and Azilian – was abandoned after c. 9000 BP. The (scanty) evidence for re-occupation of the montane interior does not come until c. 6900 BP at Los Canes in the shadow of the Picos de Europa in eastern Asturias and 5800 BP at Tarrerón near El Mirón – the latter not long before the (much delayed) appearance of megaliths and the Neolithic (Arias 1991). Of course, it is always possible that new discoveries (like Los Canes), could change the panorama and again make viable the hypothesis of functional complementarity between inland sites emphasizing hunting coastal sites emphasizing fishing and mollusc gathering (see also Clark 1989).

One of González Morales' more interesting observations is the distinction between the classic Asturian of eastern Asturias and western Cantabria and the coastal middens of eastern Cantabria and Vizcaya. The very different malacofaunas (dependent on different littoral substrates) may be the reason why Asturian picks (for limpet gathering, apparently) are not present in the sites of the Santoña or Guernica estuaries, which are rich in oysters. A remaining problem I have with the Asturian question, however, has to do with the nature of the complete human subsistence during the period between 9000-5500 BP, before the arrival of domesticates. There are relatively small amounts of terrestrial mammal bones (and no marine mammal remains) in the Asturian *concheros*. It is simply very difficult for me to imagine that people made most of their living from limpets and topshells, supplemented by small numbers of crabs, sea urchins, fish etc., even if they were also eating considerable amounts of plant foods (roots, acorns, hazelnut etc.), as I argued in 1979. The Asturian *concheros* still seem to me (and to Clark) to represent only part of the full gamut of annual human subsistence activities. Of course, the excavation and analysis of Mazaculos Cave (eastern Asturias) by González Morales show that there was initially an actual occupation of this cave mouth before it became a shell dump, but even that basal occupation surface is poor in indications of hunting. Yet hunting of red and roe deer, boar, ibex, chamois and smaller game must have been important to overall human diets. The Mazaculos work does go a long way toward resolving the question of the seasonality of the Asturian sites, first addressed by the La Riera Palaeoecological Project. I suspect, however, that more remains to be found and written on this and other Asturian subjects.

Much of what was written about the Asturian between the work of the Conde de la Vega del Sella and that of González Morales was produced by Geoffrey Clark (Chapter 15) who put the Asturian back on the Mesolithic map and agenda after a murky period of misclassification and misunderstanding in the 1950's-early 1960's. Clark's work (1976, 1983) – especially the radiocarbon dating and complete collection analyses of classic concheros – was truly seminal and led, among other things, to the La Riera re-excavation and to González Morales' new excavations. It is apparent that Clark (like I) had not read the Bailey/ Craighead manuscript before he wrote his; otherwise he would have replied to it in his essay. More on that below.

Clark's chapter shows how Mesolithic research in (essentially coastal) Iberia has exploded since his pioneering work in the late 1960's. It shows how much of the Anglo-American agenda (processualist, functionalist, ecological) has been adopted in Iberian Mesolithic research over the course of the 1970s and 1980s for a variety of reasons (research by Americans, Britons and Canadians in Cantabria, Levante, and Portugal, training and academic visits of Portuguese and Spaniards in USA and Britain). Based to a large extent on Straus (1979), I will not reiterate Clark's arguments on functional complementarity between coastal and interior sites. Obviously I subscribe to the notion that we must take into consideration fundamental facts of lithology, technology and typology, as well as tool and site function, in interpreting interassemblage variability, not only in the Mesolithic (in Cantabria as elsewhere), but also in the Upper Paleolithic. There is a fundamental, enduring distinction between 'leptolithic' artifacts made mostly on flint and heavy duty artifacts made on quartzite. Beyond this fact, it is equally clear that there are major trends (in Cantabria and in other Iberian regions) toward subsistence intensification that began long before the Mesolithic (Clark 1987). Several factors probably contributed to causing these trends; regional population packing was certainly one of them and an important one at that. Relative population density interacted with Europe-wide climatic factors (i.e., the onset of the Last Glacial Maximum) to cause the first great pulse of subsistence intensification in the Solutrean, as the human range contracted and population eventually concentrated in the favorable southern refugia. The second great pulse of intensification in Cantabria also resulted from an interaction of environment and population density – rising sea level combined with the growth of dense forest in the mountains appears to have restricted human settlement to a narrow coastal zone, where territorially constrained humans had to further diversify their food quest especially in terms of marine resource exploitation. I concur with Clark that as data become available from other regions of Iberia (Levante, Andalucia and Portugal), they tend to point in the same directions as does the Cantabrian record in terms of specific environmental changes and increases in regional population combining to produce punctuated trends in subsistence intensification through both situational specialization and overall diversification. I also see the widespread west European ideology, that no doubt lay behind the largely zoocentric rupestral and portable art of the late Last Glacial, as having supported (didactically, socially, territorially, magically, etc.) an adaptive system closely geared toward surviving the rigorous conditions of the time (heavily dependant on gregarious herbivores, but also on secondary food resources where available). When those conditions (cold, stormy, intensely seasonal, open vegetation, etc.) changed or subsided, the ideology and art systemically linked to them ceased to have meaning and quickly disappeared, to be replaced in the Azilian and Mesolithic by new ideas, worldviews and means of expression (Barton et al. 1994, Clark et al. 1996).

One of the most interesting aspects of the record from the Atlantic (and Baltic) Façade is the fact of locally late adoption of food production economies, making Clark's notion of 'Neolithization as a last resort' particularly apt. The cases of Portugal and Cantabrian Spain are instructive examples of increasingly circumscribed, coastal-based foragers finally being absorbed into the Neolithic system. The exact causes and mechanisms for this absorption are still really unknown, but probably included some combination of local population pressure in the Mesolithic 'enclaves', increasing contacts (trade, raid, intermarriage, ceremonial, etc.) with farmer/herder 'neighbors' (and relatives?), land pressure from those 'neighbors' (and shrinkage of hunting territory for the foragers), the prestige of Neolithic symbols and goods (ceramics, etc.), and so forth. When the economic-social-ideological 'switch' occurs, it can do so swiftly and dramatically, as in the case of the appearance of megaliths prominently marking the heights above many Cantabrian valleys around 5500-5000 BP (Diez 1996, González Morales & Serna 1991). Foraging lifeways continued in the rich, diverse habitats of Atlantic Europe long after food production had not only swept across the Mediterranean world and much of the rest of Europe, but also after it had become established in close proximity to the last full-time hunter-fisher-gatherers. In the case of Cantabrian Spain, not only was the Mesolithic enclave propitious for foraging, but it was also ecologically unsuited to unmodified Mediterranean agriculture – making for a considerable time lag in the leap across the mountains, whether the Pyrenees, or the Cantabrian Cordillera. The nature of those first 'Neolithic' adaptations (whether merely a grafting of new cultigens and especially domesticated animals onto the existing mixed foraging economy or a radical substitution of farming for foraging) remains to investigated. Hopefully research projects, such as our current work in El Mirón Cave will help shed light on this process – as on so many others (both Paleolithic and Chalcolithic).

I turn now to Geoff Bailey's latest attempt (Chapter 16) to discredit our view that people intensified their food quest at least partially in response to regional population pressure. This essay is written with Alan Craighead, who recently reanalyzed materials from La Riera Cave, including samples we had collected for a pioneering attempt at obtaining seasonality information by oxygen isotope analysis in the laboratory of N.J. Shackleton in Cambridge. Bailey is responsible for correctly demonstrating (with reference to the Australian ethnographic record) that the molluscs found in shellmiddens (such as those of the Asturian) represented only small proportions of the overall, annual diet of their collectors. Yet, seasonally, Clark and I believe they may have represented critical tiding-over resources. This is one of the things we wanted to investigate at La Riera. From our point of view (and that of Leslie Freeman [1973]), it is quite obvious that molluscs were more than trivial food sources as early as the Solutrean and Lower Magdalenian in Asturias and Cantabria, given the density of large-size limpet shells at sites such as La Riera, Cova Rosa, Altamira and El Juyo. Bailey and Craighead try to minimize the importance of shellfish in the late Upper Paleolithic, but it is worth noting that people were transporting these shells at least two hours from the Pleniglacial/Late Glacial shore (probably having also eaten some at the beach) and that this was done in the context of the first really systematic, large-scale hunting of red deer and ibex in the long prehistory of the region (Freeman's 'wild harvesting'). This itself constitutes further evidence for subsistence intensification under stress. To argue that temperatures were as warm as today and that the coastline was even slightly closer to La Riera in Solutrean and Magdalenian periods than somewhat earlier (as a half-hearted explanation for the presence of many seashells), as Bailey and Craighead do, flies in the face of all known evidence for the Last Glacial Maximum and Dryas I and casts considerable doubt on the credibility of their results.

In 'testing' our hypothesis of temporally vectored progressive overexploitation of shellfish, Craighead remeasured small samples of whole shells from La Riera, including only five limpets from Asturian level 29 (sevenfold fewer than were measured by the LRPP project malacologist, J.A. Ortea, in his original La Riera study [1986]). Ortea's measurements of 15,418 specimens of *Patella vulgata* (not 11,286 as stated by Bailey and Craighead) clearly show a consistent average diameter range of c. 40-45 mm for the Solutrean and early-middle Magdalenian levels, but then a consistent 35-25 mm average diameter range for the terminal Magdalenian, Azilian and Asturian levels. This size decrease occurred before the major Holocene warming of the Cantabrian Sea. To concentrate only on the small sample from Level 29 (and then to give it an average diameter value nearly 60% greater than Ortea's value), as Bailey and Craighead do, is to overlook the clear, robust trend in these data. They also totally ignore the fact that Ortea demonstrates that, under conditions of non-exploitation in the current day in Asturias, *P. vulgata* achieves an average diameter of 45.1 mm – like those in the Solutrean and earlier Magdalenian levels – whereas along stretches of shore heavily exploited today it reaches an average diameter of only 32.8 mm – similar to average sizes in the terminal Magdalenian, Azilian and Asturian of La Riera. *P. intermedia* behaves similarly, although with a lesser difference than *P. vulgata* in the modern control samples. If anything, the total limpet sample from La Riera measured by Ortea shows a size decrease even earlier than the *P. vulgata* sample – mid Magdalenian times. This demonstration was based on random collecting of large samples by a professional shell fisherman.

Based often on very small samples, Bailey and Craighead claim to show by the experimental aragonite/calcite method that limpet size at La Riera tracked ocean temperature changes. Yet their results are unconvincing. For example, they show a sea surface temperature of 21.3° C at c. 20,500 BP and several temperatures of c. 9° C for Holocene aged Azilian and Asturian levels. Even if the so-called 'Laugerie' oscillation did exist *(pace* Sanchez-Goñi 1991), no one has ever claimed that sea water temperature conditions on the verge of the Last Glacial Maximum exceeded those of the Preboreal and Boreal. Similar incredibly warm temperatures are given by Bailey and Craighead for many of the other Solutrean and early Magdalenian levels without explanation. How these data are meant to show that intensive exploitation of shellfish was not the cause of size decrease is frankly unclear to me. On the other hand, if Craighead's estimates (and extrapolations) of average limpet age based on ring counts are correct, there does seem to be an argument against a long-term trend toward overexploitation. However, it should be noted that his samples are very small and some of the youngest limpets were collected in Azilian levels 26 and 28, with no information given for the Asturian levels. They do concede "a slight but progressive increase in exploitation pressure".

Finally, Craighead's seasonality results are very interesting, as they extend the data back into the Upper Paleolithic, while conflicting (in a very limited way and based only on *P. depressa [intermedia]* measurements from Level 29) with Deith and Shackleton's (1986) conclusion (based on oxygen isotope analyses of *P. vulgata, P. intermedia,* and *Monodonta lineata)* that Asturian mollusc gathering was a cold-season (mainly winter) activity at La Riera. Craighead's results indicate multi-seasonal collecting, and would seem to lend credence to a similar overall conclusion by González Morales (this volume) based on all available evidence for the Asturian of nearby Mazaculos (i.e., warm-season deer kills), where Deith had also found that shellfish were collected during the cold season (Deith had found evidence only of winter mollusc collection in several other Asturian sites). Yet Altuna (1986) had also found one red deer each killed in summer and winter in the La Riera Asturian. Taken together, these results imply that the Asturian *concheros* may not only represent limited winter season activity involving shellfish gathering, but also warm-season occupation involving at least some hunting.

Abundant evidence (summarized by González Sainz [1989]) points to massive hunting of ever younger red deer and ibex over the course of the late Magdalenian and Azilian – yet another possible indicator of a human population under resource stress. Fawns are also particularly well represented among the admittedly rather small Asturian faunal assemblages from the region. Over the long term these people seem to have endangered their local deer food resources by this short-sighted behavior, but maybe they had few other choices without adopting labor intensive domestication economies.

The last paper in the volume, by Arias and Alvárez (Chapter 17) consists of an exhaustive survey of evidence for funerary ritual on the Iberian Peninsula. Although Upper Paleolithic burials are extremely scarce and, for the most part equivocal, the Mesolithic of Spain and (esp.) Portugal offers an important source of information that rivals that of Scandinavia, Karelia and Poland. The authors identify the 6th millennium cal BC as a period of accelerated social change in mortuary customs, possibly related to increasing population densities in the Tagus and Sado estuaries (Portugal) and to the earliest domestication economies along the Mediterranean coast (Spain). The first cemeteries also date to this interval, perhaps signaling increasingly territorial behavior in a few, exceptionally productive environments that (regardless of possible contact with agropastoral communities) probably allowed for greater sedentism, thus setting in motion well-documented decreases in birth spacing and consequent increases in the rate of population growth. In Iberia, as elsewhere, decreases in residential mobility, increases in population density, and the appearance of cemeteries appear to be associated with one another, although the definition of more precise causal chains must await more modern-quality excavations at known sites, and the discovery of new ones. Arias and Alvárez make the point that the Iberian late Mesolithic affords an important corpus of evidence on mortuary practices that has previously been underacknowledged, or overlooked altogether.

By Way of Conclusion

Without repeating detailed arguments going back to the mid-1970s (Straus 1977, 1983, 1987; Straus et al. 1980, 1981; Straus & Clark 1986), I remain convinced that there are two central facts concerning late Paleolithic and Mesolithic adaptations in Cantabrian Spain (apparently mirrored in other regions of Iberia). These are (1) that numbers of sites (and in a general, relative way, presumably regional population) increased dramatically in the Solutrean coincident with the Last Glacial Maximum and abandonment of northern Europe. Then they seem to have leveled off or increased slowly throughout the Magdalenian, and finally appear to rise again, but now along a very narrow coastal strip coincident with Holocene sea level rise and dense forestation of the abandoned interior; and, (2) that human subsistence intensified to more efficiently exploit resources that had already long been utilized in the region (i.e., large-scale, specialized red deer hunting) and to diversify the food quest, adding difficult (chamois, ibex – for which highly specialized, large-scale hunting developed in Solutrean times) and dangerous (boar – with reforestation) game and going down the food chain to exploit more intensively marine molluscs and other shellfish (crustaceans, sea urchins); anadromous, riverine, estuarine and eventually marine fish, and birds. The correlation seems apparent. We argue, further, that the relationship must ultimately have been causal under these specific, geographically circumscribed conditions.

Environmental factors played a role in circumscribing human populations, first within the southern refugia of Upper Pleniglacial Europe c. 21-17 kya, and then, within Cantabrian Spain per se, in a narrowing coastal strip. The net effect, c. 20 kya and again at c. 9 kya, was to increase the numbers of mouths to feed with available wild resources per unit area in shrinking territories. In each case, local human populations may have eventually stabilized at new levels compatible with available resources and food-gathering technologies, strategies and tactics. Yet with new changes in critical parameters (including sea level, vegetation, mammalian biomass, etc.), imbalances in the relationship between people and the land would occur again, occasioning changes in the food quest (hence, for example, the Asturian phenomenon). What specific ecological imbalances (social and territorial, as well as demographic and environmental) led in combination to the adoption of food production economies in Cantabria are as yet unknown, but I, with Clark, am certain that the end of the foraging lifeway indeed came as 'a last resort' and a major break with a multimillennial trend of extracting ever more of the available wild subsistence resources of the land, the sea and even the air. The Atlantic Façade had never been a Potemkin Village, but it too eventually was swept away by the tide.

Postscript (July 2003)

Since this overview chapter was written many years ago, there have been several developments of relevance in Cantabrian Spain. These concern limited evidence for Mesolithic 'incursions' into the mountainous interior, for a fairly early transition from the Epipaleolithic into the Neolithic in the upper Ebro drainage to the south of the Cantabrian Cordillera, and for a relatively late appearance of Neolithic cultural/economic traits along the Atlantic Façade of northern Spain—apparently including a 'package' of domesticated ovicaprines and cattle, limited cereals (notably common wheat) and megalithic monuments by 5800-5500 radiocarbon years BP. Evidence from our excavations at El Miron Cave (Ramales, Cantabria) is relevant to the first and last of these developments, and together with other material, is preliminarily presented in Straus and González Morales (2003) and in González Morales, Straus, Díez and Ruiz (n.d.). Evidence of the early yet not discontinuous transition from the Epipaleolithic to the Neolithic in the interior Basque Province of Alava has been developed as the result of recent excavations, notably by Alfonso Alday (2002a,b with references). I have further reviewed this and other aspects of Mesolithic research in Spain and Portugal in a review chapter in a book edited by Bailey and Spikins (Straus n.d.). Debates continue to rage on the Iberian Peninsula as to whether there was cultural *and* genetic continuity *or* rupture between the Mesolithic and Neolithic systems and inhabitants – reminiscent of the debate over the fate of the Neandertals on the Peninsula – and, more generally, throughout Europe. Work by the Canadian team of David Lubell and Mary Jackes, now in collaboration with Pablo Arias and associates in northern Spain as earlier in Portugal – continues to be in the forefront of biocultural research into the terminal Pleistocene/initial Holocene situation.

Endnotes

1. Not to be confused with the modern political region of Cantabria, formerly the province of Santander, a portion of Old Castile that extended to the sea in the center of the northern coast of Spain.

References

Alday, A.

2002a Los últimos cazadores-recolectores de la Iberia interior: la Alta-Media Cuenca del Ebro de la Meseta del Norte. *Munibe* 54: 79-101.

2002b Las unidades industriales mesolíticas en la Alta-Media Cuenca del Ebro. *Complutum* 13: 19-50.

Altuna, J.

1986 The mammalian faunas from the prehistoric site of La Riera. In *La Riera Cave* (L. Straus & G. Clark, eds.), pp. 237-274, 421-480. Tempe: Arizona State University Anthropological Research Papers No. 36.

Arias, P.

1991 *De Cazadores a Campesinos*. Santander: Universidad de Cantabria.

Barton, C.M. & M. Neeley

1996 Phantom cultures of the Levantine Epipaleolithic. *Antiquity* 70: 139-147.

Barton, C.M., G. Clark & A. Cohen

1994 Art as information: explaining Upper Paleolithic art in western Europe. *World Archaeology* 26: 185-207.

Clark, G.A.

1976 *El Asturiense Cantábrico.* Madrid: Bibliotheca Praehistorica Hispana No. 13.

1983 *The Asturian of Cantabria – Early Holocene Hunter-Gatherers in Northern Spain.* Tucson: Anthropological Papers of the University of Arizona No. 41.

1987 From the Mousterian to the Metal Ages: long-term change in the human diet of Cantabrian Spain. In *The Pleistocene Old World: Regional Perspectives* (O. Soffer, ed.), pp. 293-316. New York: Plenum.

1989 Site functional complementarity in the Mesolithic of northern Spain. In *The Mesolithic in Europe* (C. Bonsall, ed.), pp. 589-603. Edinburgh: John Donald Publishers, Ltd.

1996 Plus français que les Français. *Antiquity* 70: 138, 139.

Clark, G.A. & L.G. Straus

1983 Late Pleistocene hunter-gatherer adaptations in Cantabrian Spain. In *Hunter-Gatherer Economy in Prehistory* (G. Bailey, ed.), pp. 131-148. Cambridge: Cambridge University Press.

Clark, G.A., C.M. Barton & A. Cohen

1996 Explaining art in the Franco-Cantabrian refugium: an information exchange model. In *Debating Complexity* (D. Meyer et al., eds.), pp. 241-253. Calgary: Archaeological Association of the University of Calgary.

Deith, M. & N.J. Shackleton

1986 Seasonal exploitation of marine molluscs. In *La Riera Cave* (L. Straus & G. Clark, eds.), pp. 299-314. Tempe: Arizona State University Anthropological Research Papers No. 36.

Diez Castillo, A.

1996 *Evolución del Poblamiento Prehistórico en los Valles del Deva y Nansa.* Ph.D dissertation, Departmento of Prehistory, Universidad de Cantabria.

Elston, R. & S. Kuhn (Editors)

2003 *Thinking Small: Global Perspectives on Microlithization.* Arlington: Archeological Papers of the American Anthropological Association No. 13.

Fischer, A.

1995 *Man and Sea in the Mesolithic.* Oxford: Oxbow Monograph No. 53.

Freeman, L.G.

1973 The significance of mammalian faunas from Paleolithic occupations in Cantabrian Spain. *American Antiquity* 38: 3-44.

González Morales, M. & M. Serna

1991 Cuestiones sobre la prehistoria de Cantabria: los primeros productores, el megalitísmo y el arte esquemático. *Actas del XX Congreso Nacional de Arqueología*, pp. 137-169. Zaragoza: Universidad de Zaragoza.

González Morales, M., L. G. Straus, A. Díez & J.Ruíz

n.d. Postglacial coast and inland: Epipaleolithic-Mesolithic-Neolithic transitions in Vasco-Cantabrian Spain. *Munibe* 55 (in press).

González Sainz, C.

1989 *El Magdaleniense Superior-Final de la Región Cantábrica.* Santander: Universidad de Cantabria.

Léotard, J-M., M. Otte & L. G. Straus

n. d. L'Abri du Pape. Liège: ERAUL (in preparation).

Lewthwaite, J.

1986 From Menton to the Mondego in three steps. *Arqueologia* 13: 95-119.

Lubell, D., M. Jackes, H. Schwarcz, M. Knyf & C. Meiklejohn

1994 The Mesolithic-Neolithic transition in Portugal: isotopic and dental evidence of diet. *Journal of Archaeological Science* 21: 201-216.

Marambat, J. & J. Roussot-Larroque

1989 Paysage végétal et occupations humaines sur la côte atlantique: l'exemple de la Lède du Gurp. *Bulletin de l'Association Française pour l'Etude du Quaternaire* 1989: 73-89.

Neeley, M.P. & C.M. Barton
1994 A new approach to interpreting late Pleistocene microlith industries in southwest Asia. *Antiquity* 68: 275-288.

Ortea, J. A.
1986 The malacology of La Riera Cave. In *La Riera Cave* (L. Straus & G. Clark, eds.), pp. 289-298. Tempe: Arizona State University Anthropological Research Papers No. 36.

Otte, M. & L.G. Straus
1997 *La Grotte du Bois Laiterie.* Liège: ERAUL No. 80.

Penalva, C. & L. Raposo
1986 Palheirões do Alegra. *Informaçao Arqueológica* 7: 16-19.

Radovanovic, I.
1996 *The Iron Gates Mesolithic.* Ann Arbor: International Monographs in Prehistory No. 11.

Roche, J.
1972 Les amas coquilliers mésolithiques de Muge. *Fundamenta* A-3: 72-107.

Sanchez Goñi, M.F.
1991 On the Last Glaciation and the interstadials during the Solutrean. *Current Anthropology* 32: 573-575.

Straus, L.G.
1977 Of deerslayers and mountain men: Paleolithic faunal exploitation in Cantabrian Spain. In *For Theory Building in Archaeology* (L. Binford, ed.), pp. 41-76. New York: Academic Press.

1979 Mesolithic adaptations along the northern coast of Spain. *Quaternaria* 21: 305-327.

1983 Terminal Pleistocene faunal exploitation in Cantabria and Gascony. In *Animals and Archaeology: Hunters & their Prey* (J. Clutton-Brock & C. Grigson, eds.), pp.209-225. Oxford: British Archaeological Reports, International Series No. 163.

1987 Hunting in late Upper Paleolithic western Europe. In *The Evolution of Human Hunting* (M. Nitecki & D. Nitecki, eds.), pp. 147-176. New York: Plenum.

1988 New chronometric dates for the prehistory of Portugal. *Arqueologia* 20: 73-76.

1991a The Mesolithic-Neolithic transition in Portugal: a view from Vidigal. *Antiquity* 65: 899-903.

1991b Human geography of the late Upper Paleolithic in western Europe. *Journal of Anthropological Research* 47: 259-278.

1991c Southwest Europe at the Last Glacial Maximum. *Current Anthropology* 32: 189-199.

1995 *Les Derniers Chasseurs de Rennes du Monde Pyrénéen: L'Abri Dufaure.* Paris: Mémoires de la Société Préhistorique Française No. 22.

n.d. The Mesolithic of Atlantic Iberia. In *Mesolithic Europe* (G. Bailey & P. Spikins, eds.). Leicester: Leicester University Press (in press).

Straus, L.G., J. Altuna, G. Clark, M. González Morales, H. Laville, Arl. Leroi-Gourhan, M. Menéndez de la Hoz, J. Ortea
1981 Paleoecology at La Riera. *Current Anthropology* 22: 655-682.

Straus, L.G., J. Altuna, D. Ford, L. Marambat, J. Rhine, H. Schwarcz & J.-L.Vermet
1992 Early farming in the Algarve. *Trabalhos de Antropologia e Etnologia* 32: 141-162.

Straus, L.G., J. Altuna, M. Jackes & M. Kunst
1988 New excavations in Casa da Moura and at Bocas, Portugal. *Arqueologia* 18: 65-95.

Straus, L.G., J. Altuna & B. Vierra
1990 The concheiro at Vidigal: a contribution to the late Mesolithic of southern Portugal. In *Contributions to the Mesolithic in Europe* (P. Vermeersch & P. Van Peer, eds.), pp. 463-474. Leuven: University of Leuven Press.

Straus, L.G. & G.A.Clark
1986 *La Riera Cave – Stone Age Hunter-Gatherer Adaptations in Northern Spain.* Tempe: Arizona State University Anthropological Research Papers No. 36.

Straus, L.G., G.A.Clark, J. Altuna & J.A.Ortea
1980 Ice Age subsistence in northern Spain. *Scientific American* 242: 142-152.

Straus, L.G. & M. González Morales
1996 Preliminary excavations in El Mirón Cave. *Old World Archaeology Newsletter* 20: 14-18.

2003 The Mesolithic in the Cantabrian Interior: fact or fantasy. In *Mesolithic on the Move* (L. Larsson, H. Kindgren, K. Knutsson, D. Loeffler & A. Akerlund, eds.), pp. 359-368. Oxford: Oxbow Books.

Straus, L.G., M. Otte, A. Gautier, P. Haesaerts, I. López-Bayón, P. Lacroix, A. Martínez, R. Miller & J. Orphal
n.d. Late Quaternary prehistoric investigations in southern Belgium. *Préhistoire Européenne* (in press).

Tavares da Silva, C.
1989 Novos dados sobre o Neolitico antigo do Sul de Portugal. *Arqueologia* 20: 24-32.

Tavares da Silva, C., J. Soares, J. Cardoso, C. Cruz & C. Reis
1986 Neolitico da Comporta. *Arqueologia* 14: 59-82.

Vierra, B.

1992 *Subsistence Diversification and the Evolution of Microlithic Technologies: a Study of the Portuguese Mesolithic.* Ph.D dissertation, Department of Anthropology, University of New Mexico. Ann Arbor: University Microfilms.

1995 *Subsistence and Stone Tool Technology: an Old World Perspective.* Tempe: Arizona State University Anthropological Research Papers No. 47.

Woodman, P.

1985 *Excavations at Mount Sandel 1973-77.* Belfast: Northern Ireland Archaeological Monographs No. 2.

Zilhão, J.

1992 *Gruta do Caldeirão: O Neolitico Antigo.* Lisbon: Trabalhos de Arqueologia No. 6.

1993 The spread of agro-pastoral economies across Mediterranean Europe. *Journal of Mediterranean Archaeology* 6: 5-63.

List of Contributors

Estéban Alvárez-Fernández
Departamento de Ciencias Historicas
Universidad de Cantabria
Avenida de los Castros, s/n
39005 Santander SPAIN

Pablo Arias
Departamento de Ciencias Historicas
Universidad de Cantabria
Avenida de los Castros, s/n
39005 Santander SPAIN

J. Emili Aura Tortosa
Department de Arqueologia i Prehistoria
Universitat de Valencia
46020 Valencia SPAIN

Geoff Bailey
Department of Archaeology
University of Newcastle-upon-Tyne
Newcastle-upon-Tyne NE1 7RU
UNITED KINGDOM

Clive Bonsall
Department of Archaeology
University of Edinburgh
Old School Road, Infirmary Street
Edinburgh EH1 1LT SCOTLAND

Geoffrey A. Clark
Department of Anthropology
Arizona State University
Tempe, AZ 85287-2402, U.S.A.

Alan Craighead
Site 40 C12 RR No. 2
Lower Road, Gibsons
BC V0N 1V0 CANADA

Louis Detrain
Service Régional de l'Archéologie
DRAC - Aquitaine
6 bis, Cours de Gourgue
33074 Bordeaux FRANCE

Miguel-Ángel Fano Martínez
Departamento de Historia
Universidad de Deusto, Apartado 1
48080 Bilbao SPAIN

Juan Fernández-Tresguerres
Area de Prehistoria y Arqueología
Facultad de Geografía e Historia
Universidad de Oviedo
33071 Oviedo SPAIN

Pascal Foucher
Service Régional de l'Archéologie
DRAC - Poitou-Charentes
Hôtel de Rochefort, 102 Grand'Rue
86060 Poitiers FRANCE

Manuel R. González Morales
Departamento de Ciencias Historicas
Universidad de Cantabria
Avenida de los Castros, s/n
39005 Santander SPAIN

Peter Rowley-Conwy
Department of Archaeology
University of Durham
46 Saddler Street
Durham DH1 3NU
UNITED KINGDOM

Cristina San Juan
Service Régional de l'Archéologie
DRAC - Poitou-Charentes
Hôtel de Rochefort, 102 Grand'Rue
86060 Poitiers FRANCE

Lawrence G. Straus
Department of Anthropology
University of New Mexico
Albuquerque, NM 87131
U.S.A.

Christopher Tolan-Smith
Department of Archaeology
University of Newcastle-upon-Tyne
Newcastle-upon-Tyne NE1 7RU
UNITED KINGDOM

Alain Turq
Service Régional de l'Archéologie
DRAC - Aquitaine
6 bis, Cours de Gourgue
33074 Bordeaux FRANCE

José Manuel Vázquez Varela
Departament de Historia
Universidade de Santiago de Compostela
15703 Santiago de Compostela SPAIN

Bradley J. Vierra
Department of Anthropology
University of Texas
San Antonio, TX 78249, U.S.A.

Peter Woodman
Department of Archaeology
University College Cork
Cork IRELAND

João Zilhão
Max Planck Institute for Evolutionary Anthropology
Deutscher Platz, 6
B-04103 Leipzig
GERMANY

Marek Zvelebil
Department of Archaeology and Prehistory
University of Sheffield
Sheffield SI0 2TN
UNITED KINGDOM

ARIZONA STATE UNIVERSITY
ANTHROPOLOGICAL RESEARCH PAPERS

1. A CULTURAL INVENTORY OF THE PROPOSED GRANITE REEF AND SALT-GILA AQUEDUCTS, AGUA FRIA RIVER TO GILA RIVER, ARIZONA by A.E.Dittert, P. Fish and D. Simonis (1969), 26 pp., 14 figs. .... O.P.

2. PATTERN MATHEMATICS AND ARCHAEOLOGY by B. Zaslow and A.E. Dittert (1977), x + 90 pp., 39 figs., 5 tables, bibliogr. .... O.P.

3. SURFICIAL POLLEN RECORDS FOR CENTRAL ARIZONA by J. Schoenwetter and L. Doerschlag (1970), 22 pp., 1 table, bibliogr. .... O.P.

4. HECLA I: A PRELIMINARY REPORT ON THE ARCHAEOLOGICAL INVESTIGATIONS AT THE LAKESHORE PROJECT, PAPAGO RESERVATION, SOUTH-CENTRAL ARIZONA by A. Goodyear and A.E. Dittert (1973), iv + 81 pp., 15 figs., bibliogr. .... O.P.

5. PALYNOLOGY OF THE ROBINSON SITE, NORTH-CENTRAL WISCONSIN by J. Gish (1976), viii + 76 pp., 11 figs., 5 tables, app., bibliogr. .... O.P.

6. DEFINITION AND PRELIMINARY STUDY OF THE MIDVALE SITE by J. Schoenwetter, S.W. Gaines and D. Weaver (1973), viii + 173 pp., 9 maps, 22 figs., 29 tables, bibliogr. .... O.P.

7. ARCHAEOLOGICAL INVESTIGATIONS AT THE WESTWING SITE, AZ T:7:27 (ASU), AGUA FRIA RIVER VALLEY, ARIZONA by D. Weaver (1974), v + 84 pp., 24 figs., 2 app., bibliogr. .... O.P.

8. AN ARCHAEOLOGICAL SURVEY OF THE CAVE BUTTES DAM ALTERNATIVE SITE AND RESERVOIR, ARIZONA by J. Rodgers (1974), vi + 92 pp., 30 figs., 2 app., bibliogr. .... O.P.

9. HECLA II AND III: AN INTERPRETIVE STUDY OF THE ARCHAEOLOGICAL REMAINS FROM THE LAKESHORE PROJECT, PAPAGO RESERVATION, SOUTH-CENTRAL ARIZONA by A. Goodyear (1975), xvii + 401 pp., 47 figs., 50 tables, 9 app., bibliogr. $20.00

10. AN ARCHAEOLOGICAL INVESTIGATION OF BUCKEYE HILLS EAST, MARICOPA COUNTY, ARIZONA by J. Rodgers (1976), vii + 116 pp., 18 figs., 6 tables, 2 app. $10.00

11. ARCHAEOLOGY IN COPPER BASIN, YAVAPAI COUNTY, ARIZONA: MODEL BUILDING FOR THE PREHISTORY OF THE PRESCOTT REGION by M. Jeter (1977), xv + 434 pp., 47 figs., 15 tables, 8 app., bibliogr. .... O.P.

12. ARCHAEOLOGICAL INVESTIGATIONS ALONG THE GRANITE REEF AQUEDUCT, CAVE CREEK ARCHAEOLOGICAL DISTRICT, ARIZONA by J. Rodgers (1977), ix + 185 pp., 31 figs., 33 tables, 3 app., bibliogr. $15.00

13. AN ANALYTICAL APPROACH TO CULTURAL RESOURCE MANAGEMENT: THE LITTLE COLORADO PLANNING UNIT, F. Plog, Editor (1978), xiv + 293 pp., 67 figs., 96 tables, 4 app., bibliogr. .... O.P.

14. ARCHAEOLOGICAL INVESTIGATIONS AT SENECA LAKE, SAN CARLOS INDIAN RESERVATION, ARIZONA by C. R. Stafford (1978), vi + 96 pp., 31 figs., 11 tables, 3 app., bibliogr. $10.00

15. COMPUTER GRAPHICS IN ARCHAEOLOGY: STATISTICAL CARTOGRAPHIC APPLICATIONS TO SPATIAL ANALYSIS IN ARCHAEOLOGICAL CONTEXTS, S. Upham, Editor (1979), vi + 156 pp., 75 figs., 18 tables, bibliogr. $15.00

16. THE INTERPRETIVE POTENTIAL OF MOUSTERIAN DEBITAGE by P. Fish (1979), ix + 167 pp., 17 figs., 27 tables, 4 app., bibliogr. $15.00

17. ARCHAEOLOGICAL INVESTIGATIONS IN THE CAVE CREEK AREA, MARICOPA COUNTY, SOUTH-CENTRAL ARIZONA by T. K. Henderson and J. B. Rodgers (1979), xii + 198 pp., 33 figs., 34 tables, 19 plates, 4 app., bibliogr. $17.50

18. AN ARCHAEOLOGICAL TEST OF SITES IN THE GILA BUTTE-SANTAN REGION, SOUTH-CENTRAL ARIZONA by G. Rice *et al.* (1979), x + 258 pp., 65 figs., 33 tables, bibliogr. .... O.P.

19. THE NORTH BURGOS ARCHAEOLOGICAL SURVEY: BRONZE AND IRON AGE ARCHAEOLOGY ON THE MESETA DEL NORTE (PROVINCE OF BURGOS, NORTH-CENTRAL SPAIN), G.A. Clark, Editor (1979), xviii + 307 pp., 98 figs., 40 tables, 9 plates, 4 app., bibliogr. $20.00

20. SPEAKING, SINGING AND TEACHING: A MULTIDISCIPLINARY APPROACH TO LANGUAGE VARIATION (SWALLOW VIII), F. Barkin and E. Brandt, Editors (1980), vi + 482 pp., bibliogr. O.P.

21. ARCHAEOLOGICAL INVESTIGATIONS AT AZ U:6:61 (ASU), A PREHISTORIC LIMITED ACTIVITY SITE IN SOUTH-CENTRAL ARIZONA by P. Brown and A.E. Rogge (1980), viii + 83 pp., 17 figs., 17 tables, 4 plates, 1 app., bibliogr. $10.00

22. PROCEEDINGS: EIGHTH INTERNATIONAL CONGRESS FOR THE STUDY OF THE PRECOLUMBIAN CULTURES OF THE LESSER ANTILLES, S. Lewenstein, Editor (1980), xiv + 624 pp., 106 figs., 46 tables, 90 plates, bibliogr. .... O.P.

23. CURRENT ISSUES IN HOHOKAM PREHISTORY: PROCEEDINGS OF A SYMPOSIUM, D. Doyel and F. Plog, Editors (1980), iv + 281 pp., 24 figs., 19 tables, 1 app., bibliogr. .... O.P.

24. THE PROTOHISTORIC PERIOD IN THE NORTH AMERICAN SOUTHWEST, A.D. 1450-1700, D. Wilcox and W.B. Masse, Editors (1981), vi + 445 pp., 12 figs., 8 tables, 10 plates, bibliogr. .... O.P.

25. PATTERN DISSEMINATION IN THE PREHISTORIC SOUTHWEST AND MESOAMERICA by B. Zaslow (1981), vii + 57 pp., 28 figs., 2 tables, 3 app., bibliogr. $10.00

26. CHIMPANZEE VISUAL COMMUNICATION by S. Berdecio and L. Nash (1981), xv + 159 pp., 15 figs., 60 tables, 80 plates, 2 app., bibliogr. .... O.P.

27. ANTHROPOLOGICAL STUDIES IN GREAT BRITAIN AND IRELAND, M. Firestone, Editor (1982), iii + 147 pp., 17 figs., 12 tables, bibliogr. $12.00

28. GRANITE REEF: A STUDY IN DESERT ARCHAEOLOGY, P. Brown and C. Stone, Editors (1982), xvii + 443 pp., 71 figs., 69 tables, 20 plates, 4 app., bibliogr. $25.00

29. ECOLOGICAL MODELS IN ECONOMIC PREHISTORY, G.Bronitsky, Editor (1983), iii + 290 pp., 36 figs., 36 tables, bibliogr. $20.00

30. A PEASANT COMMUNITY IN CHANGING THAILAND by S. Piker (1983), ix + 157 pp., 5 figs., 19 tables, bibliogr. $15.00

31. REGIONAL ANALYSIS OF PREHISTORIC CERAMIC VARIATION: CONTEMPORARY STUDIES OF THE CIBOLA WHITEWARES, A. Sullivan and J. Hantman, Editors (1984), vi + 149 pp., 16 figs., 11 tables, 12 plates, bibliogr. $15.00

32. PREHISTORIC SUBSISTENCE AND POPULATION CHANGE ALONG THE LOWER AGUA FRIA RIVER, ARIZONA: A MODEL SIMULATION by D. Dove (1984), viii + 139 pp., 19 figs., 5 tables, 2 app., bibliogr. $15.00

33. PREHISTORIC AGRICULTURAL STRATEGIES IN THE SOUTHWEST, S. Fish and P. Fish, Editors (1984), vii + 386 pp., 61 figs., 28 tables, 2 plates, bibliogr. $25.00

34. HEALTH AND DISEASE IN THE PREHISTORIC SOUTHWEST, C. Merbs and R. Miller, Editors (1985), xix + 402 pp., 35 figs., 64 tables, 101 plates, 1 app., bibliogr. $25.00

35. POPULATION GROWTH AND SOCIAL COMPLEXITY: AN EXAMINATION OF SETTLEMENT AND ENVIRONMENT IN THE CENTRAL MAYA LOWLANDS by Anabel Ford (1986), xiii + 201 pp., 28 figs., 37 tables, 19 plates, 4 app., bibliogr., index $17.50

36. LA RIERA CAVE: STONE AGE HUNTER-GATHERER ADAPTATIONS IN NORTHERN SPAIN L.G. Straus and G.A. Clark, Editors (1986), xvii + 498 pp., 244 figs., 107 tables, 35 plates, 2 app., bibliogr. $25.00

37. POLITIES AND PARTITIONS: HUMAN BOUNDARIES AND THE GROWTH OF SOCIAL COMPLEXITY, K. Trinkaus, Editor (1987), iv + 251 pp., 43 figs., 11 tables, bibliogr. $20.00

38. COASTS, PLAINS AND DESERTS: ESSAYS IN HONOR OF REYNOLD J. RUPPÉ, S.W. Gaines, Editor (1987), xii + 282 pp., 65 figs., 48 tables, 2 plates, bibliogr. $25.00

39. NEW FRONTIERS IN THE ARCHAEOLOGY OF THE PACIFIC COAST OF SOUTHERN MESOAMERICA, F. Bove and L. Heller, Editors (1989), xvii + 292 pp., 81 figs., 40 tables, 44 plates, bibliogr. $30.00

40. PERSPECTIVES ON ANTHROPOLOGICAL COLLECTIONS FROM THE AMERICAN SOUTHWEST, A.L. Hedlund, Editor (1989), iii + 158 pp., 2 figs., bibliogr. $15.00

41. TECHNOLOGICAL CHANGE IN THE CHAVEZ PASS REGION, NORTH-CENTRAL ARIZONA, G. Brown, Editor (1990), xiv + 226 pp., 48 figs., 90 tables, 2 plates, 6 app., bibliogr. $20.00

42. THE ETHNOARCHAEOLOGY OF REFUSE DISPOSAL, E. Staski and L. Sutro, Editors (1991), vii + 92 pp., 11 figs., 13 tables, 8 plates, bibliogr. $10.00

43. AN INVESTIGATION OF IMAGE PROCESSING TECHNIQUES AT PINCEVENT HABITATION NO. 1, AN UPPER MAGDALENIAN SITE IN NORTHERN FRANCE by S. A. Lang (1992), vii + 110 pp., 44 figs., 2 tables, 13 app., bibliogr. $25.00

44. SHAMANS, PRIESTS AND WITCHES: A CROSS-CULTURAL STUDY OF MAGICO-RELIGIOUS PRACTITIONERS by M. J. Winkelman (1992), viii + 191 pp., 11 figs., 4 tables, 3 app., bibliogr. $27.50

45. MIDDLE PALEOLITHIC ASSEMBLAGE AND SETTLEMENT VARIABILITY IN WEST-CENTRAL JORDAN by J. M. Potter (1993), v + 57 pp., 23 figs., 23 tables, bibliogr. $10.00

46. EXPLORING SOCIAL, POLITICAL AND ECONOMIC ORGANIZATION IN THE ZUNI REGION, T. Howell and T. Stone, Editors (1994), ix + 126 pp., 39 figs., 28 tables, 3 plates, bibliogr. $17.50

47. SUBSISTENCE AND STONE TOOL TECHNOLOGY: AN OLD WORLD PERSPECTIVE by B.J. Vierra (1995), xiv + 283 pp., 113 figs., 95 tables, 2 app., bibliogr. $30.00

48. INTERPRETING SOUTHWESTERN DIVERSITY: UNDERLYING PRINCIPLES AND OVERARCHING PATTERNS, P. Fish and J. Reid, Editors (1996), ix + 320 pp., 61 figs., 20 tables, bibliogr. $35.00

49. WANDERING VILLAGERS: PIT STRUCTURES, MOBILITY AND AGRICULTURE IN SOUTHEASTERN ARIZONA, P. A. Gilman (1997), xii + 216 pp., 53 figs., 54 tables, 2 app., bibliogr. ... $25.00

50. THE ARCHAEOLOGY OF THE WADI AL-HASA, WEST-CENTRAL JORDAN, VOLUME 1: SURVEYS, SETTLEMENT PATTERNS AND PALEOENVIRONMENTS, N.R. Coinman, Editor (1998), v + 228 pp., 83 figs., plates; 37 tables, bibliogr. $25.00

51. MIGRATION AND REORGANIZATION: THE PUEBLO IV PERIOD IN THE AMERICAN SOUTHWEST, K.A. Spielmann, Editor (1998), x + 301 pp., 100 figs., 49 tables, 4 plates, 3 app., bibliogr. .... O.P.

52. THE ARCHAEOLOGY OF THE WADI AL-HASA, WEST-CENTRAL JORDAN, VOLUME 2: EXCAVATIONS AT MIDDLE, UPPER AND EPIPALEOLITHIC SITES, N.R. Coinman, Editor (2000), xiv + 391 pp., 143 figs., plates; 167 tables, bibliogr. $45.00

53. SOCIAL INTEGRATION IN THE ANCIENT MAYA HINTERLANDS: CERAMIC VARIABILITY IN THE BELIZE RIVER AREA by L. J. Lucero (2001), ix + 88 pp., 20 figs., 20 tables, bibliogr. $15.00

54. BOUNDARIES AND TERRITORIES: PREHISTORY OF THE U. S. SOUTHWEST AND NORTHERN MEXICO, M. E. Villalpando, Editor (2002), xi + 186 pp., 15 maps, 46 figs., 12 tables, bibliogr., index $25.00

55. THE MESOLITHIC OF THE ATLANTIC FAÇADE: PROCEEDINGS OF THE SANTANDER SYMPOSIUM, M.R.González Morales and G.A.Clark, Editors (2004), xii + 260 pp., 83 figs., plates; 20 tables, bibiogr. $30.00

Out-of-print (OP) numbers in the *Arizona State University Anthropological Research Papers* are available upon request in the form of unbound xerox copies at $.15/page. Requests should be directed to the Editor, *Anthropological Research Papers*, Department of Anthropology, Arizona State University, Tempe, AZ 85287-2402 [Tel: (480) 965-6213, 965-7596; fax: (480) 965-7671; email: gaclark@asu.edu].